WORKSHOP ON SPACE CHARGE PHYSICS IN HIGH INTENSITY HADRON RINGS

WORKSHOP ON SPACE CHARGE PHYSICS
in
HIGH INTENSITY HADRON RINGS
supported by

WORKSHOP ON SPACE CHARGE PHYSICS IN HIGH INTENSITY HADRON RINGS

Shelter Island, New York May 1998

EDITORS
A. U. Luccio
W. T. Weng
Brookhaven National Laboratory

AIP CONFERENCE PROCEEDINGS 448

American Institute of Physics Woodbury, New York

Editors:

Alfredo U. Luccio and Wu-Tsu Weng
Alternating Gradient Synchrotron (AGS)
Building 911B
Brookhaven National Laboratory
Upton, NY 11973-5000

Email: luccio@bnl.gov
weng@bnl.gov

Authorization to photocopy items for internal or personal use, beyond the free copying permitted under the 1978 U.S. Copyright Law (see statement below), is granted by the American Institute of Physics for users registered with the Copyright Clearance Center (CCC) Transactional Reporting Service, provided that the base fee of $15.00 per copy is paid directly to CCC, 222 Rosewood Drive, Danvers, MA 01923. For those organizations that have been granted a photocopy license by CCC, a separate system of payment has been arranged. The fee code for users of the Transactional Reporting Service is: 1-56396-824-X/ 98 /$15.00.

© 1998 American Institute of Physics

Individual readers of this volume and nonprofit libraries, acting for them, are permitted to make fair use of the material in it, such as copying an article for use in teaching or research. Permission is granted to quote from this volume in scientific work with the customary acknowledgment of the source. To reprint a figure, table, or other excerpt requires the consent of one of the original authors and notification to AIP. Republication or systematic or multiple reproduction of any material in this volume is permitted only under license from AIP. Address inquiries to Office of Rights and Permissions, 500 Sunnyside Boulevard, Woodbury, NY 11797-2999; phone: 516-576-2268; fax: 516-576-2499; e-mail: rights@aip.org.

L.C. Catalog Card No. 98-87648
ISBN 1-56396-824-X
ISSN 0094-243X
DOE CONF- 980567

Printed in the United States of America

CONTENTS

Preface ix
Participants xi

PLENARY TALKS

Observation I: Space Charge in Proton Linacs 3
    T. P. Wangler, F. Merrill, L. Rybarcyk, and R. Ryne
Observation II: Observation and Interpretation of Space Charge Phenomena
for High Intensity 15
    I. Hofmann, R. Bär, O. Boine-Frankenheim, and G. Rumolo
Theory I: Early Work on Space Charge Effects in Particle Accelerators 26
    B. W. Zotter
Theory II: Space Charge Dominated Beams in Synchrotrons and Linac 38
    S. Y. Lee
Theory III: Betatron Resonances with Space Charge 56
    R. Baartman
Simulation I: Simulation of Space Charge Effects in a Synchrotron 73
    S. Machida and M. Ikegami
Simulation II: Simulation with Space Charge 85
    C. R. Prior
Existing Facilities I: Status and High Intensity Performance of ISIS 104
    C. M. Warsop
Existing Facilities II: PSR Experience with Beam Losses, Instabilities, and
Space Charge Effects 116
    R. J. Macek
Existing Facilities III: Performance and Measurements of the Fermilab
Booster 128
    M. Popovic and C. Ankenbrandt
Existing Facilities IV: High Intensity Performance and Upgrades
at the Brookhaven AGS 135
    T. Roser
New Facilities I: Progress Report on the European Spallation Source,
May 1998 141
    C. R. Prior
New Facilities III: SNS Accumulator Ring Design and Space Charge
Considerations 152
    W. T. Weng

CONTRIBUTED PAPERS: OBSERVATIONS

Longitudinal Space Charge Compensation at PSR 171
    F. Neri

Some Issues on the RF System in the 3 GeV Fermilab Pre-Booster 178
    K. Y. Ng
Studies of Space-Charge Physics in Beams for Advanced Accelerator
Applications. 189
    J. G. Wang, S. Bernal, P. Chin, R. A. Kishek, Y. Li, M. Reiser, M. Venturini,
    W. W. Zhang, Y. Zou, T. Godlove, D. Kehne, I. Haber, and R. C. York
Space Charge Effect and the Booster to AGS Transfer 198
    S. Y. Zhang

CONTRIBUTED PAPERS: THEORY

An Exactly Solvable Model of Transverse Stability in Bunched Beams 213
    M. Blaskiewicz
Emittance Growth in Heavy Ion Rings Due to the Effects of Space Charge
and Dispersion 221
    J. J. Barnard, G. D. Craig, A. Friedman, D. P. Grote, B. Losic, and S. M. Lund
Micromap Approach to Space Charge in a Synchrotron................... 233
    G. Franchetti, I. Hofmann, and G. Turchetti
Longitudinal Halo in Beam Bunches with Self-Consistent 6-D Distributions ... 245
    R. L. Gluckstern, A. V. Fedotov, S. S. Kurennoy, and R. D. Ryne
An RMS Particle Core Model for Rings................................ 254
    J. A. Holmes, J. D. Galambos, D. K. Olsen, and S. Y. Lee
Achromat with Linear Space Charge for Bunched Beams 271
    D. Raparia, J. G. Alessi, Y. Y. Lee, and W. T. Weng
Dispersion and Space Charge.. 278
    M. Venturini, R. A. Kishek, and M. Reiser
The Space-Charge Impedance of RF-Shielding Wires 286
    T.-S. F. Wang
Tune Shift Due to Non-Linear Space Charge Effect....................... 298
    J. Beebe-Wang

CONTRIBUTED PAPERS: SIMULATION

Analytical and Numerical Treatment of Halo-Free Beam Transport.......... 317
    Y. K. Batygin
Overview of the WARP Code and Studies of Transverse Resonance Effects ... 329
    A. Friedman
Progress Towards Understanding Transverse Space Charge Effects on SNS
Ring Injection 332
    J. Galambos, J. Holmes, D. Olsen, J. Whealton, M. Blaskiewicz, A. U. Luccio,
    and J. Beebe-Wang
Space Charge Calculations in Rings for Uniform Focusing, FODO, and
Doublet Lattices 344
    J. A. Holmes, J. D. Galambos, J. H. Whealton, D. K. Olsen, M. Blaskiewicz,
    A. U. Luccio, J. Beebe-Wang, and S. Y. Lee

A Hybrid Fast-Multipole Technique for Space-Charge Tracking with Halos 359
      F. W. Jones

PIC Code Simulations of the Space-Charge-Dominated Beam in the University of Maryland Electron Ring 371
      R. A. Kishek, I. Haber, M. Venturini, and M. Reiser

Simulation of Longitudinal Micro-Bunch Dynamics 379
      P. R. Knaus

A Particle Simulation Code for a High Intensity Accelerator Ring 390
      A. U. Luccio, J. Beebe-Wang, M. Blaskiewicz, J. Galambos, J. Holmes, and D. K. Olsen

Nonlinear δ-F Simulation Studies of High-Intensity Matched-Beam Propagation in Periodic-Focusing Transport Systems 408
      P. H. Stoltz, W. W. Lee, and R. C. Davidson

Nonlinear Self Consistent High Resolution Beam Halo Algorithm in Homomorphic and Weakly Chaotic Systems 420
      J. H. Whealton, R. J. Raridon, D. K. Olsen, J. D. Galambos, and J. A. Holmes

SUMMARIES OF WORKING GROUPS

Summary of the Observation Group 431
      R. Cappi

Summary of the Theory Working Group 439
      R. Baartman

Summary of the Working Group on Simulation 441
      A. U. Luccio

Author Index 447

PREFACE

The main purpose of the Workshop was to review recent progresses and the current state of understanding of space charge related phenomena in high intensity hadron synchrotrons and accumulator rings, with special attention to new machines being considered (such as the SNS, ESS, JHF, µ-Collider Driver...). Attendance was by invitation.

The following colleagues kindly agreed to set up a Program Committee

S. Chattopadhyay	LBL
T-S. Wang	LANL
S-Y. Lee	Indiana University
A.U. Luccio (vice-chair)	BNL
D.K. Olsen	ORNL
M. Reiser	University of Maryland
W-T. Weng (chair)	BNL

Some answers we were trying to find are:
- What are the experimental observations of high intensity effects.
- What are the theoretical explanations and predictions.
- What computer simulation codes are available.

The Workshop took 3 and 1/2 days. In the first two days we had plenary presentations on: (1) Review of experimental results (2) Review of the theory, (3) Review of simulation computer codes, (4) Existing High Intensity Rings, (5) New Proposed Facilities. Then, Working Groups were formed on the general topics 1,2,3.

In choosing a location for the Workshop we wanted a place both not too far from Brookhaven and a secluded location. Shelter Island came to mind, as it has been the location of a number of conferences through the years, not the least being the very famous conference of 1947. That conference many of us remember, and more have read about, was, arguably, the most productive conference in the history of physics. It may be worthwhile to recall those great times.

"The attendees' bus was escorted by police from the AIP Headquarters in Manhattan to the Island and the participants set the ground for Quantum Electrodynamics (QED). Prior to that time it was well-known that there were infinities in QED, but it was thought that--somehow--these infinities could be removed and the results would be zeros (like, for example, the self-energy of the electron). Here, however, for the first time, results were presented which showed that QED must give non-zero and finite results if it was to explain the new experiments. The new experiments, from Columbia University, the I.I. Rabi group, were the shift of the 2P from the 2S state in hydrogen as disclosed by Lamb and Retherford and the magnetic moment of the electron as disclosed by Kusch and Foley. Both of these effects were not predicted by the Dirac theory and therefore required a finite QED evaluation.

Subsequent to the conference, actually on the train going back to Ithaca, there was a first calculation of the Lamb shift by Bethe which showed that it was very likely that QED would give the correct result (if cut-off properly). Subsequently there were proper calculations by Lamb and Kroll, French and Weisskopf, Schwinger, and Feynman. And, of course, all of QED was developed by Schwinger and Feynman (and independently by Tomonaga). But, one can say that it all started at that very famous conference in 1947".

We had a long history of operations of high intensity proton synchrotrons from the U.S., Europe, and Japan. However, in terms of average beam power, they are all in the range of a few to two hundred KW. The new generation of applications both in the spallation neutron source and µ-collider drivers requires average beam power from a few MW to 10 MW. This presents a new challenge in the precision of accelerator analysis and control of beam losses. To help workshop discussions and focus on the present issues for the design of new high power facilities, we assigned specific tasks to the three Working Groups:

Questions for Beam Observations Group.
  1. What are the ways to observe emittance growth, their reliability and precision.
  2. What are the ways to observe particle losses, their reliability and precision.
  3. Observation and experience of particle losses in existing facilities.
  4. Suggestion of methods to be adopted for high intensity synchrotrons in emittance and particle loss observation.
  5. What is the best strategy to design and place collimators in rings to reduce halo?

Questions for Theory Group
  1. Relationship between single particle and coherent picture of resonances in synchrotrons.
  2. Methods and accuracy of emittance growth and particle losses calculation by resonance models.
  3. Mechanisms of halo generation in synchrotrons.
  4. Methods and accuracy of emittance growth calculation by particle-core models.
  5. How to minimize halo generation in synchrotrons.

Questions for Simulation Group
  1. Characteristics, strength and weakness of existing 6-D, space-charge dominated tracking codes.
  2. What are the specific recommendations to improve them, algorithms, speed, convergence, relevance, ...?
  3. How to validate a tracking code, self-consistency, reliability, convergence, calibration with observations and other codes.
  4. Is it possible to build a code everybody agrees to use?

Our Workshop was held on May 4-7, 1998 at the Pridwin Hotel on the waterfront. The attendance was great and the weather collaborated. We had many discussions, sometimes heated, in the real spirit of a Workshop, and in conclusion we learned a lot.

W-T Weng
A.U. Luccio
Brookhaven National Laboratory
Upton, NY 11973-5000, USA

Participants

Jose Alonso
    Oak Ridge National Laboratory
    FEDC - 10H Union Valley Rd, Oak Ridge, TN 37831-8218
    alonsojr@ornl.gov

Chuck Ankenbrandt
    Fermilab MS-220
    POBox 500, Batavia, IL 60510
    ankenbrandt@fnal.gov

Rick Baartman
    TRIUMF
    4004 Wesbrook Mall, Vancouver BC V6T2A3, Canada
    krab@triumf.ca

John J. Barnard
    Lawrence Livermore National Laboratory
    L-645, Livermore, CA 94550
    jjbarnard@llnl.gov

Yuri K. Batygin
    RIKEN
    2-1 Hirosawa, Wako-shi, Saitama 351-01, Japan
    batygin@rikaxp.riken.go.jp

Joanne Beebe-Wang
    Brookhaven National Laboratory
    AGS Dept. 911A, Upton, NY 11973-5000
    bbwang@ad1.ags.bnl.gov

Mike Blaskiewicz
    Brookhaven National Laboratory
    AGS Dept. 911B, Upton, NY 11973-5000
    mmb@bnl.gov

Roberto Cappi
    CERN, PS Division
    CH-1211 Geneva 23, Switzerland
    roberto.cappi@cern.ch

Weiren Chou
    Fermilab
    POBox 500, Batavia, IL 60510
    chou@fnal.gov

Ronald L. Davidson
    Princeton Plasma Physics Lab
    Box 451, Princeton, NJ 08543
    rdavidson@pppl.gov

Vadim Dudnikov
    8 Beverly Commons 34
    Beverly, MA 01915
    dudnikov@erols.com

Mikhail D'Yachkov
    TRIUMF
    4004 Wesbrook Mall, Vancouver BC V6T2A3, Canada
    dyachkov@triumf.ca

Giuliano Franchetti
    Gesellschaft für Schwerionenforschung mbH
    HSB Group, Planckstrasse 1, D-64291, Darmstadt, Germany

g.franchetti@gsi.de
Alex Friedman
    Lawrence Livermore National Laboratory
    L-645, POBox 808, Livermore, CA 94550
    af@llnl.gov
John D. Galambos
    Oak Ridge National Laboratory
    FEDC - 10H Union Valley Rd, Oak Ridge, TN 37831-8218
    jdg@ornl.gov
Robert L. Gluckstern
    University of Maryland
    Physics Department, College Park, MD 20742
    rlg@quark.umd.edu
Terry F. Godlove
    FM Technologies
    9713 Manteo Court, Ft. Washington, MD 20744
    tgodlove@gmu.edu
Ingo Hofmann
    Gesellschaft für Schwerionenforschung mbH
    Planckstrasse 1, D-64291, Darmstadt, Germany
    i.hoffman@gsi.de
Jeffrey Holmes
    Oak Ridge National Laboratory
    FEDC, Oak Ridge, TN 37831-8218
    jzh@ornl.gov
Dong-O Jeon
    Indiana University Cyclotron Facility
    2401 Milo B. Sampson Lane, Bloomington, IN 47405
    jeon@iucf.indiana.edu
Frederick Jones
    TRIUMF
    4004 Wesbrook Mall, Vancouver BC V6T2A3, Canada
    fwj@triumf.ca
Rami A. Kishek
    University of Maryland, IPR
    Energy Research Bldg, College Park, MD 20742-3511
    ramiak@ipr.umd.edu
Patrick Knaus
    CERN, PS Division
    CH-1211 Geneva 23, Switzerland
    patrick.knaus@cern.ch
Ioanis Kourbanis
    Fermilab
    Beam Division MS 341, POBox 500, Batavia, IL 60510
    ioanis@fnal.gov
S-Y Lee
    Indiana University
    Dept. of Physics, SW 117, Bloomington, IN 47405
    shylee@indiana.edu
W.W. Lee
    Princeton Plasma Physics Lab
    Theory Dept., Princeton, NJ 08543
    wwlee@pppl.gov
Y-Y Lee

Brookhaven National Laboratory
AGS Dept. 911B, Upton, NY 11973-5000
yylee@bnl.gov

Derun Li
Lawrence Berkeley Laboratory
One Cyclotron Rd, MS-71-259, Berkeley, CA 94720
dli@lbl.gov

Alfredo U. Luccio
Brookhaven National Laboratory
AGS Dept. 911B, Upton, NY 11973-5000
luccio@bnl.gov

Steven M. Lund
Lawrence Livermore National Laboratory
L-645, POBox 808, Livermore, CA 94550
lund@hif.llnl.gov

Robert J. Macek
Los Alamos National Laboratory
LANSCE-DO, MS H848, PO Box 1663, Los Alamos, NM 87545
macek@lanl.gov

Shinji Machida
KEK
3-2-1 Midori-cho, Tanashi-shi, Tokyo, 188-8501, Japan
shinji,machida@kek,jp

Sig A. Martin
Institute for Nuclear Physics
Forschungszentrum Juelich, D-52425 Juelich, Germany
s.martin@fz-juelich.de

Michel Martini
CERN, PS Division
CH-1211 Geneva 23, Switzerland
michel.martini@cern.ch

Filippo Neri   LANCE-1, MS H808
Los Alamos National Laboratory
Los Alamos, NM 87545
fneri@lanl.gov

David Neuffer
Fermilab MS-220
POBox 500, Batavia, IL 60510
neuffer@fnal.gov

King Y. Ng
Fermilab MS-220
POBox 500, Batavia, IL 60510
ng@fnal.gov

Milorad Popovic
Fermilab MS-220
POBox 500, Batavia, IL 60510
popovic@fnal.gov

David K. Olsen
Oak Ridge National Laboratory
FEDC - 104 Union Valley Rd, Oak Ridge, TN 37831-8218
dkp@ornl.gov

Michael Plum
Los Alamos National Laboratory
LANSCE-2, MS H838, PO Box 1663, Los Alamos, NM 87545

       plum@lanl.gov
Milorad Popovic
    Fermilab
    Beam Div/Proton Source, Batavia, IL 60510
    popovic@fnal.gov
Chris Prior
    Rutherford Appleton Lab
    ISIS Dept., Chilton, Didcot Oxon OX11 0QX, UK
    crp@axpr12.rl.ac.uk
Deepak Raparia
    Brookhaven National Laboratory
    Bldg, Upton, NY 11973-5000
    raparia@bnl.gov
Martin Reiser
    University of Maryland, Institute for Plasma Research
    Energy Research Facility #, College Park, MD 20742-3511
    mreiser@glue.umd.edu
Thomas Roser
    Brookhaven National Laboratory
    Bldg 911B, AGS Dept., Upton, NY 11973-5000
    roser@bnl.gov
Andrew Sessler
    Lawrence Berkeley Laboratory
    One Cyclotron Rd, MS-71-259, Berkeley, CA 94720
    tbalbl@lbl.gov
Peter Stoltz
    Princeton Plasma Physics Lab
    Box 451, Princeton, NJ 08543
    pstoltz@pppl.gov
Marco Venturini
    University of Maryland
    Energy Research Facility #, College Park, MD 20742-3511
    mreiser@glue.umd.edu
J.G. Wang
    University of Maryland
    Institute for Plasma Research, College Park, MD 20742
    jgwang@plasma.umd.edu
Tai-Sen F. Wang
    Los Alamos National Laboratory
    LANSCE-1, MS H808, PO Box 1663, Los Alamos, NM 87545
    twang@lanl.gov
Thomas Wangler
    Los Alamos National Laboratory
    LANSCE-1, MS H817, PO Box 1663, Los Alamos, NM 87545
    twangler@lanl.gov
Chris Warsop
    Rutherford Appleton Lab
    ISIS Dept., Chilton, Didcot Oxon OX11 0QX, UK
    c.m.warsop@rl.ac.uk
Wu-Tsung Weng
    Brookhaven National Laboratory
    Bldg. 911B, Upton, NY 11973-5000
    weng@bnl.gov
John Whealton

Oak Ridge National Laboratory
Bldg. 9108, Oak Ridge, TN 37831-8088
z3w@ornl.gov

Mark Wilson
US Dept. of Energy
ER-224, 19901 Germantown Rd., Germantown MD 20874-1290
mark.wilson@oer.doe.gov

S. Y. Zhang
Brookhaven National Laboratory
Bldg. 911B, Upton, NY 11973-5000
syzhang@bnl.gov

Bruno W. Zotter
CERN, PS Division
CH-1211 Geneva 23, Switzerland
bruno.zotter@cern.ch

PLENARY TALKS

SPACE CHARGE IN PROTON LINACS

T.P.Wangler, F. Merrill, L. Rybarcyk, and R. Ryne

Los Alamos National Laboratory

1. Introduction

There are at least two reasons why we may be interested in space-charge effects in proton linacs when discussing the physics of circular machines. First, we can expect that there are areas of commonality in the space-charge physics of linacs and circular machines. Second, a linac delivers the input beam to a circular machine, so understanding the linac physics helps to explain the limitations for the input beam quality to a ring. This presentation is divided into three parts. First, we discuss space-charge effects from the linac point of view. Second, we discuss practical methods of calculation of linac beam dynamics that include space-charge forces. Finally, we summarize the status of experimental studies of the beam performance in the LANSCE linac including space-charge effects.

2. Space Charge from the Linac Viewpoint

We begin by reviewing the characteristics of space-charge forces in accelerator beams. A smooth space-charge potential is generally used to replace the Coulomb interactions of the particles in the beam. This is a good approximation when the number of particles in a sphere with radius equal to the Debye length is much larger than unity, and this is usually true for beams in linacs and rings; in high current linacs the number of particles in a Debye sphere is typically of order 10^6. Even with this approximation, the space-charge force does not have simple properties. In general it is nonlinear, time dependent, coupled in x, y, and z, and exhibits effects that are common in plasma physics such as collective oscillations and Debye shielding.

In a proton linac the space-charge force can have major consequences on the beam dynamics. First, the beam current is limited by the space-charge force in both the longitudinal and transverse directions. The value of the matched rms beam size is determined by a balance between the defocusing effects of emittance and space charge, and the focusing effect of the external forces. If the beam current increases and all other parameters remain constant, the matched rms beam size will increase. In the transverse direction the aperture limits the growth of the beam size and determines a transverse current limit. In the longitudinal direction the beam size is limited by the phase width of the rf separatrix, which sets the longitudinal current limit. Assuming that the bunch density is uniform, the space-charge force is a linear function of displacement from the bunch center, and analytic formulas can be derived for these current limits.

If the bunch density is nonuniform, the space-charge fields are nonlinear. The nonlinear space-charge fields can produce filamentation of the phase-space distribution, resulting in growth of the rms emittances. In addition if the rms size of the beam is mismatched to the periodic focusing lattice, the rms beam size will oscillate. These oscillations combined with the nonlinearity can produce an extended beam halo and a large growth of the rms emittance. The halo formation can be understood as a resonant interaction of individual particles traveling through the beam core with the time oscillating space-charge field of the core, and occurs when the particle frequency is about half the core frequency.

The importance of the space-charge force in a linac can be measured by the tune-depression ratios, one for each plane. To define the tune depression for a given plane, we first define σ_0 as the phase advance per focusing period of single particles undergoing betatron motion in the limit of zero beam current; σ_0 is a measure of the strength of a linear external focusing force. Another important quantity σ is defined as the phase advance per focusing period including the beam current. If defined in this simple way, for a nonuniform density beam there would not be a unique value for σ. A useful definition of σ that provides a unique value for the beam is obtained by referring to an equivalent uniform beam, obtained by replacing the real beam with an equivalent uniform beam with the same second moments as the real beam. Then, the tune-depression ratio is defined as σ/σ_0. The extreme values $\sigma/\sigma_0 = 0$ and 1, represent the space-charge and emittance-dominated limits, respectively. For a high current proton linac, the tune-depressions typically range between about $0.5 \leq \sigma/\sigma_0 \leq 0.9$.

In circular machines it is customary to measure the importance of space charge by a tune shift $\Delta\nu = \nu - \nu_0$, rather than the tune ratio that we use in linacs. This makes sense because in circular machines the current is limited by machine resonances that depend on the tune ν. To compare the parameters σ and ν, we note that $\nu_0 = N_p \sigma_0 / 2\pi$, where N_p is the number of focusing periods in the ring. The quantity ν includes space charge, and if this is defined as $\nu = N_p \sigma / 2\pi$, where σ is defined in terms of the equivalent uniform beam, we find that $\sigma / \sigma_0 = 1 - |\Delta \nu| / \nu_0$. Suppose we substitute $|\Delta \nu| = 0.25$, which is a conservative rule of thumb for a maximum space-charge tuneshift, and choose a value for the zero-current tune $\nu_0 = 5.75$ that is close to the design value for the SNS ring. Then we find that $\sigma/\sigma_0 = 0.95$, which is not much tune depression by linac standards. This shows that the beam in a ring is in the emittance-dominated regime, and suggests that most direct space-charge effects that are observed in linacs should be relatively less important for beams in rings. Another way to look at this is to say that it takes very little space charge in the ring before you encounter a resonance. One other implication of this emittance-dominated result is that the equilibrium beam distribution in the ring will be expected to have a Gaussian-like profile, characteristic of an emittance-dominated beam where the Debye length, which is the length scale over which the beam density falls to zero, is large compared with the rms beam size.

Next let's return to the subject of space-charge effects in the linac. In a proton linac the beam quality is degraded by space-charge-induced emittance growth. These emittance-growth effects can be separated into four general categories. First is the charge redistribution effect, which is the main cause of emittance growth in rms-matched nonequilibrium beams when there is a change in the focusing lattice. For this mechanism the emittance grows very rapidly within one-quarter plasma period with an associated transfer of space-charge field energy to particle kinetic energy as the charge redistributes to balance the forces within the beam. Second is kinetic energy transfer between planes, which is caused by a tendency of beams to reduce large thermal asymmetries between planes. This mechanism usually has a time scale of tens of plasma periods. Third is emittance growth caused by mismatch of the rms beam ellipse to the acceptance of the periodic focusing lattice. This mechanism derives its free energy for the initial potential energy associated with imbalance between the external focusing force and the defocusing forces associated with space charge and emittance; a typical time scale is tens of plasma periods. There is a fourth category of emittance growth caused by envelope instability in a periodic channel when the focusing force is too strong. This mechanism can be avoided by requiring that $\sigma_0 < 90°$.

In practice the rms mismatched case supplies the most free energy for emittance growth and this growth can lead to a significant extended beam halo. The beam halo that is observed in 2D phase-space projections may appear either as a phase-space structure in the form of filaments or rings, or it may be a relatively diffuse and featureless extension of the beam density. In 2D transverse phase-space projections of a 3D bunch the filaments that form from the nonlinear space-charge force often project into a smooth featureless halo distribution. The physical mechanism of emittance growth and halo formation for an rms mismatched beam can be explained as resulting from the space-charge field of an oscillating beam core, which can resonate with the particles undergoing betatron oscillations. This mechanism has been studied using the particle-core model in which one assumes a linear, uniform focusing channel with a round continuous uniform density beam that is mismatched to excite an azimuthally symmetric breathing mode (see Fig.1). Test particles are launched to represent individual particles passing through the core. The equations of the model include the round-beam envelope equation to represent the breathing oscillations of the core, and the single particle equation of motion in the field of a uniform density beam for a particle which passes back and forth through the core. One finds that a parametric resonance occurs when the particle oscillation frequency is half the core oscillation frequency [1]. Although the resonant amplitude is self limiting because of the amplitude dependence of the space-charge force outside the beam, the resonance is nevertheless effective at driving some particles to much larger amplitudes. For a given value of a mismatch parameter, defined as the ratio of the initial beam size to the matched beam size, the resonant amplitudes have a maximum value for a given mismatch, which is identified by a separatrix that is observed in a stroboscopic plot of the phase-space motion [2] (see Fig. 2). By solving the model equations numerically, one finds that the maximum particle amplitude normalized to the rms beam size is insensitive to the tune depression over a wide range of tune-depression values. Although the tune depression does not have a strong influence on the normalized maximum amplitude, it does influence the growth time of the halo so that for a beam in the emittance-dominated limit, the time for a resonantly-driven particle to move from a point near the inner separatrix where its amplitude is smallest to the outer separatrix where its amplitude is greatest is much larger than for a beam in the space-charge limit. The model predicts that as the mismatch is turned on, the maximum amplitude changes very strongly at a value of the mismatch parameter equal to unity (see Fig 3), so that according to the model even a few percent of mismatch would result in observable beam halo even for an emittance-dominated beam. Thus, the model predicts that the beam halo from mismatch may be hard to avoid in a real accelerator.

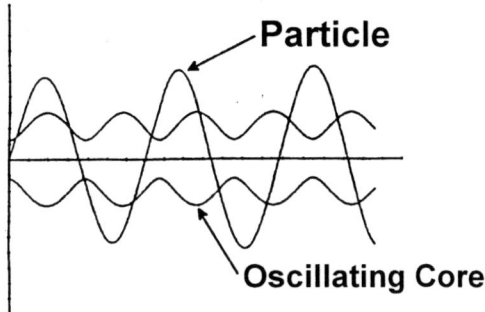

Fig. 1 Transverse oscillations versus time for a particle and an oscillating beam core as represented in particle-core model.

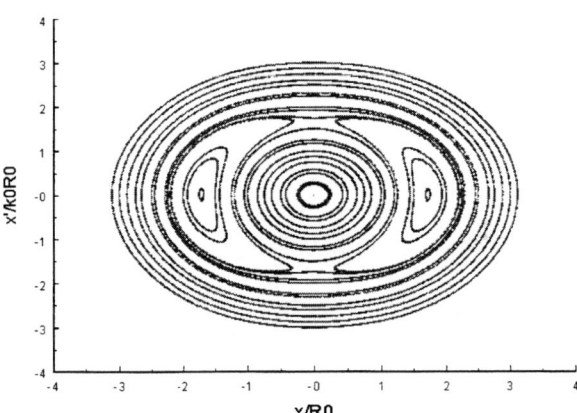

Fig. 2 Stroboscopic plot from the particle-core model of the phase space motion of 32 initial particle coordinates for a tune depression ratio of 0.95 and an initial mismatch (ratio of initial core size to matched core size) of 0.9, showing the central core region, and the resonance region centered on the two stable fixed points. The phase-space motion is strobed each time the core is at its minimum size. The intersection of the outer separatrix with the horizontal axis gives the maximum amplitude for the assumed tune-depression ratio and mismatch value.

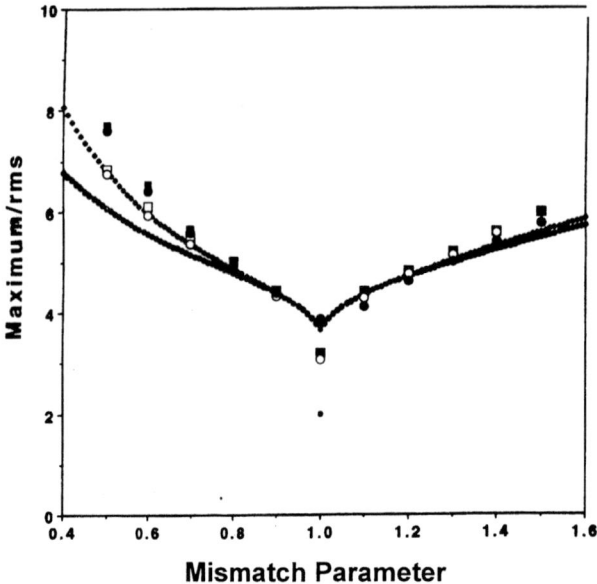

Fig. 3. Ratio of maximum amplitude to rms size versus the mismatch parameter from the particle-core model. The curves show the predictions of the particle core model for two tune depression ratios 0.5 and 0.9. The points show the results from numerical simulation for the same two tune ratios and two different initial distributions, Gaussian (squares) and thermal (circles).

The numerical solution of the model shows that the maximum amplitude is proportional to the core radius which is proportional to the rms core size, which is approximately the rms beam size. An analytic solution exists that relates the rms beam size to the beam parameters [2], and from this we obtain a scaling formula for the maximum amplitude of resonantly driven halo particles. The scaling formula predicts that for a given initial mismatch, the maximum amplitude increases with increasing initial rms emittance, and decreases with increasing bunch length, increasing bunch frequency, and increasing focusing strength.

The design principles for a high-current proton linac are strongly influenced by the need to control the space-charge effects that we have discussed. For pulsed linac injectors to rings, the main requirement is usually the need to provide linac output emittances that are safely within the acceptance of the ring. For CW linacs the requirement is usually based on the need to limit the beam losses in the linac. For energies above about 100 MeV the beam losses should be limited to at least a few watts per meter of lost beam power. As a general rule strong focusing should be provided in all three planes. It is desirable to avoid abrupt changes in the focusing strength that could lead to beam mismatch. If transitions such as to a new focusing lattice or a new frequency can not be avoided, beam matching should be provided. Beam matching requires adjusting the focusing strengths to make the Courant-Snyder parameters of the injected beam equal to the matched values for the periodic focusing lattice. The matching can be made nearly independent of beam current if σ_0/L, where L is the focusing period, is the maintained the same on both sides of the transition. Choosing higher frequency can reduce space-charge effects by distributing the total beam charge over more bunches and by increasing the longitudinal focusing strength. The frequency choice is generally limited by transverse and longitudinal acceptances.

Finally, we summarize some characteristics of proton beams in rf linacs that are often different than for beams in circular machines. Bunches in linacs are nearly spheroidal with aspect ratios that are not far from unity; linac bunches are usually longer than their transverse sizes. The bunch dimensions in rf linacs are usually small compared with the aperture radius, and as long as the beam is relatively well aligned, the image forces from charges induced in the walls are usually small in linacs compared with the direct space-charge force. With the possible exception of beam-steering elements of beam transport lines external to the linac, the linac beam passes through no dipole magnets, and linac beams are not subject to resonances caused by periodic sampling of the same errors as in circular machines. Consequently, more highly depressed tunes are possible in linacs. The linac beam spends only a few microseconds in the linac, so linac beams usually do not attain thermal equilibrium.

3. Practical Calculation of Linac Space Charge Forces

Because of the complexity of the space-charge force, calculation of space charge effects by analytic methods are of limited value and computer codes are necessary. Fortunately, the rapid progress in computer capability in the past 20 years has increased the number of particles that can be traced through the linac from 10^3 to 10^7. The simplest space-charge programs are those which replace the actual beam distribution with an equivalent uniform beam that has the same second moments as the real beam. This type of code traces the beam envelope and is especially useful for beam-matching optimization. However, to represent the full nonlinear space-charge force that is responsible for emittance growth and halo formation, multiparticle simulation methods are generally used. Although it is not practical to model the interparticle Coulomb collisions for the full 10^9 particles in a real linac bunch, representing these collisions by a smoothed space-charge potential and tracking a smaller number of macroparticles is a good approximation. We use the particle-in cell (PIC) method, in which at each step a mesh is superimposed on the bunch to smooth the space-charge fields. This reduces the effects of artificially large forces that would otherwise be caused by binary encounters between macroparticles. Computer simulation runs are carried out with up to about 10^5 macroparticles on personal computers, and 10^7 macroparticles on high performance parallel processor computers. SCHEFF, a 2D PIC code, is still the most commonly used space-charge routine for proton-linac simulations. Recently we have begun using a 3D PIC code.

The objective of the simulation runs with numbers as large as 10^7 particles is to obtain better statistical precision especially for the outer parts of the beam distribution. Running 10^7 particles using 128 processors on a **64X64X128** grid takes about 5.5 hours. Comparison was made between the SCHEFF and 3D PIC simulations of the 1.7-GeV proton-linac design for accelerator production of tritium. Excellent agreement for both the rms sizes and emittances and for the maximum particles was obtained for runs of 10^5 macroparticles using a **20X40** grid for SCHEFF and a **32X32X64** grid for the 3D PIC subroutine.

4. Experimental Studies of the LANSCE Linac

Next we describe the study of the LANSCE (formerly known as LAMPF) linac carried out by Garnett, Mills, and Wangler in 1989 [3], which has been continued and improved by recent work of F. Merrill and L. Rybarcyk[4]. A block diagram of the linac is shown in Fig.4. The studies involved simulation of the LANSCE H^+ beam and comparison of the simulation results with beam measurements. The objective was to compare the predictions of the simulation code with experimental results. Perhaps the most important part of the study was to determine the 6D input beam distribution at the drift-tube linac input. Measurements of beam current, transverse beam profile and transverse emittance were available in the low-energy beam transport (LEBT) upstream of the DTL, and these measurements were used to determine the transverse beam characteristics. The longitudinal properties of the injected beam were determined by simulation of the bunching through the fields of the buncher cavities. Typical results are shown in Fig.5. The space-charge force causes a noticeable distortion of the phase-space distributions at the input. Partial neutralization of the beam charge occurs through collisional ionization of the residual gas, and the effective beam current was determined by choosing the value that produced the best fits between the measured and calculated rms beam sizes. Table 1 shows the comparison of simulations with measurements for the transverse emittances at 100 and 800 MeV, the rms beam size at 100 MeV, and the bunch width at 100 MeV. The width is determined by sweeping the phase of the cavity to move the longitudinal rf separatrix through the beam bunch and uses an absorber and collector to measure the energy gain difference between the accelerated and non-accelerated beam. Finally the integrated particle loss along the coupled-cavity linac (CCL) was compared with the integrated loss obtained from the current monitors. The error in the measured emittance at 100 MeV is obtained from the difference between the two measurements, one from a wire-scan method and one from a slit and collector method. Error estimates for the other measurements were not available. Errors applied to the simulated quantities were based on the differences between results for a range of different assumptions about the charge neutralization and component parameter values in the LEBT. With the exception of the particle loss, the simulation results were in fairly good agreement with the measurements. We found that the magnitude of the particle loss prediction from simulation had a large uncertainty, which we believe is a result of the uncertainties associated with the population of the tail in longitudinal phase space that is caused by the bunching process.

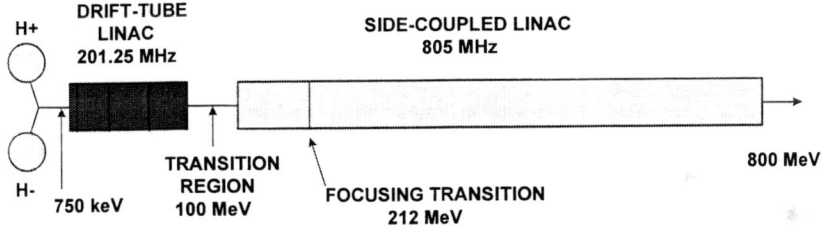

Fig. 4. Block diagram of the LANSCE proton linac at Los Alamos.

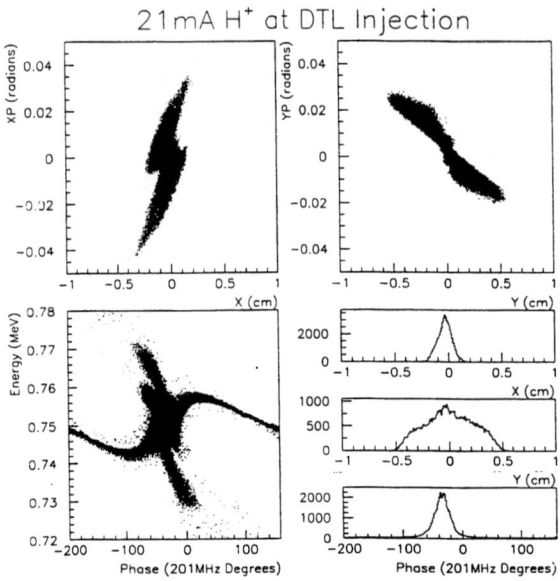

Fig. 5. Phase-space distributions and projections at the drift-tube linac input from simulation of the bunching process in the LANSCE linac.

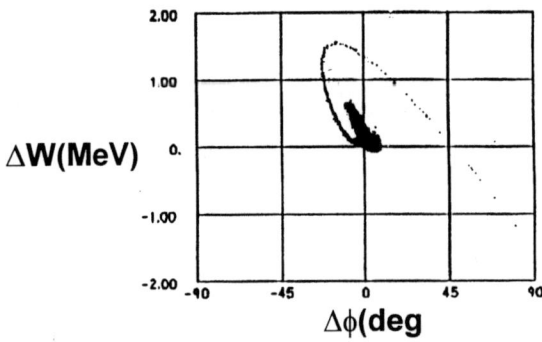

Fig. 6. Longitudinal phase space at 100 MeV from a simulation of the LANSCE linac.

Table 1 Comparison of Measurements with Simulation of the LANSCE Linac for a 15-mA H⁺ Beam.

Quantity	Measurement	Simulation
$\varepsilon_{rms,n}$ (mm-mrad) at 100 MeV	0.40±0.14	0.29±0.03
$\varepsilon_{rms,n}$ (mm-mrad) at 800 MeV	0.71	0.65±0.02
x_{rms} (mm) at 800 MeV	5.4	7.0±1.0
FWHM Phase width (deg) at 100 MeV	16	20±3
Particle loss (%) at > 100 MeV	~0.1	0.9±0.5

Beam losses above about 100 MeV are the main concern at LANSCE because of the induced radioactivation of the accelerator, which inhibits hands-on maintenance on the machine. The experimental distribution of losses along the CCL can be approximately obtained from activation measurements made after the run, together with a model that relates activation to beam losses versus energy. The locations of the two main beam loss peaks in the CCL near 100 and 212 MeV were well reproduced by the simulations. From the simulations we are able to explain the beam-loss mechanisms. A first peak near 100 MeV is caused by inadequate matching and transverse acceptance of the focusing lattice. A second peak at about 212 MeV occurs where the transverse acceptance drops at a transition where the focusing period doubles. The beam losses at that location are mostly caused by 100-MeV particles that were not captured in the 805-MHz longitudinal buckets at the 100-MeV injection into the CCL. We refer to these as longitudinal losses, and the lost particles are those that are included in the tail in longitudinal phase space caused by the incomplete bunching of the beam injected into the DTL. Other simulation studies were made of the LANSCE linac with an RFQ replacing the first DTL tank. These studies have shown that the longitudinal tails would be almost completely removed by the superior bunching characteristics of the RFQ. A proton linac with an RFQ front end would not be expected to have beam losses from the longitudinal tail, but beam halo could still be produced by mismatch. Detailed and systematic measurements of the outer beam halo after the linac were carried out by H. Koziol [5] in 1975, using a movable plate and beam-loss scintillation monitors to detect the scattered radiation from the plate.

However, machine time was never made available to make comparable halo measurements in the linac.

5. Conclusions

In this paper we have discussed space-charge effects from the linac viewpoint. The space-charge force in a linac produces limiting currents, emittance growth that degrades the beam quality, and beam halo. We have discussed the impressive advances in computing capability that have allowed us to simulate space-charge effects in a linac; the number of macroparticles that can be traced through the linac has increased in recent years from about 10^3 to 10^7. Also, excellent agreement is found between the main space-charge routines. This simulation capabilility gives us more confidence to describe the details of the beam distribution including the beam halo for a linac design that includes realistic errors. Experimental studies of the LANSCE linac reveal the complexity of the effects in a real linac and shows how they affect the beam distribution. We find that there is good agreement between beam simulation and LANSCE beam measurements for the rms quantities, the maximum beam size and the relative beam-loss distribution. Further progress can be made if we can obtain better experimental measurements of beam halo in the linac itself.

References

[1] R. L. Gluckstern, Phys. Rev. Lett. 73(1994) 1247.

[2] T. P. Wangler, E. R. Gray, S. Nath, K. R. Crandall, and K. Hasagawa, "New High Power Linacs and Beam Physics", Presented at the 1997 Particle Accelerator Conference, Vancouver, May 12-16, 1997.

[3] R. W. Garnett, R. S. Mills, and T. P. Wangler, " Beam Dynamics Simulation of the LAMPF Linear Accelerator", Proc. Of 1990 Linear Accelerator Conf., Sep. 10-14, 1990, Los Alamos Report LA-12004-C, pp 347-350.

[4] Frank Merrill and Lawrence J. Rybarcyk, " Transverse Match of High Peak-Current Beam into the LANSCE DTL Using PARMILA", Proc. of XVIII International Linear Accelerator Conf., Aug. 26-30, 1996, CERN report CERN 96-07, pp 231-233.

[5] H. Koziol, "Halo Measurements on the LAMPF 800 MeV Beam", Los Alamos MP-Division Report MP-3-75-1.

Observation and Interpretation of Space Charge Phenomena for High Intensity

I. Hofmann, R. Bär, O. Boine-Frankenheim, G. Rumolo

GSI, Planckstr.1, 64291 Darmstadt, Germany

Abstract. We discuss measurements on space charge dominated beams in circular accelerators, which could be of interest for high-current machines. In the transverse phase plane a crucial issue for high intensity is the direct observation of space charge, which was demonstrated recently at LEAR and the SIS by means of quadrupolar pick-ups. The issues of uniform and non-uniform space charge densities as well as mismatch oscillations are of interest in the context of halo formation and beam loss. For the longitudinal plane the self-bunching instability due to a resistive impedance has been measured in detail in the ESR and found to confirm very well the linearized theory; in the nonlinear phase of simulations modeling the experiment no overshoot was found, in contrast with theoretical predictions for beams without space charge.

I INTRODUCTION

Observation of space charge phenomena at high intensities and phase space densities is of increasing interest in light of the very demanding requirements on low beam loss in high power spallation sources and other applications. The issues in circular accelerators are quite distinct from those in linear accelerators, but certain features remain the same. In particular the mechanisms for beam halo formation are believed to be relatively insensitive to the actual strength of space charge forces [1,2]. Direct observation of space charge induced frequency shifts is possible in circular machines by recording signals over a large number of revolutions (applied here to the quadrupolar oscillations). As far as measurements in the longitudinal phase space an interesting question is whether the threshold of instability can be exceeded without significant deterioration of the phase space density (overshoot). Space charge acting as a broad-band impedance can lead to nonlinear behavior with features that differ from the case of relativistic beams with negligible space charge, which have been studied extensively in the past.

In this paper we focus mainly on experiments performed at the heavy ion synchrotron SIS (18 Teslameter) and the storage ring ESR (10 Teslameter) with electron cooling, both at GSI. Computer simulation has been found a necessary tool for interpreting results from observation of space-charge-dominated beams. At GSI a diversity of codes is available based on different methods and space charge calculation techniques. We have developed and used several particle-in-cell (PIC) codes: SCOP-2D(x,y), PATRIC(r,z) and SCOP-3D(x,y,z). PATRIC(r,z) is the most versatile code containing RF options on different harmonics, an r-z Poisson solver, cavity impedances and Schottky noise analysis. More recently we have employed for longitudinal simulation a direct Vlasov integrator which avoids the noise inherent to PIC simulation. This code is very useful to study long-term evolution and resolve fine structures in phase space. The PATRIC-code is presently extended to include diffusion processes and cooling forces (electron cooling, laser cooling). Nonlinear resonances are studied with a micro-map code using a Fourier transform based space charge calculation [3].

II OBSERVATION OF TRANSVERSE SPACE CHARGE DENSITY

During injection and RF capture a control of the transverse density profile is required. It is generally accepted that both the shape of the density profile as well as envelope or higher order mismatch play an important role for space-charge-induced halo and beam loss. Direct measurement of the transverse density profile by wire scans is common in proton circular machines. This technique gives sufficiently well resolved density profiles, but unfortunately it cannot be used at very high intensities. The use of beam ionization profile monitors is also likely to be practically limited by the effect of the beam space charge potential.

A Direct Observation by Quadrupolar Oscillations

A method which works particularly well for high intensity without interfering with the beam is based on measuring the frequency shift of quadrupolar (envelope) oscillations induced by space charge. In this method (already proposed by Hardt [4]) the coherent space charge tune shift of quadrupolar oscillations is used to determine the incoherent tune shift, which is difficult to measure directly. There is a theoretical relation between the coherent frequency shift and the incoherent tune shift, which can be derived from Sacherer's rms envelope equations. The result can be written approximately as

$$Q_{coh,1} - 2Q_{0,x} = -(3 - a_x/(a_x + a_y))\Delta Q_{inc,x}/2 \approx -5/4\Delta Q_{inc,x} \tag{1}$$

A second coherent frequency $Q_{coh,2}$ is obtained by simply interchanging x and y. $Q_{0,x}$, $Q_{0,y}$ are the bare machine tunes that are identical with the dipole tunes if image charge effects are negligible. It should be noted that Eq. 1 also describes the coherent resonance shift for crossing of half-integer resonances (see Refs. [5,6]).

In the limiting case of round beams and symmetric focusing the two eigenfrequencies are those of the round "breathing mode" and the "quadrupolar mode" discussed in papers about beam halo (see, for instance, Ref. [1]). These relationships can be used, in combination with the formulas relating the incoherent tune shifts with the emittances, to determine in each plane the incoherent shifts and thus the emittances. Since Eq. 1 is a result of the rms envelope equations valid for all kinds of distribution functions the measured tune shifts only depends on rms quantities, hence we strictly obtain the rms emittances by this method.

We have implemented this method in the heavy ion synchrotron SIS (for details see Ref. [7]) after it was successfully demonstrated at LEAR [8]. The beam is excited with a signal sweeping over the envelope oscillation frequencies which are shifted from $2Q_{0,x,y}$ due to space charge. The coherent response signal on a quadrupolar pick-up is Fourier analyzed. Both the exciter and pick-up must have quadrupolar symmetry. The LEAR-experiment with electron cooled protons has given very interesting insight into the behavior near a half-integer resonance due to the combined use of a beam ionization profile monitor. One of the findings of this experiment was that for an uncorrected resonance the working point could not be shifted (from above) through the half integer by means of increased space charge. The beam responded with emittance blow-up if the zero-current tune was too close to the half-integer.

The scope of the SIS experiment in particular was to carry out a detailed comparison with computer simulation (with SCOP-2D), determine geometry factors and clarify the relationship between coherent and incoherent signals. In our computer simulation we have found, for instance, that in spite of a spread in incoherent frequencies typical for a water-bag distribution, there is no visible broadening of the coherent quadrupolar line (which was claimed for the LEAR experiment).

In the SIS we have used existing equipment consisting of four plates connected appropriately. We have taken the difference from the sum signals on the vertical rsp. horizontal plates, which suppresses the dipole and longitudinal modes (see Fig. 1 where different side bands of these coherent lines are shown). First successful measurements have been carried out with a coasting Ne^{10+} beam of maximum 25 mA at the injection flat-top (11.5 MeV/u) and a typical sweep time of ≈ 1 s. The thus determined ΔQ_{inc} have been used to determine the rms emittances and the space charge shift of the working point (see Fig. 2).

The method works best for high intensity and beam sizes not too small compared with the pick-up spacing, since quadrupolar signals drop with distance. Shorter observation times can be reached with higher intensity, since the needed frequency resolution is reduced accordingly and the signal to noise ratio improves. It should be noted that the method should in principle also work for bunched beams (where a spread of quadrupolar frequencies exists along the bunch) if the quadrupolar signals can be separated from all other coherent bunch lines including synchrotron side bands.

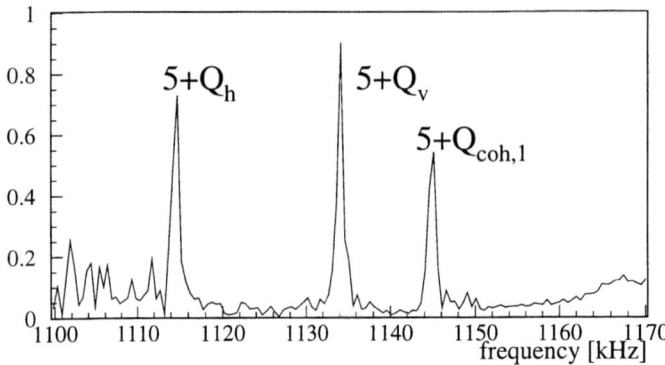

FIGURE 1. Measured spectra on quadrupolar pick-up in SIS showing sideband of one of the quadrupolar modes (horizontal) and two betatron sidebands (dipole mode).

B Space Charge Shifts for Second and Higher Order Resonances.

Besides describing the rms emittance the quadrupolar tune shifts also determine the shift of the resonance condition for half-integer resonances due to space charge. Eq. 1 implies that the actually dangerous resonance condition is not that of the incoherent tune crossing the half-integer, but that of the coherent tune. The latter happens for an 8/5 times increased incoherent tune shift, which allows a corresponding intensity increase. Hence, space charge introduces the important distinction between incoherent and coherent resonance crossing. For the latter the driving force is the sum of the machine error and the space charge force of the perturbed beam. The resonance condition therefore cannot be identical with that of single particles (see also Refs. [5,6]. A similar phenomenon is discussed in a contribution to this workshop [3] for the space charge effect in crossing of a skew resonance.

For higher order resonances there are similar coherent tune shifts. Such higher order modes of originally round beams can be excited by mismatch oscillations, which moves particles out of the core into the halo and helps to determine the absolute intensity in the halo (see Ref. [9]). The frequency shifts for non-round anisotropic KV-distributions can be calculated analytically for modes of third order (sextupolar) and fourth order (octupolar) symmetry [10]. These coherent shifts generally are a function of the tune ratio and the emittance ratio, besides depending on intensity. For a resonance to be satisfied the corresponding incoherent shifts can be increased by factors which are above unity, but not exceeding the 8/5 of the second order (quadrupolar) mode, which appears to be the largest coherent tune

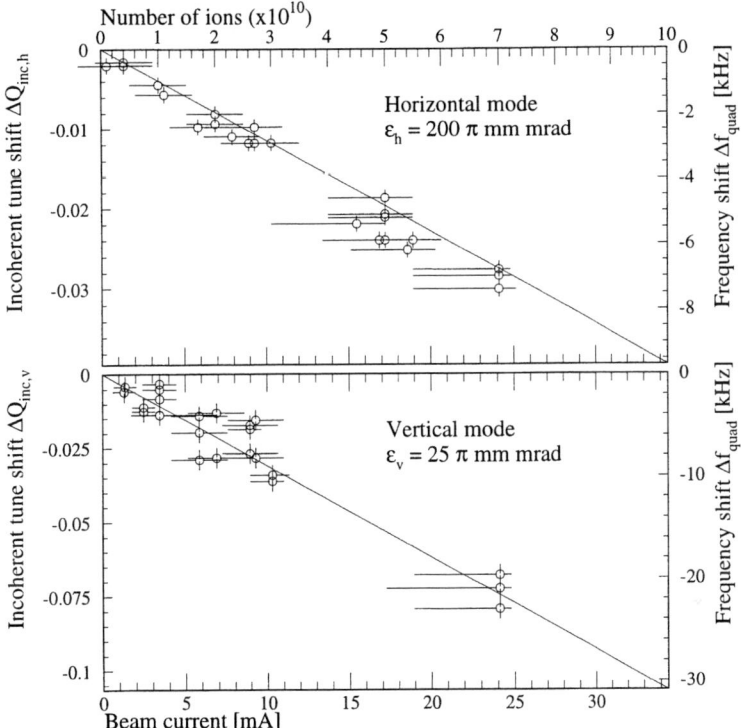

FIGURE 2. Measured space charge shifts of quadrupolar oscillation frequencies in the SIS as a function of intensity.

shift.

Obviously it is desirable - with respect to halo formation - to measure the frequency shifts of these higher order coherent modes and compare them with the theoretical values.

III OBSERVATION OF COHERENT LONGITUDINAL EFFECTS

One of the challenging questions is whether a high-current machine must necessarily operate under conditions where the beam is linearly stable. This issue has been addressed in some detail in the context of heavy ion inertial fusion drivers, where the requirement of longitudinal stability cannot easily be reconciled with the small momentum spread needed for final focusing. It seems acceptable to start outside the region of linear stability provided that either the residence time in the machine is short enough, or the instability saturates with a favorable distribution function. This question will be discussed in the following.

A Linearized Regime

During storage of high currents beams are subject to the longitudinal instability if a resistive impedance component is present at some multiple of the revolution frequency and the momentum spread is below a threshold. The problem was studied experimentally already years ago in the ISR [11] and other machines to check if the theory developed in the sixties (see Refs. [12,13]) was right. These early and the following measurements have been made above transition energy, where the space charge impedance is negligible.

A detailed study of the *space charge dominated regime* has recently been undertaken in the time domain in a linear resistive electron channel at the University of Maryland [14]. The experiment has confirmed the growth rates from linearized theory for fast and slow waves, which could be launched separately by appropriate initial conditions.

In the ESR and SIS we have investigated this mode in frequency and also in time domain. The ESR electron cooling allows to carry out such experiments in the vicinity as well as far away from the stability boundary. We have cooled a C^{6+} beam at 250 MeV/u and 0.3 mA current in order to obtain a very small longitudinal momentum spread near 10^{-5}. The frequency of the RF cavity was then shifted from an initially strongly de-tuned value towards the beam revolution frequency to obtain the expected unstable behavior. A small but finite gap voltage of 300 V was present to control the cavity impedance. The linearized growth rates were found in excellent agreement with the results from computer simulation, which also included the finite gap voltage (see Ref. [15]). Interestingly, the finite gap voltage had the effect of reducing the e-folding growth time.

B Nonlinear Regime and Overshoot

One of the most interesting issues in this context is the question of overshoot. A very simplified picture of nonlinear saturation is that of a beam which self-stabilizes after it has gained enough momentum spread due to the instability. The overshoot theorem then claims that the finally achieved momentum spread is actually larger than the threshold momentum spread necessary to avoid onset of the instability [16].

In the ESR experiments we have found that the real picture is more complicated. In order to evaluate the full history of the instability we have recorded the line density synchronously with the revolution period (mountain range diagram of Fig. 3).

The diagram shows the initially exponential growth of the slow wave (moving to the right, with time increasing from bottom to top), nonlinear saturation and decay into a fast wave moving to the left. The self-bunching effect is generally not

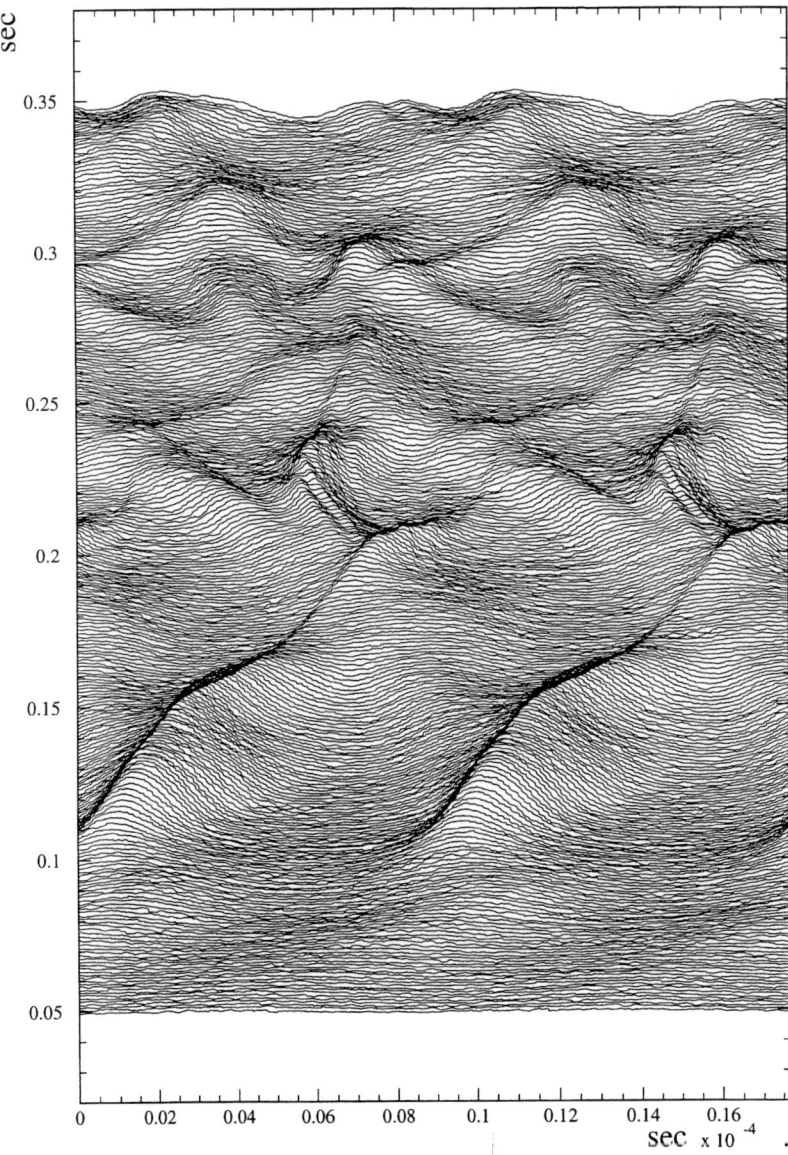

FIGURE 3. Exponential growth and nonlinear saturation phase of longitudinal resistive instability of cooled coasting beam in the ESR driven by the RF cavity on the first harmonic. The mountain range plot shows subsequent time traces from bottom to top over 350 ms (each trace is the current profile over 2 subsequent turns).

exceeding 50% of the coasting beam current, which is in excellent agreement with our computer simulation. The appearance of significantly higher harmonic signals at some later time (0.2 seconds in Fig. 3) can be noted, which are not present if the instability on the fundamental harmonic is absent. We assume this results from a loss of Landau damping in the filamented phase space distribution of the saturated instability.

An interesting nonlinear effect is that the strong coherent signals persist for at least as long as a second and are not Landau damped. This is confirmed by the longitudinal Vlasov code simulation. The working point in the experiment (and likewise the simulation) was far in the unstable region pertaining to the Gaussian momentum distribution (see dot in Fig. 4 which is due to the combined effect of the cavity and space charge impedance).

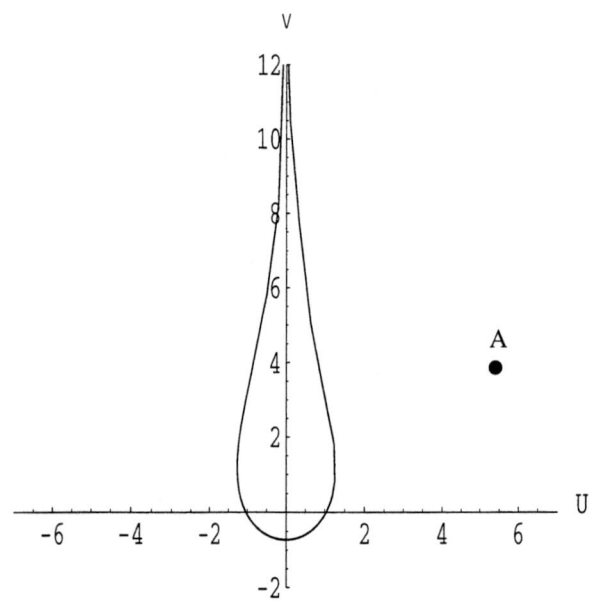

FIGURE 4. Stability diagram with unstable working point (A) due to RF cavity impedance for the Gaussian distribution in ESR experiment.

The Vlasov simulation for the working point of Fig. 4 shows that the long-term behavior is characterized by phase space "bubbles" trapped by the nonlinear wave originating at the lower momentum edge (Fig. 5). These "bubbles" continuously travel through momentum space, which explains the experimental observation of continuing oscillations of the line density.

The initial e-folding time for this case is about 60 ms. The time evolution of the fwhm momentum spread shows that it saturates after roughly five e-foldings at a value which corresponds to the threshold momentum spread, hence no overshoot as is shown in Fig. 6. At later times there is slow growth of the momentum spread.

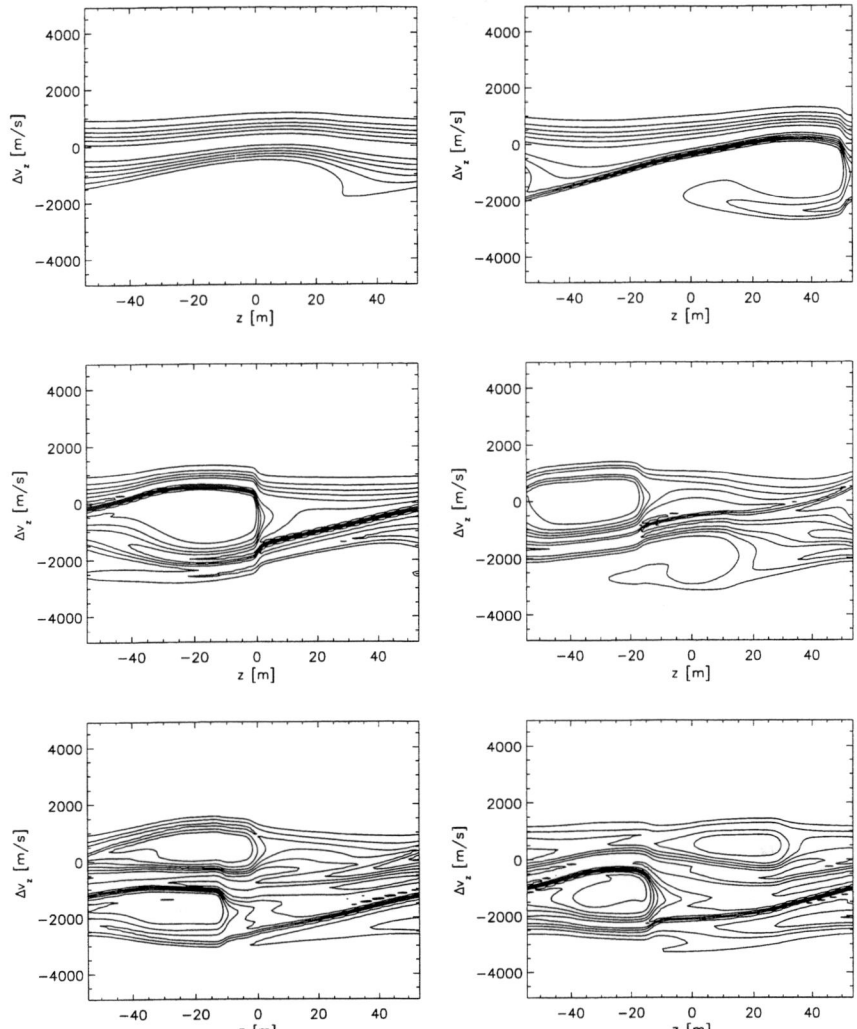

FIGURE 5. Computer simulation (Vlasov) of longitudinal instability at working point for parameters of the ESR experiment. Shown are longitudinal phase space density contours at equidistant times starting from the linear phase into the nonlinear region until time 850 ms.

The beam gradually looses energy, since the circulating energy is the only source of energy available for the resistive heating. The leveling-off at the threshold momentum spread permits the interpretation that the beam settles down near the stability margin, hence fluctuations (phase space "bubbles") are practically undamped and persist for a long period of time. This behavior is obviously in contrast with the

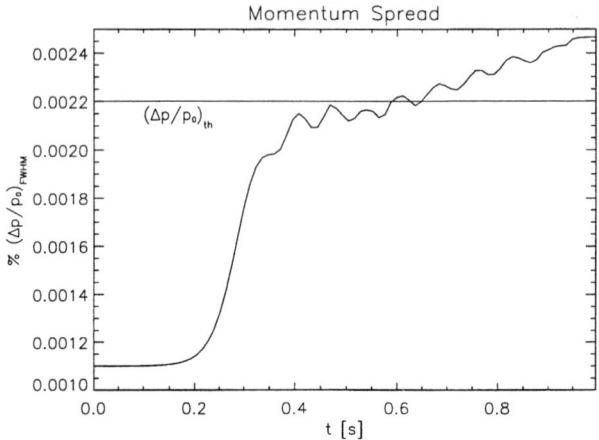

FIGURE 6. Evolution of the fwhm momentum spread showing the absence of overshoot at saturation of the instability.

different forms of the overshoot theorem, which have all been derived in the absence of space charge impedance. The theorem by Dory [16] predicts that

$$\Delta p/p_{final} \Delta p/p_{initial} \geq \Delta p/p_{thresh}^2, \qquad (2)$$

hence a doubling of the momentum spread with respect to the threshold value. Other overshoot criteria suggest weaker effects [17,18], but none of them describes our findings with practically no overshoot. We assume that the difference in behavior is due to the significant space charge broad band impedance present in our case.

An even more extreme case was claimed in earlier simulation work, where the working point in the stability diagram was chosen close to the imaginary axis, but far in the unstable region (see Ref. [19]). The result was that self-stabilization by a thin low-momentum tail occurred, whereas the bulk of the momentum distribution had a width much smaller than the threshold value. In view of the present new high-intensity projects the subject of nonlinear saturation obviously deserves further study both on the experimental and theoretical side.

Acknowledgment: The authors are thankful for comments by Michel Chanel on the LEAR measurements.

REFERENCES

1. J.S. O'Connell, T.P. Wangler, R.S. Mills and K.R. Crandall, Proc. 1993 Part.Accel.Conf., Washington DC, 3651 (1993)
2. R.L. Gluckstern, *Phys. Rev. Lett.*, **73**, 1247 (1994)
3. G. Franchetti et al., these proceedings
4. W. Hardt, CERN internal report, ISR/Int 300 GS/ 66.2 (1966)
5. S. Machida, these proceedings
6. R. Baartman, these proceedings
7. R. Bär et al. to be published in Nucl. Instr. Meth. **A**, 1998
8. M. Chanel, Proc. European Particle Accel. Conf., Sitges, June 10-14, 1996, p. 1015
9. R.L. Gluckstern, W-H. Cheng and H. Ye, *Phys. Rev. Lett.*, **75**, 2835 (1995)
10. I. Hofmann, Phys. Rev. E **57**, 4713 (1998)
11. B. Zotter and P. Bramham, IEEE Trans. Nucl. Sci **NS-20**, 830 (1973)
12. L.J. Laslett, V.K. Neil and A.M. Sessler, Rev. Sci. Instr. **32**, 276 (1961)
13. A.G. Ruggiero and V.C. Vaccaro, CERN-Report ISR-TH/68-33 (1968)
14. J.G. Wang, H. Suk and M. Reiser *Phys. Rev. Lett.*, **79**, 1042 (1997)
15. G. Rumolo et al. to be published in Nucl. Instr. Meth. **A**, 1998
16. R.A. Dory, MURA-Report 654 (1962)
17. Y. Chin and K. Yokoya, Phys. Rev. D **28**, 2141 (1983)
18. S.A. Bogacz and K.-Y. Ng, Phys. Rev. D **36**, 1538 (1987)
19. I. Hofmann, Laser and Part. Beams **3**, 1 (1985)

Early Work on Space Charge Effects in Particle Accelerators

Bruno W Zotter, CERN

1) Introduction

Some time ago Bill Weng asked me to review work on space charge effects in the sixties, claiming "since I was there!"

This is not really true as I joined CERN - and accelerator physics - only in 1968. Much analytical work - and even some computations - had already been done then, mostly in connection with the design of large proton machines such as AGS (BNL), ZGS (ANL), PS and ISR (CERN), and even earlier at MURA where much of accelerator theory was initiated. Several fundamental papers had already been published or were being prepared:

- Detuning due to direct space charge of beams with elliptical cross section, *Lee Teng* ANLAD-59 (1960);

- Detuning due to induced or *image charges* in vacuum chamber walls and magnetic pole pieces, *Jackson Laslett*, BNL Summer Study 1963, pp. 324 ff;

- Beam envelope oscillations in *Frank Sacherer's* thesis, UCRL 18454 (1968) and *Bob Gluckstern*, Proc. Linac Conf. 1970, pp. 811 ff;

- A lecture series on space charge effects given by *Pierre Lapostolle* in CERN Academic training program 1968, unfortunately before I came!

2) Direct Space Charge Fields of Charged Particle Beams

The simplest case is that of an unbunched, cylindrical beam of *circular cross section* with radius a, moving with constant velocity $v = \beta c$ in the direction of its axis. For a beam current I the volume charge density is $\rho = I/\pi a^2 v$.

The electric field has only a radial component, which can be obtained from Gauss' law. Inside the beam ($r \leq a$) it is linear:

$$E_r = \frac{\rho r}{2\varepsilon_0} = \frac{I}{2\pi\varepsilon_0 \beta c}\frac{r}{a^2},\qquad(1)$$

and exerts a force directed *radially outwards*. The magnetic field is azimuthal and Stoke's law yields:

$$B_\phi = \frac{\beta \rho r}{2\varepsilon_0 c} = \frac{\beta E_r}{c},\qquad(2)$$

corresponding to a magnetic force $\mathbf{v}\times\mathbf{B} = -\beta^2 E_r$ directed *radially inwards*. The total Lorentz force is multiplied by $1-\beta^2 = 1/\gamma^2$ and is thus *strongly reduced* in *relativistic beams* with $\gamma \gg 1$. Several effects can disturb this cancellation, such as neutralization by opposite charges or curvature of the beam trajectory, as discussed below.

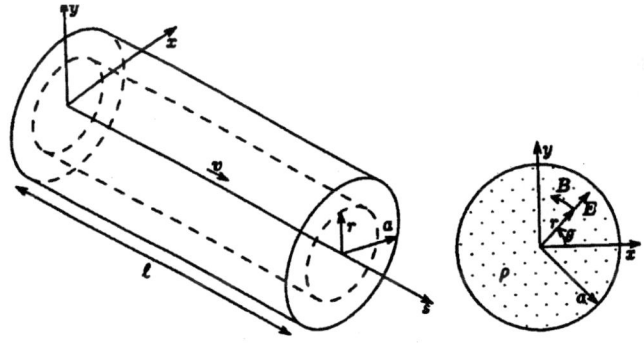

Figure 1: Calculation of electric field with Gauss law.

For a beam with *elliptic cross section*, half axes a_x and a_y, the electric field components in horizontal and vertical directions are still given by Eq. (1), but the factor a^2 in the denominator should be replaced by $a_{x,y}(a_x+a_y)/2$.

For beams with transversely non-uniform charge density, the EM forces are no longer linear, but may depend on particle amplitudes. For a *Gaussian beam*, the linear part obtained by replacing the beam radius a with $\sigma\sqrt{2}$.

For a *bunched beam*, also longitudinal electric fields exist - however, they are negligible for beams with relativistic velocities $\beta \approx 1$. In that case the transverse field strength is modulated along the beam proportional to the local charge density.

3) Induced and Image Fields

The unavoidable presence of metallic or ceramic vacuum chamber walls, as well as that of magnetic material near a charged particle beam leads to modifications of the EM fields which may shift the oscillation frequencies and may even endanger stability.

Mathematically the simplest case is that of a single, infinite, perfectly conducting plate, parallel to the beam at a distance d. The particle beam induces charges and currents such that the electric field becomes normal to the plate surface. These EM fields can be calculated by putting an *image* of opposite charge at the distance d behind the plate. The name *image fields* is commonly used for these *induced fields*, although this really refers only to a mathematical technique used to solve the problem.

For a circular cylindrical vacuum chamber surrounding a beam, which may be displaced by a distance d from its axis, a single image is sufficient to fulfil the boundary condition $E_\parallel = 0$ at the wall. It should be located at the distance b^2/d in the direction of the beam displacement, which we take as x for simplicity. At the beam location, the electric field then has only an x component:

$$E_x^{im} = \frac{\lambda}{2\pi\varepsilon_0} \frac{d}{b^2} . \tag{3}$$

The image technique can be generalized to the more realistic case of two plates, one on either side of a beam, but then also images of images have to be taken into account which leads to an infinite series. Although it can be summed explicitly, it is actually easier to express the problem directly with the proper boundary conditions.

Another useful technique for finding fields in perfectly conducting, uniform beam pipes is *conformal mapping*, used by *Laslett* to compute tune shift coefficients bearing his name for beams on the axis of elliptic chambers or between parallel plates.

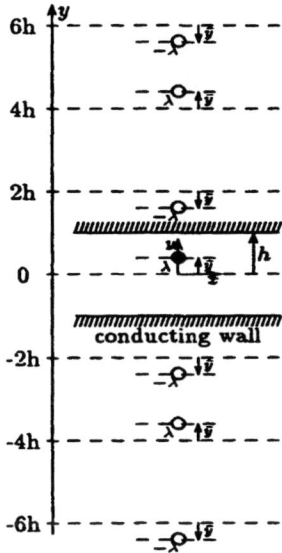

Figure 2: Images of charged beam between 2 parallel plates.

4) Coherent and Incoherent Tune Shifts

A beam can perform *coherent oscillations* either as a whole, i.e. (*rigid* or *dipole oscillations*, or considerable parts of it oscillate with a fixed phase relationship (*higher modes*). In a 2-D phase plane, one can distinguish *azimuthal modes*: quadrupole, sextupole etc., and *radial modes*. However, only the lowest modes are usually important for stability. Oscillations of single particles inside a beam, without fixed phase relationship, are called *incoherent* and may cover a range of frequencies called *frequency spread*.

Longitudinal oscillations in *energy* or *density* are usually called *synchrotron oscillations* for bunched beams. Transverse or *betatron oscillations* may occur in 2 directions - e.g. horizontal and vertical. However, when the betatron motions are coupled, these directions should be replaced by those of the *normal modes*.

In general, the forces due to direct space charge and due to induced fields may change the oscillation frequencies or *tunes*, i.e. their ratios to the revolution

Figure 3: Incoherent and coherent motion of a swing.

frequency[1]. Direct space charge forces move with the beam and thus do **not** influence the coherent dipole frequency. The induced fields are centered on the axis of the beam pipe and therefore may change both coherent and incoherent tunes.

The geometry effect can be expressed conveniently by the *Laslett coefficients* $\varepsilon_{1,2}, \xi_{1,2}$, to which we add the *direct space charge coefficients* ε_{sc} as shown in table 1 for several common beam pipe cross sections.

The vertical tune shifts of an **unbunched beam**, in a vacuum chamber of half-height h, between magnetic pole pieces at a distance $\pm g$ from axis are

---

[1]Unfortunately tune is abbreviated ν in US and Q in Europe, giving rise to confusion.

Table 1: **Laslett Coefficients for Simple Geometries**

Coeff.	circular	elliptic cross section	par. plate	comments
$\varepsilon_{1V/H}$	0	$\pm\frac{h^2}{12d^2}\left[(1+k'^2)\left(\frac{2K}{\pi}\right)^2 - 2\right]$	$\pm\pi^2/48$	coherent electric
$\varepsilon_{2V/H}$	-	-	$\pm\pi^2/24$	coherent magnetic
ξ_{1V}	1/2	$\frac{h^2}{4d^2}\left[\left(\frac{2K}{\pi}\right)^2 - 1\right]$	0	incoherent electric
ξ_{1H}	1/2	$\frac{h^2}{4d^2}\left[1 - \left(\frac{2Kk'}{\pi}\right)^2\right]$	$\pi^2/16$	
ξ_{2V}	-	-	0	incoh. magnetic
ξ_{2H}	-	-	$\pi^2/16$	
ε_{scV}	1/2	$b/(a+b)$	-	direct space charge
ε_{scH}	1/2	$b^2/a(a+b)$	-	

given by the equations:

$$\Delta\nu_{inc} = -\frac{2r_0}{ec}\frac{RI_0\langle\beta\rangle}{\beta^3\gamma}\left[\frac{\varepsilon_1}{h^2}(1-\eta) + \beta^2\frac{\varepsilon_2}{g^2} + \frac{\varepsilon_{sc}}{2b^2}(1-\beta a^2-\eta)\right] \quad (4)$$

$$\Delta\nu_{coh} = -\frac{2r_0}{ec}\frac{RI_0\langle\beta\rangle}{\beta^3\gamma}\begin{cases}\left[\frac{\xi_1}{h^2}(1-\eta) + \frac{\beta^2\xi_2}{g^2}\right], \\ \left[\frac{\beta^2\varepsilon_1}{h^2} + \frac{\xi_1}{\gamma^2 h^2} + \frac{\beta^2\varepsilon_2}{g^2}\right].\end{cases}$$

Here $\langle\beta\rangle$ is the average beta function (R/ν), e the charge, and r_0 the classical radius of a beam particle. The first expression for the coherent tune shift, called *integer formula*, is valid for very thin walls through which the AC magnetic fields can penetrate. The second one, or *half-integer formula*, is valid for the usual case of thicker walls where these fields do not penetrate. For machines with vacuum chambers consisting of sections with unequal dimensions, the tune shifts of each section should be added in proportion to their lengths.

The **neutralization** η is defined as ratio of the number of charges of opposite sign to the total number. It reduces the nearly perfect cancellation of electric and magnetic forces in relativistic beams and thus should be kept small, e.g. by clearing electrodes or by extremely good vacuum.

For **bunched beams**, several terms in the expressions for the tune shifts are proportional to peak rather than average current. This can be taken into account by division with the *bunching factor* $B < 1$, defined as the ratio of average to peak current:

$$\Delta \nu_{inc} = -\frac{2r_0}{ec} \frac{RI_0 \langle \beta \rangle}{\beta^3 \gamma} \left[\left(\beta^2 + \frac{1-\beta^2-\eta}{B} \right) \frac{\varepsilon_1}{h^2} + \right.$$
$$\left. + \beta^2 \frac{\varepsilon_2}{g^2} + \frac{1-\beta^2-\eta}{B} \frac{\varepsilon_{sc}}{b^2} \right] \quad (5)$$

$$\Delta \nu_{coh} = -\frac{2r_0}{ec} \frac{RI_0 \langle \beta \rangle}{\beta^3 \gamma} \times$$

$$\times \begin{cases} \left[\left(\beta^2 + \frac{1-\beta^2-\eta}{B} \right) \frac{\xi_1}{h^2} + \beta^2 \frac{\xi_2}{g^2} \right], \\ \left[\beta^2 \frac{\varepsilon_1}{h^2} + \frac{1-\beta^2-\eta}{B} \frac{\xi_1}{h^2} + \beta^2 \frac{\varepsilon_2}{g^2} \right]. \end{cases}$$

Neutralization in bunched beams usually weaker than in unbunched ones, and can be further reduced by leaving a *gap* in the filling pattern.

5) Experimental Evidence for Tune Shifts and Spreads.

In the ISR, a coasting beam was *stacked* in momentum space. A fraction of the beam was always lost - presumably on a resonance - whenever a certain current level was reached. This was called *brickwall effect*, and could be avoided by applying periodic tune corrections during stacking which kept the tune of the whole beam inside a resonance free region.

The corrections were calculated with *impedance model*, originally obtained from theoretical estimates and from measurements of transfer function. Injection was interrupted every 3 A, and the working line corrected with both quadrupoles and multipoles (pole face windings). This permitted to stack record currents of over 50 A.

The impedance calculations were verified with a special *experimental cavity* with variable shunt impedance. Unfortunately, even its strongest damping turned out to be insufficient, and the cavity itself was a major current limitation! Short-circuit jaws had to be added to hide the cavity from the beam during operation. Later the tune corrections were based on measurements of coherent tunes across a beam with rf excitation, and of incoherent tunes with Schottky scans. Here we show the calculation and the measurements of the tunes, across a stacked beam in the ISR, on which the tune corrections were based.

Also in bunched beams betatron tunes are often measured as function of current in order to determine the transverse impedance. However, since they are proportional to the *effective impedance* and hence depend also on bunch length and shape, the relevant parameters should be carefully recorded.

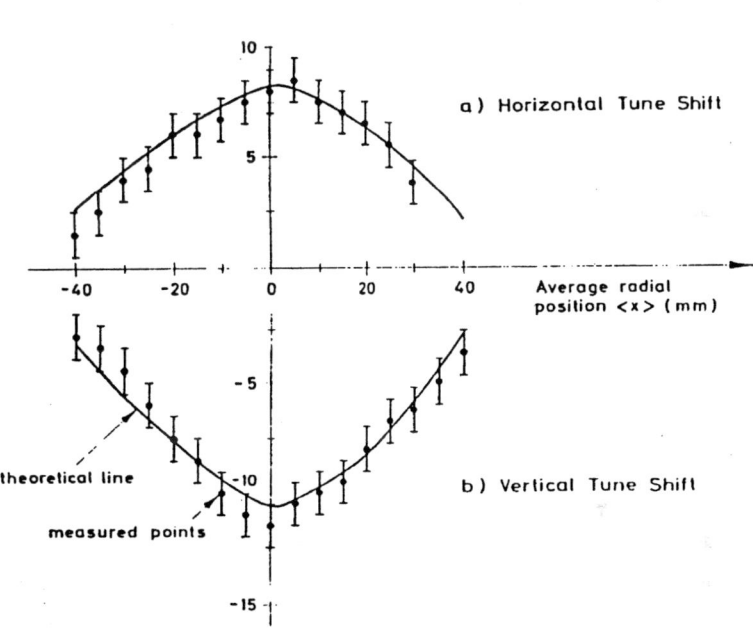

Figure 4: Tunes across a stacked ISR beam

6) Space Charge Impedance.

The longitudinal stability of a particle beam against exponentially growing oscillations can be expressed by a *dispersion relation* which was originally written $(U - iV)I(\omega) + 1 = 0$. Here U and V are the *dispersion relation coefficients*, the *dispersion integral* is

$$I(\omega) = \int_{-\infty}^{\infty} dW \frac{df_0(W)}{dW} \frac{dW}{\omega - n\dot{\theta}} , \qquad (6)$$

where n is the number of wavelengths around the circumference, $\dot{\theta} \approx \omega_0 + k_0 W$ the angular velocity of beam, $W = 2\pi(p_\theta - p_0)$ the canonical angular momentum, and $f_0(W)$ the unperturbed distribution function.

The integration around the pole can be performed using the *Landau prescription*, i.e. along an infinitesimal semi-circle in the complex plane, which yields an imaginary part $i\pi df_0(W)/dW$ in addition to the (real) principal value.

For a circular cylindrical beam of radius a in a perfectly conducting, concentric tube of radius b, the so-called *g-factor* is $1 + 2\ln(b/a)$ and

$$U = \frac{Ne^2}{\gamma^2} \frac{n}{R} g , \qquad (7)$$

proportional to beam current, while V vanishes. Normalized coefficients U' and V' were introduced, but were soon replaced by the *coupling impedance* defined as the (negative) ratio of the induced voltage per turn to the beam current at a particular frequency[2]

$$Z_\|(\omega) = \frac{V - jU}{eI} . \qquad (8)$$

Because of the form of the resistive wall impedance, the transverse dispersion relation was written $\nu\omega_0[U_T + (1+i)V_T]I_T(\omega) = 1$. The *transverse dispersion relation coefficients* U_T and V_T have the dimensions s^{-1}, the same as frequency. The transverse dispersion integral is

$$I_T(\omega) = \int_{-\infty}^{\infty} dW \frac{f(W)}{\omega - (n - \nu)\omega_0} . \qquad (9)$$

The *transverse impedance* $Z_\perp$ is usually defined as the transverse gradient of the deflecting voltage on the axis, divided by the beam current and thus has the dimension Ω/m.

$$Z_\perp(\omega) = jZ_0 \frac{e\nu\gamma}{r_0 I}[V_T + j(U_T + V_T)] . \qquad (10)$$

---

[2]Usually the *physics convention* $exp(-i\omega t)$ is then replaced by the *electro-magnetic convention* $exp(j\omega t)$, this can be obtained by simply putting $i = -j$.

7) Envelope Equations

The beam radius a of a uniformly charged, one-dimensional (axially symmetric) beam is given by the self-consistent equation:

$$\frac{d^2a}{ds^2} + K(s)a - \frac{E^2}{a^3} - \frac{\lambda r_0}{\gamma^3 a} = 0 , \tag{11}$$

where $K(s)$ is the external focusing function (including gradient errors), E the beam emittance, and λ the line density.

In terms or the dimensionless ratio $x = a/a_0$, where a_0 is the radius of the unperturbed, matched beam, and the betatron phase $\phi = \int ds/\nu\beta$, the unperturbed envelope equation becomes

$$\frac{d^2x}{d\phi^2} + \nu^2 x - \frac{\nu^2}{x^3} = 0 . \tag{12}$$

It has the solution $x^2 = \sqrt{1 + A^2} + A\sin(2\nu\phi + \alpha)$, where A and α are constants. The matched solution $x = 1$ is obtained for $A = 0$, other solutions oscillate about it with frequency 2ν. Including the tune shifts $\Delta\nu_s$ due to gradient errors and $\Delta\nu_{sc}$ due to space charge, the equation becomes:

$$\frac{d^2x}{d\phi^2} + [\nu^2 + 2\nu\Delta\nu_s \cos(n\phi)]x - \frac{\nu^2}{x^3} - 2\nu\Delta\nu_{sc} = 0 . \tag{13}$$

Replacing the amplitude function β by its average R/ν in the (small) last term, it becomes independent of ϕ. In the absence of gradient errors, there is then a constant solution $x = 1 + \Delta\nu_{sc}/2\nu$. Oscillations about it, with small amplitude ξ, are given by

$$\frac{d^2\xi}{d\phi^2} + 2(2\nu^2 - 3\nu\Delta\nu_{sc})\xi = -2\nu\Delta\nu_s \cos(n\phi) . \tag{14}$$

Resonances will occur when $\Delta\nu_{sc} = (4/3)[\nu - n/2]$. Due to the variation of beam size, the space charge limit is thus by 1/3 larger than the (usual) value for constant beam size.

This analysis can be generalized to 2 dimensions. The equations of motion for a beam half widths $a_{x,y}$ become

$$\frac{d^2 a_{x,y}}{ds^2} + K_{x,y}(s)a_{x,y} - \frac{E^2_{x,y}}{a^3_{x,y}} - \frac{2\lambda r_0}{\gamma^3(a_x + a_y)} = 0 . \tag{15}$$

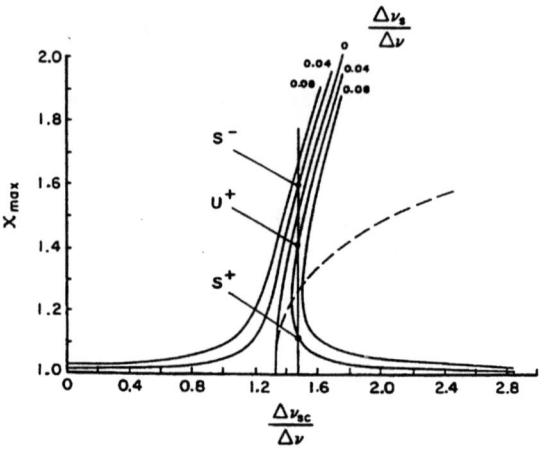

Figure 5: Detuning of a resonance with amplitude.

A self-consistent solution of these equations has first been given by I. Kapchinski and V. Vladimirsky (Proc. Conf. CERN 1959, p. 274) as a shell in 4-dimensional phase space. Unfortunately an extension to 3 dimensions is not possible.

8) Emittance Growth

Due to the non-linearity of the envelope equations, the beam emittance may increase when resonances are crossed, e.g. during injection or acceleration, depending on the speed and also the sense of crossing.

The worst case is a slow, adiabatic change of parameters with increasing $\Delta\nu_{sc}$ - then the beam may blow up indefinitely. On the other hand, due to the detuning of resonances with amplitude, oscillations about the matched value increase only to some finite maximum value when a beam is accelerated (decreasing $\Delta\nu_{sc}$). If the vacuum chamber is large enough to contain these oscillations, they will again become smaller as the acceleration continues.

The maximum amplitude can be found directly from Fig. 6, which shows the relative amplitude for various values of tune shift due to gradient errors over detuning $\Delta\nu_s/\Delta\nu$.

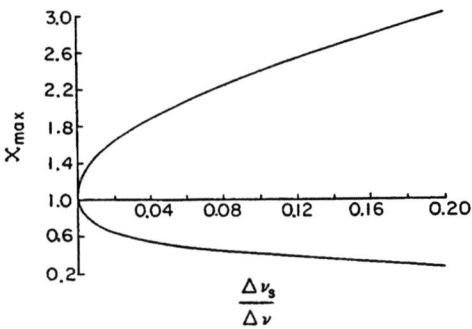

Figure 6: Minimum and maximum amplitude of oscillation.

9) Conclusions

The calculation of space charge effects in the sixties and early seventies concentrated mainly on analytic estimates of tune shifts and emittance growth due to crossing of resonances during acceleration, bunching, or current increase of the beam. Only uniformly distributed impedances - such as due to direct space charge or wall resistivity - were considered in these days, localized impedances such as rf or other cavities were not yet important. The particle beams were assumed to be either unbunched or to have rather long bunches.

Nevertheless, the compensation of tune shifts during stacking in the CERN ISR led to record currents of over 50 A, and was initially based entirely on theoretical impedance models.

References

R. Gluckstern, Proc. Linac Conf. 1970, pp. 811 ff

P. Lapostolle, Notes CERN Academic training lectures 1968 (unpubl.)

J. Laslett, BNL Summer Study 1963, pp. 324 ff

F. Sacherer, thesis, report UCRL 18454 (1968)

L. Teng, Report ANLAD-59 (1960)

Space Charge Dominated Beams in Synchrotrons and Linac

S.Y. Lee

Department of Physics, Indiana University, Bloomington, IN 47405

Abstract.
We employ the Kapchinskij-Vladimirskij envelope Hamiltonian to describe the envelope evolution, and the particle Hamiltonian to describe particle motion in a space charge dominated beam. In a uniform focusing channel, particle motion can encounter parametric resonances arising from the envelope oscillations of a mismatched beam. Large amplitude envelope oscillation can cause global chaos near the vicinity of the beam core, and lead to enhanced halo formation. In a periodic focusing channel, enhancement of halo formation may also arise from structure resonances, i.e. envelope-lattice and particle-lattice resonances. The onset of global chaos exhibits a first order phase transition like behavior when the amplitude of envelope oscillations for a mismatched beam is larger than a critical value. The equations of motion for space charge dominated beams in synchrotron are derived. We find that the space charge force generates an identical defocusing function to the betatron motion and dispersion function. The self-consistent envelope equation obeys the Kapchinskij-Vladimirskij type equation similar to that of the linear transport system. We employ these results to analyze the stability of crystalline beams, and discuss the implication on the high intensity proton driver for the neutron spallation sources. Possible experiments are suggested.

I INTRODUCTION

Intense charged particle beams have many applications. Some of these applications are proton drivers for spallation neutron sources, energy amplifiers, secondary beam sources such as muons, pions, and kaons, and heavy ion beam drivers for fusion energy [1,2]. Thus the stability of the space charge dominated beam is an important topic in beam physics.

For a space charge dominated beam, Kapchinskij and Vladimirskij (KV) have constructed a self consistent equilibrium distribution function, which obeys the KV equation governed by external focusing force and the force due to self beam charge and current [3]. When the KV envelope function and the particle motion experience a time dependent modulation, nonlinear parametric resonances may cause beam instability. There are two methods commonly used in the study of the stability of space charge dominated beams. The linearized Vlasov equation method studies the threshold of an equilibrium distribution perturbatively [4], while the particle-core model studies the core and test-particle stability using the envelope equation and Hill's equation [5–7]. The linearized Vlasov equation approach has been shown to provide accurate description of the threshold behavior of collective modes. On the other hand, the particle core model has been successfully used to describe the halo formation using nonlinear parametric resonances. Numerical simulations have been found to agree well with both the linearized Vlasov equation theory and the particle core model in linear transport systems.

This paper reviews the stability of space charge dominated beams based on the particle-core model. In Sec. II we will show that the equation of motion for a space charge dominated beam in a synchrotron is identical to that of a linac transport channel. The stability of envelope and particle Hamiltonians of space charge dominated beams will then be discussed in Secs. III and IV. The conclusion is given in Sec. V.

II EQUATION OF MOTION IN SYNCHROTRON

In the past, space charge dominated beams were studied mainly in linacs, where the beam energy is low, and thus the space charge force is important [1,8]. Since the synchrotron can accumulate linac beams and attain a much higher line density, the space charge force can be as important. Deriving an envelope equation for a space charge dominated beam in synchrotron is complicated by the dispersion function. In the Frenet-Serret coordinate system with $(\mathbf{x}, \mathbf{s}, \mathbf{z})$ as unit basis vectors, the particle phase space coordinates are $(x, p_x, z, p_z, t, -E)$ [9], where the synchrotron phase space coordinates (t, E) are the time and the energy of the particle, x, z are respectively the betatron coordinates, and p_x, p_z are the corresponding conjugate coordinates. We consider only transverse force, where we choose $A_x = A_z = 0$ for the vector potential. The Hamiltonian up to the second order in p_x, p_z is

$$\tilde{H} \approx -\left(1 + \frac{x}{\rho}\right)p + \left(1 + \frac{x}{\rho}\right)\left(\frac{p_x^2 + p_z^2}{2p}\right) - eA_s, \tag{1}$$

where e and p are the charge and the momentum of the particle, ρ is the radius of curvature, and $A_s = (1 + x/\rho)\mathbf{A}\cdot\mathbf{s}$ is a component of the vector potential with $B_z = [1/(1 + x/\rho)](\partial A_s/\partial x)$, $B_x = -[1/(1 + x/\rho)](\partial A_s/\partial z)$. We expand the momentum about the reference value p_0 and obtain

$$\Delta p = p - p_0 \approx \frac{\Delta E}{\beta_0 c} - \frac{1}{2p_0}\left(\frac{\Delta E}{\beta_0 c \gamma_0}\right)^2 - \frac{eV_{sc}}{\beta_0 c}, \tag{2}$$

where $\Delta E = E - E_0$ is the energy deviation, and β_0 and γ_0 are the Lorentz factors of the reference particle. Here V_{sc} is the self-consistent space charge scalar potential that satisfies the Poisson equation

$$\left(\frac{\partial^2}{\partial x^2} + \frac{\partial^2}{\partial z^2}\right)V_{sc} = -\frac{e}{\epsilon_0}\int\int\int dp_x dp_z d(-\Delta E) f(x, z, p_x, p_z, -\Delta E; s), \tag{3}$$

where the distribution function f obeys the Vlasov equation $df/ds = 0$. The equilibrium distribution function in a synchrotron must also be a periodic function of s, and is generally a function of the effective Hamiltonian that includes the space charge mean field potential. For a few special distributions, self consistent space charge potential can be expressed in analytic form, e.g., the Kapchinskij-Vladimirskij (KV) distribution [3] in a linear transport system.

Now we consider the case of a coasting beam without synchrotron motion so that the longitudinal rf electric field is zero. The vector potential is

$$A_s = B_0 x + \frac{B_0}{2\rho} x^2 + \frac{1}{2} B_1 (x^2 - z^2) + A_{sc}, \tag{4}$$

where $B_1 = \partial B_z / \partial x$ is the focusing function evaluated at the reference orbit, and A_{sc} is the vector potential due to the space charge force. Since the magnetic force is equal to $-\beta_0^2$ times the electric force, we obtain $A_{sc} = \beta_0^2 e V_{sc} / \beta_0 c$. Substituting the vector potential into the Hamiltonian, one obtains

$$H_0 = -p_0 \frac{\Delta E}{\beta_0^2 E_0} + p_0 \frac{1}{2\gamma_0^2} (\frac{\Delta E}{\beta_0^2 E_0})^2 - p_0 \frac{\Delta E}{\beta_0^2 E_0} \frac{x}{\rho}$$
$$+ \frac{p_x^2 + p_z^2}{2 p_0} + \frac{p_0}{2} \left(K_x x^2 + K_z z^2 \right) + \frac{e}{\beta_0 c \gamma_0^2} V_{sc}. \tag{5}$$

where $K_x = 1/\rho^2 - B_1/B\rho$ and $K_z = B_1/B\rho$ are the focusing functions for betatron motion, and we have used the identity $B_0 = -p_0/e\rho$, that signifies the expansion of x around the reference orbit.

The transformation of the coordinate system onto the closed orbit for the particle at energy E is accomplished by using the generating function [10]

$$F_2(x, \bar{p}_x, t, \Delta E) = (x - D_x \frac{\Delta E}{\beta_0^2 E_0}) \bar{p}_x + (z - D_z \frac{\Delta E}{\beta_0^2 E_0}) \bar{p}_z - (E_0 + \Delta E) t$$
$$+ x \frac{D_x'}{\beta_0 c} \Delta E - \frac{1}{2} D_x D_x' p_0 (\frac{\Delta E}{\beta_0^2 E_0})^2 + z \frac{D_z'}{\beta_0 c} \Delta E - \frac{1}{2} D_z D_z' p_0 (\frac{\Delta E}{\beta_0^2 E_0})^2.$$

Since the equilibrium distribution is a function of x and z, the mean field Coulomb potential is

$$V_{sc} \approx V_{sc,0} + \frac{1}{2} V_{sc,xx} \left(\bar{x} + D_x \frac{W}{\beta_0 c} \right)^2 + \frac{1}{2} V_{sc,zz} \left(\bar{z} + D_z \frac{W}{\beta_0 c} \right)^2$$
$$+ V_{sc,xz} \left(\bar{x} + D_x \frac{W}{\beta_0 c} \right) \left(\bar{z} + D_z \frac{W}{\beta_0 c} \right) + \cdots, \tag{6}$$

where $V_{sc,xx} = \partial^2 V_{sc}/\partial \bar{x}^2$, $V_{sc,zz} = \partial^2 V_{sc}/\partial \bar{z}^2$, $V_{sc,xz} = \partial^2 V_{sc}/\partial \bar{x} \partial \bar{z}$ are partial derivatives of the space charge potential evaluated at the reference orbit, $V_{sc,0}$ is a constant, and $W = \Delta E/p_0$. Eliminating the cross terms in the Hamiltonian, the equations for the dispersion functions are

$$D_x'' + \left(K_x + \frac{eV_{sc,xx}}{\beta_0 c p_0 \gamma_0^2}\right) D_x + \frac{eV_{sc,xz}}{\beta_0 c p_0 \gamma_0^2} D_z = \frac{1}{\rho}, \tag{7}$$

$$D_z'' + \left(K_z + \frac{eV_{sc,zz}}{\beta_0 c p_0 \gamma_0^2}\right) D_z + \frac{eV_{sc,xz}}{\beta_0 c p_0 \gamma_0^2} D_x = 0. \tag{8}$$

Note that the space charge mean field reduces the focusing strength and may introduce a linear coupling to the equations of motion. With $\tilde{p}_x = \bar{p}_x/p_0$ and $\tilde{p}_z = \bar{p}_z/p_0$, the new Hamiltonian becomes

$$\tilde{H}_2 = \frac{1}{2}\left[\tilde{p}_x^2 + \left(K_x + \frac{eV_{sc,xx}}{\beta_0 c p_0 \gamma_0^2}\right)\tilde{x}^2 + \tilde{p}_z{}^2 + \left(K_z + \frac{eV_{sc,zz}}{\beta_0 c p_0 \gamma_0^2}\right)\tilde{z}^2 + 2\frac{eV_{sc,xz}}{\beta_0 c p_0 \gamma_0^2}\tilde{x}\tilde{z}\right]$$
$$- \frac{W}{\beta_0 c} + \frac{1}{2}\left(\frac{W}{\beta_0 c}\right)^2 \left[\frac{1}{\gamma_0^2} - \frac{D_x}{\rho}\right].$$

The synchrotron equation of motion is

$$\frac{d(\Delta\tilde{t})}{ds} = \frac{1}{\beta_0 c}\left(\frac{D_x}{\rho} - \frac{1}{\gamma_0^2}\right)\frac{W}{\beta_0 c}, \tag{9}$$

where $\Delta\tilde{t} = \tilde{t} - s/\beta_0 c$ is the relative time. The corresponding momentum compaction factor for the space charge dominated beam

$$\alpha_{c,sc} = \frac{1}{C}\oint \frac{D_x}{\rho} ds, \tag{10}$$

where C is the circumference of the synchrotron, has an identical form as that of the emittance dominated beams except that the dispersion function is modified by the space charge mean field. The equations of betatron motion are

$$\tilde{x}'' + \left(K_x + \frac{eV_{sc,xx}}{\beta_0 c p_0 \gamma_0^2}\right)\tilde{x} + \frac{eV_{sc,xz}}{\beta_0 c p_0 \gamma_0^2}\tilde{z} = 0, \tag{11}$$

$$\tilde{z}'' + \left(K_z + \frac{eV_{sc,zz}}{\beta_0 c p_0 \gamma_0^2}\right)\tilde{z} + \frac{eV_{sc,xz}}{\beta_0 c p_0 \gamma_0^2}\tilde{x} = 0. \tag{12}$$

We observe that Hill's equations for the betatron motion have an identical focusing function as that of the dispersion functions. Furthermore, the space charge force may introduce linear coupling to the betatron motion and gives rise to the vertical dispersion.

We now apply our formalism to analyze the crystalline beams in a storage ring as they are the ultimate form of space charge dominated beams [11]. By using a properly tailored "tapered" cooling force, an ordered state of the crystal beam can be obtained by the molecular dynamics numerical simulations [12]. When the crystalline state is formed, the normalized temperatures, defined in Ref. [12], will be less than 10^{-4} in all degrees of freedom. Figure 1 shows the dispersion function

x_{co}/δ, and the vertical closed orbit $z_{co}/\langle z_{co}\rangle$ in one period of the lattice. The existence of an unique dispersion function shown in the upper curve of Fig. 1 indicates that the horizontal closed orbit of each particle is related to the dispersion function. The fact that $z_{co}/\langle z_{co}\rangle \approx 1$ for all particles in the crystalline beam indicates that (1) the space charge force has almost fully compensated the quadrupole focusing force, and (2) there is no vertical dispersion function and no inhomogeneous term in Eq. (12). Thus the linear coupling due to the space charge force is small, i.e., $V_{sc,xz} \approx 0$.

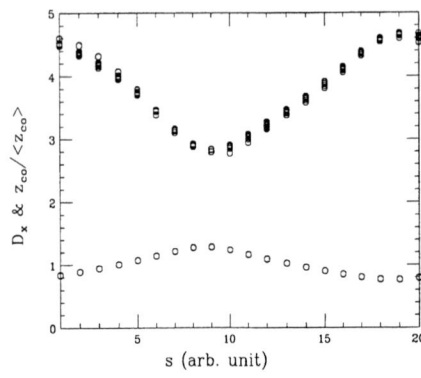

FIGURE 1. The horizontal orbit x_{co}/δ (normalized unit) and the vertical orbit $z_{co}/\langle z_{co}\rangle$ of 20 particles Obtained from the molecular dynamics calculations are plotted in one superperiod of simple FODO lattice 10 superperiods and betatron tunes $Q_x = Q_z = 2.34$. Note that the horizontal closed orbit of the crystalline particle arises from dispersion function, and the vertical closed orbit is obtained from the homogeneous Hill's equation of Eq. (12) with a focusing force almost fully balanced by the Coulomb mean field.

Neglecting the linear coupling and nonlinear contribution from the space charge potential, the betatron Hamiltonian becomes

$$\tilde{H}_3 = \frac{1}{2}\left[\tilde{p}_x^2 + \tilde{p}_x^2 + \left(K_x + \frac{eV_{sc,xx}}{\beta_0 c p_0 \gamma_0^2}\right)\tilde{x}^2 + \left(K_z + \frac{eV_{sc,zz}}{\beta_0 c p_0 \gamma_0^2}\right)\tilde{z}^2\right], \quad (13)$$

and the KV distribution is a self consistent solution [3]. In the KV model, we obtain

$$V_{sc,xx} = -\frac{eN}{\pi\epsilon_0}\frac{1}{a(a+b)}, \quad V_{sc,zz} = -\frac{eN}{\pi\epsilon_0}\frac{1}{b(a+b)}, \quad (14)$$

where N is the number of particles per unit length, a and b are the horizontal and the vertical betatron beam envelopes. For a KV beam, the envelope equations become

$$a'' + K_x a - \frac{\epsilon_x^2}{a^3} = \frac{2K_{sc}}{a+b}, \quad b'' + K_z b - \frac{\epsilon_z^2}{b^3} = \frac{2K_{sc}}{a+b}, \quad (15)$$

where ϵ_x and ϵ_z are the beam emittances, $K_{sc} = 2Nr_0/\beta_0^2\gamma_0^3$ is the space charge perveance, and r_0 is the classical radius.

III THE ENVELOPE HAMILTONIAN

With the longitudinal distance s as the time coordinate, the KV Hamiltonian for a transport channel with paraxial symmetry is $H_e = \frac{1}{2}P_b^2 + \tilde{V}_{KV}(R_b)$, where (P_b, R_b) are the conjugate *envelope phase space* coordinates, $R_b = a = b$ is the equilibrium radius of the beam, $P_b = R_b'$, and the KV potential is

$$\tilde{V}_{KV} = \frac{1}{2}k_f(s)R_b^2 - K_{sc}\ln R_b + \frac{\epsilon^2}{2R_b^2}. \qquad (16)$$

Here $k_f(s)$ is the external focusing field, and ϵ is the emittance of the beam. The space charge perveance parameter of the beam is $K_{sc} = 2Nr_0/\beta^2\gamma^3$, where r_0 is the classical radius of the particle, β and γ are the relativistic factors of the beam, and N is the number of particles per unit length. For a uniform focusing channel, $k_f(s)$ is constant. For a periodic focusing channel, $k_f(s)$ is periodic, i.e. $k_f(s) = k_f(s+L)$, where L stands for the cell length of the focusing field.

We transform the KV Hamiltonian to a *dimensionless* form [7] with parameters $\theta = 2\pi s/L$ for time variable, $k(\theta) = L^2 k_f(s)$ for the focusing strength in a periodic cell, $K = LK_{sc}/\epsilon$ for the effective space charge perveance parameter, and $R = R_b/\sqrt{\epsilon L}$ and $P = \sqrt{L/\epsilon}P_b$ for conjugate envelope phase space coordinates. Here L is the basic period of the accelerator structure. The parameter $\frac{K}{4\pi} = \frac{LK_{sc}}{4\pi\epsilon}$ signifies the linear space charge tune shift in circular synchrotrons. The envelope Hamiltonian becomes

$$H_e = \frac{1}{4\pi}P^2 + \frac{1}{4\pi}k(\theta)R^2 - \frac{1}{2\pi}K\ln R + \frac{1}{4\pi R^2}. \qquad (17)$$

A Uniform Focusing Channel

For a uniform focusing channel with $k(\theta) = \mu^2$, where μ is the phase advance of transverse betatron oscillations. In terms of new phase space coordinates, the envelope Hamiltonian becomes

$$H_{e0} = \frac{1}{4\pi}P^2 + V_0(R) = \frac{1}{4\pi}P^2 + \frac{1}{4\pi}\mu^2 R^2 - \frac{1}{2\pi}K\ln R + \frac{1}{4\pi R^2}. \qquad (18)$$

The relevant focusing parameter in the uniform focusing channel is the phase advance per unit length $\mu_f = \mu/L$.

The envelope radius of a matched beam, given by $dV_0(R)/dR = 0$ at $R = R_0$, is

$$R_0 = \left(\frac{1}{\mu}(\sqrt{\kappa^2+1}+\kappa)\right)^{1/2}, \qquad (19)$$

where the effective space charge parameter $\kappa = K/2\mu$ is the only scaling factor of the dynamical system. Because the Hamiltonian in Eq. (17) is integrable, the envelope radius for a mismatched beam will follow an envelope torus of the Hamiltonian flow. The action of a given envelope torus is

$$J_e = \frac{1}{2\pi} \oint P dR = \frac{1}{\sqrt{\pi}} \oint (E_e - V_0(R))^{1/2} dR, \qquad (20)$$

where the "energy" E_e of the envelope Hamiltonian can be expanded in action as

$$H_e = E_e(J_e) = \nu_e J_e + \frac{1}{2} \alpha_e J_e^2 + \cdots. \qquad (21)$$

The envelope tune is

$$Q_e(J) = \frac{dE_e}{dJ_e} = \nu_e + \alpha_e J_e + \cdots, \qquad (22)$$

where the nonlinear detuning parameter

$$\alpha_e = \frac{3}{32\pi^3 R_0^4 \nu_e^2}(K + \frac{10}{R_0^2}) - \frac{5}{192\pi^5 R_0^6 \nu_e^4}(K + \frac{6}{R_0^2})^2 + \cdots \qquad (23)$$

can easily be obtained by the canonical perturbation method, and

$$\nu_e = 2\frac{\mu}{2\pi}[1 - \kappa(\sqrt{\kappa^2 + 1} - \kappa)]^{1/2} \qquad (24)$$

is the tune of small amplitude envelope oscillations. The phase advance of the envelope function in one period is 2μ at a zero space charge limit, and $\sqrt{2}\mu$ at the infinite space charge limit, i.e. $Q_e \in [\nu_e, \frac{\mu}{\pi})$. Therefore, the nonlinear detuning α_e arises solely from the space charge alone. Figure 2 shows a typical example of $2\pi Q_e/\mu$ as a function of the envelope amplitude for the parameters $R_0 = 1.4109$ and $\kappa = 2.1913$.

Using the generating function

$$F_2(R, J_e) = \int_{\hat{R}}^{R} P dR, \qquad (25)$$

where $\hat{R}$ is the maximum amplitude of an envelope torus, the conjugate angle coordinate is

$$\psi_e = \frac{\partial F_2}{\partial J_e} = 2\pi \frac{\partial E_e}{\partial J_e} \int_{\hat{R}}^{R} \frac{dR}{P}. \qquad (26)$$

Hamilton's equations motion are $\dot{J}_e = 0$, i.e. J_e is invariant, and $\dot{\psi}_e = Q_e(J)$, i.e. $\psi_e = Q_e(J)\theta + \psi_{e0}$. The envelope oscillation is approximately sinusoidal with an amplitude dependent tune.

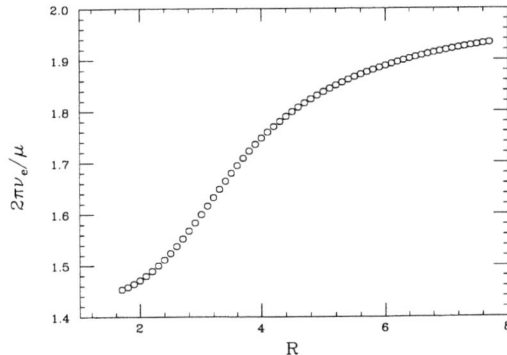

FIGURE 2. The envelope tune Q_e normalized to $\frac{\mu}{2\pi}$ vs the maximum amplitude $R = \hat{R}$ for an envelope torus is shown for parameters $R_0 = 1.4199$ and $\kappa = 2.1913$. Note that the envelope tune depends quadratically on $\hat{R} - R_0$ for a small $\hat{R} - R_0$. Asymptotically, we have $2\pi Q_e/\mu \to 2$ as $R \to \infty$. For a space charge dominated beam with $K/2\mu \gg 1$, we have $2\pi Q_e/\mu \to \sqrt{2}$ as $R \to R_0$.

B Small amplitude envelope oscillations

For a weakly mismatched beam, we can expand the envelope function around the average radius, i.e. $R = R_0 + Y$. The resulting envelope Hamiltonian is $H_e \approx \frac{1}{4\pi}P^2 + \pi\nu_e^2 Y^2 + \cdots$. Thus the envelope function of a weakly mismatched beam is $R \approx R_0 + (J_e/\pi\nu_e)^{1/2}\cos\nu_e\theta$, where the action is $J_e \approx \pi\nu_e(\hat{R} - R_0)^2$ for small amplitude oscillations. Since the tune Q_e, shown in Fig. 2, depends quadratically on $\hat{R} - R_0$ up to about $\hat{R} \approx 3$, the expansion of Eq. (22) is a good approximation within the same range. We define a mismatch parameter M as

$$M = \frac{R_0 - R_{\min}}{R_0} = \frac{1}{R_0}\sqrt{\frac{J_e}{\pi\nu_e}}, \tag{27}$$

where $R_{\min}$ is the minimum of the envelope radius of a mismatched beam.

C Periodic focusing channel

Making Floquet transformation to the envelope Hamiltonian (17), we obtain

$$H_e = H_{e0} + \frac{1}{4\pi}(k(\theta) - \mu^2)R^2, \tag{28}$$

where μ is the tune. The matched envelope radius (closed orbit or the betatron amplitude function) can be obtained by solving the linearized equation. The second term in Hamiltonian (28 provide the driving terms for envelope structure resonances [6,7]. The envelope structure resonance can enhance halo formation in space charge dominated beams [7].

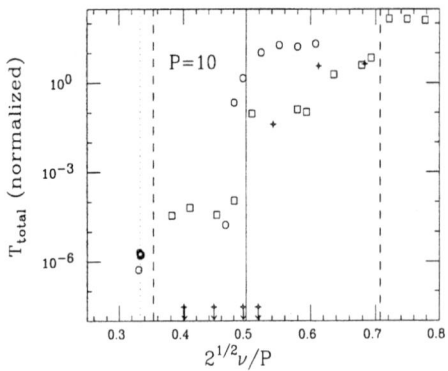

FIGURE 3. The total beam temperature $T_{\text{total}} = T_x + T_z + T_s$ (in the normalized unit [12]) obtained from molecular dynamics calculations with different lattices vs betatron tune. The solid line at $\sqrt{2}\nu = P/2$ is the envelope stopband of a cold beam and dashed lines at $2\nu = P/2, P$ are the stopbands of a hot beam. The dotted line shows the third order envelope stopband.

D Half integer stopbands and Envelope Resonances

When the envelope tune encounters a resonance at $mQ_e = nP$, where m and n are integers, P is the superperiod, the envelope motion will be strongly perturbed. In particular, the Mathieu instability $Q_e = P$ is related to the half integer stopband, that can be excited by linear quadrupole errors.

To test the stability condition, we study the stability of crystalline beams in synchrotron. When the horizontal and vertical betatron tunes for the crystalline beam lattice are equal, The envelope tune of Eq. (22) or Eq. (24) can encounter half-integer stopbands occur at $P/2, P, 3P/2, \cdots$. Therefore, in order to maintain a crystalline beam, the lattice must satisfy $\sqrt{2}\nu \leq P/2$ [12]. Figure 3 shows the beam temperature, obtained from the molecular dynamics calculations, vs $\sqrt{2}\nu/P$. When the betatron tune of a cold beam reaches the envelope stopband at $\sqrt{2}\nu/P = 1$, the betatron envelope for all off-momentum particles becomes unstable, and the temperature of the beam increases suddenly. When the betatron tune of a hot beam encounters the envelope stopband at $2\nu/P = 1$, the temperature of the beam increases further. Besides the linear Mathieu instability, nonlinear systematic stopbands occur at $\sqrt{2}\nu = P/m$, $(m = 3, 4, \cdots)$ [7]. Fortunately, higher order stopbands have zero width, and these stopbands can be easily suppressed by the cooling force.

IV THE PARTICLE HAMILTONIAN

Using the longitudinal coordinate s as the time variable, the *particle Hamiltonian* for transverse oscillations in a paraxial symmetry transport channel is

$$H_p = \frac{1}{2}(p_x^2 + p_z^2) + \tilde{V}_p(x, z), \tag{29}$$

where (x, p_x, z, p_z) are transverse phase space coordinates in the Larmor precessing frame. The transverse focusing potential, $\tilde{V}_p(x, z)$, that includes self-fields of the beam in the KV model is

$$\tilde{V}_p = \frac{1}{2}k_f(s)r^2 - \frac{K_{sc}}{2R_b^2}r^2\Theta(R_b - r) - \frac{K_{sc}}{2}(1 + 2\ln\frac{r}{R_b})\Theta(r - R_b), \quad (30)$$

where $r = \sqrt{x^2 + z^2}$. Using the generating function

$$F_2(x, z, p_r, p_\varphi) = p_r\sqrt{x^2 + z^2} + p_\varphi \arctan\frac{z}{x}, \quad (31)$$

where p_r and p_φ are new momenta with conjugate coordinates (r, φ) respectively, the rotationally symmetric Hamiltonian becomes

$$H_p = \frac{1}{2}(p_r^2 + \frac{p_\varphi^2}{r^2}) + \tilde{V}_p(r). \quad (32)$$

Since the Hamiltonian is independent of the angle variable φ, the momentum p_φ is invariant, which corresponds to the conservation of angular momentum [7,4]. The system reduces to a one dimensional (1D) equation of motion. In this section, we will study a restricted 1D motion with $p_\varphi = 0$.

Using y_b to stand for either x or z coordinates, and p_b for either p_x or p_z, θ for the time variable, and the normalized coordinates

$$y = \frac{y_b}{\sqrt{\epsilon L}}, \quad p = \sqrt{\frac{L}{\epsilon}}p_b, \quad (33)$$

for the conjugate phase space coordinates, the Hamiltonian becomes

$$H_p = \frac{1}{4\pi}p^2 + \frac{1}{4\pi}k(\theta)y^2 - \frac{K}{4\pi R^2}y^2\Theta(R - |y|) - \frac{K}{4\pi}(1 + 2\ln\frac{|y|}{R})\Theta(|y| - R), \quad (34)$$

where R follows the KV Hamiltonian flow discussed in Sec. III, and Θ is the step function used to describe space charge potential inside or outside the core given by $\Theta(\zeta) = 1$ for $\zeta \geq 0$ and 0 otherwise. Hamilton's equations of motion for the beam in the uniform focusing channel are

$$\dot{y} = \frac{1}{2\pi}p, \quad \dot{p} = -\frac{1}{2\pi}\mu^2 y + \frac{K}{2\pi R^2}y\Theta(R - |y|) + \frac{1}{2\pi}\frac{K}{y}\Theta(|y| - R). \quad (35)$$

For a weakly mismatched beam, the envelope function is $R = R_0(1 - M\cos\nu_e\theta)$. Expanding the radius R about R_0, we obtain the particle Hamiltonian as $H_p = H_{p0} + \Delta H_p$, where

$$H_{p0} = \frac{1}{4\pi}p^2 + \frac{1}{4\pi}\mu^2 y^2 - \frac{K}{4\pi R_0^2}y^2\Theta(R_0 - |y|) - \frac{K}{4\pi}[1 + 2\ln\frac{|y|}{R_0}]\Theta(|y| - R_0), \quad (36)$$

$$\Delta H_p \approx -\frac{K}{2\pi R_0^2}\left[\frac{Y}{R_0}(y^2 - R_0^2) + \frac{3Y^2}{2R_0^2}(y^2 - \frac{1}{3}R_0^2) + \cdots\right]\Theta(R_0 - |y|), \quad (37)$$

and $Y = R_0 - R = MR_0\cos\nu_e\theta$. Note here that we have thrown away many time dependent terms, which do not depend on the phase space variables (y, p). These terms contribute to the fluctuation of Hamiltonian values without giving rise to resonance phenomena. Since the Coulomb force depends only on the total charge inside the envelope radius, the envelope oscillations do not perturb particle motion outside the envelope radius, and the perturbing potential exists only inside the envelope radius.

The action for a torus of the unperturbed Hamiltonian is

$$J_y = \frac{1}{2\pi} \oint p \, dy. \tag{38}$$

The scale transformation of Eq. (33) transforms the actual action of a beam particle by the scaling factor of the emittance, i.e. $J_b = \epsilon J_y$. With the generating function $F_2(y, J_y) = \int_{\hat{y}}^{y} p \, dy$, the conjugate angle variable is

$$\psi_y = \frac{\partial F_2}{\partial J_y} = 2\pi \frac{\partial E_p}{\partial J_y} \int_{\hat{y}}^{y} \frac{dy}{p}. \tag{39}$$

The energy of the unperturbed Hamiltonian is a function only of the action, i.e. $E_p(J_y)$. The particle tune is $Q_y(J_y) = \partial E_p / \partial J_y$.

A Properties of the unperturbed Hamiltonian flow

For a beam with a matched space charge envelope, the beam radius is R_0. Particle motion, governed by the Hamiltonian (36), can be divided into two regions.

1. For particles inside the equilibrium KV envelope radius, we have $E_p \leq E_{p0}$ with $E_{p0} = \pi \nu_y^2 R_0^2 = \frac{1}{2}\nu_y$. The betatron tune is

$$\nu_y = (\mu/2\pi)[\sqrt{\kappa^2 + 1} - \kappa]. \tag{40}$$

The action is related to the energy by $E_p = \nu_y J_y$. Thus the tune of particle motion is constant, i.e. $Q_y = \nu_y$ with $\nu_y \sim \frac{\mu}{4\pi\kappa}$, when $\kappa \gg 1$. Since all particle inside the core has an identical tune ν_y, the space charge force causes a *constant incoherent* tune shift by $\Delta\nu_y = \frac{\mu}{2\pi} - \nu_y$. The particle motion can be described by

$$y = \sqrt{\frac{J_y}{\pi\nu_y}} \cos\nu_y\theta, \quad p = -\sqrt{4\pi\nu_y J_y} \sin\nu_y\theta. \tag{41}$$

The maximum action is $\hat{J}_y = \pi\nu_y R_0^2 = \frac{1}{2}$. The torus for particles on the matched envelope ellipse is $y_e^2/R_0^2 + R_0^2 p_e^2 = 1$.
2. In the limit of a large action $J_y \to \infty$, the space charge force is not important, and we have $Q_y(J_y \to \infty) \to \frac{\mu}{2\pi}$. In general, we have $Q_y \in [\nu_y, \frac{\mu}{2\pi})$.

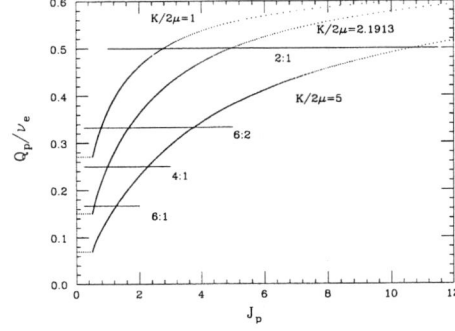

FIGURE 4. The particle tune Q_y, normalized to $\frac{\mu}{2\pi}$, is plotted vs the particle action $J = J_y$ for $(R_0, \kappa) = (1,1), (1.4199, 2.1913), (2, 5)$ respectively. Asymptotically, we have $\frac{2\pi Q_y}{\mu} \to 1$ as $J_y \to \infty$.

Fig. 4 shows the particle tune vs J_y with parameters $(R_0, \kappa) = (1,1), (1.4199, 2.1913)$, and $(2.0, 5.0)$. The sharp rise (cusp) of the particle tune near $\hat{J}_y = \frac{1}{2}$ will bear important implications to parametric resonances when the system is perturbed harmonically. Such a harmonic perturbation can be generated by wake fields of vacuum chamber impedances, or by the space charge force resulting from mismatched envelope oscillations.

B Parametric resonances

Now the task is to expand the perturbation in action-angle variables of the unperturbed Hamiltonian. For example, the term linear in M can be obtained with with

$$(y^2 - R_0^2)\Theta(R_0 - y) = \sum_{n=-\infty}^{\infty} G_n(J_y)e^{in\psi_y}, \tag{42}$$

where

$$G_n(J_y) = \frac{1}{2\pi}\int_{-\pi}^{\pi}(y^2 - R_0^2)\Theta(R_0 - y)e^{-in\psi_y}d\psi_y, \tag{43}$$

with $G_{-n} = G_n^*$ and γ_n is the phase of G_n. Because the Hamiltonian of Eq. (36) is symmetric with respect to $y \to -y$, all odd harmonics vanish, i.e. $G_n = 0$ for $n = $ odd. The Hamiltonian can then be expressed as

$$H_p = E_p(J_y) + \frac{K}{4\pi R_0^2}\sum_{m=1}^{\infty}\sum_{n>0}^{\infty}(m+1)M^m|G_{n,m}| \times$$
$$\times [\cos(n\psi_y - m\nu_e\theta + \gamma_n) + \cos(n\psi_y + m\nu_e\theta + \gamma_n)] + \cdots, \tag{44}$$

where $G_{n,1} = G_n$ of Eq. (43). Here only terms that are important to particle dynamics are explicitly shown. Terms with $m > 1$ are usually not important for a weakly mismatched beam. The resonance strength function $G_{n,m}$ can be obtained similar to that of Eq. (43) by replacing integrand with an appropriate perturbation term, e.g $(3y^2 - R_0^2)/2$ for $m = 2$, etc.

The Hamiltonian of Eq. (44) expressed in action-angle variables exhibits clearly parametric resonances. A coherent perturbation to a Hamiltonian torus occurs when the stationary phase condition, $n\dot\psi_y \approx m\nu_e$, is satisfied. Such a resonance is called a $n{:}m$ primary resonance. To understand the effect of a resonance condition on the Hamiltonian flow, we perform another canonical transformation to the resonance rotating frame using the generating function

$$F_2 = (\psi_y - \frac{m}{n}\nu_e\theta + \frac{\gamma_n}{n})I_y. \tag{45}$$

Here $I_y = J_y$ and $\phi_y = \psi_y - \frac{m}{n}\nu_e\theta + \frac{\gamma_n}{n}$ are conjugate phase space variables. Neglecting the time dependent terms, which are averaged to zero, we obtain the time averaged Hamiltonian in the resonance rotating frame as

$$\langle \tilde{H}_p \rangle \approx E_p(I_y) - \frac{m}{n}\nu_e I_y + h_{n,m}(I_y)\cos n\phi_y, \tag{46}$$

where the effective resonance strength $h_{n,m}$ is

$$h_{n,m} = \frac{(m+1)M^m K}{4\pi R_0^2}|G_{n,m}(I_y)|. \tag{47}$$

The fixed points of the time averaged Hamiltonian is $\sin n\phi_{p,\text{FP}} = 0$, and

$$nQ_y(I_{p,\text{FP}}) - m\nu_e \pm nh'_{n,m}(I_{p,\text{FP}}) = 0, \tag{48}$$

where the prime corresponds to the derivative with respect to I. There are thus n unstable fixed points (UFPs) and n stable fixed points (SFPs) in the particle phase space for the $n{:}m$ resonance.

We would like to investigate the parametric resonance before the resonance strength calculation. Roughly speaking with the condition that $h'_{n,m} \approx 0$, the $n{:}m$ resonance of Eq. (48) is approximately given by $Q_y/\nu_e \approx m/n$, where $Q_y/\nu_e \in [\frac{\nu_y}{\nu_e}, \frac{\mu}{2\pi\nu_e})$. Since ν_y/ν_e and $\mu/2\pi\nu_e$ depend only on the effective space charge parameter κ, the $n{:}m$ parametric resonance condition depends only on κ. Figure 5 shows ν_y/ν_e and $\mu/2\pi\nu_e$ as a function of the parameter κ as solid curves, where the lower curve, for ν_y/ν_e, is the tune of particles inside the KV envelope with $J_y \in [0, \frac{1}{2}]$, the upper curve, for $\frac{\mu}{2\pi\nu_e}$, depicts the limit that $J_y \to \infty$, and the shaded region corresponds to particles with action $J_y \in (\frac{1}{2}, \infty)$. Horizontal lines, $\frac{1}{2}, \frac{1}{4}, \frac{1}{6}, \cdots$ corresponds to 2:1, 4:1, 6:1 $\cdots$ resonances.

Here we note that the 2:1 resonance condition is satisfied for all values of κ. This 2:1 resonance bifurcate at $J_y \leq 1/2$ as $\kappa \to 0$, which becomes the linear Mathieu instability. Since the linear Mathieu's instability occurs only in the limit of zero space charge parameter, where the resonance strength is zero, it is *not* important. Using results of Sec. IV A, the stable fixed point for the 2:1 resonance can be estimated to be $R_{\text{SFP}} \sim 1.8R_0$ for large κ.

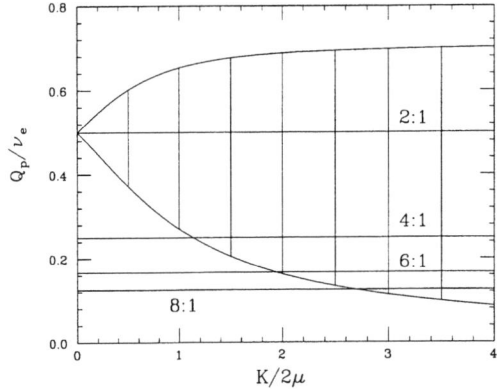

FIGURE 5. The ratio of the particle tune to the envelope tune is plotted as a function of the effective space charge strength κ. The shaded region corresponds to the allowable region of the particle tune. Horizontal lines mark parametric resonances due to the envelope modulation. Note here that the 2:1 resonance can exist at all value of the effective space charge parameter. A higher order resonance exists only when the horizontal line intersect the shaded area.

The 4th and the 6th order resonances bifurcate at $\kappa \approx \frac{3}{\sqrt{7}}$ and $\kappa \approx \frac{8}{\sqrt{17}}$ respectively. In general, the n:1 resonance can exist only

$$\kappa \geq (n^2 - 4)/\sqrt{8(n^2 - 2)}. \tag{49}$$

The condition for the n:m resonance can be obtained by replacing n with $\frac{n}{m}$ in the above equation, e.g. the 8:3 resonance bifurcate at $\kappa = 0.4865$ and the 6:2 resonance bifurcate at $\kappa = 0.6682$. Although the higher order parametric resonance strength functions become zero at the envelope radius (see next next section), particles near the envelope core can percolate into these resonance islands in numerical particle-in-cell calculations. When parametric resonances are located just outside the envelope core, beam particles can become more sensitive to errors and noise.

The horizontal lines in Fig. 4 mark the occurrence of parametric resonance condition. Note that the 2:1 resonance is always further apart from high order resonances, where *halo particles are defined as beam particles which orbit about 2:1 resonance islands*. In fact, as the parameter κ increases, the fixed point of the 2:1 resonance moves away from the core with $J_{\text{FP};2:1} \sim (\sqrt{\kappa^2 + 1} + \kappa)$. This process makes the core less susceptible to the halo formation. In reality, the radial extension of a beam with a large κ will be correspondingly larger, thus the halo formation will depend on the actual particle distribution in the tail region. The intersection of the horizontal line m/n with Q_y/ν_e curve in Fig. 4 ($Q_y = Q_p$) corresponds to the n:m resonance action, $J_{\text{FP};n:m}$. Based on our numerical simulations, we find a scaling property with $J_{\text{FP};n:m}(\kappa) \approx C_{n:m}\kappa$, where $C_{2:1} \approx 2.15, C_{4:1} \approx 0.43, C_{6:1} \approx 0.24, \cdots$. This means that all resonances in this dynamical system expand almost uniformly as a function of the parameter κ.

C Halo formation and global chaos

Within the KV model, all particles with actions less than $1/2$ will remain inside the envelope function for a matched or mismatched beam. A mismatched envelope oscillates at the tune of Q_e and all particles inside the envelope oscillate at a tune of ν_y. *Since the core envelope remains intact and the motion of particles inside the envelope is linear, Hamiltonian tori inside the envelope can be distorted but not destroyed.* The torus, which follows envelope oscillations with an action $J_y = \frac{1}{2}$, is the envelope torus.

The Hamiltonian flow outside the envelope torus has a very different structure. The 2:1 resonance exists for all values of κ. The SFP and UFP of the 2:1 resonance are always outside the envelope torus. When the parameter κ increases, we find that the SFP action of 2:1 resonance is approximately given by $J_{\text{FP},2:1} \approx \sqrt{\kappa^2 + 1} + \kappa$.

Since the envelope torus does not corrupt, how can halo be generated? The answer lies in the fact that particle distribution of finite temperature has a diffusive tail. When global stochasticity develops, particles outside the artificial KV envelope can become halo particle. *The diffusion process produces tail distribution and the resonance generates halo.* With this physical picture in mind, we investigate the dependence of the critical mismatch parameter M_c, which is defined as the envelope mismatch for the onset of global chaos, on the effective space charge parameter κ.

The evolution of the mis-matched envelope ellipse is $R^2(\theta)p_e^2 + y_e^2/R^2(\theta) = 1$, where $R(\theta)$ describes the envelope oscillations in the transport channel. The Hamiltonian value (energy) of an envelope particle oscillates with time, but the action, $J_{ye} = \frac{1}{2}$, and the Poincaré surface of section are invariant. Since we always consider the Poincaré surfaces of section at the minimum radius locations. the maximum energy at $y_e = 0$ is defined as the Poincaré energy for an envelope particle given by $E_{p,P} = 1/[4\pi(1-M)^2 R_0^2] = \nu_y/[2(1-M)^2]$.

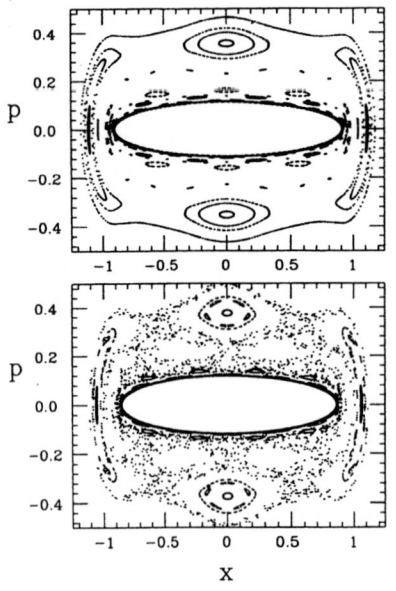

FIGURE 6. Poincaré surfaces of section in particle phase space (x,p), where $x = y$, for $(R_0, K) = (1.005, 5)$ with $M = 0.1$ and 0.15 are shown in the upper and lower plots. The lower plot is a clear example of the bounded local chaos. Here, a 10% tail distribution within the bounded local chaotic region can produce a factor of two increase in emittance. Since higher order resonances usually stay close to the core, local chaos may enhance emittance dilution without generating halo provided that the initial beam distribution does not extend very far from the core.

The Poincaré energy of a test particle outside the envelope ellipse is larger than $E_{p,P}$. We now consider the motion of a test particle outside the core with Poincaré energy $E_t = \eta E_{p,P}$ with $\eta > 1$. For a given mismatch parameter M, there is a critical number η_c such that all test particles with $\eta > \eta_c$ will orbit about the 2:1 resonance islands and become halo particles. Although particles with $\eta < \eta_c$ may encounter local chaos due to higher order primary or secondary resonances, they are bounded by a Hamiltonian torus (see Fig. 6), which separate the 2:1 resonance from the rest of higher order resonances.

Figure 7 shows the η_c vs the mismatch parameter M for $\kappa = 2$ and 2.5 respectively. Here, a unique feature is a relatively "smooth dependence" of η_c vs M for $\kappa = 2$ and a "first order phase transition like" character for $\kappa = 2.5$. Numerical simulations indicate that a sharp transition in η_c occurs always when $\kappa \geq 2.2$. The critical mismatch parameter M_c can then be defined as the mismatched parameter where a sudden jump of the critical Poincaré energy occurs as shown on Fig. 7 reaching the envelope core, where $\eta_c \approx 1$. For example $M_c \approx 0.2975 \pm 0.0025$ at $\kappa = 2.5$. The physics of this sharp transition can be understood as follows.

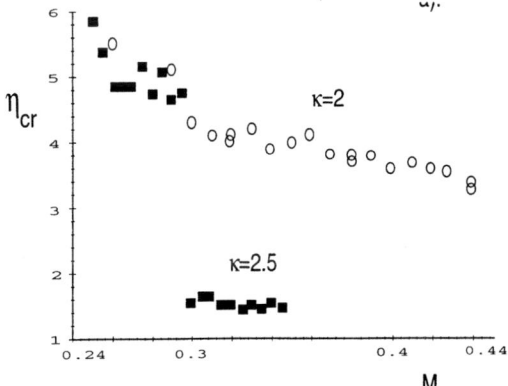

FIGURE 7. The minimum Poincaré energy of a test particle that orbits the 2:1 resonance is shown as a function of the mismatch parameter for $\kappa = 2$ and 2.5 respectively. The critical mismatch parameter corresponds to the sudden jump of the Poincaré energies. In this example, $M_c \approx 0.2975$. Note that the smooth transition of the minimum Poincaré energy shown for $\kappa = 2$ has become a first order phase transition like behavior shown for $\kappa = 2.5$.

When κ is small, there are few resonances near the core (see Figs. 5 and 4). Since the width of the 2:1 parametric resonance, or equivalently the corresponding Melnikov integral, varies smoothly with the parameter M, η_c will decrease smoothly as the parameter M increases. Along with the smooth decrease in η_c, a small stepwise decrease is expected arising from the overlapping 2:1 resonance with high order resonances one by one like a staircase. This is the case for $\kappa = 2$, shown as open circles in Fig. 7. Because there are relatively few resonances in the vicinity of the core, a small stepwise decrement in η_c appears when the 2:1 resonance overlaps with the 4:1 resonance around $M \approx 0.3$. The chaotic region produced by the overlapping 2:1 and 4:1 resonances is still relatively far away from the core. Therefore η_c continue to decrease smoothly with respect to an increasing mismatch parameter M.

When κ is large, there are many primary and secondary resonances near the envelope core. This appears to be the case when $\kappa \geq 2.2$. As the mismatch parameter is varied, local chaos is formed near the envelope core for a small mismatch parameter. Once the stochastic layer of the 2:1 resonance overlaps with the boundary of the local chaos, mainly the 4:1 resonance, global chaos occurs. A sudden jump in the critical Poincaré energy shown in Fig. 7 for $\kappa = 2.5$ provides us a sensitive determination of the critical mismatch parameter M_c.

V CONCLUSION

We find that the space charge defocusing forces for the equation of motion of the betatron coordinate $\tilde{x}$ and the dispersion function D_x are identical. The KV beam is also a self-consistent distribution for the space charge dominated beams in synchrotrons. For a KV beam, the envelope equations of motion are identical to that of linear transport channels. Our theory is found to be consistent with the numerical results obtained from a molecular dynamics simulation for the crystalline beam.

The space charge dominated beam transport problem in a linac channel is studied in a one-way self consistency model by solving the KV envelope and Hill's equations in Hamiltonian dynamics. Parametric resonances of the particle Hamiltonian can be generated by a mismatched envelope oscillations. The resonance condition is found to depend only on a single effective space charge parameter κ, i.e. the ratio of the space charge perveance parameter to the phase advance of the focusing field. Large amplitude beam halo is generated by the 2:1 resonance arising from the mismatched beam envelope oscillation. This resonance occurs at all space charge perveance parameter. Existence of envelope and particle structure resonances in periodic focusing channel can enhance halo formation. Our method provides analytic solutions to space charge transport problems.

For high density beams in synchrotrons, the space charge parameter is designed to satisfy the condition that the Laslett tune shift parameter $\Delta\nu_{\text{laslett}} = \nu\kappa \leq 0.4$. The corresponding space charge parameter is small in comparison with that of the linac beams or crystalline beams. However, the beam particles stay in the synchrotron for a long time, the accumulated effect can be as important. Here the systematic and random half integer stopband for the envelope equation may play an essential role in the stability of the space charge dominated beams. Comparison of numerical simulations with the theory presented in this work would be valuable.

acknowledgments

The work is supported in part by a grant from the Oak Ridge National Laboratory, NSF PHY-9512832 and DOE DE-FG02-92ER40747. I thank Drs. J. Holmes, D. Olsen, and W.T. Weng for many useful discussions concerning the design of the spallation neutron source.

REFERENCES

1. J.D. Lawson, *Physics of charged particle beams*, 2nd ed. (Oxford Univ. Press, N.Y. 1988); R.C. Davidson, *Physics of non-neutral plasma*, (Addison-Wesley, reading, 1990); See also *Space Charge Dominated Beams and Applications of High Brightness Beams*, AIP Conference Proceedings 377, S.Y. Lee, ed., (AIP, N.Y. 1996)
2. R.A. Jameson, in *Proc. of the 1993 Part. Accel. Conf.* p.3926 (IEEE, Piscataway, 1993); Los Alamos Report No. LA-UR-93-1209 (1993) and LA-UR-94-3753 (1994) (unpublished).
3. I.M. Kapchinskij and V.V. Vladimirskij, *Proceedings of the International Conf. on High Energy Accelerators*, p. 274 (CERN, Geneva, 1959); P.M. Lapostolle, IEEE Trans. Nucl. Sci. **NS-18**, 1101 (1971); F.J. Sacherer, ibid. 1105 (1971); J.D. Lawson, P.M. Lapostolle, and R.L. Gluckstern, Part. Accel. **5**, 61 (1973); E.P. Lee and R.K. Cooper, ibid. **7**, 83 (1976).
4. R.L. Gluckstern, Proceedings of the 1970 Linear Accelerator Conference, 811 (1970); unpublished; I. Hofmann et al., Part. Accel. **13**, 145 (1983).
5. R.L. Gluckstern, Phys. Rev. Lett. **73**, 1247 (1994).
6. C. Chen and R.C. Davidson, Phys. Rev. E**49**, 5679 (1994); Phys. Rev. Lett. **72**, 2195 (1994).
7. S.Y. Lee and A. Riabko, Phys. Rev. E **51**, 1609 (1995); A. Riabko et al., Phys. Rev. E **51**, 3529 (1995); S.Y. Lee and A. Riabko, Chinese Journal of Physics (Taiwan), **35**, 387 (1997).
8. J.S. O'Connell, T.P. Wangler, R.S. Mills and K.R. Crandall, in *Proceedings of the Particle Accelerator Conf.* edited by J. Bisognano, p.3657 (IEEE, Piscataway, NJ, 1993). J.M. Lagniel, Nucl. Inst. and Meth. in Phys. Res. **A345**, 46 (1994), ibid. 405 (1994). I. Hofmann, L.J. Laslett, L. Smith, and I. Haber, particle Accelerators, **13**, 145 (1983); J. Struckmeier and M. Reiser, Particle Accelerators, **14**, 227 (1983).
9. E.D. Courant and H.S. Snyder, Ann. Phys. (N.Y.) **3**, 1 (1958).
10. T. Suzuki, Part. Accel. **12**, 237 (1982).
11. See e.g., *The crystalline beams and related issues*, edited by D.M. Maletic and A.G. Ruggiero (World Scientific Pub. Co., Singapore, 1996).
12. J. Wei, A. Drawscke, A.M. Sessler, and X.P. Li, p. 229 in Ref. [11]; and references therein. As the density of a space charge dominated beam becomes higher, the intrabeam scattering becomes more important, and the heating rate increases. However, when a crystalline structure is formed, the random scattering vanishes. The crystalline state corresponds to a state with with vanishing small transverse and longitudinal emittances.

Betatron Resonances with Space Charge

RICHARD BAARTMAN

TRIUMF, 4004 Wesbrook Mall, Vancouver B.C., V6T 2A3, Canada

Abstract. The point is made that betatron resonances do not occur at the incoherent value of the tune, but rather at the frequencies of the appropriate collective modes. This has important implications not only for the design of high intensity, low energy proton synchrotrons, but also for the interpretation of machine studies at existing synchrotrons of this type.

I INTRODUCTION

In many papers, in proceedings of accelerator schools, and even in textbooks on accelerator physics, we read that the linear part of the space charge force is added to the linear equation of motion, leading to a tune shift, which if large enough can place individual particles on low-order betatron resonance lines. This picture, though in some sense compelling, is misleading and inhibits understanding the transverse intensity limit in low energy proton synchrotrons.

A Few Historical Notes: The fact that the incoherent tune is irrelevant when investigating integer resonances was first emphasized in 1962 by Morin [1]. It was re-emphasized by Lapostolle [2] in 1963. However, neither of these two mentioned at the time that the same sort of reasoning applied to higher order resonances. L. Smith [3], in a seminal paper also published in 1963, first pointed out that half-integer, or 'quadrupole' resonances do not occur at the incoherent tune either, and that "a machine designed conservatively regarding aperture and injector emittance could easily handle more beam than the usual space charge limit". However, his analysis was only for round isotropic beams (i.e. the same tune and emittance in each transverse direction). His student F. Sacherer in his Ph.D. thesis [4] (1968) extended the envelope analysis to non-round, anisotropic beams. As well, Sacherer treated the general case for any order resonance in one dimension, for the idealized distribution which gives exactly linear space charge force. This was extended to two-dimensional round isotropic beams by R. Gluckstern [5] (1970). It was further extended to non-round, anisotropic beams by I. Hofmann [8] (1998). These last two did not use the approximation that the tune shift be small compared with the tune, and so are applicable to space charge dominated beams in linacs and transport lines as well.

CP448, *Workshop on Space Charge Physics in High Intensity Hadron Rings*
edited by A. U. Luccio and W. T. Weng
© 1998 The American Institute of Physics 1-56396-824-X/98/$15.00

The case of general distribution is a very difficult problem even for small tune shift. Many people have proposed non-self-consistent models which ignore the non-linear forces inherent in the (non-KV) stationary distribution, and therefore miss an essential feature, namely, Landau damping. Recently, S. Lund and R. Davidson [6] have made good theoretical progress in this direction.

Naturally, the problem can also be studied with computer simulations. This was done by I. Hofmann [9] (1985), who simulated integer and 1/4-integer resonances. He found that neither of these occurred at the incoherent tune. As well, simulations by Machida have firmly established the non-relevance of the maximum incoherent tune shift to the half-integer (quadrupole) resonance [10] (1991) and to the third-integer (sextupole) resonance [19] (1998).

At least two machines have been investigated recently with a view to understanding the space charge limits, namely, the CERN PS [21] and the PSR [22]. In both of these cases, the space charge limits were found to exceed the condition that the incoherent tunes lie far from the lowest order resonances.

Lastly, as emphasized by Rees and Prior [20], effects of images can cloud the issue by contributing driving terms which make it appear that the beam is responding to incoherent resonances.

II THEORY

In the following analysis, we first ignore the forces due to image charges, and include only the 'direct' space charge term. In terms of locating resonance frequencies, this approximation is valid in proton synchrotrons whose injection energy is less than a few GeV. For example, the approximation is good in the AGS right up to the extraction energy of 30 GeV, since as energy increases, making images relatively more important, the beam size also shrinks, making the direct space charge relatively more important.

We will also be ignoring synchrotron motion. In principle, the results are correct only for coasting beams, but in practice they are applicable if the synchrotron tune is small compared with the space charge tune shift.

Incorrect theory: One often thinks of transverse space charge as just another force F_{sc}:

$$x'' + \nu_{0x}^2 x + F_{sc} = F(x,\theta) \tag{1}$$

where ν_{0x} is the unperturbed tune and $F(x,\theta)$ represents lattice errors. Then for linear space charge, $F_{sc} = -\alpha x$ and

$$x'' + (\nu_{0x}^2 - \alpha)x = F(x,\theta) \tag{2}$$

or,

$$x'' + \nu_{ix}^2 x = F(x,\theta) \tag{3}$$

and the new tune ν_{ix} is shifted downward from the bare tune by the *space charge tune shift* $\Delta\nu_x = \alpha/(2\nu_{0x})$. (We use the sign convention where space charge tune shifts are positive quantities and so the incoherent tune is $\nu_i = \nu_0 - \Delta\nu$.) If desired, nonlinear space charge terms can be added to $F(x,\theta)$, to treat the case of non-KV distributions. The resulting equation of motion can be analyzed using the well-known tools that have been developed for betatron resonances.

But this approach is incorrect. Forces arising from the beam itself are not the same as external forces. As will be shown, any theory which treats the two types of forces in the same way is incorrect and will make incorrect predictions. One often hears it said that although this approach is not self-consistent, it is at least correct for the onset of a betatron resonance. This also is incorrect. It proceeds from the false notion that particles react instantaneously according to their particular (incoherent) frequencies. Actually, particles with different tunes react differently only after a number of turns approximately equal to the reciprocal of the tune difference. Over short time scales, particles respond together; and since they do, the looked-for incoherent motion never gets started. Therefore, the response of a beam to a perturbation proceeds from the more collective to the less collective, not the other way around. In other words, it is the frequencies of the collective modes which are relevant, not the incoherent frequencies.

A Integer Resonance

One of the things that equation 2 leaves out is the fact that space charge forces are centred on the beam, not on the reference orbit through the optics elements. Therefore,

$$x'' + \nu_{0x}^2 x = \alpha(x - \bar{x}) + F(\theta) \qquad (4)$$

where $\bar{x}$ is the x-coordinate of the centre of charge and $F(\theta)$ is now independent of x, as appropriate for the integer resonance case. We take the average and find

$$\bar{x}'' + \nu_{0x}^2 \bar{x} = F(\theta). \qquad (5)$$

Thus **the motion of the centre of charge is not affected by space charge** self-forces. This is called the 'coherent' motion. Clearly, this coherent motion will be unstable if the coherent tune, ν_{0x} (equal here to the bare tune since we are neglecting image charges), is equal to an integer. Or, another way of saying the same thing is that as the coherent tune approaches an integer, the closed orbit distortion increases without limit.

Now to find the 'incoherent' motion, we simply subtract eqn. 5 from eqn. 4.

$$(x - \bar{x})'' + \nu_{0x}^2(x - \bar{x}) = \alpha(x - \bar{x}) \qquad (6)$$

or

$$(x - \bar{x})'' + \nu_{ix}^2(x - \bar{x}) = 0 \tag{7}$$

where, as before, ν_{ix} is the incoherent tune. Notice that the incoherent equation of motion contains no driving terms for the integer resonance. Therefore, **incoherent motion is not affected by dipole errors**. This means that the incoherent tune can be equal to an integer, with no adverse effects. Conceptually, this is the biggest hurdle to leap. It is therefore worth re-emphasizing in different words: A particle which is shifted by direct space charge to a tune of exactly an integer, turn after turn sees the same dipole errors at the same betatron phase, and yet is not even slightly affected compared with other particles which do not have integer tune. This is not due to space charge stabilizing the resonance, as claimed in ref. [11], since in this completely linear case there is no incoherent tune spread to generate Landau damping. There just isn't any driving term for the incoherent motion.

Although we derived the coherent and incoherent equations of motion for linear space charge, they remain true in the general case, following directly from Newton's third law. For particle i, the equation of motion is

$$x_i'' + \nu_{0x}^2 x_i = F_{\text{sci}} + F(\theta) \tag{8}$$

$F_{\text{sci}} = \sum_j F_{ij}$, F_{ij} being the force on particle i by particle j. Since $F_{ij} = -F_{ji}$, when we sum eqn. 8 over i to obtain the coherent motion, we recover eqn. 5. Subtracting, we now find for the incoherent motion

$$(x_i - \bar{x})'' + \nu_{0x}^2(x_i - \bar{x}) = F_{\text{sci}} \tag{9}$$

Since the space charge force is now nonlinear, this cannot be reduced to a new simple harmonic equation with a shifted frequency. Nevertheless, it retains the feature that the incoherent motion does not 'see' dipole errors.

B Half-Integer Resonance

Incorrect theory: Now let us investigate errors of the type $F(x, \theta) = x f(\theta)$. The equation of motion is

$$x'' + \nu_{0x}^2 x = \alpha(x - \bar{x}) + x f(\theta) \tag{10}$$

Coherent motion is found by taking the average

$$\bar{x}'' + \nu_{0x}^2 \bar{x} = \bar{x} f(\theta), \tag{11}$$

and to find the equation for incoherent motion, we again subtract:

$$(x - \bar{x})'' + \nu_{ix}^2(x - \bar{x}) = (x - \bar{x}) f(\theta). \tag{12}$$

This looks exactly like a single particle equation of motion in the presence of a half-integer resonance driving term. Again, **this approach is incorrect**. The response of a beam to linear errors is to modulate in size, but the beam size is contained in the space charge term α. The correct self-consistent approach is to formulate the problem directly in terms of the beam size, or 'envelope'.

In the following analysis, we first assume that the beam has the correct distribution to give only linear space charge forces. However, the results are not restricted to such special distributions, as will be pointed out later in our discussion. No originality is claimed for the analysis, as it can all be found in Sacherer's thesis [4].

1 One Dimension

We start with the 1-dimensional case, since it exhibits the features we want to emphasize, unencumbered by other considerations such as beam aspect ratio and tune split. For this case the distribution which gives linear space charge is uniform in configuration space between the beam edges at $\pm \hat{x}$.

It can be shown that $\alpha \propto 1/\hat{x}$ and so from the equation of motion (10), it is clear that

$$\alpha = 2\nu_0 \Delta \nu a/\hat{x} \tag{13}$$

where $\Delta \nu$ is the incoherent tune shift, and a is $\sqrt{\beta \epsilon}$, the unperturbed beam size. In terms of the normalized beam size $\tilde{x} = \hat{x}/a$, it can be shown that the envelope equation is

$$\tilde{x}'' + \nu_0^2 \tilde{x} - \frac{\nu_0^2}{\tilde{x}^3} = 2\nu_0 \Delta \nu + f(\theta)\tilde{x}. \tag{14}$$

Keeping in mind that $\Delta \nu \ll \nu_0$, we see that the stationary solution to this equation is

$$\tilde{x} = 1 + \frac{\Delta \nu}{2\nu_0} \tag{15}$$

and this can be seen as a change in the β-function due to space charge: $\beta_x = \beta_{x0} \nu_0/(\nu_0 - \Delta \nu)$.

Rewriting the envelope equation in terms of small perturbations δ with respect to the stationary solution, i.e. $\tilde{x} = 1 + \frac{\Delta \nu}{2\nu_0} + \delta$, we find

$$\delta'' + (4\nu_0^2 - 6\nu_0 \Delta \nu)\delta = f(\theta). \tag{16}$$

Clearly, this is unstable for the nth Fourier component of $f(\theta)$ when $4\nu_0^2 - 6\nu_0 \Delta \nu = n^2$, i.e., when

$$\frac{n}{2} = \nu_0 - \frac{3}{4}\Delta \nu = \nu_\mathrm{i} + \frac{1}{4}\Delta \nu. \tag{17}$$

So the incoherent tune $\nu_\mathrm{i} = \nu_0 - \Delta \nu$ can be depressed beyond the half-integer by a quarter of the space charge tune shift! Again, this is not due to a 'stabilizing' or 'self-limiting' effect of space charge, but to the fact that betatron resonance occurs at the appropriate collective mode frequency and not at the incoherent frequency.

2 Two Dimensions

Linear space charge is provided by the KV distribution: this is a uniformly-filled ellipse in any 2-dimensional projection, and an ellipsoidal shell in the 4 phase space dimensions. Kapchinsky and Vladimirsky [12] showed that the space charge force coefficient α is

$$\alpha_x \propto \frac{1}{\hat{x}(\hat{x}+\hat{y})}, \text{ and } \alpha_y \propto \frac{1}{\hat{y}(\hat{x}+\hat{y})}. \tag{18}$$

By comparison with eqn. 10, it is therefore clear that in terms of incoherent tune shifts,

$$\alpha_x = 2\nu_{0x}\Delta\nu_x \frac{a}{\hat{x}}\frac{a+b}{\hat{x}+\hat{y}} \tag{19}$$

$$\alpha_y = 2\nu_{0y}\Delta\nu_y \frac{b}{\hat{y}}\frac{a+b}{\hat{x}+\hat{y}} \tag{20}$$

where $a = \sqrt{\beta_x \epsilon_x}$ and $b = \sqrt{\beta_y \epsilon_y}$ are the unperturbed beam sizes.

The envelope equations are the well-known KV equations, here written in terms of tune shifts and the normalized beam sizes $\tilde{x} = \hat{x}/a$, $\tilde{y} = \hat{y}/b$:

$$\tilde{x}'' + \nu_{0x}^2 \tilde{x} - \frac{\nu_{0x}^2}{\tilde{x}^3} = 2\nu_{0x}\Delta\nu_x \frac{a+b}{a\tilde{x}+b\tilde{y}} \tag{21}$$

$$\tilde{y}'' + \nu_{0y}^2 \tilde{y} - \frac{\nu_{0y}^2}{\tilde{y}^3} = 2\nu_{0y}\Delta\nu_y \frac{a+b}{a\tilde{x}+b\tilde{y}}. \tag{22}$$

For clarity, we have dropped the gradient error driving terms. As a check, we find by direct substitution that to first order in smallness of $\Delta\nu/\nu$, $\tilde{x} = 1 + \Delta\nu_x/(2\nu_{0x})$ and $\tilde{y} = 1 + \Delta\nu_y/(2\nu_{0y})$, again reflecting the incoherent effect of space charge on the β-functions.

As before, we can linearize for small perturbations δ_x and δ_y, i.e. $\tilde{x} = 1 + \frac{\Delta\nu_x}{2\nu_{0x}} + \delta_x$ and $\tilde{y} = 1 + \frac{\Delta\nu_y}{2\nu_{0y}} + \delta_y$.[1] The result is two coupled simple harmonic oscillators. We can find the eigenfrequencies ('eigentunes') of the system, but the general expression is complicated and unenlightening. For the case of a round beam ($a = b$, $\nu_{0x}\Delta\nu_x = \nu_{0y}\Delta\nu_y$), the frequencies are

$$\nu^2 = 2\nu_{0x}^2 + 2\nu_{0y}^2 - 5\nu_{0x}\Delta\nu_x \pm \sqrt{(2\nu_{0x}^2 - 2\nu_{0y}^2)^2 + (\nu_{0x}\Delta\nu_x)^2}. \tag{23}$$

If the tune split is small, i.e. $|\nu_{0x} - \nu_{0y}| \ll \Delta\nu_x/4$, we find two distinct eigenmodes:

$$\nu^2 = \begin{cases} 4\bar{\nu}^2 - 4\nu_{0x}\Delta\nu_x \\ 4\bar{\nu}^2 - 6\nu_{0x}\Delta\nu_x \end{cases} \tag{24}$$

---

[1] This is not as straightforward as Sacherer implies, since the expansion must be made with the correct order of smallness, namely, $1 \gg \Delta\nu/\nu \gg \delta$.

$(2\bar{\nu}^2 = \nu_{0x}^2 + \nu_{0y}^2)$, while if the tune split is not small compared with the tune shift, the two modes are

$$\nu^2 = \begin{cases} 4\nu_{0x}^2 - 5\nu_{0x}\Delta\nu_x \\ 4\nu_{0y}^2 - 5\nu_{0x}\Delta\nu_x \end{cases} \quad (25)$$

The physical interpretation is straight-forward. In the small-split case, the two transverse motions are tightly coupled together, and a gradient error in either transverse plane can drive either mode. One mode is symmetric, envelope modulations are in phase so the beam 'breathes' in both directions together; the other mode is antisymmetric, envelope modulations are 180° out of phase. In the large-split case, the envelope modulations in x and y cannot stay in phase, so they are essentially decoupled and act independently.

Resonance occurs for integer values of the eigentune, i.e. $\nu = n$ where n corresponds to the Fourier component of the driving gradient error. All the resonance conditions can be cast into the form

$$\nu_{0x,y} - C\Delta\nu_{x,y} = n/2. \quad (26)$$

For the small-split case, $C = 1/2, 3/4$ (resp. for symmetric and anti-symmetric), and for the large-split case, $C = 5/8$. The analysis can be repeated for non-round beams. For example with $a = 2b$, $C_x = 7/12$ and $C_y = 2/3$ in the large-split case, and $C = 0.273, 0.684$ (sym., anti-sym., $\Delta\nu_{x,y}$ set equal to the larger of the two incoherent tune shifts) in the tightly-coupled case $\nu_{0x} \approx \nu_{0y}$. In the incorrect theory based upon the incoherent tune, the resonance condition in eqn. 26 is $C = 1$.

Those more familiar with formulas for space charge modes as they have been derived for linacs and transport lines (see e.g. Ref. [7, eqn. 10]) will remember the following envelope mode frequencies for isotropic focusing:

$$\nu^2 = \begin{cases} 2\bar{\nu}^2 + 2\nu_i^2 \\ \bar{\nu}^2 + 3\nu_i^2 \end{cases} \quad (27)$$

With the substitution $\nu_i = \bar{\nu} - \Delta\nu$, these are seen to be consistent with eqn. 24. In fact, these are more accurate than eqn. 24, as they do not require $\Delta\nu \ll \nu_0$.

The foregoing analysis is not applicable to coupling resonances, since the envelope equations assume no coupling i.e. no xy, xy', etc. correlations. The general case has been treated by Hofmann [8]. There are 4 resonance modes: these can be identified with the betatron resonances $\nu_x = N/2$, $\nu_y = N/2$, $\nu_x - \nu_y = N$, and $\nu_x + \nu_y = N$.

C Higher Order Resonances

1 One Dimension

We have seen that integer resonances can be investigated using the equations of motion of the first moments, and half-integer resonances can be investigated using the equations of motion of the second moments. In general, higher order resonances require the simultaneous solution of the equations of motion of the corresponding higher order beam moments. These equations are not easily solved. However, for the simple case of 1-D motion and linear space charge, the general Vlasov equation can be solved.

For the case of longitudinal motion in a bunch whose length is large compared with the conducting beam pipe, the eigenfrequencies are well-known [17]. In that case, the internal space charge force is proportional to $\hat{x}^{-3}$. In the present case, the space charge force is proportional to $\hat{x}^{-1}$, and so the eigenfrequencies are not the same. However, all other aspects, such as the doubly-infinite number of modes, and the qualitative shapes of these modes, are similar. Sacherer's findings [4] can be summarized in the resonance condition

$$n/m = \nu_0 - C_{mk}\Delta\nu, \qquad (28)$$

with the matrix C given in Table 1. There are two indices: m for azimuthal and k for radial. For example, the $(1,3)$ eigenmode is the first harmonic azimuthally (dipole) and has 2 nodes radially. It is the first non-rigid dipole mode, since only odd-k modes exist for any odd m, and only even for even. In general, the mode (m,k) is driven by error driving terms proportional to $x^{k-1}\cos(n\theta)$ where $n \approx m\nu$.

TABLE 1. Coherent mode coefficients C_{mk} for 1-D transverse space charge

	m=1	2	3	4	5
k=1	0				
2		3/4			
3	9/8		7/8		
4		17/16		59/64	
5	65/64		133/128		121/128

This agrees with the results already found for the rigid dipole and quadrupole modes in 1-D, namely, $C_{11} = 0$ and $C_{22} = 3/4$.

Remarkably, the frequencies of the non-rigid modes ($k > m$) are shifted *past* the incoherent frequency. However, as explained below, these modes are not seen in simulations, and there is good reason to believe they are not important. Realistic distributions never have perfectly linear space charge and the frequency spread brings with it Landau damping. It is known that in the longitudinal case, Landau thresholds depend mostly upon k [18]; for example, if the lowest sextupole mode is Landau damped, the first non-rigid dipole mode is as well. Recent work by Lund and Davidson [6] using a warm fluid model for the transverse case indicates that these modes are associated with the unphysical aspects of the ideal linear space charge distribution.

2   Two Dimensions

As already stated, Hofmann [8] has treated the general case up to octupole. It is difficult to make a concise summary of his results because of the additional parameters; tune split and emittance ratio. However, he gives simplified formulas for the case of round beam, equal tunes in both planes. The coefficients C_{mk} extracted from these formulas are shown in Table 2.

TABLE 2. Coherent mode coefficients C_{mk} for 2-D round beam

	m=1	2	3	4
k=1	0			
2		1/2, 3/4		
3	3/4, 5/4		3/4, 11/12	
4		7/8, 5/4		13/16, 7/8, 31/32

Lund and Davidson [6] (LD) give a formula for the simplest of these round beam modes, namely those corresponding to round (symmetric) perturbations with $m = k$. Cast into the form of eqn. 28, their eqn. 103 becomes:

$$C_{mm,\text{sym}} = 1 - \frac{2}{m^2} \tag{29}$$

Since this is for symmetric perturbations, it only applies to even m. We see that it agrees with those of Table 2 for $m = 2, 4$. As with the 1-D case, some of the modes in Table 2 are largely irrelevant for realistic distributions. LD argue that the non-appearance of the mode $(m, k) = (2, 4)$ with $C = 5/4$ in their warm-fluid model is due to the non-physical nature of the KV distribution. Hofmann mentions that this mode, which is intrinsically unstable for $\nu/\nu_0 < 0.24$, changes the phase space density of the beam by only a negligible amount beyond this threshold.

Of the other modes in Table 2, it is not clear at this point which are the important ones. Likely, only the diagonal ($m = k$) ones are.

D   Nonlinear space charge

The foregoing was for the case of ideal distributions which give linear space charge. Originally, it seemed that the general case could only be treated with the Vlasov equation. However, in 1971 Sacherer [13] proved that the envelope equations apply to any distribution, provided that the beam parameters are replaced by the appropriate statistical quantities: the beam boundary $\hat{x}$ becomes the rms size and the emittance becomes the rms emittance. Although the resulting equations are still correct, they are no longer closed, since rms emittance is not a conserved quantity in the presence of nonlinear forces. In practice, however, this turns out to only cause difficulty if one wishes to study the evolution of an unstable distribution. Starting with an arbitrary stationary distribution, and approaching a betatron resonance, the envelope equations still correctly predict thresholds. In fact, Sacherer began to study the possibility of more general KV equations because simulations performed by Lapostolle [14] indicated that these equations very accurately described the evolution of rms beam sizes even for non-KV distributions.

This concept of 'linear-space-charge-equivalent beam' (= 'KV-equivalent beam' in 2-D) has been used in studies of higher order resonances as well. It is well-established in the case of space charge dominated beams in transport channels. (See for example, Struckmeier et al. [15].) On the other hand, it has been largely ignored in studies of space charge effects in synchrotrons.

Nonlinear space charge implies that the tune shift is no longer a single quantity, but is a function of amplitude; particles of smaller amplitude have in general a larger tune shift than large-amplitude particles. An important, but often overlooked, implication of the universality of the envelope equations is that the relevant tune shift is not the maximum shift but rather the tune shift of the linear-space-charge-equivalent beam. For example, in the 2-D case, the beam envelope of the KV distribution is twice the rms size and the 100% emittance is 4 times the rms emittance. Therefore, to obtain the tune shift relevant for half-integer resonances, one can use the usual (incoherent) tune shift formula [16]

$$\Delta \nu_x = \frac{r_\mathrm{p} N G}{2\pi \beta^2 \gamma^3 \epsilon_x}, \qquad (30)$$

provided that the form factor G is set equal to 1, and the emittance is interpreted as 4 times the rms emittance, i.e. $\epsilon_x = 4\sqrt{\overline{x^2}\,\overline{x'^2} - \overline{xx'}^2}$. It is easily shown that with this definition of emittance, the value of G for maximum incoherent tune in the Gaussian case is 2. Therefore, with quadrupole errors, the maximum incoherent tune shift can exceed that needed for coincidence with a half-integer resonance by as much as a factor of 3.2; a factor of 2 for the Gaussian beam and a further factor of 8/5 to take into account the location of the envelope resonance in the large-split case.

III SIMULATIONS

A One Dimension

We performed simulations with up to 50,000 particles and the following equation of motion;

$$x'' + \nu_0^2 x = \alpha x^{m-1} \cos(n\theta) + F_\mathrm{SC}. \qquad (31)$$

The derivatives are with respect to θ; m and n are integers.

For no space charge, the particle experiences resonant growth when

$$m\nu_0 \approx n. \tag{32}$$

The smaller is the strength α, the more accurately does this resonance condition have to be met to see an effect.

The space charge force on the ith particle in the simulations is simply equal to an intensity parameter multiplied by the difference between the number of particles to its left and to its right. Physically, this corresponds to particles being in the form of planes in free space. A distribution which results in exactly linear space charge force is that which is uniform in x. This is the analogue of the 'Kapchinsky-Vladimirsky' distribution for 2-D.

GIF animations of the simulations can be found at

http://decu10.triumf.ca:8080/ht/

In each of the animations the intensity is changing slowly to sweep first the incoherent tune through the resonance, and then the coherent tune. This closely mimics the situation in rings which accumulate a large tune shift over many turns. Since animation snapshots are taken at intervals of m turns, it is easy to verify that nothing happens when the incoherent tune is equal to n/m: for the stationary distribution, all the particles look as though they have stopped rotating in phase space and, even though the error driving term is significant, there is no hint of amplitude growth. However, when the frequency of the coherent mode m is equal to n/m, one can see that the m-fold distortion of the beam boundary is stationary. In that case, with an appropriate driving term, the amplitude of the distortion grows steadily. For example, for $m = 2$, $\nu_i = \frac{n}{2} - \frac{1}{4}\Delta\nu$ (i.e. $\nu_0 = \frac{n}{2} + \frac{3}{4}\Delta\nu$), one sees a growing mismatch which is at a constant orientation in phase space, and individual particles moving around the edge of the mismatched phase space ellipse at a rate equal to the difference between the coherent quadrupole mode tune and the incoherent tune ($= \frac{n}{2} - \nu_i = \frac{1}{4}\Delta\nu$).

We have not been able to find the coherent modes with $k \neq m$.

For any other distribution, particles at different betatron amplitudes rotate in phase space at different rates. We have run simulations for the Gaussian distribution. As the incoherent frequency of the small amplitude particles reach the resonance, nothing happens. As the intensity continues to increase, a stage is reached where there is a barely discernable amplitude growth at an amplitude which contains roughly 50% of the particles. For larger intensities yet, the incoherent tune of the particles in the tail of the Gaussian coincide with the resonance, and show a dramatic emittance increase along with an appearance of an m-fold symmetric island structure. In summary, we see the core is affected only by coherent core modes, and the tail affected by incoherent resonance. The reason for this behaviour is that halo particles feel space charge forces from the core of the beam, and these act as external forces because they depend mainly upon the number of particles in the core and not so much on the shape of the core. This picture qualitatively bridges the two (seemingly contradictory) concepts of coherent space charge modes on the one hand, and the core-halo modes on the other.

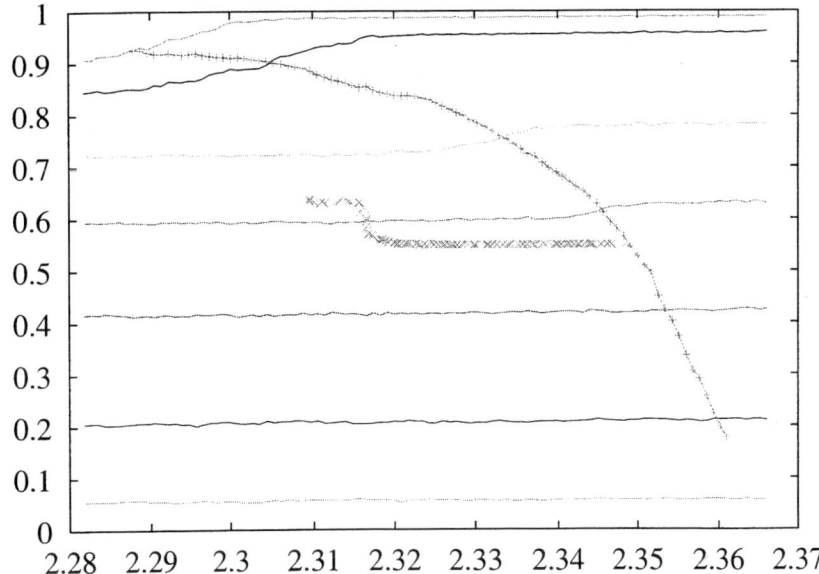

FIGURE 1. The plot contains 3 types of curves all plotted against the incoherent tune of the equivalent stationary distribution (ITESD). The data are from simulations in which the bare tune is 2.45 and the intensity is slowly raised to bring the incoherent tune past the 7/3 resonance; so the curves should be thought of as proceeding from right to left. The thick curve in the centre is the rms size of the stationary distribution. A threshold at the coherent mode frequency 2.3167 is evident. The horizontal curves are the fraction of particles inside a fixed emittance for the Gaussian distribution. They step downwards as particles are driven to larger amplitude. The curve composed of connected + symbols is the emittance at which the incoherent tune is on resonance. So for example, when the ITESD is 2.36, small-amplitude particles have tune equal to 7/3; when the ITESD is 2.345, the emittance containing 60% of the particles has a tune of 7/3.

An example is shown in Fig. 1, where we have plotted for the gaussian distribution the fraction of particles inside a given betatron amplitude as a function of the incoherent tune of the stationary beam of same rms size. The resonance is $3\nu = 7$, the bare tune is 2.45, so the value of the incoherent tune of the equivalent stationary beam that puts the coherent $(m, k) = (3, 3)$ mode on resonance is $(2.45 - (2.45 - 2.333)/C_{33} =) 2.31667$. The rms beam size of the stationary distribution has also been plotted, and there is clearly a sharp threshold at the expected location. On the other hand, in the Gaussian case, the effect is much broader, but still centred upon the same location. This verifies that the relevant comparison between beams of different distribution is not one where the peak incoherent tune is the same, but where the rms beam size is the same.

B Two Dimensions

Machida [10] has performed simulations for quadrupole resonance in the SSC LEB. With $\nu_{0x} = 11.87$, ν_{0y} was varied in the range 11.95 to 11.55. The beam distribution was Gaussian, the maximum incoherent tune shift was kept at $\Delta\nu_y = 0.33$, and the half-integer stopband was 0.02. Machida found the threshold for emittance growth to be approximately 11.63, or, in other words, with an incoherent tune shift of 0.33, the bare tune could be placed as close as 0.13 from the 1/2-integer. Equation 23 predicts that the resonance $2\nu_y = 23$ causes an envelope instability when $\nu_{0y} = 11.61$. Agreement is even better when the stopband width is taken into account.

Hofmann [9] has simulated the case for an octupole driving term near a 1/4-integer resonance. He found that emittance growth occurred when the incoherent frequency was 10% beyond the resonance. This is within his uncertainty of $1 - C_{mm,\text{sym}} = \frac{1}{8}$ of Table 2.

IV EXPERIMENT

A number of observers have noted that the incoherent space charge tune shift can exceed the distance of the bare tune to the nearest low order resonance. For example, in the CERN PS with vertical tune 6.22, Cappi et al. [21] observed losses when the tune shift was larger than 0.27. When they moved the vertical tune to 6.28, losses occurred only for tune shift larger than about 0.35.

Experiments performed on the LANL PSR (Nueffer et al. [22]) test the theory more quantitatively since both beam sizes and beam distributions were accurately measured. However, the calculated peak incoherent tune shifts contained in the report are a factor of 2 too small, since their tune shift formula (1) is incorrect [23]. In fact, their formula gives the tune shift of the KV-equivalent beam, which is a factor of 2 smaller than the peak tune shift of the Gaussian beam. Their Table I should therefore read as follows.

TABLE 3. LANL PSR data $\nu_{0x} = 3.155$, varying ν_{0y}

ν_{0y}	$N/10^{13}$	$2\sigma_y$	$2\Delta p/p$	ϵ_x	ϵ_y	$\Delta\nu_{yi}$	$\Delta\nu_{y\text{KV}}$
2.193	0.6	8.5	0.38	20.1	8.6	0.150	0.075
	1.18	9.9	0.41	22.4	11.6	0.230	0.115
	2.3	13.3	0.38	26.6	20.0	0.296	0.148
2.142	0.6	8.4	0.41	19.7	8.4	0.150	0.075
	1.18	11.5	0.45	16.7	15.7	0.190	0.095
	2.3	15.5	0.41	25.4	28.6	0.226	0.113
2.100	0.6	9.5	0.47	15.5	10.7	0.122	0.061
	1.18	14.5	0.45	16.4	25.0	0.136	0.068
	2.3	21.5	0.31	44.0	55.0	0.128	0.064
2.059	0.6	12.6	0.46	15.0	18.9	0.086	0.043
	1.18	20.6	0.45	16.0	50.5	0.078	0.039

Note that the experiments were performed on bunched beams. Therefore, $\nu_0 > 2$ with $\nu_i < 2$ indeed means that some slice of the beam bunch has an incoherent tune of exactly 2. Resonances of all orders are expected at $\nu_y = 2$, since there was no correction of the driving terms. The order m of the intensity-limiting resonance was not known. An experiment at low intensity had been performed in 1987 [24], in which ν_y was lowered towards 2. It was found that losses began at $\nu_y \approx 2.03$. The second last entry in the table indicates that the coherent mode responsible for emittance growth is shifted by $((2.059 - 2.03)/0.043 =) 0.67$ of the KV-equivalent tune shift. We guess that the mode responsible is the quadrupole mode $2\nu_y = 4$, since $\Delta\nu_c/\Delta\nu_{KV} = 5/8$ (for round beams), and the stopband of the quadrupole mode is likely the largest.

As well, there is a qualitative result from the PSR experiments in support of the coherent mode theory. According to the incoherent resonance theory, only those particles whose tunes are on resonance will experience amplitude growth. In the particular case of a Gaussian beam approaching a resonance from above, this would imply that the smallest-amplitude particles would grow in amplitude, thus de-populating the centre of the Gaussian, making it flatter. Fig. 1c of [22] (which corresponds with the third-last line of Table 3) shows that in at least one case, the opposite occurred.

V IMAGES

As mentioned above, a signature of the integer resonance $\nu = n$ is an n-fold closed orbit distortion. As the tune approaches the integer, the steering of the orbit around the ring must more and more accurately correct the nth Fourier component. According to the theory emphasized in the present report, the integer resonance is not approached by increasing the incoherent tune shift while leaving the bare tune unchanged, since the bare tune is the frequency of the coherent mode. Nevertheless, it was observed at ISIS [20] that with a tune above 4, the 4th harmonic of the closed orbit needed to be corrected as the intensity was raised. Rees et al. concluded that this was due to the (usually neglected) closed-orbit-dependent terms of the image force expansion. Let us briefly derive this expansion.

For a parallel plate configuration, we simply extend the derivation first given by Laslett [25, Appendix B]. The origin is midway between the plates, which are situated at $y = \pm h$. A line charge λ is located at $\bar{y}$ (y_1 in Laslett's notation). Then the potential at other locations y is given by

$$U = -2\lambda \log \left| \frac{\sin\left(\frac{\pi y}{2h}\right) - \sin\left(\frac{\pi \bar{y}}{2h}\right)}{1 + \cos\left(\frac{\pi(y+\bar{y})}{2h}\right)} \right|. \tag{33}$$

From this we subtract the direct line charge field $-2\lambda \log \left|\frac{\pi(y-\bar{y})}{4h}\right|$ and expand up to 4th order in y and $\bar{y}$, to get:

$$-\frac{U_{\text{image}}}{\lambda} = \frac{\pi^2}{24\,h^2}\left(y^2 + 4y\bar{y} + \bar{y}^2\right) + \frac{\pi^4}{11520\,h^4}\left(7y^4 + 32y^3\bar{y} + 42y^2\bar{y}^2 + 32y\bar{y}^3 + 7\bar{y}^4\right) \tag{34}$$

We can find the force by differentiating w.r.t. y. However, since the beam is held at $y = \bar{y}$ by external dipole errors, and we are interested in perturbations about this point, we express the result in terms of $\check{y} = y - \bar{y}$:

$$\frac{E_{y\text{image}}}{4\lambda} \approx \frac{\pi^2}{48}\frac{\check{y}}{h^2} + \frac{\pi^2}{16}\frac{\bar{y}}{h^2} + \frac{\pi^4}{192}\frac{\bar{y}^3}{h^4} + \frac{\pi^4}{128}\frac{\check{y}\bar{y}^2}{h^4} + \frac{\pi^4}{256}\frac{\check{y}^2\bar{y}}{h^4} + \frac{7\pi^4}{11520}\frac{\check{y}^3}{h^4} \tag{35}$$

The factor of 4 on the left is in order to make the usual image coefficient as defined by Laslett [25] appear explicitly in the expansion. We recover the usual 'incoherent' space charge image coefficient $\epsilon_1 = \frac{\pi^2}{48}$ and the 'coherent' image coefficient $\xi_1 = \frac{\pi^2}{16}$.

The case of the circular boundary (radius h) is much simpler mathematically, since there is only one image:

$$\frac{E_{y\text{image}}}{4\lambda} = \frac{1}{2}\frac{\bar{y}}{h^2 - y\bar{y}} \approx \frac{1}{2}\frac{\bar{y}}{h^2} + \frac{1}{2}\frac{\bar{y}^3}{h^4} + \frac{1}{2}\frac{\check{y}\bar{y}^2}{h^4}, \tag{36}$$

and this yields as usual, $\epsilon_1 = 0$ and $\xi_1 = \frac{1}{2}$.

Let us introduce new 'higher order' image coefficients κ for the general boundary:

$$\frac{E_{y\text{image}}}{4\lambda} = \epsilon_1 \frac{\check{y}}{h^2} + \xi_1 \frac{\bar{y}}{h^2} + \kappa_{30}\frac{\bar{y}^3}{h^4} + \kappa_{21}\frac{\check{y}\bar{y}^2}{h^4} + \kappa_{12}\frac{\check{y}^2\bar{y}}{h^4} + \kappa_{03}\frac{\check{y}^3}{h^4} + \cdots \tag{37}$$

The κ_{21} term was investigated by Rees and Prior [20]: it represents a quadrupole, or half-integer, driving term whose strength is proportional to the square of the closed orbit distortion. Therefore, if ν_y is depressed toward n, and there is an n-fold distortion of the closed orbit, there will be a $2n$-fold quadrupole driving term, and a $2\nu_y = 2n$ stopband. We expect this to occur at the quadrupole coherent mode (not at the incoherent tune as suggested in [20]), and this has been confirmed in simulations [26]. Similarly, the κ_{12} term will open a sextupole stopband for n-fold distortions of the closed orbit where $3\nu_y = n$. As well, the modulations of the vacuum chamber can drive octupole (κ_{03}) and quadrupole (ϵ_1) resonances.

For large space charge tune shift, the higher order image terms are not small compared with driving terms due to lattice errors, and they can have a significant impact. It is important to realize that they are all intensity-dependent and so correction schemes derived by experimenting with a low intensity beam will not work as intended at high intensity.

VI CONCLUSION

The main point of the present paper is that the practice of restricting the incoherent tune spread to lie between the lowest order betatron resonances is too conservative. Conversely, the assumption that in a well-tuned machine, the incoherent tune spread does not overlap any low-order resonances, is also not warranted. The error incurred is especially large for centrally-peaked distributions. For example, the Gaussian distribution in 2 dimensions can have a peak incoherent tune shift over twice as large as the distance of the bare tune from the low-order resonance.

ACKNOWLEDGEMENTS

The author would like to thank the TRIUMF accelerator division for supporting this work, and M. D'yachkov for coding the simulations and constructing the animations. He would also like to thank the organizers of the Shelter Island workshop: interactions with other attendees resulted in a more complete paper.

REFERENCES

1. D.C. Morin, *Transverse Space Charge Effects in Particle Accelerators* MURA Report 649 (Ph.D. thesis, University of Wisconsin, 1962).
2. P. Lapostolle, *Oscillations in a Synchrotron Under Space Charge Conditions* Proc. Int. Conf. on High Energy Acc. (Dubna, 1963) p. 1235.
3. L. Smith, *Effect of Gradient Errors in the Presence of Space Charge Forces* Proc. Int. Conf. on High Energy Acc. (Dubna, 1963) p. 1232.
4. F. Sacherer, *Transverse Space Charge Effects in Circular Accelerators* Lawrence Rad. Lab Report UCRL-18454 (Ph.D. thesis, University of California, 1968).
5. R. Gluckstern, *Oscillation Modes in Two Dimensional Beams* Proc. Linac Conf. (Fermilab, Batavia, 1970) p. 811.
6. S. Lund and R. Davidson, *Warm-Fluid Description of Intense Beam Equilibrium and Electrostatic Stability Properties* to be published in Physics of Plasmas (1998).
7. J. Struckmeier and M. Reiser *Theoretical Studies of Envelope Oscillations and Instabilities of Mismatched Intense Charged-Particle Beams in Periodic Focusing Channels* Part. Acc. **14** (1984) p. 227.
8. I. Hofmann, *Stability of Anisotropic Beams with Space Charge* Phys. Rev. E **57** (1998) p. 4713.
9. I. Hofmann and K. Beckert, *Resonance Crossing in the Presence of Space Charge* IEEE Trans. Nucl. Sci. **NS-32** (PAC 1985) p. 2264.
10. S. Machida, *Space Charge Effects in Low Energy Proton Synchrotrons* Nucl. Inst. Meth. **A309** (1991) p. 43.
11. W.T. Weng, *Space Charge Effects – Tune Shifts and Resonances* AIP Conf. Proc. **153** (1987) p. 348.
12. I.M. Kapchinsky and V.V. Vladimirsky, *Limitations of Proton Beam Current in a Strong Focusing Linear Accelerator Associated with the Beam Space Charge* Proc. Int. Conf. on High Energy Acc. (CERN, 1959) p. 274.
13. F. Sacherer, *RMS Envelope Equations with Space Charge* IEEE Trans. Nucl. Sci. **NS-18** (PAC 1971) p. 1105. See also the longer report of the same title, CERN-SI-Int.-DL/70-12 (Nov. 1970).

14. P. Lapostolle, *Quelques Propriétés Essentielles des Effets de la Charge d'Espace dans les Faisceaux Continus* CERN-ISR-DI/70-36.
15. J. Struckmeier, J. Klabunde and M. Reiser *On the Stability and Emittance Growth of Different Particle Phase-Space Distributions in a Long Magnetic Quadrupole Channel* Part. Acc. **15** (1984) p. 47.
16. K.H. Reich, K.H. Schindl, H.O. Schönauer, *An Approach to the Design of Space-Charge Limited High Intensity Synchrotrons* Proc. Int. Conf. on High Energy Acc. (Fermilab, 1983) p. 438.
17. G. Besnier, B. Zotter, *Oscillations Longitudinales d'une Distribution Elliptique,...* CERN-ISR-TH/82-17.
18. G. Besnier *Contribution à la Théorie de la Stabilité des Oscillations Longitudinales d'un Faisceau Accélère en Régime de Charge d'Espace* Ph. D. thesis (B-282-168) Université de Rennes, France (1978).
19. S. Machida, *Space Charge Effects in Low Energy Proton Synchrotrons* Nucl. Inst. Meth. **A309** (1991) p. 43.
20. G. Rees and C. Prior, *Image Effects on Crossing an Integer Resonance* Part. Acc. **48**(1998) p. 251.
21. R. Cappi, R. Garoby, S. Hancock, M. Martini, J.P. Rinaud, *Measurement and Reduction of the Transverse Emittance Blow-up Induced by Space Charge Effects* Proc. PAC93 p. 3570.
22. D. Neuffer, D. Fitzgerald, T. Hardek, R. Hutson, R. Macek, M. Plum, H. Thiessen, T.-S. Wang, *Observation of Space-Charge Effects in the Los Alamos Proton Storage Ring* Proc. PAC91 p. 1893.
23. R. Macek, private communication (1998).
24. E. Colton, talk given at TRIUMF (1987).
25. L.J. Laslett, *On the Intensity Limitations Imposed by Transverse Space-Charge Effects in Circular Accelerators* Proc. 1963 Summer Study on Storage Rings, Accelerators and Experimentation at Super-High Energies p. 324
26. S. Machida, private communication (1998).

Simulation of Space Charge Effects in a Synchrotron

Shinji Machida
KEK-Tanashi
Midori-cho, Tanashi-shi, Tokyo, 188-8501 JAPAN, E-mail: shinji.machida@kek.jp
and
Masanori Ikegami
JAERI
Tokai-mura, Naka-gun, Ibaraki-ken, 319-11 JAPAN, E-mail: ikegami@linac.tokai.jaeri.go.jp

Abstract. We have studied space charge effects in a synchrotron with multi-particle tracking in 2-D and 3-D configuration space (4-D and 6-D phase space, respectively). First, we will describe the modelling of space charge fields in the simulation and a procedure of tracking. Several ways of presenting tracking results will be also mentioned. Secondly, it is discussed as a demonstration of the simulation study that coherent modes of a beam play a major role in beam stability and intensity limit. The incoherent tune in a resonance condition should be replaced by the coherent tune. Finally, we consider the coherent motion of a beam core as a driving force of halo formation. The mechanism is familiar in linac, and we apply it in a synchrotron.

INTRODUCTION

Space charge force is one of major sources of emittance increase of beam core, halo formation, and beam loss. Because of its nature of electromagnetic force proportional to $1/\gamma^2$, it is significant only at low energy. In a ring accelerator, especially in a synchrotron, space charge effects are measured with tune shift as a result of its defocusing force against focusing one by an external magnetic lattice. A typical magnitude of tune shift in a synchrotron at the low energy end is around -0.1 to -1.

Space charge force is a function of beam size in transverse and longitudinal planes as well as number of particles in a bunch. That makes the analysis of space charge effects difficult. Once the beam emittance growth occurs due to the effects, the beam is no longer affected by the same space charge force as before. In addition, at the low energy end of synchrotron, the number of particles is gradually increased as a function of turn number throughout an injection process. The transverse emittance is sometimes manipulated by the so-called phase space painting at the same time. The longitudinal bunch length and peak intensity strongly depend on an rf voltage pattern. Therefore, it seems that simulation study is the only way to model the effects and analyze beam behavior in a self-consistent way.

We have been developing a simulation code named Simpsons (Simulation of proton synchrotrons). Simpsons tracks multi particles and a single particle in a synchrotron with space charge effects. There are two versions; one models a beam in 6-D phase space with acceleration and the other assumes a coasting beam and tracks only transverse 4-D phase space coordinates. Its ultimate goal is to simulate beams self-consistently in proton (or ion) synchrotrons where space charge effects play a significant role. It also includes time dependent parameters such as rf voltage, bending fields, injection orbit (for phase space painting), power supply ripple, etc.

In this paper, we first explain the way of simulation. Secondly, we discuss two subjects as a demonstration of Simpsons. One is a coherent motion and its role in beam stability. The other is halo formation due to coherent oscillations of the core.

MODELLING OF SPACE CHARGE EFFECTS

Field Calculation

A calculation of space charge fields is based on a particle-in-cell (PIC) method (1). As a coordinate system, cylindrical one for a 3-D beam and polar one for a 2-D beam are taken. Fractional charge of each macro particle is allocated to grid points nearby according to area weighting. Poisson equation is then solved with a boundary condition of circular cross sectional beam pipe. A contribution from the magnetic field is simply taken into account with a $1/\gamma^2$ factor.

Under the Lorentz gauge, a scalar potential satisfies,

$$\Delta\phi - \varepsilon_0\mu_0 \frac{\partial^2\phi}{\partial t^2} = -\frac{\rho}{\varepsilon_0} \tag{1}$$

We introduce the ordering (2),

$$\frac{\partial^2}{\partial z^2}\phi \sim \varepsilon_0\mu_0 \frac{\partial^2\phi}{\partial t^2} \ll \frac{1}{r}\frac{\partial}{\partial r}\left(r\frac{\partial}{\partial r}\phi\right) + \frac{1}{r^2}\frac{\partial^2}{\partial\varphi^2}\psi \tag{2}$$

Then,

$$\frac{1}{r}\frac{\partial}{\partial r}\left(r\frac{\partial}{\partial r}\phi\right) + \frac{1}{r^2}\frac{\partial^2}{\partial\varphi^2}\phi = -\frac{\rho}{\varepsilon_0} \tag{3}$$

Once a charge distribution at grid points is obtained, it is Fourier transformed in the azimuthal direction.

$$\phi = \sum_m \phi_m \exp(im\varphi) \tag{4}$$

$$\frac{\rho}{\varepsilon_0}(\equiv n) = \sum_m n_m \exp(im\varphi) \tag{5}$$

with the inverse transform,

$$\phi_m = \frac{1}{2\pi}\int_0^{2\pi}\phi(r,z,\varphi)\exp(-im\varphi)d\varphi \tag{6}$$

$$n_m = \frac{1}{2\pi}\int_0^{2\pi}n(r,z,\varphi)\exp(-im\varphi)d\varphi \tag{7}$$

Note that, $n_{-m} = n_m^*$. Then,

$$n_m(r,z) = \frac{1}{r}\frac{\partial}{\partial r}\left(r\frac{\partial}{\partial r}\phi_m(r,z)\right) - \frac{m^2}{r^2}\phi_m(r,z) \tag{8}$$

The general solution to the equation ($m \geq 0$) is,

$$\phi_m = \alpha\left(\frac{r}{b}\right)^m + W(r) - \left(\frac{r}{b}\right)^m W(b) \tag{9}$$

On the boundary condition, $\phi_m = 0$ at $r = b$ (b is the beam pipe radius), it becomes

$$\phi_m = W(r) - \left(\frac{r}{b}\right)^m W(b) \tag{10}$$

where

$$W(r,z) = \int_0^r \ln\frac{r}{r'} n_m(r',z) r'\, dr' \qquad m = 0 \tag{11}$$

$$= \frac{r^m}{2m}\int_0^r r'^{(1-m)} n_m(r',z) dr' - \frac{r^{-m}}{2m}\int_0^r r'^{(1+m)} n_m(r',z) dr' \qquad m \neq 0 \tag{12}$$

where $\phi_{-m} = \phi_m^*$ is used to compute ϕ_m for $m < 0$.

We replace the integral with summation at grid points. The electric fields are

$$E_r = \sum_m \frac{\partial}{\partial r}\phi_m \exp(im\varphi) \tag{13}$$

$$E_\varphi = \sum_m \frac{1}{r}\frac{\partial}{\partial \varphi}\phi_m \exp(im\varphi) \tag{14}$$

$$E_z = \sum_m \frac{\partial}{\partial z}\phi_m \exp(im\varphi) \tag{15}$$

The differentiation is done analytically beforehand except with z.

Tracking Procedures

The independent variable is *time* so that a snapshot of a beam in configuration space is obtained. At each time step, space charge fields are calculated and applied to each macro particle as a impulse kick. Then, the particle position is advanced with newly calculated momentum. Those two computations are repeated until as many turns as specified initially.

The size of grids is recalculated at each time step so that all the macro particles are included with the minimum required grid size. Once a particle amplitude becomes larger than the pipe radius, it is regarded as a lost particle. The number of grid points is optimized with some tests. For example, radial grid points are fixed as 40 in order to reproduce a smooth field curve.

The number of longitudinal grids is determined in such a way that a half FODO cell is divided into at least 5 grids. Since a beam stretches over several transverse focusing periods and transverse beam size is accordingly modulated (3), it is necessary to divide a beam in longitudinal direction such that harmonic components driven by transverse space charge are included. Once the longitudinal grid size is determined, a time step is calculated as,

$$dt \sim \frac{dz}{\beta c} \qquad (16)$$

The number of azimuthal modes is usually taken up to four, which is enough to model octupolar shape in configuration space.

The space charge defocusing force modifies the lattice functions and an optical matching condition. If the force is linear, all the particles feel the same focusing force and therefore the modified matching condition with space charge applies to a whole beam. In reality, however, a distribution is not uniform and nonlinear space charge force cannot make the whole beam matched. In the simulation, the rms envelope equations including space charge force are solved to find the matched optical condition for the rms emittance. In the 3-D simulation, a peak current is assumed to solve the envelope equations so that only the center of a bunch is matched.

As an initial particle distribution, one can specifies several kinds, such as K-V, waterbag, parabolic, and gaussian for transverse planes, and uniform, parabolic, and gaussian for longitudinal one. We adopt a concept of *equivalent beam* so that the rms emittance is fixed when we compare two or more distributions.

Presentation of Simulation Results

Simulation results are presented in several ways. Plotting of macro particles in phase space every n turns (n should be reasonably large unless a particle distribution changes very quickly) is most common one. With its projection to position and gradient axes, it shows the particle distribution and its evolution in the most primitive way.

Some statistical quantities, such as second or higher order moments, can be derived from the particle ensemble. An example is the rms emittance defined by the second moments and their cross term. When a machine has its physical acceptance, we can define a fraction of particles which fits in the acceptance. A fraction of particles outside of the acceptance can be regarded as particle loss. Then, one can trace particle loss as a function of time or turn.

Instead of dealing with all the macro particles, we also see a single particle trajectory, so-called Poincare map. Several particle trajectories with different initial coordinates reveal phase space structure. In a synchrotron, it is common to look at the single particle coordinates at one fixed point in a ring. As a tracking study in linac, however, it is also possible to plot coordinates with other periodicity such as the one introduced by initial mismatch or a coherent mode oscillations.

A single particle trajectory gives a tune of individual particle. It is defined as a frequency at which the following quantity has a peak as shown in Fig. 1 (left) (4).

$$m_j^{klm}(v) = \frac{1}{n_{max}} \sum_n^{n_{max}} x_j^k(n) y_j^l(n) z_j^m(n) \exp(-2\pi i v n) \qquad (17)$$

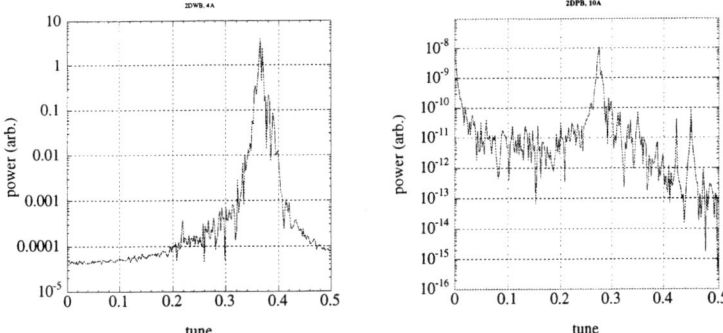

FIGURE 1. Fourier spectrum of single particle motion (left) and of second moment (right).

Each individual particle has its own incoherence tune. That results in an incoherent tune spread in a tune diagram, called 'necktie diagram'.

A characteristic frequency of the whole beam is also calculated. One example is a coherent frequency defined in the following. First, a moment if the particle coordinates is calculated every turn,

$$M_{coh}^{klm}(n) = \frac{1}{N}\sum_{j}^{N} x_j^k(n) y_j^l(n) z_j^m(n) \tag{18}$$

Then, the frequency spectrum of the moment is

$$M_{coh}^{klm}(\nu) = \frac{1}{n_{max}}\sum_{n}^{n_{max}} M_{coh}^{klm}(n)\exp(-2\pi i \nu n) \tag{19}$$

The frequency at which the above quantity has a peak is the coherent tune as shown in Fig. 1 (right). The sum of $k+l+m$ defines the order of the coherent modes.

Validation of a Code

The code was tested in several ways in the 2-D simulation. First, the necessary number

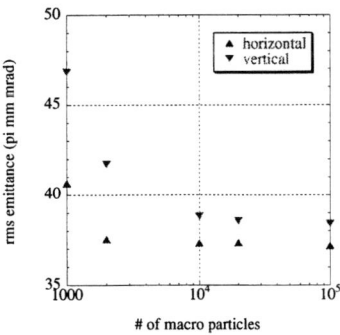

FIGURE 2. Number of macro particle dependence on the results in 2-D calculation. When it is more than 10^4, the rms emittance converges.

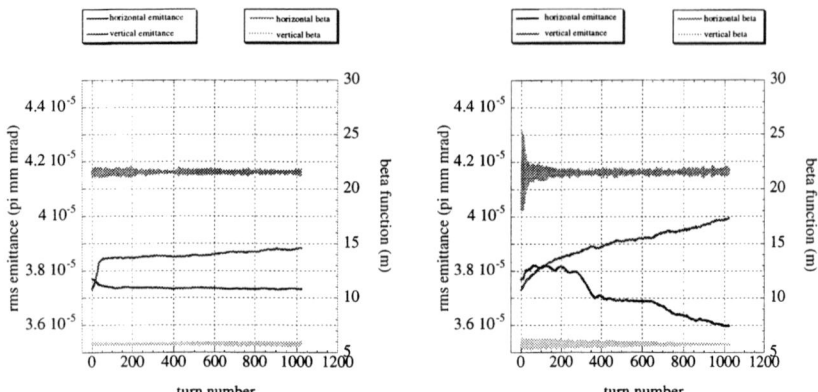

FIGURE 3. Evolution of rms emittance and β-functions with initial distribution of K-V. With (left) and without (right) rms matching at injection. The bare tune is (6.84,5.81) and incoherent tune shift is -0.81.

of macro particles are examined as shown in Fig. 2. When one chooses more than 10^4 macro particles, tracking results should not depend on the number. Another check of the code is tracking of the uniform (K-V) distribution. As shown in Fig. 3 (left), the rms emittance and β-functions stay constant as long as the initial matching condition is satisfied. The initial mismatch introduces oscillations of β-functions. The rms emittance growth occurs since the K-V distribution cannot remain the K-V (Fig. 3 (right)).

When the beam intensity is increased, even the matched K-V distribution becomes unstable as shown in Fig. 4. It is not clear if it becomes unstable due to numerical (artificial) reasons or it is a physical phenomenon.

For the 3-D simulation, validation of the code is not established. We need further study.

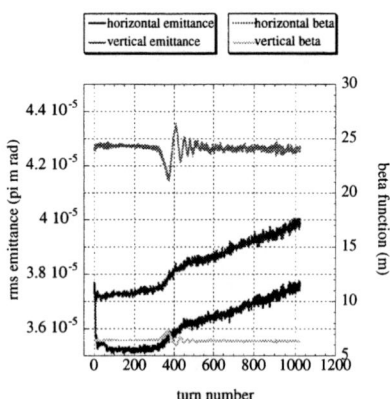

FIGURE 4. Unstable K-V beam. For first 300 turns, the beam is stable. Then, suddenly envelopes start to oscillate and rms emittance growth occurs. The bare tune is (5.84,5.81) and incoherent tune shift is -1.22. It is not clear if it is due to numerical reasons or a physical phenomenon.

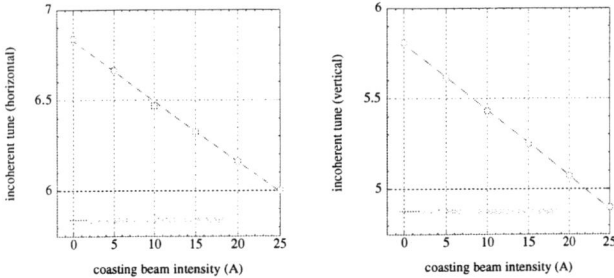

FIGURE 5. Incoherent tune as a function of beam intensity. Left figure shows horizontal result and right figure does vertical. Particle distribution is uniform (K-V) so that all the particles have the same tune.

SIMULATION OF COHERENT MODE

A study of beam stability with the envelope equations had been made by L. Smith and F. Sacherer in 1960s (5,6). Their essential statement is that amplitude growth occurs when an envelope mode frequency coincides with quadrupole field harmonics. Although their analysis was based on the uniform (K-V) distribution when it first appeared, Sacherer himself found in 1971 that the same equations should be valid if one substitutes rms beam size for sharp edge beam size of the K-V distribution (7). Extension of their analysis to non-uniform particle distributions and also to nonlinear error fields are not trivial,

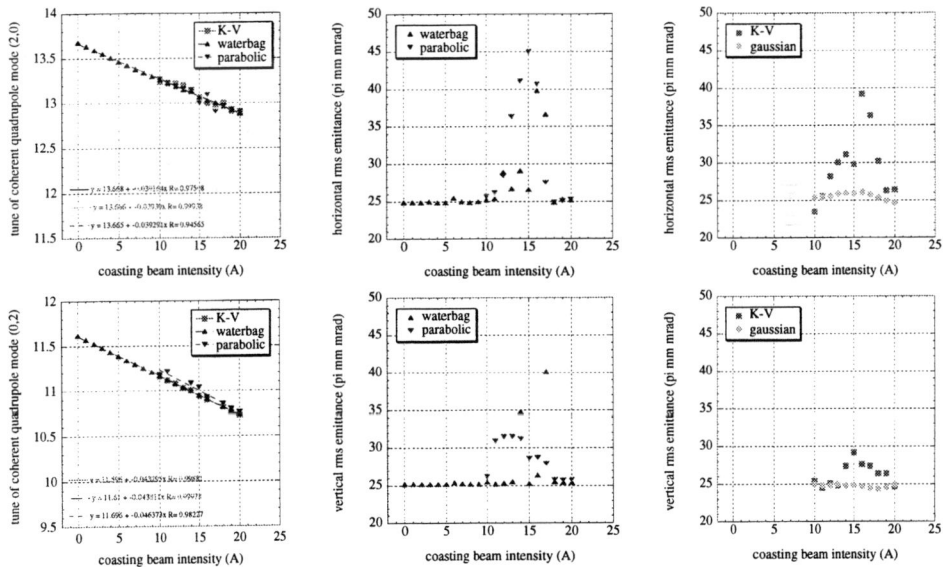

FIGURE 6. Tune of coherent quadrupole mode (left) and rms emittance at 512 turns after injection (center and right). Upper figures show horizontal results and lower ones vertical. In both planes, rms emittance growth is observed when the coherent quadrupole tune becomes integer.

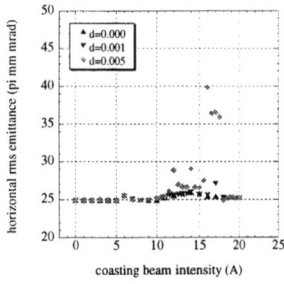 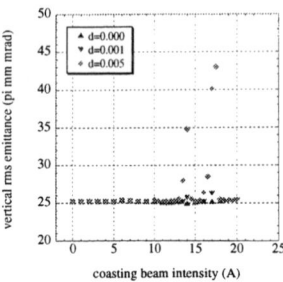

FIGURE 7. RMS emittance growth before and after harmonic correction. The growth is suppressed with the correction. Horizontal (left) and vertical (right) planes.

but can be done numerically.

Let us start from the coherent quadrupole mode, whose answer we knew by Sacherer. Figure 6 (left) shows the intensity dependence of the coherent quadrupole mode. At zero current, it is just twice as much as the bare tune, that is 6.84 for horizontal and 5.81 for vertical. When the intensity is increased, the coherent tune decreases linearly.

If one compares the slope of the quadrupole mode with the slope of the incoherent tune of Fig. 5, the quadrupole tune shift is about 60% of twice of the incoherent tune shift, that agrees with the analytical prediction by Sacherer, namely 63% (or 5/8).

In addition to the uniform (K-V) distribution, the equivalent beams with different distributions; waterbag and parabolic, follow the same intensity dependence (Fig. 6 (left)). In either particle distribution, the emittance growth and particle loss is observed when the coherent mode frequency becomes integer where quadrupole error harmonics is excited. Since the incoherent tune is already well below the half integer when the coherent one hits the integer, one can conclude that the resonance occurs when the coherent mode frequency satisfies the resonance condition. The emittance growth can be suppressed by harmonic correction of quadrupole error fields (Fig. 7). The beam loss is shown in Fig. 8.

Now, we look at higher order modes and their resonance behavior with external magnetic error fields. Figure 9 (left) shows the coherent mode frequency of sextupole as a function of intensity. At zero current, the coherent mode frequency is three times as much

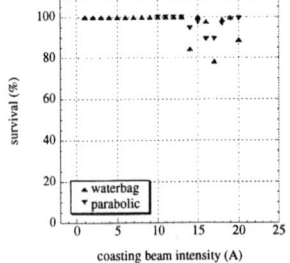 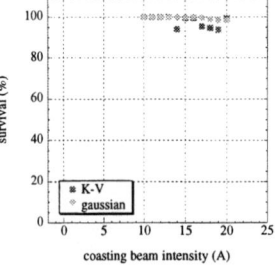 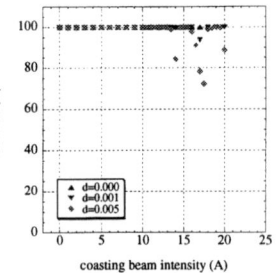

FIGURE 8. Beam loss due to resonance of coherent quadrupole modes. Results of several initial distributions (left and center) and those with harmonic correction (right).

80

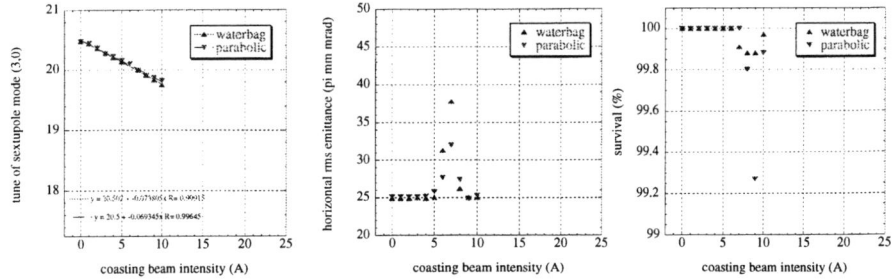

FIGURE 9. Tune of coherent sextupole mode (left), rms emittance growth (center), and beam loss.

as the bare tune. We assume non-uniform initial distributions: waterbag and parabolic.

When the intensity is increased, the coherent mode frequency decreases. The slope of intensity dependence of the coherent tune is about 72% of the three times of the incoherent tune. With sextupole error fields, the rms emittance growth and beam loss is observed at the intensity where the coherent mode frequency of sextupole becomes integer. Similar to the quadrupole mode, those resonance is explained by the resonance condition of the coherent sextupole mode. Those can be suppressed by harmonic correction of sextupole error fields (Fig. 10). Similarly the coherent octupole mode is shown in Fig. 11.

Until here, image charges play a minor role. Once a beam travels off-center of a beam pipe, however, it becomes resonance driving force similar to the external error fields (8). Figure 12 shows that the emittance growth and beam loss occur when the coherent quadrupole mode becomes integer. Although there is no external field error, the closed orbit distortions destroy the symmetry of image charges and excite the resonance.

BEAM HALO FORMATION IN A SYNCHROTRON

One of trends in linac beam dynamics for last several years is a study of beam halo formation based on the particle-core model (9). The particle-core model assumes time dependent oscillations of a beam core. In a linac case, initial mismatch is regarded as a major source of the core oscillations. When the frequency of core oscillations becomes as twice as that of single particle oscillations, a parametric resonance occurs and the particle at the tail may have a large amplitude, becoming halo.

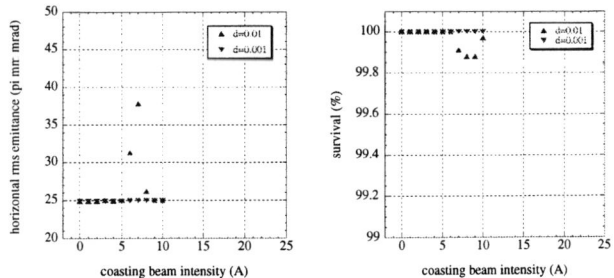

FIGURE 10. RMS emittance growth (left) and beam loss before and after harmonic correction.

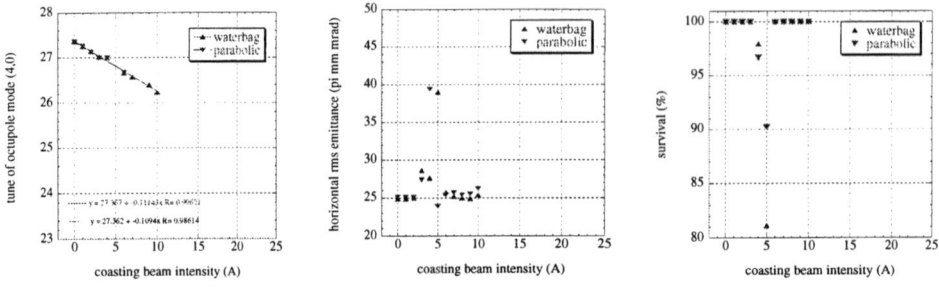

FIGURE 11. Tune of coherent octupole mode (left), rms emittance growth (center), and beam loss.

Although the injection mismatch may be minimized in a synchrotron, there are other sources which introduce time dependent core oscillations. The coherent mode oscillations discussed above is one of them. We have investigated the possible halo formation mechanism in a synchrotron based on the particle-core model and Simpsons.

Let us first look at how the coherent mode oscillations can excite parametric resonance of tail particles. As we have mentioned, the coherent tune shift is smaller than the twice of the incoherent tune shift. Figure 13 (left) schematically depicts the intensity dependence of coherent (quadrupole in this example) and incoherent tune as a function of intensity. Now let us plot both tune as a function of particle amplitude (Fig. 13 (right)). Here we assume a 2-D uniform (K-V) distribution. In the beam core region, the incoherent tune is constant and increasing outside approaching the bare tune. If one compares that with the coherent quadrupole tune in the core, the incoherent tune is always smaller than the half of the coherent one. In the limit of large amplitude, however, the incoherent tune is larger than the half of the coherent one. Therefore, there is a certain amplitude where the incoherent tune is exactly the half of the coherent one.

In order to see the phase space structure in more detail, the particle-core model is applied with the JHF booster parameters. The ring consists of 28 FODO cells. The horizontal and vertical bare tune are 6.8. The beam intensity are chosen so that the coherent quadrupole tune is 13. Since there is no external field error, beam blow up should not

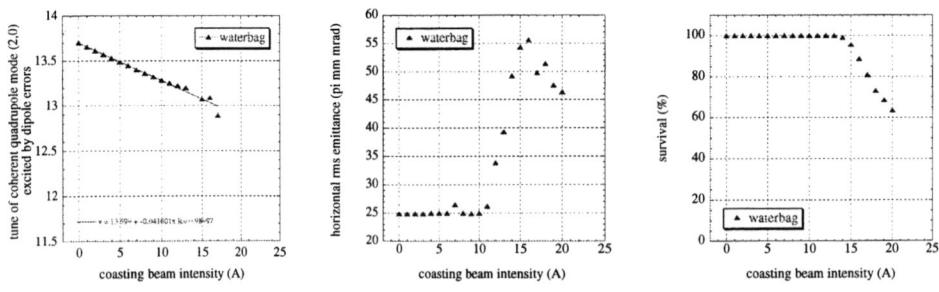

FIGURE 12. Tune of coherent quadrupole mode (left), rms emittance growth (center), and beam loss. There is no quadrupole errors. However, the image charge due to closed orbit distortion excites the quadrupole mode.

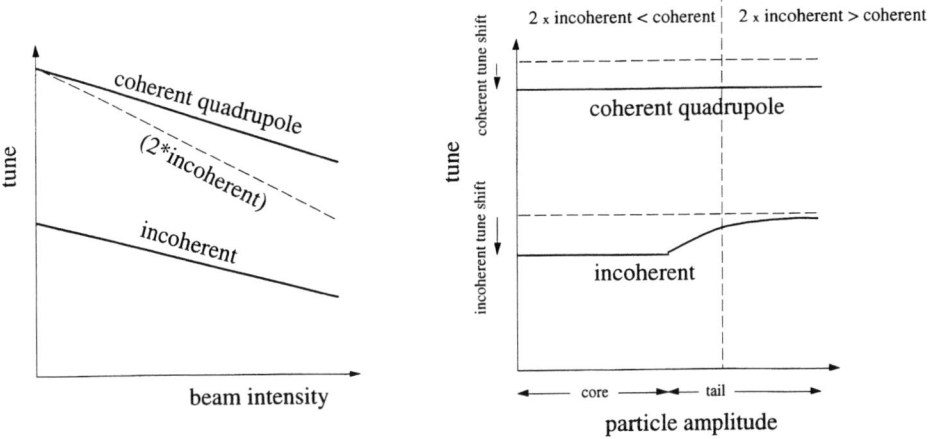

FIGURE 13. Incoherent tune in the core and coherent quadrupole tune as a function of beam intensity (left), and as a function of particle amplitude (right). Since twice of the incoherent tune is always smaller than the coherent tune in the core (left), there should be an amplitude outside of the core, where 2 x incoherent tune is equal to coherent one. Fixed point of the parametric resonance locates there.

occur. The coherent tune is adjusted at integer simply because Poincare map can be obtained by looking at particle trajectories at a fixed point in a ring. Otherwise, one has to look at it at an interval according to the periodicity of the coherent mode, which may not coincide with the lattice periodicity.

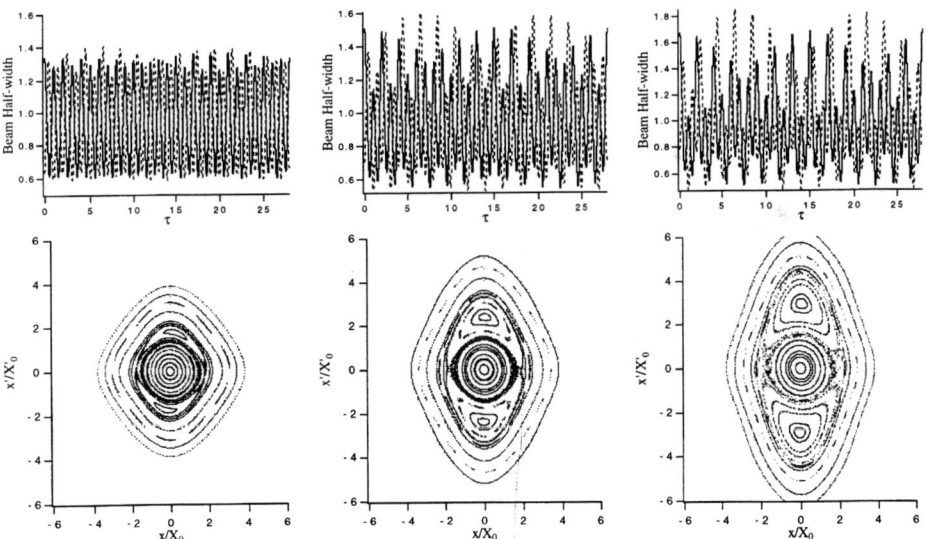

FIGURE 14. Beam envelope modulations (upper) and phase space structure (lower) based on the particle-core model. Modulation amplitude is 2-5% (left), 15% (center), and 27% (right). We assume that a ring lattice consists of 28 FODO cells and the envelope oscillates 13 times per ring.

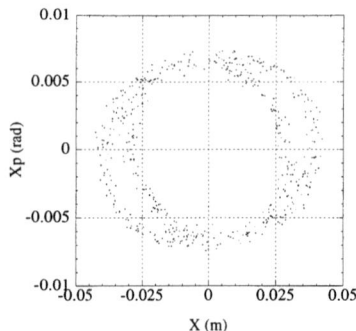

FIGURE 15. Phase space trajectory of a particle initially located near the edge of the core. When it is observed, the coherent quadrupole frequency of the core is almost 13 and the incoherent tune of the particle is 6.5.

An amplitude of the coherent oscillations is an unknown parameter. That should depend on the strength of external error fields, distance to a resonance, beam intensity, etc. We have looked at three cases; 2-5%, 15%, and 25%. Phase space structure together with a modulation of horizontal and vertical beam size along the ring are shown in Fig. 14. With relatively low intensity compared with linac (but with typical magnitude in a synchrotron), the particle-core model predicts beam halo due to the coherent oscillations.

We, then, tracked multi particles to see if we can observe the same resonance island. Now, we need to setup external quadrupole error fields to excite the coherent quadrupole mode at $\nu=13$. At the same time, the strength of errors should be small enough so that the oscillations of core should not become unstable in a time scale of the halo formation.

In Fig. 15, we show a single particle trajectory which indicates fixed point in the phase space. The particle is initially located near the edge of the beam core. When the island is observed, the coherent tune is almost 13 and incoherent tune of that particle is 6.5. In conclusion, the parametric resonance and halo formation induced by the coherent quadrupole oscillations are indeed observed in the multi-particle tracking.

REFERENCES

1. For example, Birdsall, C. K. and Langdon, A. B., *Plasma Physics via Computer Simulation*, New York: MacGraw-Hill Book Company.
2. This was suggested by Sloan, M. L.
3. That is true for spallation source type of a synchrotron, where rf harmonic number is a few, whereas the number of focusing period is a few tens.
4. Orlov, Y. and Soffer, A., "Fourier Analysis of Higher Order Coherent and Incoherent Resonances in Beam-Beam Interaction", CLNS 92/1178.
5. Smith, L, "Effects of Gradient Errors in the Presence of Space Charge Forces", Proc. of International Conference on High Energy Accelerators, 897 (1963).
6. Sacherer, F. J., "Transverse Space-Charge Effects in Circular Accelerators", Ph.D. Thesis, 1968.
7. Sacherer, F. J., "RMS Envelope Equations with Space Charge", IEEE Trans. Nucl. Sci. **18**, 1105 (1971).
8. Baartman, R., in this proceedings.
9. The following reference summarizes the literature on the particle-core model, Okamoto, H. and Ikegami, M., "Simulation Study of Halo Formation in Breathing Round Beams", Physical Review E **55**, No.4, 4694 (1997).

SIMULATION WITH SPACE CHARGE

C.R. Prior

Rutherford Appleton Laboratory, Chilton, Didcot, U.K.

I INTRODUCTION

A developing involvement in accelerators in which space charge effects play a prominent rôle has prompted studies at the Rutherford Appleton Laboratory into computational tools to treat theoretical problems of beam transport by means of simulation. The main emphasis has been on storage rings and the modelling of injection using stripping foils or septa, but other aspects of beam dynamics have also been investigated. By-products of the tracking codes have been analytical computer programs which generate basic sets of working parameters and, through linear theory, are able to achieve fairly high degrees of optimisation. The simulation codes are then used to confirm these results and explore non-linear effects.

As with all particle tracking codes, the calculation of space charge forces occupies a considerable slice of the programming and, usually, the major part of the CPU run time. Computer speed and core were major restrictions on the kind of modelling which could be achieved 10-15 years ago and many ingenious ideas were developed in order to produce worthwhile results. Nowadays, storage is not a problem, and computing speeds have increased and the cost of CPU time been dramatically reduced so that we think little of running continuously for several days in order to produce results that would have been unthinkable only a few years ago. Earlier ideas for saving on storage and time are, however, not wasted: they now release space and provide opportunities for more complicated simulations using greater numbers of macro-particles, and allow greater accuracy in the type of calculations which, for example, depend on the refinement of a mesh or the size of an integration step length.

In this paper, the general approach to simulation and the fundamental details behind the structure of the RAL codes are described. A range of illustrations is included covering modelling of the H$^-$ injection scheme for the European Spallation Source (ESS), the tilted septum method of injecting heavy ions for the European inertial confinement fusion study (HIDIF), the proposals for an upgrade to ISIS using a method of dual harmonic RF acceleration, and a recent investigation to determine the extent to which octupoles and higher order elements might be used to adapt particle distributions and transverse beam shapes into forms suitable for spallation and other problems involving targets.

II PRINCIPLES OF SIMULATION

The basic ideas behind the simulation of particle beams in accelerators are mathematically simple but require care in their implementation. A model beam of typically 10^4 to 10^6 macro-particles is used to represent a real beam, which might have the order of $\sim 10^{14}$ particles in a bunch. The degree to which this is a genuine representation, and predicted effects are not consequences of the reduced number of particles, is related to problems of a statistical nature, and other difficulties, such as rounding and interpolation errors generated in computers, which have finite word-length and work to a fixed precision, also need to be countered.

In the absence of space-charge, recent applications of mathematical theories such as differential and Lie algebras have led to fast and accurate codes based on symplectic transfer maps working to very high order of accuracy. Examples are COSY INFINITY [1] and MARYLIE [2]. Study is underway to include the effects of self-fields, though, since the principles behind the codes are somewhat different to normal tracking techniques, this can probably only be achieved, at least in the first instance, by approximation methods.

Conventional particle tracking codes rely on the generation of an input model beam distribution in phase-space, giving the macro-particles for the subsequent simulation. Packages are readily available for most standard distributions and can quickly be written in other cases. A convenient technique is the "ratio of uniform deviates" method [3], where a density function $f(x)$ is created by uniformly filling the region

$$0 < u < \sqrt{f\left(\frac{v}{u}\right)}$$

of the two-dimensional (u, v)-plane, with a random number generator. The random variable $x = \dfrac{v}{u}$ has the desired density function.

Progress of the beam through the accelerator is then determined by integrating the appropriate equations of motion. A suitable formalism, based on a standard Serret-Frenet frame of unit vectors moving with the beam, and the ensuing Hamiltonian development, is described in [4]. Various methods of integration may be employed, from linear Euler (forward) difference methods to more complicated Runge-Kutta approaches. However the higher the order of accuracy, the longer the computing time taken to integrate forward in time and it is usually best to adopt the traditional leapfrog method, giving accuracy to second order in the step-length, and use more steps of shorter length. At each stage the external forces (from the accelerating elements) need to be calculated as well as the internal, space charge, fields. One assumes some kind of coding to read the details of the accelerator into the program, and this can range from direct, but often crude, methods to more sophisticated ideas. At RAL, we adopt an approach which allows the definition of structures and substructures of elements to any number of levels, and this generally permits the details of some quite complicated accelerating systems to be read in in a short and fairly simple way.

III CALCULATION OF SPACE CHARGE FORCES

A large proportion of computing time in a simulation run is occupied by solving for the space charge forces. Several different approaches are adopted. Older codes simply evaluated the Coulomb forces between charges, a process which is $O(n^2)$ and is operable only with relatively small numbers of particles n. There are problems with this method when particles move too close together, so some kind of cut-off distance (typically the Debye length) has to be introduced and used whenever particles come within this limit. Also the method is susceptible to rounding errors, particularly for particles near the axes where the perpendicular force will, from symmetry considerations, often be small: adding together many fairly large terms whose sum should actually approach zero is clearly prone to unavoidable computational inaccuracies. For example, in the simplest case of a round beam in two-dimensions with angular symmetry, 50,000 simulation particles generated so as to model a uniform distribution - a problem for which the analytical answer can be written down - gave less than 60% of the space charge forces within a 10% error band; and, of course, the calculations took some minutes to complete. Other approaches to space-charge may involve approximations to save computing time, such as those used in ACCSIM, for instance [5], but the general approach to using a set number of macro-particles to simulate the behaviour of a much larger number of beam particles follows the techniques of computational fluid mechanics and uses finite difference or finite element techniques. It may be argued that the latter is the more flexible approach and offers considerable advantages, although the full degree of sophistication is often not needed in accelerator physics. Finite element methods are readily adapted to different boundary shapes and different boundary conditions at the accelerator walls; codes may be written in such a way that different orders of accuracy may be used without major modification; calculations may readily be switched between different types of coordinates, for example two-dimensional cartesians (x,y) to three-dimensional cylindrical polars with axisymmetry (r,z); much theoretical work has been done to save on computer storage and time; and several techniques exist to complete the calculations once the problem has been put into standard form. For these reasons, but especially because of the advantage of being able automatically to handle conducting walls and the associated "image" charges, finite element techniques are used at RAL. Essentially one transforms (relativistically) into the beam frame, where the particles may be treated as at rest, and Poisson's equation is solved, then the answers are converted back to the laboratory frame for integration forward by one time step. The space charge potential in the beam frame is the function ϕ which gives a stationary value to

$$\pi(\phi) = \tfrac{1}{2} \int_V |\nabla \phi|^2 \, dV + \frac{1}{\epsilon_0} \int_V \rho \phi \, dV - \int_{\partial V} \bar{\phi}_n \phi \, dS. \tag{1}$$

This variational problem is equivalent to

$$\nabla^2 \phi = \frac{\rho}{\epsilon_0} \quad \text{in } V \tag{2}$$

with

$$\phi = \bar{\phi}, \qquad \frac{\partial \phi}{\partial n} = \bar{\phi}_n \qquad \text{on} \quad \partial V. \qquad (3)$$

In two-dimensions, V would be the cross-section of the accelerator channel with ∂V the bounding wall; in three-dimensions, V would be the (topologically) cylindrical region occupied by the beam in the accelerator, with appropriate (say periodic) boundary conditions imposed longitudinally.

Taking two-dimensions as illustration, the variational problem is solved by imposing a triangular mesh across V and fitting to each element a polynomial function for $\phi(x,y)$ of degree n. To avoid extra calculations which do not actually enhance accuracy, one looks for a complete set of functions, which means that ϕ contains $N = \frac{1}{2}(n+1)(n+2)$ unknowns: for example, for $n = 2$ the unknowns are the six coefficients a_i in $a_0 + (a_1 x + a_2 y) + (a_3 x^2 + a_4 xy + a_5 y^2)$. This expression can be restructured in terms of the values of ϕ at specific points of the mesh (the "nodes"). The so-called serendipity method of allocation may be used and an illustration of the nodes for $n = 3$ would then be as shown below.

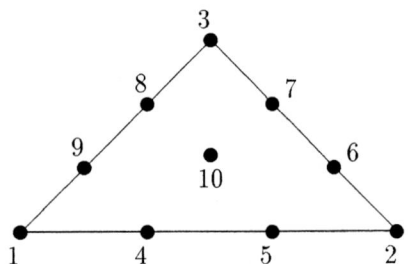

FIGURE 1. Node arrangement for triangular mesh with $n = 3$ potential fit.

Introducing standard areal coordinates ξ_i, $i = 1, 2, 3$, for each triangular element ($\sum \xi_i = 1$), we can write

$$\phi(x,y) = \sum_{i=1}^{N} f_i(\xi_j)\phi_i \qquad (4)$$

where f_i are the shape functions and $\{\phi_i\}$ the potentials at the nodes. For $n = 1$, $f_i = \xi_i$, and for $n = 2$, we would use $f_i = \xi_i(2\xi_i - 1)$ for $i = 1, 2, 3$, $f_4 = 4\xi_1\xi_2$, $f_5 = 4\xi_2\xi_3$, and $f_6 = 4\xi_3\xi_1$.

The source terms $\frac{1}{\epsilon_0} \int \rho \phi \, dV$ depend on a model for the charge density ρ. The simplest is to treat the macro-particles as δ-functions:

$$\rho(\mathbf{x}) = \sum_{\text{particles, } i} q_i \delta(\mathbf{x} - \mathbf{x}_i).$$

Alternatively, one could introduce an arbitrary spreading of charge by, for example, treating each macro-particle as a small Gaussian distribution of pre-determined size. The δ-function formalism under (1) is in fact equivalent to assuming a uniform spread of charge over a disc, provided it is contained entirely within a single element of the mesh.

Substituting these formulae into (1), the variational problem reduces to the equivalent condition $\partial \pi / \partial a_i = 0$, which results in a set of simultaneous equations for the $\{\phi_i\}$. These are adjusted to account for the boundary conditions (3), giving an equation involving only internal nodes:

$$K_{ij}\phi_j = Q_i. \tag{5}$$

Here the "stiffness" matrix K_{ij} depends only on the mesh and Q_i are the source terms. Equation (5) is solved using techniques such as the conjugate gradient method or the method of triple factoring. The matrix K is extremely sparse and there are well-known methods for handling such linear problems. With a few exceptions, only the non-zero elements are stored and this number can be minimised by appropriate global numbering of the mesh nodes. Space charge forces are then found from

$$\nabla \phi = \sum_{i,j} \frac{\partial f_i}{\partial \xi_j} \phi_i \nabla \xi_j.$$

Programming for any level of polynomial modelling is relatively simple but the degree of accuracy of the answers depends on the refinement of the mesh and the order of fit. Taking 10,000 simulation particles, for $n = 1$ (linear potential fit) roughly 3000 triangular mesh elements are needed to give about 90% of space charge forces within a 10% error band for the circular, uniform beam mentioned above. For $n = 3$ this can be reduced to about 400 and would give 95% accuracy (the errors being along the axes and actually very small, though greater than 10% in relative terms).

Meshes used in the RAL codes are formed of triangular elements generated by radial and circular lines based on the beam centre, with different (continuous) radial scalings over the beam and the external region. As the beam moves, the mesh adjusts internally, while the boundary wall is fixed (infinite boundaries are simulated by "super-elements" mapping the potentials onto $\ln r$ terms (2D) for large r). An example of an off-centre beam in a square conducting chamber is shown in Figure 2. If the beam varies slowly, the mesh need only be re-generated every 5-10 time steps. In such cases the matrix K_{ij} may be pre-inverted and stored (using the same storage locations), so that only the source terms Q_i need be re-calculated as the particles change their positions; the $\{\phi_i\}$ are then quickly computed from $\phi_i = K_{ij}^{-1} Q_j$.

Extension to three-dimensions is both simple and natural. The region is split longitudinally into planes covered by triangular meshes. Pairs of elements in adjacent planes are joined to form tetrahedra, in each of which volume coordinates ξ_i, $i = 1, 2, 3, 4$, $(\sum \xi_i = 1)$ are used. With four vertices, the simplest polynomial fit is

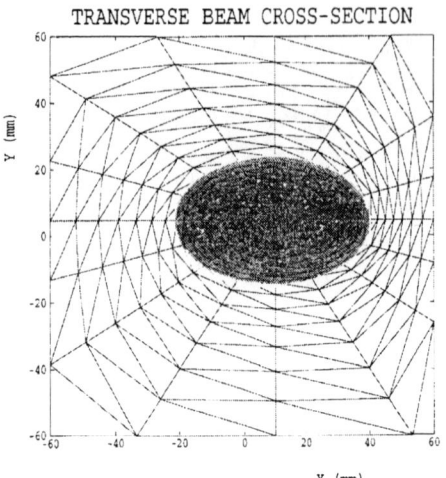

FIGURE 2. Finite element space charge mesh showing an off-centre particle beam in a rectangular conducting chamber.

$\phi = a_0 + a_1 x + a_2 y + a_3 z$; the quadratic form requires six extra nodes, which are positioned at the mid-points of each of the edges. The procedure described above applies and equations (1) to (5) remain valid in 3D.

Note that a slight adaptation of the 2D x-y code will give a 3D r-z code with axisymmetry. Starting with a rectangular mesh, a linear potential fit $\phi = a_0 + a_1 r + a_2 z + a_3 rz$ would correspond to nodes at the four vertices, and a 12-node fit (up to cubic terms plus terms $r^3 z$ and $z^3 r$) would include two nodes per mesh edge and give better accuracy. Then all that is needed is the change $dx\, dy \to r\, dr\, dz$ in the integrals of (1), which are in general performed numerically, and the relevant re-interpretation of the answers.

IV 1D LONGITUDINAL TRACKING CODES

A code which has proved particularly useful in recent years is the RAL program TRACK1D[1], which simulates the injection, acceleration and trapping of bunches in a synchrotron. The basic physics is governed by the equations [6]

$$\frac{d}{dt}\Delta\phi = \frac{h\omega_0 \eta}{\beta_s^2 E_s} \Delta E \qquad (6)$$

$$\frac{d}{dt}\frac{\Delta E}{\omega_0} = \frac{e}{2\pi}[V(\phi) - V(\phi_s) + V_{sc}(\phi)] \qquad (7)$$

---

[1] RAL tracking codes are noted for the complete lack of originality in their names.

where variations ΔE and $\Delta\phi$ in energy and phase from synchronous values are used as coordinates. $V(\phi)$ is the applied RF waveform, h is the harmonic number, ω_0 the revolution frequency and $\eta = 1/\gamma_t^2 - 1/\gamma^2$. In this one-dimensional model, the space charge forces are given in terms of the particle line density, λ, by [7]

$$V_{sc}(\phi) = e\frac{h}{R}\frac{d\lambda}{d\phi}\left[\frac{Rg_0}{2\epsilon_0\gamma^2} - L\beta^2 c^2\right]. \tag{8}$$

L is the total inductance per turn of the reactive wall (usually small and ignored), R is the average machine radius, and for a beam of circular cross-section of mean radius a in a pipe of radius b, $g_0 = 1 + 2\ln(b/a)$. (This formula for g_0 is valid only for long bunches with half-length $z \gtrsim 10a$; for shorter bunches, the linac formula

$$g_0 = \begin{cases} \dfrac{1}{1-\zeta^2} - \dfrac{\zeta\cos^{-1}\zeta}{(1-\zeta^2)^{3/2}} & \text{for } \zeta < 1 \\ \dfrac{\zeta\cosh^{-1}\zeta}{(\zeta^2-1)^{3/2}} - \dfrac{1}{\zeta^2-1} & \text{for } \zeta > 1 \\ \dfrac{1}{3} & \text{for } \zeta = 1 \end{cases}$$

where $\zeta = z/a$, is more appropriate.) From an initial model bunch of $N_P \sim 10^4$ simulation particles in ΔE-$\Delta\phi$ space, a repetitive process of calculating the applied voltages and space charge terms and updating the coordinate in intervals of one time step is applied. For such a simple model, it may be surprising that implementing a suitable code is quite difficult. The main problem is that the line density has first to be determined from the particle distribution and this will inevitably suffer from statistical fluctuations owing to the relatively small N_P. Trying to calculate the derivative, $\dfrac{d\lambda}{d\phi}$, from such a λ will lead to substantial errors which will propagate and most likely grow with time. The solution is to introduce a smoothing process to iron out the statistical effects while at the same time retaining the main features of the distribution. Fourier analysis with up to approximately 10 harmonic terms tends to follow the noise too closely. However a satisfactory approach has been found at RAL using cubic splines in which an automatic node allocation identifies the real trends in λ and minimises random errors [8].

TRACK1D contains other unusual features which have been developed from a need to simulate the injection of many turns into longitudinal phase space. A standard approach to injecting N_t turns would be to use N_P/N_t simulation particles per turn, but this number can be unreliably small or forces N_P to be awkwardly large (for $N_t = 1000$, one would need $N_P \geq 10^6$ for example). TRACK1D therefore adopts a process during injection of building up charge at selected points, somewhat akin to the Euler approach to fluid dynamics of identifying changes in properties at fixed locations. After injection is completed the code reverts to a more traditional Lagrangian technique of following particles as the distribution evolves. A third coordinate ("charge" or, more precisely, the number of real particles each macro-particle represents) is introduced. As each beamlet is injected, its particles are allocated to an imaginary rectangular grid of $\lesssim N_P$ nodes in ΔE-$\Delta\phi$ space according to a weighting scheme given by functions of the type $f_i(\xi_j)$ in (4). Beam already injected is similarly allocated to the grid, producing at each stage approximately N_P elements of charge, whether for the first turn or much later in the cycle. Once injected, the nodal charges are regarded as particles to generate the space charge forces for a single time step evolution according to (6) and (7) under which they will move from their grid positions. As the next turn is injected, particles in the machine are weighted back onto the imaginary grid, giving a slightly different arrangement, and the process repeated. The particles used under this procedure, therefore, all carry different amounts of charge, which change as the process is repeated and beam is built up. The specific charge allocation schemes that might be used [9] are (in order of complexity) commonly known as the nearest-grid-point method (NGP), cloud-in-cell scheme (CIC) and the triangular-shaped-cloud (TSC), given mathematically by repeatedly convoluting the "top-hat" function

$$\Pi(x) = \begin{cases} 0 & |x| > \frac{1}{2} \\ \frac{1}{2} & |x| = \frac{1}{2} \\ 1 & |x| < \frac{1}{2} \end{cases}$$

with itself. In fact, TRACK1D uses the next function up in this series (so far unnamed). Charge assignment has not only the benefit of smoothing the distribution but also the unfortunate side-effect of artificially spreading the beam, and TRACK1D includes correction techniques to counteract this effect. The process of assigning charge to the grid points is depicted in Figure 3 and examples of the use of the code are covered in section IX.

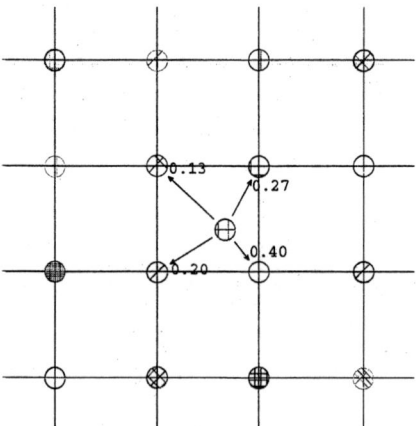

FIGURE 3. Injection mesh for TRACK1D showing the allocation of an incoming particle to the nodes (cloud-in-cell method).

V 2D TRANSVERSE TRACKING CODES

Many of the ideas contained in the longitudinal code have recently been incorporated into the two-dimensional transverse program TRACK2D [10]. This code is built in modules, using the same routines for reading in details of the beamline or accelerating structure, graphics and optimising facilities as other related codes. By these means, one is sure that elements are being handled in exactly the same way in the tracking calculations as in, say, analytical treatments used for more basic design and comparison. It is also easy to add new modules within the existing structure as new questions arise.

TRACK2D includes the finite element space charge techniques described above. The default routine uses a third order potential fit, but orders up to six are used for verification purposes. Equations of motion are either based on paraxial modelling or include third order effects of the transverse particle angles. Apart from direct simulation of a particle beam through a focusing channel (a storage ring or a beamline, such as the final focusing system for a heavy ion fusion scheme), the code will handle injection and contains an optimiser to determine design parameters for systems under non-linear effects. The optimiser is essentially the analytical code KVBL [11], used for machines with zero current or linear space charge governed by the KV-envelope equations, driving TRACK2D with the space-charge mesh in reduced form. It is best used for slight modifications to a system pre-optimised, so far as is possible, analytically.

In injection mode, the code currently still uses the conventional approach of N_P/N_t simulation particles per injection turn but work is in progress to incorporate the variable-charge scheme described in section IV. This will need to use an irregular 4-dimensional grid in phase space in order to maintain a suitable number of "particles" in real space for proper space charge calculations and beam simulation. The code is able to simulate multiturn injection using septum magnets in either or both planes, or charge-exchange injection by means of stripping foils, as, for example, in ISIS or the ESS [12]. In the latter case, details of the passage of particles through the foil can be used to study its thermal properties and likely peak temperatures [13].

Machine errors such as field strength errors, magnet alignment errors (horizontal, vertical and longitudinal) and bending field errors, may be read in in a simple way or generated internally via a Gaussian random number generator using prescribed rms values. The code also contains models of fringing fields. There was originally an intention that it could determine emittance effects in electron machines, such as synchrotron light sources, where machine errors, leading to coupling via closed orbit and dispersion, need careful control; but since space charge is negligible in these cases, the calculations can be performed much more quickly and effectively by other means [14].

By allowing most major parameters to be read in as tables of varying data, TRACK2D is also able to take into account certain longitudinal effects of bunched beams. Information from TRACK1D, such as bunching factor or changes in momentum spread $\Delta p/p$, can thereby be incorporated into the simulations. Particles carry as an extra coordinate their charge, which may be varied as beam particles move in and out of the transverse slice under consideration. This means that TRACK1D and TRACK2D work in conjunction with each other, each being run successively with appropriate modifications to data until a convergent set of parameters is achieved. Since full three-dimensional simulation under non-linear space charge is extremely slow, this is quite an effective alternative and can be carried out over a realistic timescale.

VI 3D TRACKING CODES

The three-dimensional tracking code, TRACK3D, is an extension of TRACK2D with the inclusion of a third longitudinal (position) coordinate into Maxwell's full equations of motion under the Lorentz force. Space charge is calculated using the 3D version of the finite element codes described in section III. Although written and running some years ago, this code was little used and largely untested because of restrictions on computer storage and, more especially, the availability of CPU time. In consequence, it fell behind TRACK2D in its development as current projects and new questions arose (section IX). It is now being brought up to date with a view to modelling in more detail various aspects of the ESS design [12] and ISIS. In particular, an improved method of global node numbering for the finite element space charge test needs to be discovered in order to reduce the storage space required for the sparse stiffness matrix K_{ij}. Additional use of the method known as static condensation, whereby sub-units of the mesh, rather than the whole mesh, are read into the computer and processed in turn, needs to be introduced.

In parallel, a 3D code to model beam transport through RF quadrupoles is being developed at RAL by Letchford [15]. This can handle quite complicated geometries, though the space charge solver is still based on Coulomb forces and cannot handle the effects of image charges. It is thought that the routines from TRACK3D might eventually be incorporated into this code. It will not only provide a useful test of the finite element approach but almost inevitably be faster than the current method as well as solve the problem of how to deal with conducting boundaries.

VII FORMULATION OF RESULTS

Deciding on the quantities to be output and the form of the results from a tracking code is not obvious and important calculations are often overlooked. It is not always clear how best to describe the behaviour of a highly non-linear beam. Quantities such as rms emittances and rms beam sizes are traditionally used but contain a limited amount of information, and perhaps the best approach is to store as often as possible the particle coordinates so that further calculations can be carried out as required at a later date. This can however lead to storage restrictions. Graphical output probably enables one to see best how a beam is evolving and how, for example, halos are generated. But again, different laboratories use different graphics systems, which can prevent codes being truly portable. History shows that attempts at a universal standard, such as GKS, seem inevitably to fail. Furthermore, without cross referencing, results may be misleading: for example, the RAL codes use colour or shading to represent different levels of charge but in the final plots only the topmost colours in the superposition are visible; and again regions which in terms of colour appear to have the highest charge density may in fact be outweighed by regions with apparently lower charge when the actual number of particles plotted is taken into account. This is not a major problem but emphasises that one needs to think carefully about one's results and not always accept them at face value. Interactive, graphical simulation is perhaps the most informative approach as it not only shows how a beam is evolving but indicates immediately when parameters are unsuitable, allowing termination and avoiding wastage of CPU time.

VIII BENCHMARKS AND VALIDATION

Individual calculations in tracking codes need, of course, to be thoroughly tested for accuracy. In addition, it is important to check that a code actually simulates the motion of a particle beam by agreement with accepted theory. The number of available benchmarks is limited but there are sufficiently many for suitable testing.

In the 1D longitudinal case, there is a stationary distribution found by Hofmann and Pedersen [16] (also Neuffer). This is characterised by the line density being proportional to the potential, whatever the shape of the focusing force, so that self-forces caused by space charge and inductive wall impedances are proportional to the external applied voltages, and the longitudinal particle distribution is preserved. A simulation code of the type described in section IV should show single particles in the bunch describing closed orbits in phase-space without deviation. A tracking test should cover several hundred synchrotron periods.

For 2D transverse tracking codes, the main benchmark is the KV distribution [17], where simulation results can be compared directly with numerical solutions of the KV envelope equations. A simple test might be a matched beam in a periodic focusing channel, but a more stringent challenge could be a final focusing beamline for a heavy ion fusion plant, tracking a space charge dominated beam (current $\sim 1\,\text{kA}$, $\beta \sim 0.3$) through a FODO channel, blowing it up under the effects of emittance and space charge from a radius of about 10 mm to 40 cm, and then focusing this to a target spot size of about 3 mm. Using a model based on the envelope equations, one would expect a 2D tracking code to predict at least 90% of particles hitting the target. If this is not the case, one would have little confidence in results obtained from simulations using non-linear beams, such as input Gaussian or waterbag distributions. Codes should also be able to confirm other theoretical results, such as the rms matching of non-linear beams or the basic properties of two-dimensional stationary waterbag and Gaussian distributions (circular beams in constant, axisymmetric, focusing channels [18]).

Suitable benchmarks for 3D simulations essentially reduce to three-dimensional stationary distributions for spherically symmetric bunches in constant focusing structures [18]. In some cases, results from a 3D code can be directly compared with separate longitudinal and transverse simulations where the coupling is small and good agreement might be expected.

Good working practice demands that accurate, up-to-date records are maintained of all changes made to programs, whether relatively minor corrections or more major developments. At each stage, the altered code should be checked against the standard set of benchmark tests. Where possible, independent validation should also be sought, preferably through codes held at other laboratories. ACCSIM [5] is widely available and is useful in this respect. However, it often turns out that one has specific problems in mind when developing a program or making additions and other codes may not yet have found it necessary to address these issues.

IX ILLUSTRATIONS

By way of illustration, four examples of recent use of the codes can be considered.

The first concerns the injection process for the European Spallation Source, depicted in Figure 4 [19]. 2.34×10^{14} H$^-$ ions, in bunches chopped at 60% of the linac duty cycle, are injected over 1000 revolutions into each of two accumulator rings at an energy of 1.334 GeV. A suitable longitudinal distribution is achieved by painting the phase space both by ramping the momentum of the injected turns in time and steering the RF bucket by varying the RF frequency. Repeated simulations using TRACK1D indicated that the optimum (simple) method of ramping is to increase the momentum linearly from the range $[0-2] \times 10^{-3}$ to $[2-4] \times 10^{-3}$ over the period of injection, 0.6 msec. During this time the RF frequency is held at a value equivalent to a particle orbiting with $\Delta p/p = 2 \times 10^{-3}$ and then reduced to the natural frequency at the design energy of the ring. Thus turns are injected into the lower part of the phase space bucket at the start of injection and into the upper half towards the end. This is visible in Figure 4, as is the phase space grid used to model injection as described in section IV. The resulting distribution has a bunching factor of around 0.46. In the transverse plane, injection is at a point in the ring where the normalised dispersion $\alpha_p/\sqrt{\beta} = 2.0$ and this combines with the momentum ramping to provide horizontal phase space painting and help reduce the number of times recirculating protons pass through the foil. Vertical painting is provided by vertical orbit bump magnets which are programmed to give inversely correlated oscillation orbits, large to small in the horizontal plane, small to large in the vertical plane. The resulting mean number of foil traversals per particle is 6-7 and a good transverse distribution is achieved. In these simulations TRACK1D and TRACK2D were used in alternating runs, results from the longitudinal code being linked via the ring dispersion, RF steering and variable charge method into the transverse plane, where subsequent modifications to reduce foil heating were fed back to the longitudinal plane for further optimisation. The results of the simulations have been confirmed by Ohmori [20] and, more recently, Galambos [21] using ACCSIM and other codes.

The second example, principally of the use of TRACK2D, concerns the design of the multiturn injection scheme for 10 GeV Bismuth ions into accumulator rings for the HIDIF fusion study [22]. A novel method is used with an electrostatic septum tilted at an angle of 44° so that injection is into both transverse planes simultaneously. By optimising vertical and horizontal closed orbit bumps, and carefully selecting the nominal tunes of the rings, it proves theoretically possible to inject 20 turns at 400 mA per turn with losses of less than 3%. The incoming beam is largely debunched at injection to reduce space charge effects, so that tune depressions are of the order of 0.04. Selection of the parameters is by the code MISHIF [23] for zero or small space charge, with TRACK2D exploring the more general case of a non-linear beam as shown in Figure 5. Results of the analytical model are closely confirmed.

An interesting third use of TRACK1D, reported in more detail in [24], is the development of a dual $h = 1/h = 2$ harmonic RF system for injection, acceleration and trapping in the ISIS synchrotron. The voltage waveform used here is

$$V(\phi, t) = V_0(t)[\sin \phi - \delta \sin(2\phi + \theta)]$$

where V_0, δ and θ are functions of time, varying according to a carefully calculated programme. The Hamiltonian for such a system has two stable centres about which particles oscillate (visible in Figure 6); by creating these in a continuous process under acceleration and then merging them at the end of the cycle, it should prove possible to trap at least 20% more particles and possibly as much as 50%. This is borne out by simulation, and the scheme is currently featured in plans for an ISIS upgrade, in collaboration with KEK, Japan.

Finally, the use of the optimising routines in TRACK2D is illustrated in Figure 7, which shows the results of an investigation into the extent to which a system of octupole and dodecapole magnets might be used to convert a beam with an initially Gaussian distribution into other forms. Repeated runs with a beam of 5000 macroparticles in the absence of space-charge determined in each case the strengths of the higher order elements needed to create a square beam with a uniform distribution, and round beams with two-dimensional parabolic or elliptic distributions. Some minor optimisation was then carried out in the presence of space charge (which was not great for the parameters used in the model). The octupoles have the effect of folding the tail of the Gaussian into the more central parts of the beam and, by adjusting their strengths carefully, this can give a fairly close approximation to each of the distributions considered. The method may conceivably be useful in controlling the development of halos in highly non-linear particle beams.

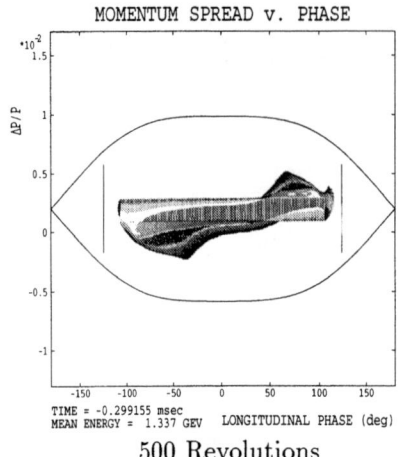
500 Revolutions

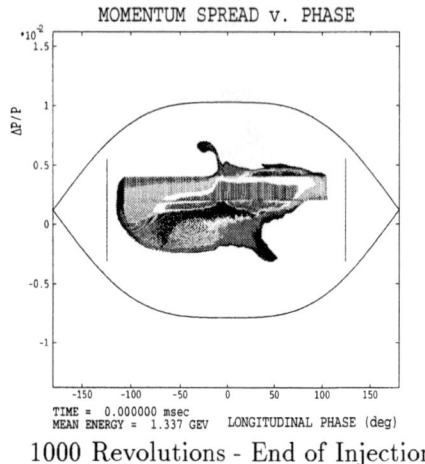
1000 Revolutions - End of Injection

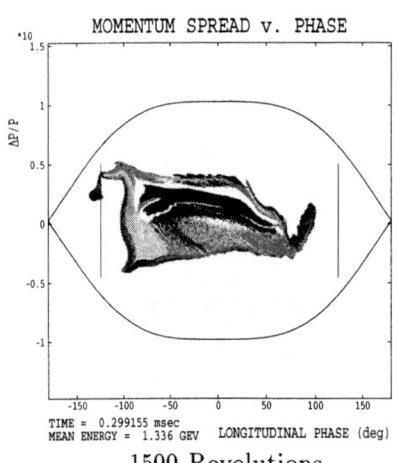
1500 Revolutions

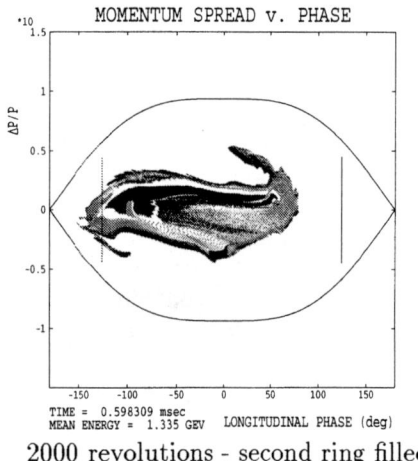
2000 revolutions - second ring filled

FIGURE 4. Longitudinal phase space plots from TRACK1D for ESS injection.

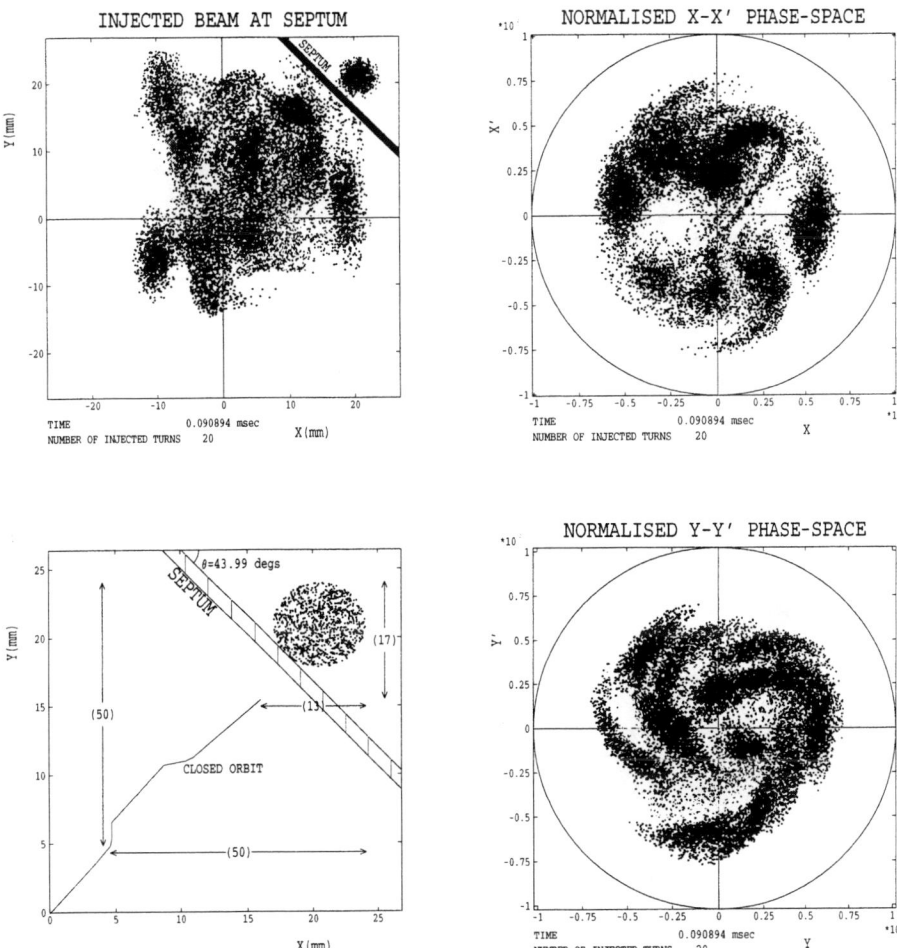

FIGURE 5. Transverse injection using a tilted electrostatic septum into the HIDIF accumulator ring. 20 turns are injected of a beam with a truncated Gaussian distribution and emittances of $4\pi\mu$ rad m in each plane to achieve a beam with emittances of $50\pi\mu$ rad m. Losses of less than 3% are predicted.

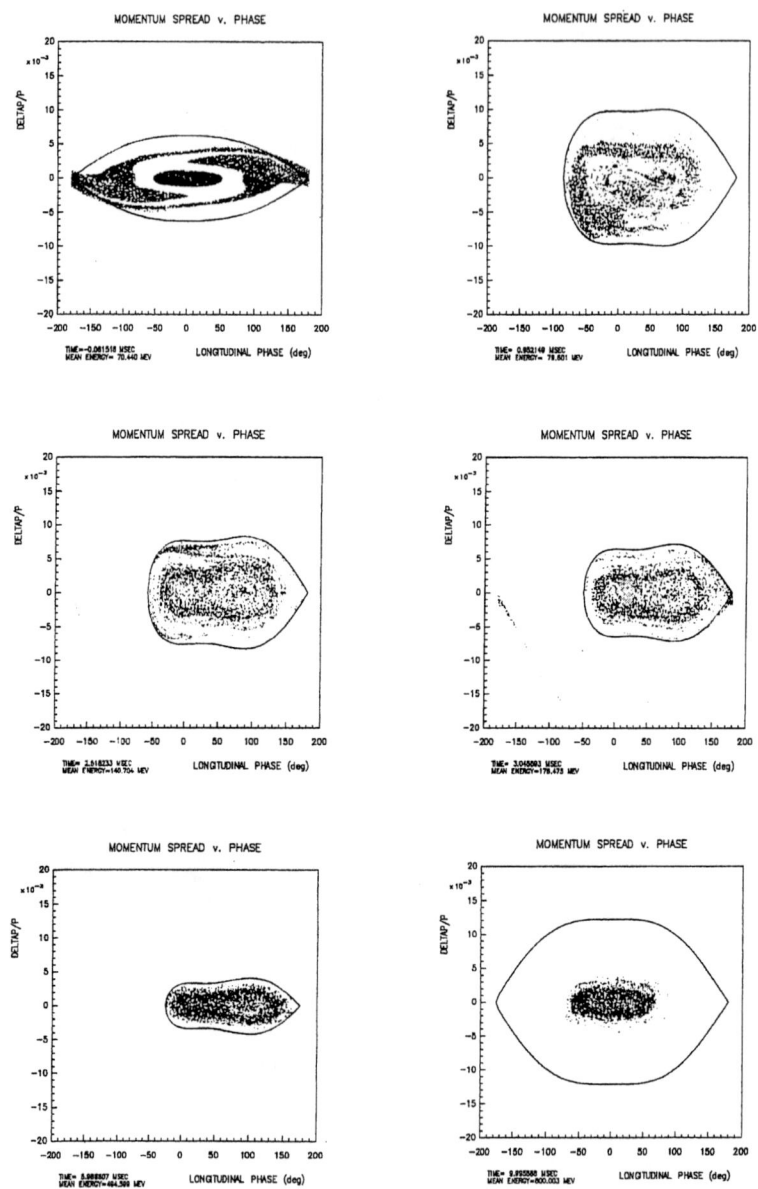

FIGURE 6. Longitudinal simulation of injection, acceleration and trapping for the proposed dual harmonic RF system on ISIS.

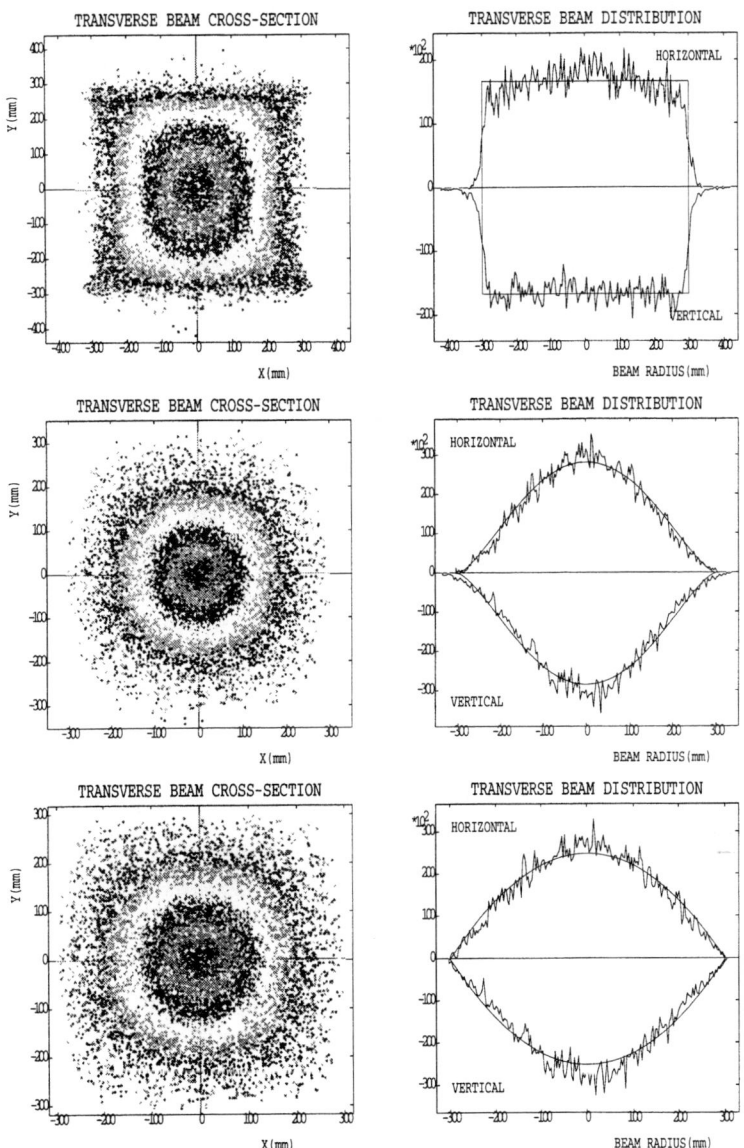

FIGURE 7. Transverse beam plots from TRACK2D showing transformation of an initially Gaussian beam using an octupole-dodecapole focusing system into a uniform, square distribution (top), a 2D parabolic distribution (centre) and a 2D elliptic distribution (bottom).

X SUMMARY

While computers have advanced dramatically and provide far greater scope for research through simulation than in the past, personal experience suggests that progress - at least in Europe - tends to be by individuals in *ad hoc* ways rather than via a co-ordinated approach to a common goal. Different laboratories have different projects and their own particular problems, and codes tend to be developed to meet these individual needs. Seldom is a code found to answer all questions as soon as they arise, and the dream of a set of all-purpose tools is as far away as ever. The manner in which technological advances are made in a highly competitive commercial field can sometimes actually handicap the user, who would prefer not to repeatedly have to update working code to fit in with the requirements of new software packages. This acts against portability and encourages duplication and individuality. Furthermore it does seem to be the case that researchers new to the modelling field or changing direction within the accelerator community often start from base level, unaware until later of what is available, and what is already known.

Simulation codes are notoriously difficult to understand, so that it may actually be the case that one may only trust the predictions when one has carried out the programming oneself. Codes developed by others then become tools for comparison and a means of generating confidence, both of which are vitally important, and too often overlooked. Validation must be carried out with a well-defined set of benchmarks, both theoretical and practical. In this respect the TRACK codes have been used to model ISIS and to predict some of the changes which led to clear improvements being made. It is because of this that there is confidence in the proposals for the dual-harmonic upgrade and in the modelling carried out for ESS.

REFERENCES

1. M. Berz: *COSY INFINITY, An Arbitrary Order General Purpose Optics Code*. Computer Codes and the Linear Accelerator Community, Los Alamos LA-11857-C:137, 1990.
2. A.J. Dragt: *MARYLIE. A Program for Charged Particle Beam Transport Based on Lie Algebraic Methods*. University of Maryland, USA.
3. A.J. Kinderman and J.F. Monahan: *Computer Generation of Random Variables Using the Ratio of Uniform Deviates*. ACM Transactions on Mathematical Software, Vol. 3, No. 3, 1979.
4. G. Ripken: *Non-linear Canonical Equations of Coupled Synchro-Betatron Motion and their Solution within the Framework of a Non-Linear 6-Dimensional (Symplectic) Tracking Program for Ultra-Relativistic Protons*. DESY Report 85-084, 1985.
5. F.W. Jones, G.H. Mackenzie and H. Schönauer: *ACCSIM - A program to Simulate the Accumulation of Intense Proton Beams*. Proc. 14th International Conference on High Energy Accelerators, Particle Accelerators 31, p 199, 1990.
6. B.W. Montague: *Single-Particle Dynamics - RF Acceleration*. Proceedings of the First Course of the CERN International School of Particle Accelerators, Erice, 1976.

7. A. Hofmann: *Single-beam Collective Phenomena - Longitudinal*. Proceedings of the First Course of the CERN International School of Particle Accelerators, Erice, 1976.
8. M.J.D. Powell: *Curve Fitting by Cubic Splines*. Harwell Report T.P. 307, 1967.
9. R.W. Hockney and J.W. Eastwood: *Computer Simulation Using Particles*. McGraw Hill, 1981.
10. C.R. Prior: *The Multi-Particle Code TRACK2D: A Guide for Users*. CLRC, Rutherford Appleton Laboratory, 1993.
11. C.R. Prior: *The KV-Envelope Code KVBL: A Guide for Users*. CLRC, Rutherford Appleton Laboratory, 1990.
12. G. Bauer, T. Broome, D. Filges, H. Jones, H. Lengeler, A. Letchford, G. Rees, H. Stechemesser, G. Thomas (eds): *ESS: A next Generation Neutron Source for Europe. Vol III: The ESS Technical Study.*, November 1996.
13. J.P. Duke: *Foil Stripping Temperatures in the European Spallation Source*. Proceedings of the 5th European Particle Accelerator Conference EPAC96, Sitges, Spain, 1996.
14. C.R. Prior: *The Two-dimensional Analytical Coupling Code COUPXY*. CLRC, Rutherford Appleton Laboratory, 1993.
15. A.P. Letchford: Minutes of ESS Meeting 1, Rutherford Appleton Laboratory, April 1998.
16. A. Hofmann and F. Pedersen: *Bunches with Local Elliptic Energy Distributions.* IEEE Trans. Nucl. Sci. Vol NS-26, No. 3, 1979.
17. I.M. Kapchinskij and V.V. Vladimirskij: *Limitations of Proton Beam Current in a Strong Focusing Linear Accelerator associated with Beam Space Charge*. Proc. Int. Conf. on High Energy Accelerators, CERN, Geneva, 1959.
18. I. Hofmann and J. Struckmeier: *Generalised Three-Dimensional Equations for the Emittance and Field Energy of High-Current Beams in Periodic Focusing Structures*. Particle Accelerators, Vol 21, pp 69-98. 1987.
19. C.R. Prior: *Longitudinal Injection for the ESS*. ESS-94-7-R, November 1994.
20. C. Ohmori: *Multi-Particle Tracking Study for ESS Ring*. Private communication.
21. J. Galambos, J. Holmes and D. Olsen: *Longitudinal Injection Studies for ESS and SNS*. Private communication.
22. C.R. Prior and G.H. Rees: *Multiturn Injection and Lattice Design for HIDIF*. Proceedings of the 12th International Symposium on Heavy Ion Inertial Fusion, Heidelberg, Germany. September 1997.
23. C.R. Prior: *The Computer Code MISHIF: Information for Users*, CLRC, Rutherford Appleton Laboratory, 1996.
24. C.R. Prior: *Studies of Dual harmonic Acceleration in ISIS*. Proceedings of the Twelfth Meeting of the International Collaboration on Advanced Neutron Sources, ICANS-XII, Abingdon, UK. May 1993.

Status and High Intensity Performance of ISIS

C M Warsop

Rutherford Appleton Laboratory, UK

1. Introduction

ISIS is presently the world's most intense pulsed neutron source, regularly providing beam powers of 150-160 kW, over operational periods of about 3500 hours per year. The status and performance of the ISIS Accelerators are reported, with emphasis on the Synchrotron. A summary of operational statistics is followed by a review of high intensity effects limiting synchrotron performance. A key consideration in operating a high intensity machine is beam loss; methods of monitoring and control are described along with resulting levels of activation and their consequences. Developments of diagnostics to overcome difficulties associated with regularly tuning a high intensity machine to optimal settings are summarised. Finally a review of developments of the ISIS facility is presented.

2. The ISIS Accelerators

The ISIS accelerators consist of a Penning H⁻ ion source at 665 kV, an accelerator column, a four tank 70 MeV Alvarez type Linac and an 800 MeV 50 Hz proton Synchrotron [1]. The linac provides ~22 mA of H⁻ over the 200 µs injection interval, which is stripped to H⁺ in a 0.3 µm thick aluminium oxide foil as it enters the ring acceptance. Beam is painted in both transverse planes, over the ~130 turns of injection to minimise space charge effects, and allow accumulation of 2.8×10^{13} protons. The harmonic number two RF system bunches the initially coasting injected beam, and accelerates it to 800 MeV on the 10 ms rising edge of the sinusoidal main magnet field. There are six double-gap ferrite-tuned RF cavities which provide a peak voltage of 140 kV per beam revolution. At extraction the two 100 ns pulses, separated by 230 ns, are deflected in a single turn into the extraction line with a lumped fast magnetic kicker. The beam is then transported to the spallation target.

3. Operational Performance Statistics

ISIS has achieved and exceeded its design current of 200 µA, with levels of 205 µA being reached for prolonged periods with acceptable losses. Figure 1 summarises beam delivered in µA-hours during operational running for each year since 1989. Over 1996-97, there were 3630 hours of beam to target, with a mean current of 171 µA, and total beam of 621 mA hours. Mean current over the last 5 week cycle exceeded 200 µA. Operational availability (scheduled beam time/actual beam time) is typically 85%. Figure 2 shows beam current, the mean over each year, and the peak 24 hour average for each year.

CP448, *Workshop on Space Charge Physics in High Intensity Hadron Rings*
edited by A. U. Luccio and W. T. Weng
© 1998 The American Institute of Physics 1-56396-824-X/98/$15.00

Figure 1
Total Annual Beam Delivered 1989-1997

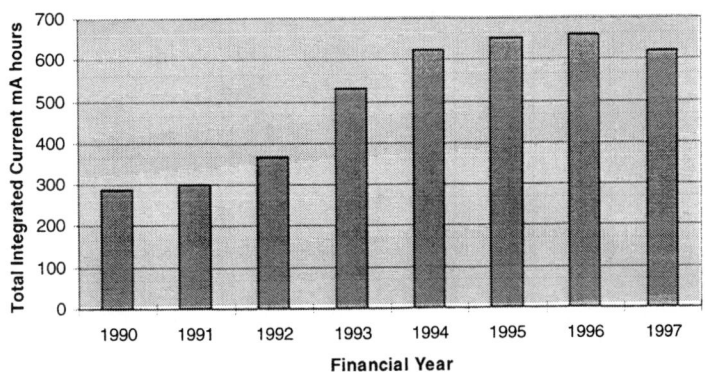

Figure 2
Operational Currents: Yearly Mean, Peak Daily Mean

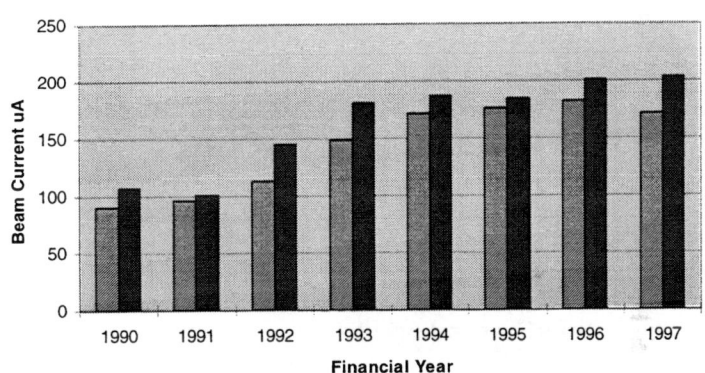

Figure 2 provides a useful indication of how well the machine operates relative to its maximum capability, comparing peak levels achieved over 24 hours with those averaged over the year. Practical constraints, including hardware reliability, will always prevent constant running at optimal levels, and generally ISIS operational and peak currents have increased together.

Although for most years the performance improves, in 1996/7 operating currents were reduced due to aperture restrictions from damaged RF shields in the ring, and this highlighted the difficulties in achieving consistent beam control at highest intensities. Tightening up on beam loss protection has done much to overcome these problems, with more recent operations returning to record performance levels. However, some substantial upgrades to diagnostics are planned, with the aim of providing more detailed information, thereby improving fault diagnosis and ability to optimally tune the machine.

4. Review of Main Loss Mechanisms

4.1 Summary of Losses

Typical losses at 200 µA in the Synchrotron are:

- 1% at injection due to foil H^0 stripping loss,
- 11% during trapping due to fast cycling, and high intensity effects
- 0.01% at extraction

Losses at injection and extraction are relatively low, and near the theoretical minimum. The dominant loss, and that having the most serious impact on activation, is during trapping of the injected beam over 0-2 ms. At low intensities, 6% of beam is lost, due principally to the non adiabatic nature of the trapping process. Percentage losses increase with intensity, reaching 11% at 200 µA. At these levels the dominant loss is still trapping loss, driven additionally by longitudinal space charge forces. However, at these levels and higher, transverse mechanisms come increasingly into play; significant effects due to incoherent space charge tune shifts and image charges are observed.

Key parameters for achieving highest intensities are:

- Q value settings at 15 time points through the cycle
- Steering magnet settings at 15 time points through the cycle
- Adjustment of parameters determining distributions during injection
- Effective control of RF system under heavy beam loading

These are referred to in detail below.

4.2 Longitudinal Losses

Longitudinal losses are due to fast bunching and space charge. These have been studied in some detail with suitable computer codes [2], and optimal RF voltage laws derived. It is found that low levels of RF volts during injection reduce later trapping losses. Studies have also led to the proposed dual harmonic upgrade, adding h=4 harmonics to the present h=2 RF system. Simulations indicate currents of ~300 µA will be possible at the same absolute loss levels [3]. The fast cycling and heavy beam loading impose many demands on the RF system which are covered extensively elsewhere [4].

4.3 Transverse Effects: Betatron Q Value Optimisation in the Synchrotron

There are 2 sets of 10 programmable trim quads in the ISIS ring, which allow fine tuning of Q values at 15 time points through the machine cycle. Optimal setting of these has been critical in achieving highest intensities. The cycle may be divided into four regions where Q values are determined by different effects: (i) injection where trim quads compensate for the natural chromaticity and the falling magnet field, (ii) ramping of Q's over 0-1 ms to minimise effects of space charge depressions during trapping, (iii) reduction in Q's over 2-4 ms to avoid transverse resistive wall instability, and (iv) lowering of Q's to avoid coupling resonances at extraction. Figure 3 shows Q values through the machine cycle as used for 200 µA running.

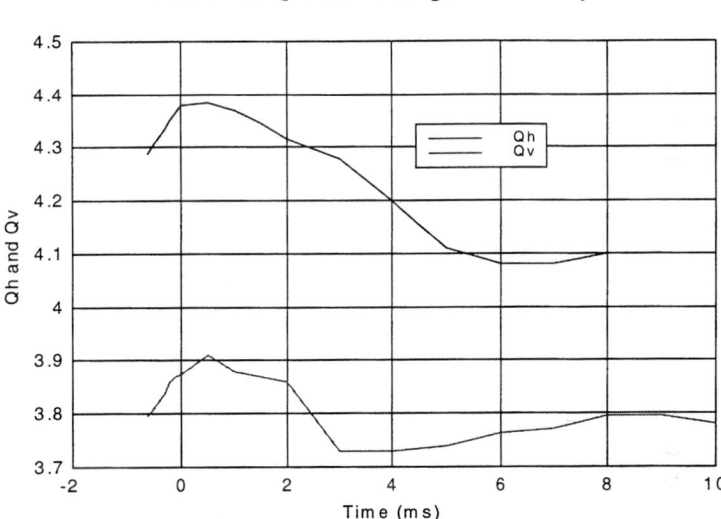

Figure 3
Variation of Q Values through Machine Cycle

Figure 4 shows the tune diagram for the synchrotron. The lattice is predominantly linear with no sextupoles or octupoles used, and minimal non linear fringe fields. The important linear resonances are shown. The dashed line outlines the path of the (Q_h,Q_v) from the start to the end of acceleration (0-10 ms). The tunes and estimated maximum incoherent tune depressions at 1 ms are also shown.

Q Shift Calculations

Settings of Q values, especially over the 0-2 ms interval, critically affect losses; raising values was instrumental in achieving design intensity. Incoherent tune depressions and associated resonance crossing can be a major cause of loss.

Detailed estimates of the incoherent space charge Q shift have been made [5], which include elliptical transverse distributions and allow for finite emittances of particles injected. Calculations indicate the maximum depression is ~-0.4 in both planes, which forces some particles over the $Q_h=4$, $2Q_h=8$, $2Q_v=7$ resonances, Figure 4. Measurements indicate coherent Q shifts at 200 µA of about -0.1 in both planes. To reduce driving terms for gradient errors, trim quads are varied azimuthally to correct 7th and 8th harmonic components.

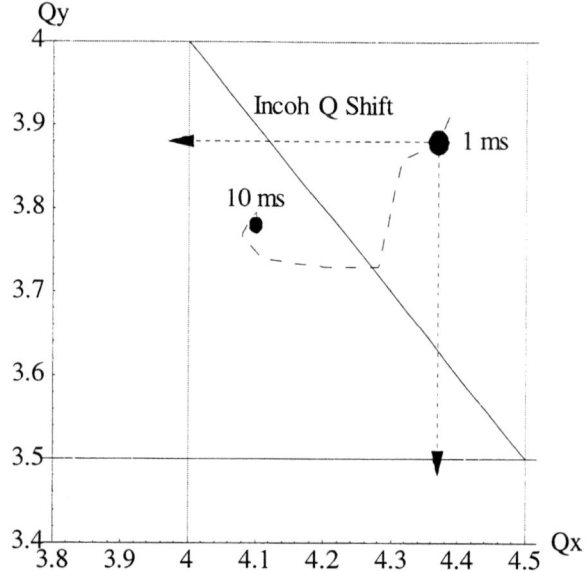

Figure 4
ISIS Synchrotron Q Diagram, Showing Main Linear Resonances,
Locus of Q values 0-10 ms, Q Values and Max Incoherent Shifts at 1 ms

Image Charge Effects

Experience has shown that highest intensities are only achieved when closed orbits during trapping are minimised, particularly the 4th harmonic components. Closed orbits are actually corrected at about 7 time points early in the cycle, and optimal settings are found by minimising beam loss. The beam typically occupies 70 % of the aperture at trapping, and image effects in the rectangular vacuum chambers significantly affect high intensity performance. Theoretical studies [6] indicate that the coherent displacement of the beam enhances variation of forces across the transverse axes, so influencing incoherent resonance crossing. A significant enhancement of transverse amplitude is predicted for some particles.

Aperture Occupation

The ring has transverse acceptances of 540 π mm mr in the horizontal and 430 π mm mr in the vertical. Collimators are usually set at about 300 π mm mr in both planes. Transverse beam size reaches a maximum at 1 ms, when it fills the available aperture. At 800 MeV, adiabatic damping would give extracted emittances of 75 π mm mr, but they are actually about 150 π mm mr.

4.4   Injection Set up

A key part of setting up the machine is optimising injection. The injection process does much to determine the initial transverse beam distributions, and therefore incoherent tune shifts. Betatron amplitudes are varied during the 120 turn painting process; vertical amplitudes decreased through injection, with horizontal motion inversely correlated. Under highest intensity conditions large vertical amplitudes have given best results. A large number of parameters in the injection line and ring determine injected betatron amplitudes, and specialised diagnostics methods are allowing tighter measurement and control.

The momentum spread at injection is also important, and can be adjusted with a debuncher cavity in the injection line. Diagnostics are available to measure initial injected momentum spread distributions. Adjustment of the initial longitudinal beam distribution, as would be expected, influences trapping loss.

Although there is a complex interplay amongst parameters at injection and trapping, generally non-peaky distributions produced by injection of many large amplitude particles, with momentum spreads enhanced with the debuncher, give best results.

4.5   Instabilities

Instabilities do not limit the high intensity performance of ISIS. Only the transverse resistive wall instability has any effect on running, and is cured by manipulating Q values. Features reducing the susceptibility of the ring to instabilities include low impedance lumped extraction kickers and profiled RF shields. A summary of instabilities observed on ISIS can be found in [7].

5. Constraints on High Intensity Operation: Beam Loss Control

Optimal operation of the machine is a compromise between maximising high intensity running whilst keeping activation and risk of machine damage to a minimum. Activation over most of the machine has to be kept low enough to allow hands on maintenance. Effective monitoring of loss distributions is essential for tuning and protecting the machine. There are two systems, toroids and beam loss monitors.

Toroids are placed at a number of locations around the machine, in the injection line, the ring, and the extraction line. They give an instantaneous measurement of current, which is used to provide a monitor of circulating intensity through the 10 ms cycle (Figure 6), and to derive efficiencies for the injection, trapping and extraction processes (Table 1). The latter are displayed and monitored continuously, with warnings or beam trips being generated as appropriate. Warnings are generated for persistent low levels of loss as well as large sudden losses.

Table 1
Typical Running Efficiencies @ 193 µA

Time	ppp @ 50 Hz	%
Injection 0 ms	2.80×10^{13}	99
Trapping 0-2 ms	2.41×10^{13}	88
Extraction 10 ms	2.41×10^{13}	100

Beam loss monitors are argon filled ionisation chambers [8], which are distributed along the length of the accelerators, with 40 around the circumference of the ring. These give the spatial distribution of loss as well as time dependency, and have high sensitivity measuring losses of < 0.01%, levels not detectable on the toroids. Each beam loss monitor has a preset trip level, which if exceeded will switch off the beam. This ensures most loss is confined to the collector regions. A typical loss distribution around the ring, summed over a whole cycle, is shown in Figure 5. Of the total loss of 13%, nearly all is localised near the collectors in superperiods 1 and 2, with the small loss in superperiod 0 at injection. Summing all the ring beam loss monitors, and displaying 'total beam loss' as a function of time, yields the signal in Figure 6. It can be seen most loss is during trapping, 0-2 ms, at energies in the range 70-100 MeV. Small losses at injection and extraction are also visible.

Figure 5
Beam Loss Distribution around Synchrotron, Summed over Machine Cycle

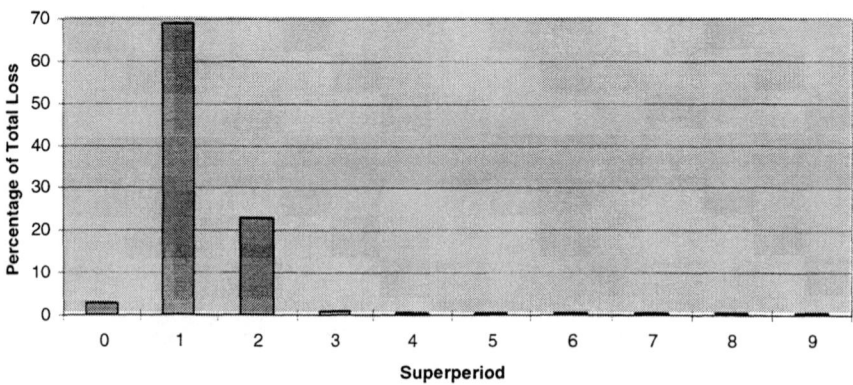

Figure 6
Beam Intensity and Beam Loss Monitor Sum Signal Over Machine Cycle

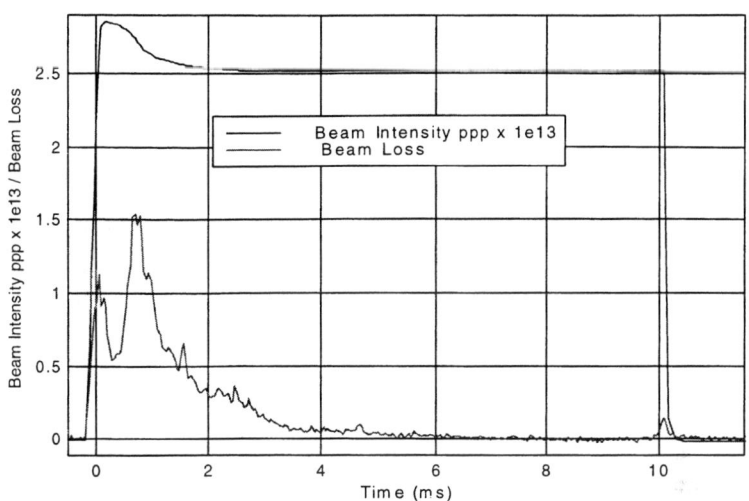

Beam Loss Collector Systems

The dominant loss during trapping is efficiently collected at a jaw situated at a high dispersion point in superperiod 1. As the field ramps, low energy untrapped beam is swept into the collector. Further horizontal collimation is provided by two downstream collectors. This system removes most lost beam (~80%). In the vertical plane there is a system of 3 jaws situated in superperiod 1. Transverse growth is also intercepted on these collector systems. The set up of collectors and closed orbits is critical for effective machine protection.

Radiation levels

Levels of activation vary markedly around the machine, with peaks at the collectors reaching levels of 300 mSv/hour on contact. However the combination of effective loss localisation, and substantial local shielding in affected areas, generally keeps typical levels on the machine during maintenance at levels of 100 µSv/hour. During in situ maintenance of equipment, dose levels are minimised with the use of mobile local shielding, including lead blankets and shutters. Working practices on ISIS are such that dose levels of all workers are ≤ 5 mSv per year.

Radiation levels, along with machine protection, are the determining factor limiting running intensity. Any upgrades are constrained by the requirement that increases in intensity must not cause a significant increase in activation.

6. Diagnostics Developments

Substantial upgrades to the synchrotron diagnostics are presently underway. Use of much improved data acquisition and computing power, along with specialist low intensity diagnostic beams, will provide more detailed information on the machine.

Essentially the upgrade consists of adding many fast digitiser channels, which allow measurement of turn by turn transverse positions of the beam centroid (trajectories) at 10 points around the ring, on the same machine pulse. Improvements to amplifiers on the capacitative position monitors allow measurement of normal high intensity beams or special low intensity *diagnostic* beams. A dedicated computer workstation provides the power needed to process data and access to all accelerator parameters via the control system. An additional long record length digitiser is incorporated for study of longitudinal bunch shapes.

Low Intensity Measurements and Injection

The utility of specially configured low intensity beam diagnostics has been exploited on ISIS for some time [9]. A low intensity beam occupying a small fraction of the transverse and longitudinal acceptances can undergo large coherent oscillations, the observation of which provides valuable information. Measurements on high intensity beams can be less useful as aperture limitations mean coherent motion is small, and interpretation is complicated by summing over many particles and space charge effects. Comparison of low and high intensity measurement can also be very useful.

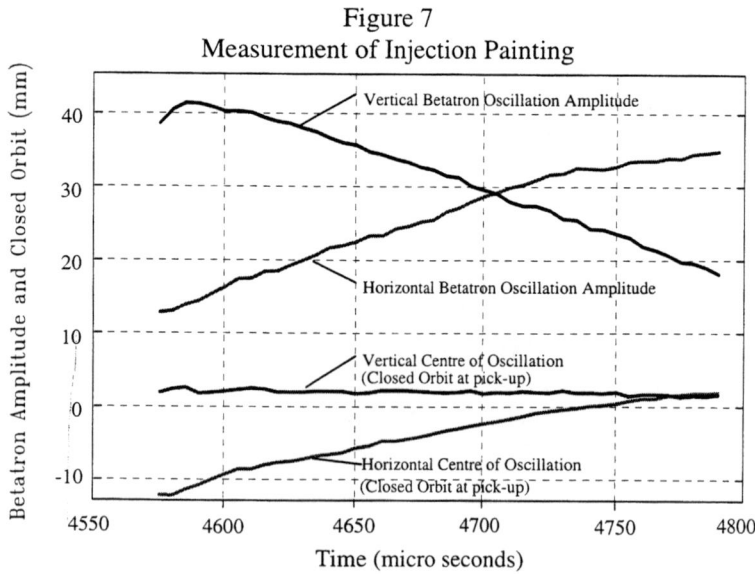

Figure 7
Measurement of Injection Painting

On ISIS, chopping the injected pulse down to 100 ns provides a beam filling a small fraction of all acceptances. Measuring the transverse motion of the injected 'chopped' beam as a function of injection time gives direct values of betatron amplitude being painted, Figure 7. Similarly, debunching of this beam pulse in the ring with RF off provides measurement of injected momentum spread. The system allows optimal injected beam distributions to be easily identified.

Transverse Information

The trajectory information has applications for aperture exploration, optimal set up of collectors, Q measurements, closed orbit correction, beta function and local phase advance measurement. Taking position measurements as a function of excitation in correction elements also gives beta functions at trim quads and steering coefficients for closed orbit correction.

Longitudinal Information

Use of long record length digitisers allows the longitudinal pulse shape from capacitative pickups to be digitised over many thousands of turns, and then analysed off line. Measurements can again be taken with low or high intensity beams. The motion of short pulse, low intensity beams, occupying a fraction of the RF bucket, provides basic longitudinal parameters. Having quick convenient access to details of bunch formation during high intensity trapping may allow finer tuning and lower loss. An example of the development of longitudinal bunch shapes at trapping is shown in Figure 8. This shows the two bunches on an AC coupled capacitative monitor. Non trapped beam is visible emerging on the right hand side of the bunches.

Practical Considerations

Most measurements can be taken during 1 in 128 of the normal 50 Hz high intensity pulses. This amounts to near negligible <1% reduction in user beam current, and allows for extensive on-line experimentation during operational running. The extremely low power associated with low intensity diagnostic beams makes them non-destructive, and ideal for trouble-shooting. Many complicated experiments can be automated, making information easily available.

As well as giving basic lattice information, these measurements can serve as useful checks of hardware. Most of these measurements have been demonstrated with low intensity diagnostic beams, but may also be possible with high intensity beams, especially at higher energies where there is spare aperture for coherent motion. It is hoped a more detailed knowledge of the high intensity effects will result.

Figure 8
Longitudinal Profile during Trapping, starting at 0 ms.
Formation of Two High Intensity Bunches at 170 µA

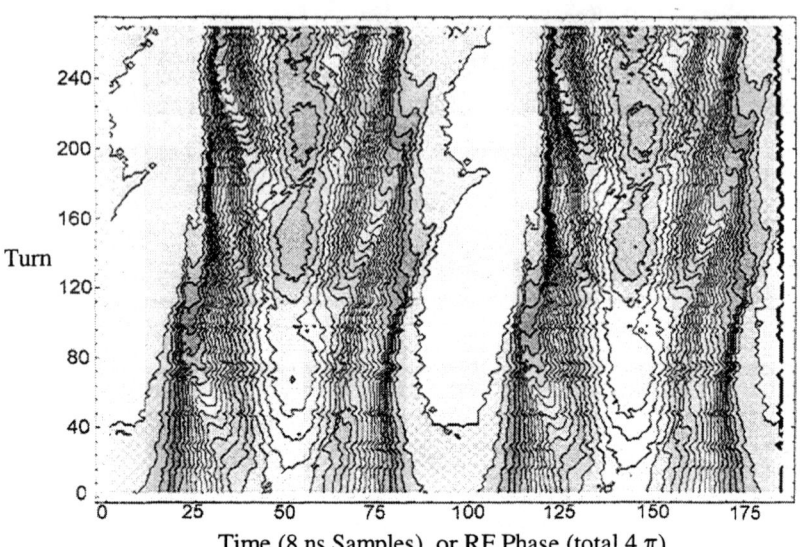

Time (8 ns Samples), or RF Phase (total 4 π)

7. ISIS Upgrades & Future Plans

Design and construction is presently underway for an RFQ to replace the 665 kV Cockcroft-Walton pre-injector, which has significantly affected performance in the past. A test stand is to be constructed to check the beam and operational characteristics of the RFQ system. Also in preparation is the dual harmonic RF upgrade for the Synchrotron, which is expected to increase currents to 300 µA levels. Experimental tests for such a system and construction of prototype higher harmonic cavities are well underway.

Detailed design studies are also in progress for major developments to the user facilities, which would make particular use of increased 300 µA currents. These are a second, lower repetition rate (10 Hz), neutron target and Sirius a large radioactive beams facility.

8. Acknowledgements

This paper reports the work of many members of the ISIS Facility. Particular thanks are due to G H Rees for many helpful comments on high intensity effects and A I Borden for information on diagnostics. Thanks are also due to D J Adams for production of Figure 8.

9. References

[1] B Boardman, Spallation Neutron Source: Description of Accelerator and Target, Rutherford Appleton Lab Report RL-82-006.
[2] S R Koscielniak, PhD Thesis, March 1987, RAL Report RALT 043
[3] C R Prior, Studies of Dual Acceleration in ISIS, Proceedings of ICANS XII, May 1993, Abingdon, UK, RAL Report RAL 94 025
[4] P J S Barratt, The ISIS Synchrotron RF System at High Intensity, Proceedings of ICANS XII, May 1993, Abingdon, UK, RAL Report RAL 94 025
[5] G H Rees, Maximum Space Charge Tune Shifts on ISIS, Internal RAL Report ISIS/ϕ/N1/92
[6] G H Rees, C R Prior, Image Effects on Crossing an Integer Resonance, Particle Accelerators, 1995, Vol.48, pp 251-257
[7] G H Rees, High Intensity Performance of the ISIS Synchrotron, Proceedings of the Workshop on Beam Instabilities in Storage Rings, 1994, Hefei, China
[8] M A Clarke Gayther, Global Beam Loss Monitoring Using Long Ionisation Chambers at ISIS, Proceedings of EPAC 94, 1994, London, UK
[9] C M Warsop, Low Intensity and Injection Studies on the ISIS Synchrotron, Proceedings of EPAC 94, 1994, London, UK

PSR Experience with Beam Losses, Instabilities and Space Charge Effects

Robert J. Macek

Los Alamos National Laboratory, Los Alamos, NM 87545

Abstract

Average current from the PSR has been limited to ~70 µA at 20 Hz by beam losses of 0.4 to 0.5 µA which arise from two principal causes, production of H^0 excited states and stored-beam scattering in the stripper foil. To reduce beam losses, an upgrade from the two-step H^0 injection to direct H^- injection is underway and will be completed in 1998. Peak intensity from the PSR is limited by a strong instability that available evidence indicates is the two-stream e-p instability. New evidence for the e-p hypothesis is presented. At operating intensities, the incoherent space charge tune shift depresses both horizontal and vertical tunes past the integer without additional beam loss although some intensity-dependent emittance growth is observed.

Introduction

The proton storage ring (PSR) at Los Alamos has operated successfully for over 10 years as an accumulator ring to drive the short-pulse-spallation-neutron source at the Lujan Center. The ring layout for 1997 operations is shown in Figure 1. Injection is a two step charge-exchange process; H^- is stripped to H^0 in a stripping magnet and drifts through a channel in the dipole iron to the stripper foil where it is converted to H^+.

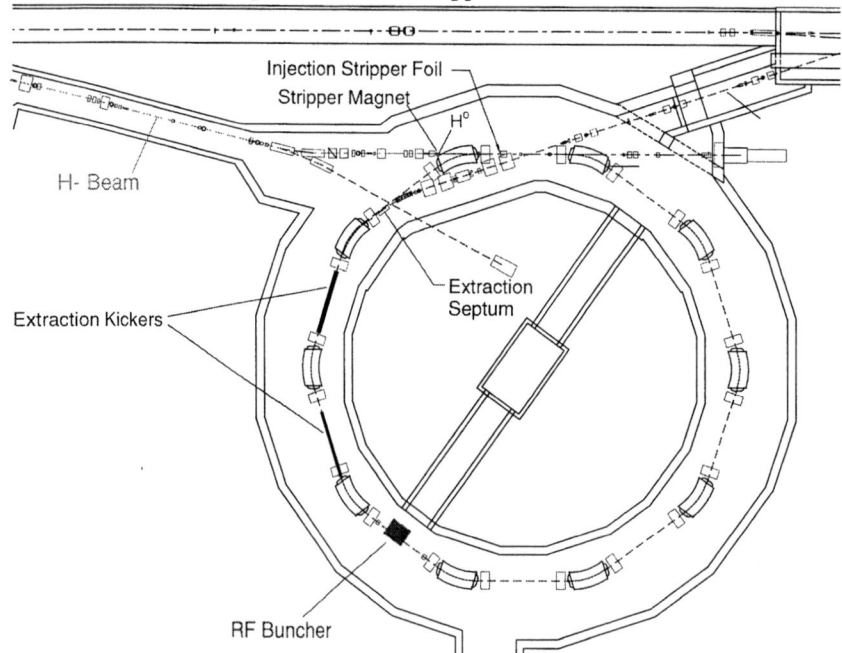

Figure 1. Layout of PSR for 1997 operations.

In this manner, about 1700 turns of beam are accumulated then extracted in a single turn by stripline kickers and transported to the neutron production target. A list of important parameters for PSR is given in Table I below.

Table I PSR Parameters 1997

Beam Kinetic Energy	T	797 MeV
Betatron tunes	v_x, v_y	3.16, 2.14
Inc. Tune Shifts @ 70 µA	$\Delta v_x, \Delta v_y$	-0.22, -0.185 calc
Chromaticities (normalized)	ξ_x	-1.22±.07 meas, -0.8 calc
	ξ_y	-1.14±.06 meas, -1.3 calc
Transition γ	γ_T	3.1
Phase Slip factor	η	-0.19
Max rf voltage	V_{rf}	12 kV
Synchrotron Tune (10kV)	v_0	0.00042
Buncher Harmonic, freq	h, f	1, 2.795 MHz
Mean Pipe Radius	b	0.05 m
Circumference	$C = 2\pi R$	90.2 m

Beam Losses

Average current from the PSR has been limited to 70 µA at 20 Hz by beam losses of 0.4 to 0.5 µA. They arise from two principal causes: 1) production of H^0 excited states that fieldstrip in the first downstream dipole and 2) nuclear plus large angle Coulomb scattering of stored beam in numerous traversals of the stripper foil. Beam losses are measured by a system of evenly spaced detectors placed around the ring on the tunnel walls opposite each dipole and half way in between. The signals are summed for a total loss signal as shown in Figure 2.

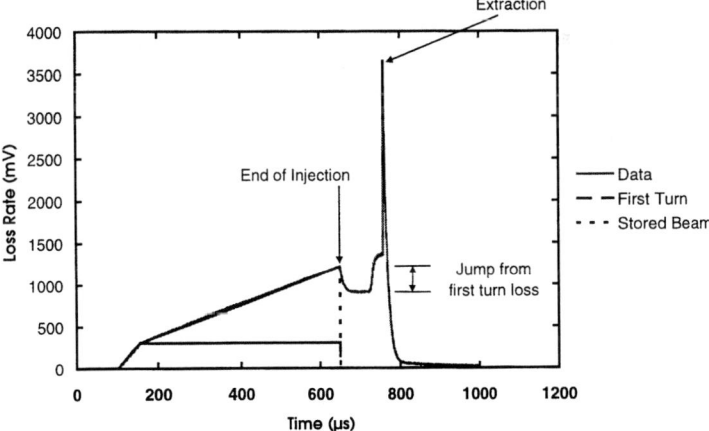

Figure 2. Loss monitor signal for 550 µs accumulation and 100 µs extraction delay.

By delaying extraction one can separate the "first turn losses" (0.2-0.3% from the field stripping of excited states) from the foil scattering losses (0.3-0.5%). The negative jump at the end of accumulation results from the cessation of excited state production. Likewise, the two components can be measured at each detector location around the ring. As expected, the first turn losses from the excited states are limited to the first quarter turn. Their level (0.2-0.3%) is in agreement with that estimated from sum of yields for n=3, 4 and 5 states, assuming that yields from an H^0 beam are one half those from an H^- beam [1],[2]. Component activation as shown in Figure 3 correlates quite well with observed loss signals.

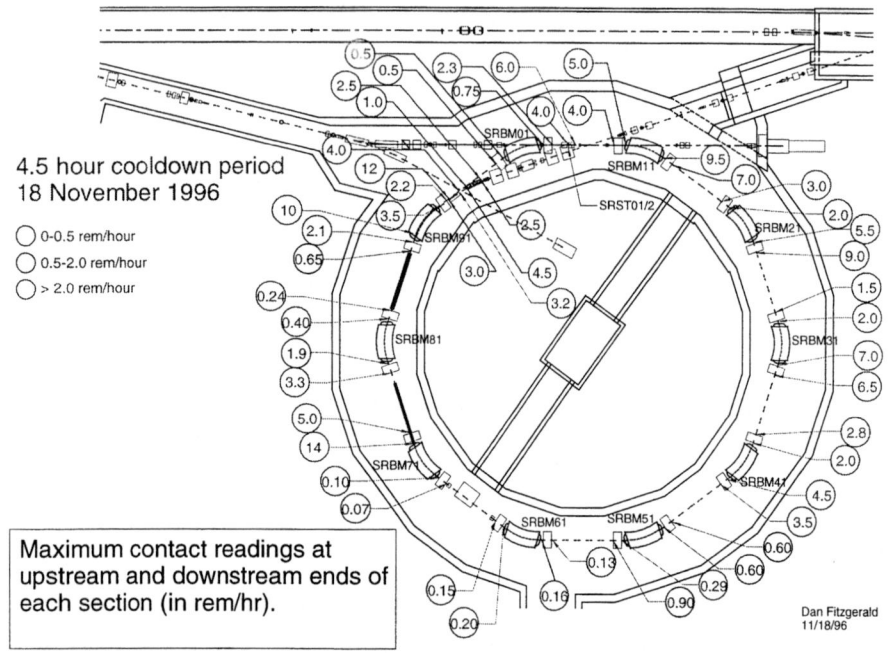

Figure 3. Late 1996 activation map.

Losses from foil scattering are large because the average proton traverses the stripper foil ~35% of the time for an average of 300 traversals (as determined from tracking simulations using ACCSIM). Reduction in the number of foil traversals is not possible with the H^0 injection system for several reasons. Fundamentally, one cannot manipulate or tailor the neutral beam to optimally match the injected beam to the ring acceptance. In addition, there is emittance growth by a factor of ~3 in the bend plane of the stripper magnet. This coupled with a large mismatch in the horizontal plane (see Figure 4 for the injection phase space) leads to a beam that fills the horizontal acceptance of the ring leaving no room for a horizontal offset to reduce foil traversals. In the vertical plane, the H^0 emittance is smaller and some offset is possible.

By implementing direct H^- injection, as shown in Figure 5, a factor of 10 reduction in foil traversals is possible. With H^- beam, emittance growth of the injected beam is avoided and the beam optics can be adjusted for minimum foil traversals. In addition, a time-programmed closed orbit bump is introduced in the vertical plane to move the stored beam off the foil. Part of the gain from reducing the foil traversals is given back in greater foil thickness (400 µg/cm^2 from 220 µg/cm^2) in a trade off between foil scattering and production of excited states. Consequently, the net reduction in the foil scattering loss rate is a factor of 5. Overall loss rates are thus reduced by a factor of about 4.5.

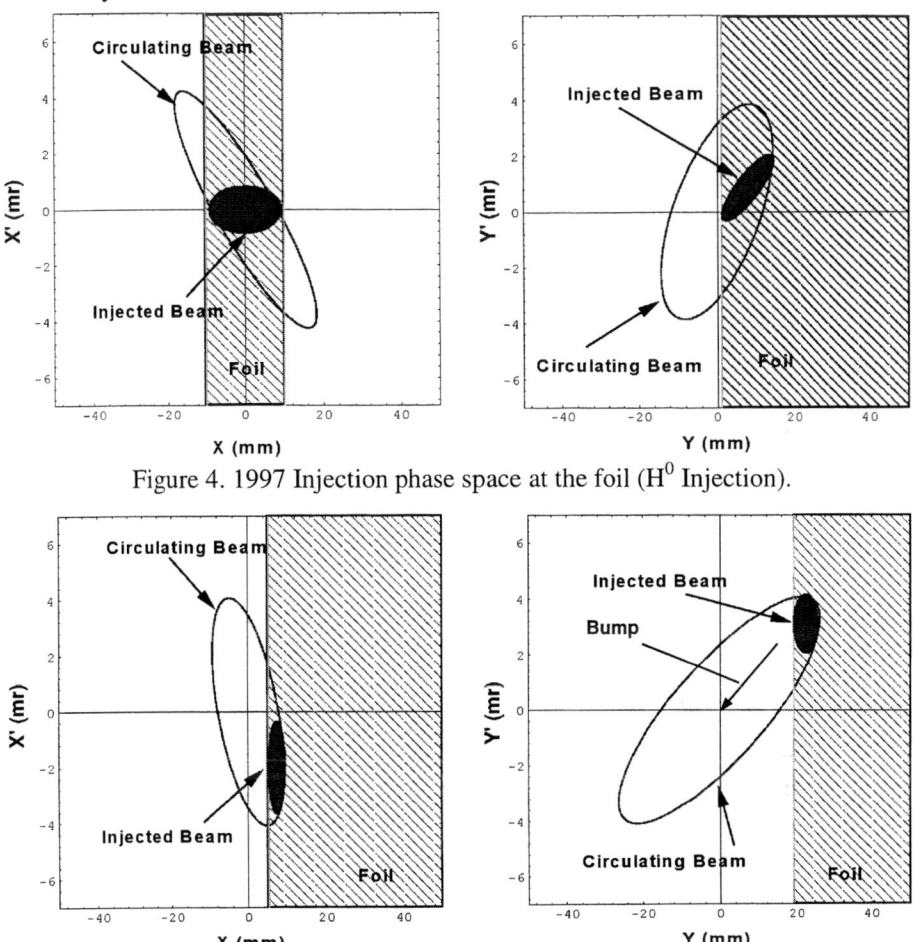

Figure 4. 1997 Injection phase space at the foil (H^0 Injection).

Figure 5. Injection phase space at the foil after the upgrade to direct H$^-$ injection.

Implementation of direct H$^-$ injection necessitated complete revision of the ring injection line and the injection region of the ring as shown in the layout of Figure 6 [3]. Skew quadrupoles were added in the injection line to remove the x-y coupling in the skewed beam line. The ring dipole down stream of the stripper foil was replaced with two large aperture C magnets to accommodate the closed orbit bump and to facilitate transport of the waste beams from the foil to a beam dump. The installation is well along and commissioning will begin in July of 1998.

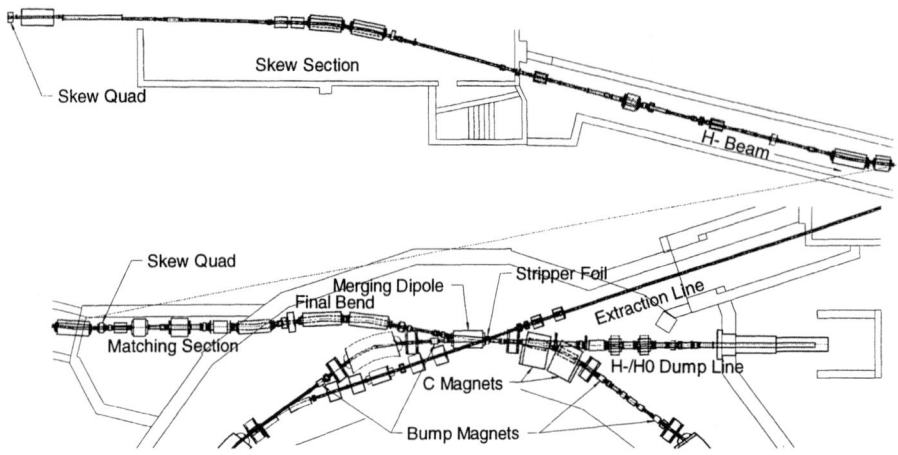

Figure 6. Layout of the direct H⁻ injection upgrade.

One of the many benefits of the injection upgrade is an increase in the size of the beam core thereby reducing the space charge tune shift. The change is shown in Figure 7 where the integral distributions of the Courant-Synder invariant for the two injection schemes are plotted. The data are from ACCSIM simulations.

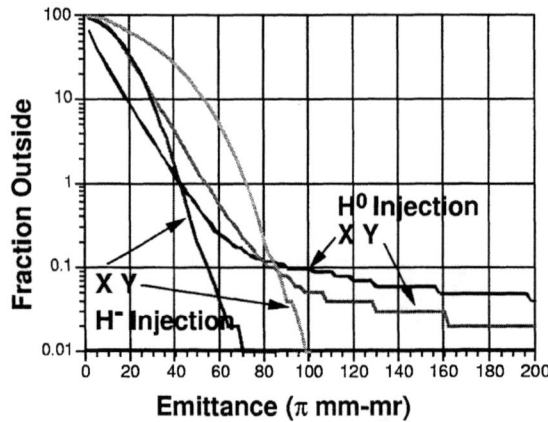

Figure 7. Integral distributions of the Courant-Synder invariant (Emittance).

Space Charge Effects

Space charge plays a role at PSR but is not a significant source of beam loss. At operating intensities the calculated maximum incoherent tunes for both x and y are depressed over the integer with no significant increase in the beam loss per proton. See the tune space diagram for PSR in Figure 8. In addition, intensity-dependent effects (presumably from space charge) have been observed but are also not a significant source of beam loss. These include intensity-dependent emittance growth and measurements of coherent tune shifts from exciting a coherent betatron oscillation with a small kick.

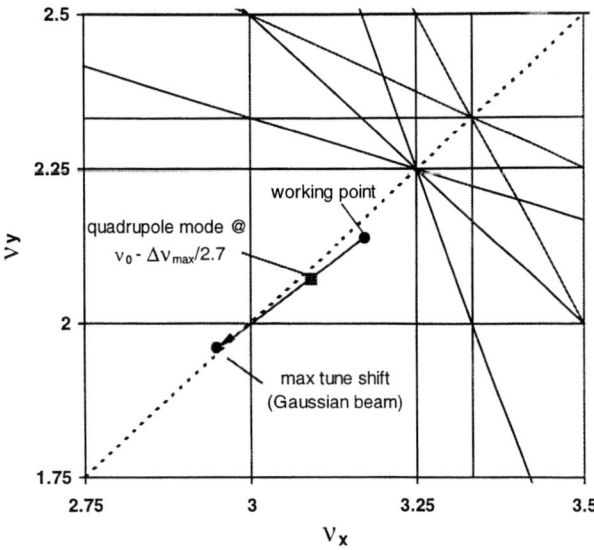

Figure 8. PSR Tune Space

The maximum incoherent tune shifts are calculated using the formulas (1) and (2) below for Gaussian beam profiles. Included are direct self-field terms plus image coefficients for centered beams. Other conditions are no neutralization and non-penetrating AC magnetic fields [4]:

$$(1) \quad \Delta v_x(\text{inc}) = -\frac{N \cdot r_p}{\pi \cdot \beta^2 \cdot \gamma} \left[\frac{\bar{\beta}_x(\text{ring})}{2 \cdot \gamma^2 \cdot B_f \cdot \sigma_x \cdot (\sigma_x + \sigma_y)} - \frac{\pi^2 \cdot \beta^2 \cdot C_2 \cdot \bar{\beta}_x(\text{bend})}{24 \cdot g^2} \right]$$

$$(2) \quad \Delta v_y(\text{inc}) = -\frac{N \cdot r_p}{\pi \cdot \beta^2 \cdot \gamma} \left[\frac{\bar{\beta}_y(\text{ring})}{2 \cdot \gamma^2 \cdot B_f \cdot \sigma_y \cdot (\sigma_x + \sigma_y)} + \frac{\pi^2 \cdot \beta^2 \cdot C_2 \cdot \bar{\beta}_y(\text{bend})}{24 \cdot g^2} \right]$$

$$(3) \quad \sigma_x^2 = \bar{\beta}_x(\text{ring}) \cdot \varepsilon_x(\text{rms}) + (\bar{\eta} \cdot \delta)^2, \quad \sigma_y^2 = \bar{\beta}_y(\text{ring}) \cdot \varepsilon_y(\text{rms})$$

In the above equations N is the number of protons, r_p the classical radius of the proton, B_f the longitudinal bunching factor, g the half height of the dipole gap, C_2 the fraction of the circumference occupied by dipole magnet iron, β and γ the usual relativistic factors, $\bar{\beta}_{x,y}$ the average beta functions around the ring or in the dipoles, σ the rms beam size for Gaussian profiles, ε the rms emittance, $\bar{\eta}$ the average momentum dispersion in the ring and δ the fractional momentum spread of the ring beam. For typical production beams we have $B_f = 0.359$, $\bar{\beta}_x(\text{ring}) = 7.61$, $\bar{\beta}_y(\text{ring}) = 8.81$, $C_2 = 0.283$, $\bar{\beta}_x(\text{bend}) = 6.79$, $\bar{\beta}_y(\text{bend}) = 9.04$, and $\bar{\eta} = 1.96$. Values for ε are obtained from profiles at the extraction wire scanners, ROWS01&2. Figure 8 shows two series of ROWS02 profiles. One series (left) is for the standard production beam and the other for a smaller initial beam size. In each series the intensity is changed in steps of a factor of 2 by injecting every pulse (count down, CD=1), or every 2nd pulse (CD=2), or every 4th pulse (CD=4). In this way, all machine parameters except intensity remain the same.

Using the data in the left series of Figure 9 (along with ROWS01 data not shown) in equations (1) and (2) provides maximum incoherent tunes shifts of $\Delta v_x = -0.22$ and $\Delta v_y = -0.185$ for an intensity of 70 µA @ 20 Hz. Tune shifts were also calculated using ACCSIM and yielded somewhat lower values of $\Delta v_x = -0.158$ and $\Delta v_y = -0.125$.

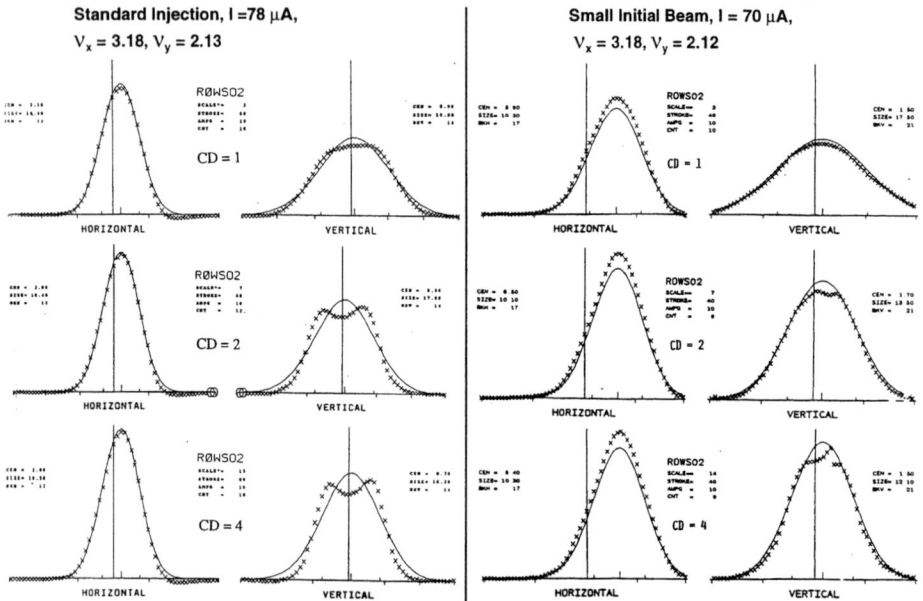

Figure 9. Beam profiles at extraction for various intensities.

Inspection of the data in Figure 9 shows emittance growth with increasing current especially in vertical plane. Curves of ε_y (rms for both series) plotted as a function of the circulating current, I_{circ}, in Figure 10 show this trend more clearly.

Coherent tune shifts can be accurately measured by observing the coherent betatron oscillation following a small kick. Measured tune shifts as a function of intensity are plotted in Figure 11 along with a linear fit to the data. The fit is good and the slope of -0.00143±0.00008 is very close to the predicted value of -0.00144/A. Coherent tune shifts are calculated using the following formulas where the h is the radius of the radius of the vacuum chamber. Other symbols have been defined earlier.

$$(4) \qquad \Delta v_x(\text{coh}) = -\frac{N \cdot r_p}{\pi \cdot \beta^2 \cdot \gamma} \left[\frac{\overline{\beta}_x(\text{ring})}{2 \cdot \gamma^2 \cdot B_f \cdot h^2} - \frac{\pi^2 \cdot \beta^2 \cdot C_2 \cdot \overline{\beta}_x(\text{bend})}{24 \cdot g^2} \right]$$

(5) $$\Delta v_y(\text{coh}) = -\frac{N \cdot r_p}{\pi \cdot \beta^2 \cdot \gamma}\left[\frac{\overline{\beta}_y(\text{ring})}{2 \cdot \gamma^2 \cdot B_f \cdot h^2} + \frac{\pi^2 \cdot \beta^2 \cdot C_2 \cdot \overline{\beta}_y(\text{bend})}{24 \cdot g^2}\right]$$

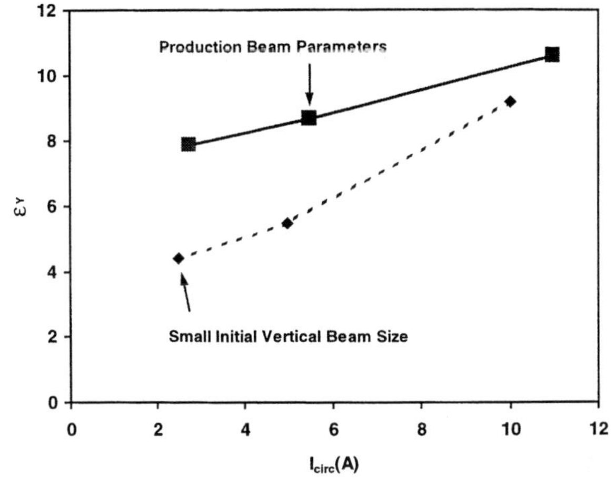

Figure 10. RMS Emittance plotted as a function of intensity (circulating current, I_{circ}).

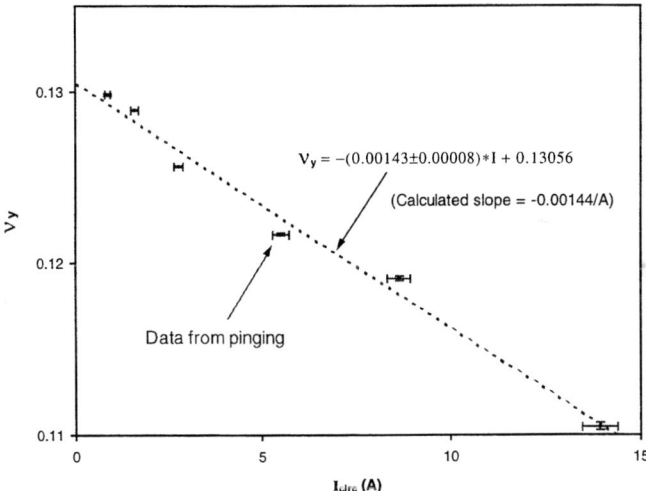

Figure 11. Coherent tune as a function of intensity.

Update on the PSR Instability

Peak intensity from PSR is limited by a strong instability that all available evidence indicates is the two-stream e-p instability. This was discussed extensively at the 1997 Sante Fe Workshop [5]. The report here will be an update on progress and new evidence obtained since the workshop. Several experimental studies were carried out in 1997 with the goal of insuring that the instability could be controlled to the level needed for 200 μA @ 30 Hz (the primary goal for the Short Pulse Spallation Source (SPSS) Enhancement project at PSR). The new studies included a systematic search for all the important control variables that affect the instability, further tests of the e-p hypothesis and, in collaboration with FNAL, tests of an inductive insert to passively compensate longitudinal space charge.

One of the most unambiguous predictions of the theory for e-p (Neuffer [6]) concerns the dependence of the frequency (of the unstable motion) on the threshold intensity, N, as shown in the formula in Figure 12. Other symbols are r_e, the classical radius of the electron, η_e, the neutralization of the proton beam, a, the half width of the beam, b, the half height and R, the mean radius of the ring. The data of Figure 12 shows that the central frequency increases by a factor of 1.4 to 1.5 for a factor of two increase in intensity which is in very good agreement with the expected factor of $\sqrt{2}$.

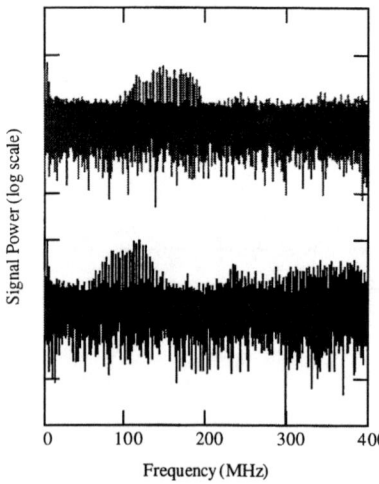

- The peak in the signal spectrum depends on the beam intensity.
- Top spectrum is twice the intensity of the bottom spectrum
- Beam conditions for the top and bottom spectra are the same except for the beam intensity and the buncher voltage.

$$f = \frac{1}{2\pi}\sqrt{\frac{2Nr_ec^2(1-\eta_e)}{\pi b(a+b)R}}$$

Figure 12. Frequency spectra from FFT of vertical motion at the onset of instability.

Neuffer's model for the e-p instability with bunched beam requires some method of trapping electrons during the passage of the ~100 ns gap. In reference [6] it was shown that trapping can occur if a small fraction of beam leaks into the gap. In order to test this idea, a controlled amount of beam was introduced in the gap by inhibiting chopping for a few turns of injection. The results are plotted in Figure 13 and show that more buncher rf voltage is needed to keep the beam stable as the beam in the gap is increased. This effect is equivalent to lowering the threshold for the instability (see Figure 16 for effect of buncher voltage on threshold). The test of adding beam in the gap was made for two different bunch widths, 220 and 150 ns, leading to two different gap widths, 140 ns and 210 ns respectively. The effect was seen in both cases.

Another expectation from the e-p model is that the threshold intensity should increase as the beam size is increased. Two factors contribute. The first factor is the dependence of the threshold condition on the bounce frequencies of the electrons and protons (which depend on beam size). This condition remains the same if the threshold intensity increases with beam size. The second factor is that the beam size affects the g parameter (see equation (6) below) in the longitudinal space charge term such that the force, which pushes beam into the gap, decreases with increasing beam size. Measurements of the effect of vertical beam size were made and the results are plotted in Figure 14. They show the trend expected from our model for the e-p instability thus adding to the evidence supporting the e-p hypothesis.

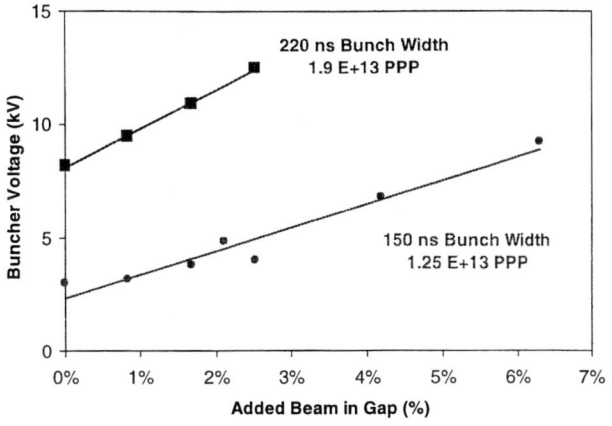

Figure 13. Effect on instability threshold from added beam in the gap.

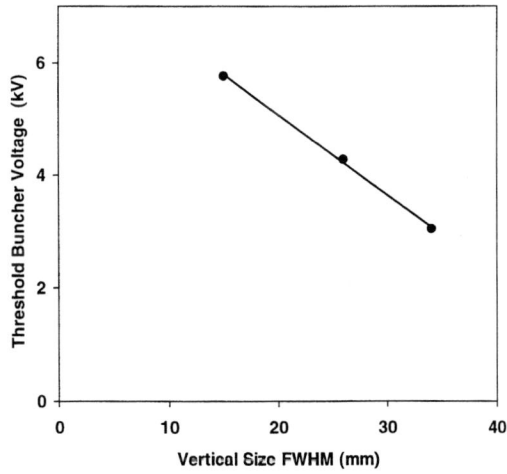

Figure 14. Effect of vertical beam size (emittance) for beam of 2.2×10^{13} ppp.

Tests of an Inductive Insert

One of the functions of the rf buncher in PSR is to overcome the longitudinal space charge force which tends to push beam into the gap and thereby lowers the e-p instability threshold. The net voltage per turn, V_s, from space charge self-voltage and inductive wall impedance (below transition) is given by:

$$\text{(6)} \quad V_s = \frac{\partial \lambda(s)}{\partial s} \left[\frac{g_0 Z_0}{2\beta\gamma^2} - \Omega_0 L \right] e\beta cR$$

where $\lambda(s)$ is the line density of charge, $g_0 = 1 + 2\ln(b/a)$, b the vacuum pipe radius, a the beam radius, $Z_0 = 377$ ohms, Ω_0 the angular frequency of revolution, L the wall inductance and R the mean radius of the ring. The idea of the inductive insert is to add sufficient ferrite to increase L until V_s vanishes.

Inductor tests at PSR addressed two main issues: a check on space charge compensation and a test of the effect on the PSR instability threshold. The check on space charge compensation was to measure bunch broadening for a narrow (50 ns) pulse and to compare to simulations as shown in Figure 15. The effect is not large but it is reproducible and agrees with the ACCSIM simulation. Note that the inductance with no bias was only 2/3 the value needed for full compensation while the value with bias on was about 1/6 of that needed for full compensation.

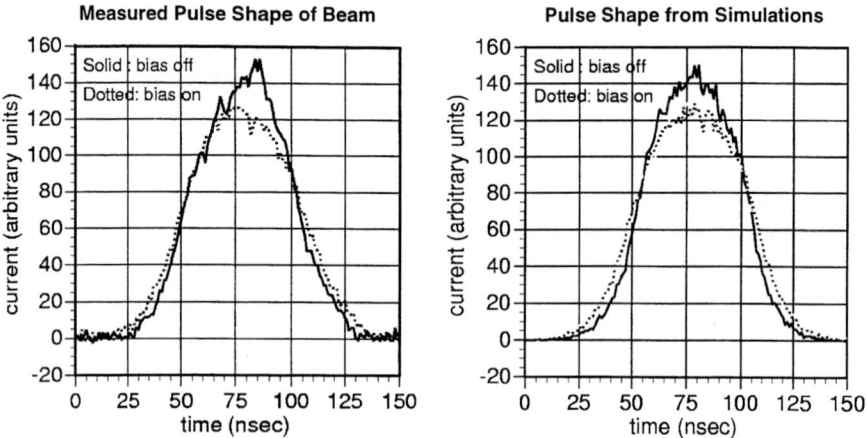

RF on @ 7.5 kV, 625 μs accumulation, 4x10^{12} protons, L ≃ 8 μH bias off, L ≃ 2 μH bias on (900A)

Figure 15. Test with 50 ns injected bunch length

The effect of the inductive insert on the e-p instability was studied by measuring the threshold intensity as a function of buncher rf voltage. The results are plotted in Figure 16 along with historical data and other 1997 data. The highest intensity point of the inductor data (triangle symbols) lies to the left of the historical trend line and indicates that less rf voltage was needed to keep the beam stable. However, it is not possible to draw a strong conclusion since other points, not involving the inductor, also lie to the left of the trend line. It is perhaps fair to say that the highest intensity point shows a two-standard deviation effect, which is encouraging but not conclusive or definitive.

Conclusions

Space charge effects are observed at PSR but to date beam losses are primarily due to stored beam scattering in the stripper foil and delayed stripping of excited states of H^0 produced in the stripping process. Space charge effects will likely increase in importance at PSR as beam current is pushed to 200 μA @ 30 Hz and beyond.

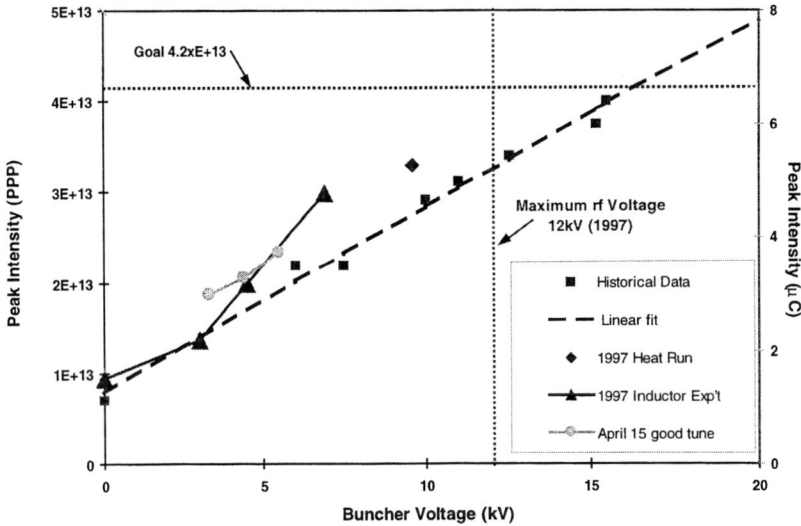

Figure 16. Threshold intensity as a function buncher rf voltage.

Evidence that the PSR instability is e-p continues to mount. One remaining concern is why efforts to collect or suppress the electrons were not effective in increasing the threshold. The possibility that many electrons are produced by beam induced multipactor needs further study.

Initial tests of an inductive insert at PSR were encouraging but not conclusive in the effect on the e-p instability. The inductor undoubtedly compensates longitudinal space charge and does not produce any unexpected adverse effects. Further study is warranted.

References

[1] R. Hutson and R. Macek, "First Turn Losses in the LANSCE Proton Storage Ring", Proceedings of the 1993 Particle Accelerator Conference, p. 363, 1993.

[2] M. Gulley et al, "Measurement of H^-, H^0 and H^+ yields produced by foil stripping of 800-MeV H^- ions", Phys. Rev. A53 (1996), p 3201-10.

[3] H. A. Thiessen et al, "Proton Storage Ring Upgrade to Direct H- Injection: Phase II Technical Design Report", LANL internal report, (1996).

[4] P. Bryant and K. Johnsen, "The Principles of Circular Accelerators and Storage Rings", (Cambridge University Press, 1993), Chapter 8, p 183.

[5] T.S. Wang and A. Jason, "Proceedings of the Santa Fe Workshop on Electrons Effects", March 5-7, 1997, LANL report LA-UR-98-1601.

[6] D. Neuffer et al, "Observations of a fast transverse instability in the PSR", NIM A321 (1992), p 1-12.

Performance and Measurements of the Fermilab Booster

M. Popovic and C. Ankenbrandt

Fermi National Accelerator Laboratory, Batavia, IL 60510[1]

Abstract. We will describe measurements of the beam in the Fermilab Booster during the first five milliseconds. Most of the particle losses in the Booster are over after the first few milliseconds. At high intensity of 4×10^{12} the transmission is 75%. Such high beam loss can be a limiting factor for future high repetition rate operation of the Booster. The evidence, although indirect, suggests that the losses are the result of incoherent space-charge effects at low energy.

INTRODUCTION

The Booster was designed [1] in the 1960's with a 200 MeV proton linac as injector. Since 1978 multi-turn H^- injection is used to build up the beam intensity in the Booster [2]. The critical parameters which limit the luminosity in the Tevatron Collider are the beam emittances and the number of particles at collisions. Thus in the Collider era Booster performance was no longer judged simply by how many protons can be accelerated but by what density of particles can be delivered. The suspected cause of the emittance growth during the first few milliseconds after injection in the Booster was the tune spread caused by space charge [3] and errors in the magnetic guide field. Thus in 1993 the Linac energy was upgraded to 400 MeV [4]. The result of the upgrade is seen all the way to the Tevatron, an increase in the phase space density at collisions, Fig.1. Improvement in beam phase space density delivered from the Booster has also relieved some of the aperture problems present in the Main Ring and allows us to accelerate and deliver a significantly larger quantity of protons onto the antiproton target. This has resulted in 50% increase in the antiproton production rate, thus increasing the luminosity in the Tevatron Collider.

---

[1] Work supported by the Universities Research Association, Inc., under contract DE-AC02-76CH00300 with the U. S. Department of Energy.

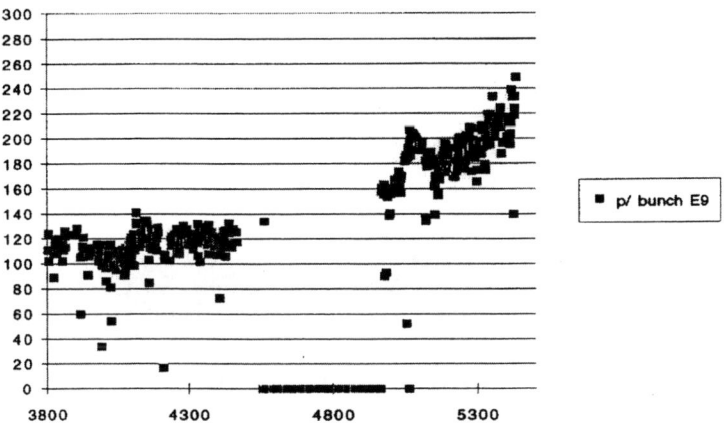

FIGURE 1. Proton intensity at collisions: The zero intensity points divide stores before and after the Linac upgrade.

OVERVIEW

The Fermilab Booster [5] is a rapid-cycling, 15 Hz, alternating-gradient synchrotron with average orbit radius of 75.47 meters. It accelerates protons from 400 MeV, the kinetic energy of the Linac beam, to 8 GeV, the nominal injection kinetic energy of the Main Ring/Main Injector. The lattice consists of 96 combined-function magnets in 24 periods. The nominal horizontal and vertical tunes are 6.7 and 6.8. The revolution time at injection is 2.2 μs. The linac delivers peak-current of 45 mA and the transfer line to the Booster usually runs with 98% transmission efficiency. Usually up to ten turns of H^- beam is injected. During injection a pulsed orbit bump magnet system is used to superimpose the trajectories of circulating and injected beam. The beam is accelerated with 17 rf cavities, transition gamma of the ring is $\gamma_t = 5.445$ and harmonic number of the ring is $h = 84$.

MEASUREMENTS

Many effects determine the behavior of the beam in the Booster at injection: the linac beam energy, the energy spread, the beam transverse emittances, and injection mismatch. In addition, the Booster has no "porch" at injection, the main magnetic field oscillates sinusoidally at 15 Hz and DC beam must be "adiabatically" captured in rf buckets fairly quickly. All these effects are mixed with effects coming from nonlinear field errors and significant remanent sextupole field at injection. In order to determine the relative contribution of all of these effects on the beam loss we have measured and vary most of these parameters one at a time around their

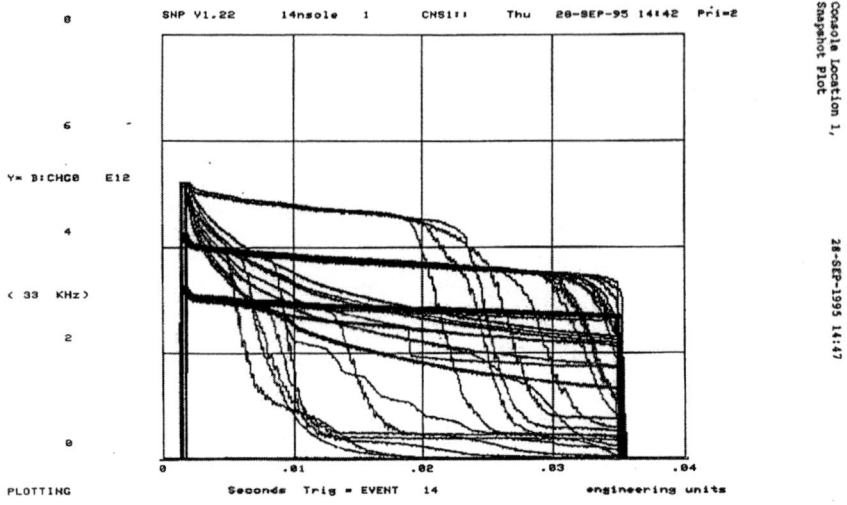

FIGURE 2. Booster DC at 400 MeV, RF on. More than 5.3E12 protons for 20 msec.

design/nominal values.

Injection Energy and Energy Spread

The Linac output energy is measured in two different ways, using a spectrometer magnet at the end of a diagnostic line and Phase-Scan method [6] which fixes the phase and amplitude of the accelerating modules to their design values. The output energy is 401.5 MeV with error less than ±1.8 MeV. The absolute value of the output energy of the linac is not a critical parameter for Booster operations because B_{min}, the minimal value of magnetic field, can be easily adjusted. The critical parameters are energy variations during the linac pulse and from pulse to pulse. Energy variations during the linac pulse are monitored in the diagnostic line after a 40 degree bend. The signal from a beam position monitor is digitized using 2 MHz quick digitizer and displayed using linac control system. Variation of the position of the beam from beginning to end as well from pulse to pulse translates to energy variation of $\frac{\Delta T}{T} = 0.2\%$. The stability of the Linac beam energy from pulse to pulse is also monitored using the phase of the beam-induced rf signal at a strip line detector 10-meters upstream of the injection to the Booster, right before the Debuncher. The role of the Debuncher is to minimize energy spread of the beam coming from the Linac and to correct for any energy variation from head to tail and from pulse to pulse of the injected beam. The phase and amplitude of the Debuncher are adjusted with a feed forward system to be flat during beam time. To measure the energy spread inside the 200-MHz bunches we have set the Booster to DC mode keeping it at 400 MeV and turning off all rf cavities. By injecting ∼ 90% of a single full turn into the Booster we were able to measure turn

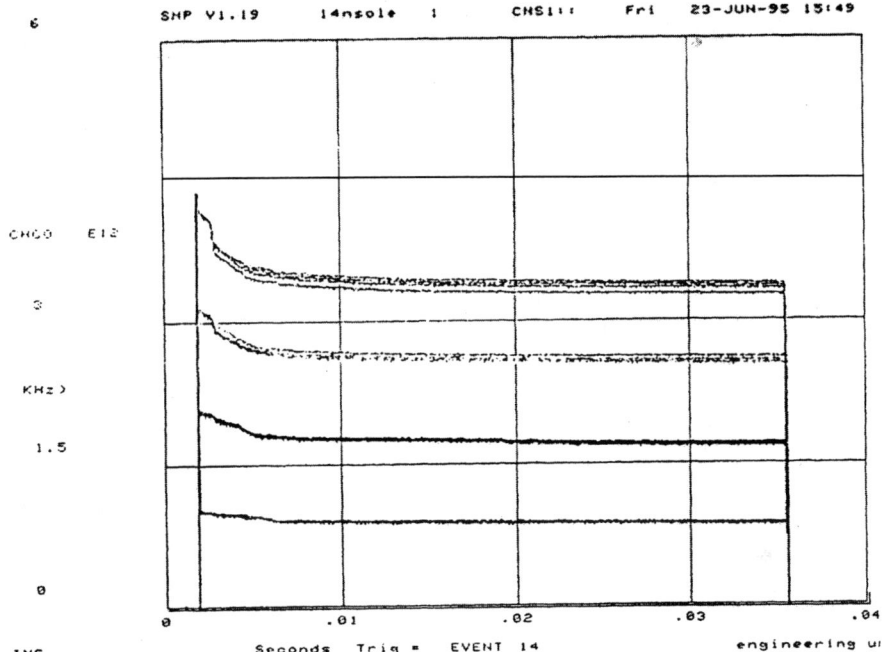

FIGURE 3. Charge vs. time(msec) for whole Booster cycle, 2, 4, 6, and 8 turns injected. There are two type of losses clearly visible in the first 5 msecs.

by turn energy spread of any particular bunch. We found that the Debuncher can be adjusted to keep total energy spread to be as low as $\frac{\Delta T}{T} < 3.0 \times 10^{-3}$ for 95% of the beam. We have measured energy spread only for the first few turns. Injected beam normally passes through the stripping Carbon foil about ten to twenty times. We have estimated that injected protons lose ~ 60 eV per foil passage and that the energy spread introduced by the foil in normal operations is negligible. Energy spread due to multiple scattering has not been measured.

Injected Transverse Emittances

All the measurements and simulations of the beam coming out from the Linac indicate that transverse emittances are less than 6 π mm-mrad. Here and throughout this paper, the normalized emittances containing 95% of the beam are quoted and units of mm-mrad are used. A Gaussian transverse distribution of rms width σ, observed at a location where the dispersion is zero and the lattice amplitude function is β_L corresponds to a normalized, 95%, emittance ϵ_N given by

$$\epsilon_N = \frac{6\pi\beta\gamma\sigma^2}{\beta_L}.$$

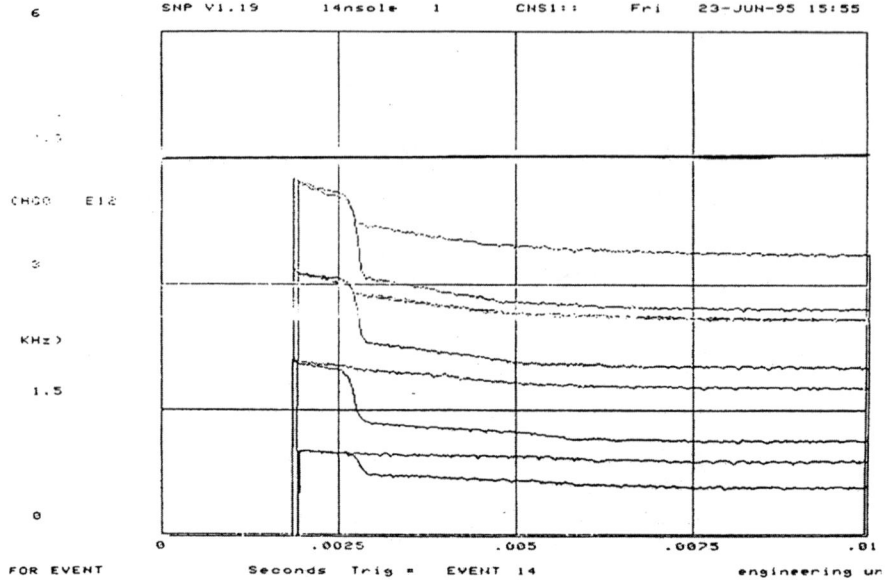

FIGURE 4. Beam intensity in the first 8 msecs for 2, 4, 6, and 8 turns injected with Debuncher on and off.

At the end of the Linac, emittances are measured using three wires with no focusing element in between. In the transport line between Linac and Booster, beam profiles are measured using multiwires for different settings of the quadrupoles upstream of each wire. The emittances are extracted using Trace3d code.

At injection the emittance growth due to the Coulomb scattering on the $\frac{200\mu g}{cm^2}$ Carbon stripping foil can be estimated from multiple-scattering theory. The calculated result is an increase of normalized emittance of 0.03π in horizontal plane and 0.11π in vertical plane per foil crossing, assuming no coupling during injection. This effect was not measured; transverse mismatch between injected beam and machine lattice hides the effect of the foil.

RF Capture Process

If the Booster accelerating cavities are all turned on at a sufficiently low gap voltage to cause good adiabatic capture, it is known that some of the cavities will suffer from electron multipactoring, subsequently inhibiting further increase in voltage. To avoid this all the cavities are turned on before injection but with pairs of cavities out of phase ("paraphased") so that the net accelerating voltage is small. When voltage is required the cavities are brought into phase by a "paraphase program" which allows optimum capture. It is apparent that this rf turn-on procedure can be done in a large variety of ways, and the manner in which the beam responds to these variations is not always obvious. Among the rapidly variable parameters that

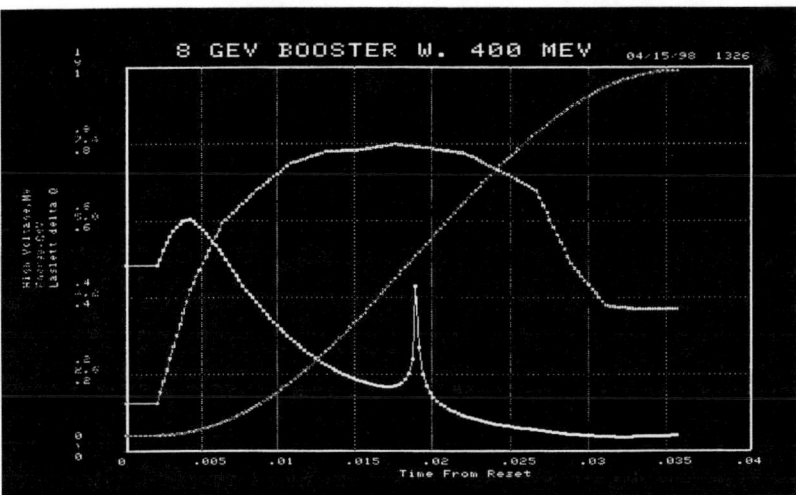

FIGURE 5. The Booster rf voltage, energy and Laslett tune shift. Note that the tune shift is larger than the injection value for the first 5 msec of beam.

affect capture are; injection time, minimum value of magnetic field, amplitude and phase of Debuncher rf voltage, the initial value of the Booster rf frequency program, the initial value of the rf amplitude program and the parameters of the paraphase program. The paraphase program was derived from the longitudinal tracking program ESME which includes longitudinal space charge effects. The starting time and the level of the starting voltage were "tuned" for best capture. Figure 3. shows the beam intensity in the Booster in the first 8 msecs, for 2, 4, 6, and 8 turns injected, with the Debuncher on and off. The first fast loss is the beam not captured in the rf bucket. This type of beam loss can be avoided with adjustment of the Debuncher. In operations, the starting time and the level of the starting voltage are "tuned" for best transmission and optimal loss pattern.

Slow Loss

A second type of losses clearly visible in Figures 2 and 3, which we call slow losses, are always present and "impossible" to tune out. We believe that these losses are related to the Laslett tune shift of the beam. Figure 5. [7] shows Booster rf voltage, energy and Laslett tune shift during the Booster cycle. The Laslett tune shift is calculated according to the formula

$$\Delta \nu_{sc} = \frac{3 r_0 N_{tot}}{2 B \beta \gamma^2 \epsilon_N}$$

where r_0 is the proton radius, N_{tot} total number of protons in the ring, ϵ_N normalized 95% emittance, and B the ratio of the average to the peak circulating current. The bunching factor, B, is calculated under the assumption that the bunch shape is Gaussian. Figure 5. shows that the Laslett tune shift stays above the tune shift at injection for more than 5 msecs after injection, about the time that the slow loss

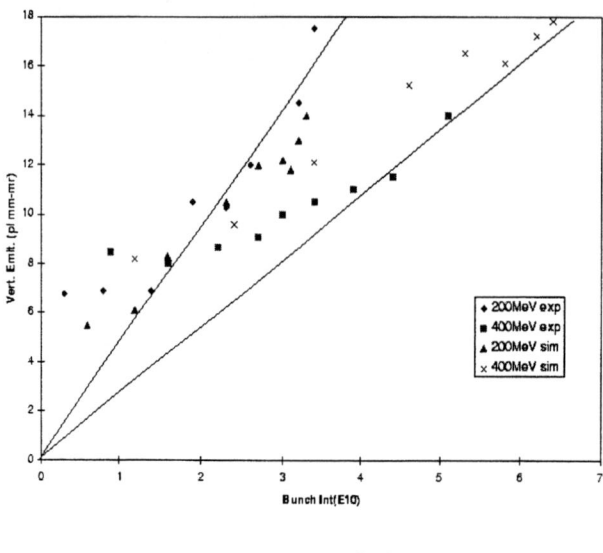

FIGURE 6. Vertical Emittance as function of bunch intensity

pattern is present. In support of this interpretation we present measurement and simulation of vertical emittances for 200 MeV and 400 MeV injection energy to the Booster. The simulation results are taken from Steven Stahl's Ph.D. thesis [8] and include effect of gradient errors, steering errors, chromaticity and space charge.

REFERENCES

1. National Accelerator Laboratory, Design Report, Second Printing, July 1968.
2. C. Hojvat, et al. Particle Accelerator Conf. 1979, IEEE Trans. Nucl. Sci. 26 (1979) pp4009-4011.
3. C. Ankenbrandt and S. Holmes, 1987 IEEE Particle Accelerator Conference Proceedings Vol. 2, pp1066.
4. Fermilab Linac Upgrade, Conceptual Design, Revision 4A, Nov. 1989.
5. E. Hubbard, et al. "Booster Synchrotron", Fermilab Technical Memo, TM-405, (1973).
6. T. Owens, et al. Particle Accelerators, 1994, Vol. 48, pp. 169-179.
7. Console program T126, written by R. Johnson.
8. Steven M. Stahl, "Beam Dynamics in the Fermilab Booster in the Presence of Space Charge". Ph.D. thesis, Northwestern University, Evanston, 1991.

High Intensity Performance and Upgrades at the Brookhaven AGS*

Thomas Roser
AGS Department, Brookhaven National Laboratory

Recent AGS High Intensity Performance

Fig. 1 shows the present layout of the AGS-RHIC accelerator complex. The high intensity proton beam of the AGS is used both for the slow-extracted-beam (SEB) area with many target station to produce secondary beams and the fast-extracted-beam (FEB) line used for the production of muons for the g-2 experiment and for high intensity target testing for the spallation neutron sources and muon production targets for the muon collider. The same FEB line will also be used for the transfer of beam to RHIC.

The proton beam intensity in the AGS has increased steadily over the 35 year existence of the AGS, but the most dramatic increase occurred over the last couple of years with the addition of the new AGS Booster[1]. In Fig. 2 the history of the AGS intensity improvements is shown and the major upgrades are indicated. The AGS Booster has one quarter the circumference of the AGS and therefore allows four Booster beam pulses to be stacked in the AGS at an injection energy of $1.5 - 1.9\,GeV$. At this increased energy, space charge forces are much reduced and this in turn allows for the dramatic increase in the AGS beam intensity.

The 200 MeV LINAC is being used both for the injection into the Booster as well as an isotope production facility. A recent upgrade of the LINAC rf system made it possible to operated at an average H$^-$ current of 150 μA and a maximum of 12×10^{13} H$^-$ per 500 μs LINAC pulse for the isotope production target. Typical beam currents during the 500 μs pulse are about 80 mA at the source, 60 mA after the 750 keV RFQ, 38 mA after the first LINAC tank (10 MeV), and 37 mA at end of the LINAC at 200 MeV. The normalized beam emittance is about $2\,\pi\,mm\,mrad$ for 95 % of the beam and the beam energy spread is about $\pm 1.2\,MeV$. A magnetic fast chopper installed at 750 keV allows the shaping of the beam injected into the Booster to avoid excessive beam loss.

---

*Work performed under the auspices of the U.S. Department of Energy

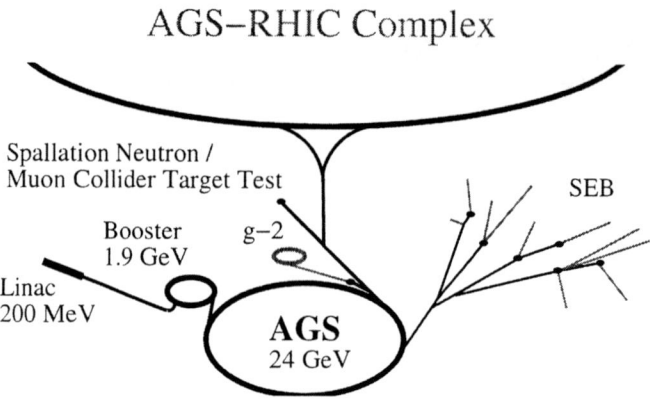

Figure 1: The AGS-RHIC accelerator complex.

The achieved beam intensity in the Booster surpassed the design goal of 1.5×10^{13} protons per pulse and reached a peak value of 2.3×10^{13} protons per pulse. This was achieved by very carefully correcting all the important nonlinear orbit resonances especially at the injection energy of 200 MeV and by using the extra set of rf cavities that were installed for heavy ion operation as a second harmonic rf system. The second harmonic rf system allows for the creation of a flattened rf bucket which gives longer bunches with lower space charge forces. The fundamental rf system operated with 90 kV and the secondary harmonic with 30 kV. The typical bunch area was about 1.5 eVs. Even with the second harmonic rf system the incoherent space charge tune shift can reach one unit right at injection (3×10^{13} protons, norm. 95 % emittance: 50 π mm $mrad$, bunching factor: 0.5). Of course, such a large tune shift is not sustainable, but the beam emittance growth and beam loss can be minimized by accelerating rapidly during and after injection. Best conditions are achieved by ramping the main field during injection with 3 T/s increasing to 9 T/s after about 10 ms. The quite large non-linear fields from eddy currents in the Iconel vacuum chamber of the Booster are passively corrected using correction windings on the vacuum chamber that are driven by backleg windings[2].

The AGS itself also had to be upgraded to be able to cope with the higher beam intensity. During beam injection from the Booster, which cycles with a repetition rate of 7.5 Hz, the AGS needs to store the already transferred beam bunches for about 0.4 seconds. During this time the beam is exposed to the strong image forces from the vacuum chamber which causes beam loss from resistive wall coupled bunch beam instabilities within as short a time as a few hundred revolutions. A very powerful feedback system was installed that senses any transverse movement

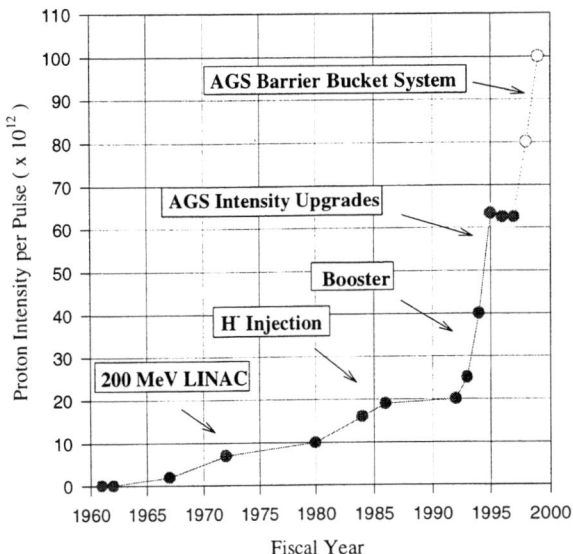

Figure 2: The history of the evolution of the proton beam intensity in the Brookhaven AGS.

of the beam and compensates with a correcting kick. This transverse damper can deliver ± 160 V to a pair of 50 Ω, one-meter-long strip-lines. A recursive digital notch filter is used in the feed-back circuit to allow for accurate determination of the average beam position and increased sensitivity to the unstable coherent beam motion. This filter design is particular important for the betatron tune setting of about 8.9 which is required to avoid non-linear octupole stopband resonance at 8.75. With an incoherent tune shift at the AGS injection energy of 0.1 to 0.2 it is still necessary, however, to correct the octupole stopband resonances to avoid excessive beam loss.

To reduce the space charge forces further the beam bunches in the AGS are lengthened by purposely mismatching the bunch-to-bucket transfer from the Booster and then smooth the bunch distribution using a high frequency 100 MHz dilution cavity. The resulting reduction of the peak current helps both with coupled bunch instabilities and stopband beam losses.

During acceleration the AGS beam has to pass through the transition energy after which the revolution time of higher energy protons becomes longer than for the lower energy protons. This potentially unstable point during the acceleration cycle was crossed very quickly with a new powerful transition energy jump system with only minimal losses even at the highest intensities. The large lattice distortions introduced by the jump system prior to the transition crossing severely limits the available aperture of the AGS in particular for momentum spread. Efforts to correct the distortions using sextupoles have been partially successful[3]. After the transition energy, a very rapid, high frequency instability developed which could only be avoided by purposely further increasing the bunch length using again the high frequency dilution cavity.

The peak beam intensity reached at the AGS extraction energy of 24 GeV was 6.3×10^{13} protons per pulse also exceeding the design goal for this latest round of intensity upgrades. It also represents a world record beam intensity for a proton synchrotron. With a 1.6 second slow-extracted beam spill the average extracted beam current was about 3 μA. This level of performance was reached quite consistently over the last few years and during a typical 20 week run a total of 1×10^{20} protons are accelerated in the AGS to the extraction energy of 24 GeV.

At maximum beam intensity about 30 percent of the beam is lost at Booster injection (200 MeV), 25 percent during the transfer from Booster to AGS (1.5 GeV), which includes losses during the 0.4 second storage time in the AGS, and about 3 percent is lost at transition (8 GeV). Although activation levels are quite high all machines can still be manually maintained and repaired in a safe manner.

Possible Future AGS Intensity Upgrades

Currently the number of Booster beam pulses that can be accumulated in the AGS is limited to four by the fact that the circumference of the AGS is four times the circumference of the Booster. This limits the maximum beam intensity in the AGS to four times the maximum Booster intensity which itself is limited to at most 2.5×10^{13} protons per pulse by the space charge forces at Booster injection. To overcome this limitation some sort of stacking will have to be used in the AGS. The most promising scheme is stacking in the time domain. To accomplish this a cavity that produces isolated rf buckets can be used to maintain a partially debunched beam in the AGS and still leave an empty gap for filling in additional Booster beam pulses. The stacking scheme is illustrated in Fig. 3. It makes use of two isolated rf buckets to control the width of this gap. Isolated bucket cavities, also called Barrier Bucket cavities, have been used elsewhere[4]. However, for this stacking scheme, a high rf voltage will be needed to contain the large bunch area of the high intensity beam. An additional important advantage of this scheme is that while the beam is partially debunched in the AGS the beam density and therefore space charge forces are reduce by up to a factor of two. A successful test of this scheme has recently been completed[5] and two 40 kV Barrier cavities are being installed in the AGS with the aim of accumulating six Booster beam pulses in the AGS to reach an intensity of about 1×10^{14}.

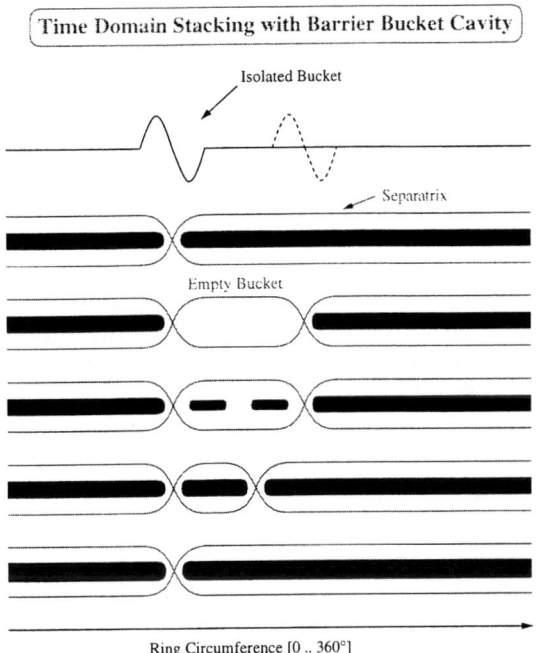

Figure 3: Time domain stacking scheme using a barrier bucket cavity. The evolution of the longitudinal beam structure during the stacking process are shown from top to bottom.

For further increases in the intensity the space charge forces at Booster injection represent the main limitation. This could be overcome by an energy upgrade of the LINAC to about 600 MeV by replacing part of the present 200 MHz cavities with higher gradient 400 MHz cavities driven by Klystrons. At 600 MeV the space charge limit at Booster injection would be 5×10^{13} protons per pulse or 2×10^{14} protons in the AGS for 4 cycles per AGS cycle.

As more Booster beam pulses are accumulated in the AGS the reduction in the overall duty cycle becomes more significant. For fast-extracted beam operation (FEB) the accumulation of four Booster pulses contributes already significantly to the overall cycle time. With the addition of a $2\,GeV$ accumulator ring in the AGS tunnel this overhead time could be completely avoided. Such a ring could be build rather inexpensively using low field magnets. The maximum repetition rate of the Linac and Booster is 10 Hz. Since the circumference of the AGS is four times that of the Booster a repetition rate of 2.5 Hz would maintain a throughput of 80 μA through the whole accelerator chain. Such an increase of the AGS repetition rate by a factor of 2.5 could be achieved by an upgrade of the AGS main magnet power supply only. The resulting beam power of 2 MW at 25 GeV corresponds to the required proton driver performance needed for a demonstration muon collider project [6]. The upgrades to the AGS complex are summarized in Fig. 4.

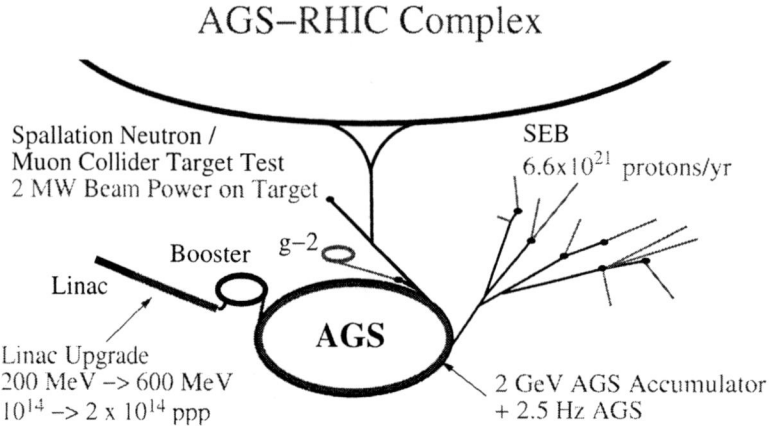

Figure 4: Summary of intensity upgrades for the AGS.

References

[1] M. Blaskiewicz et al., High Intensity Proton Operations at Brookhaven, 1995 Particle Accel. Conf. Dallas, Texas, May 1995, p. 383.

[2] G.T. Danby and J.W. Jackson, Part. Accel. **27** (1990) 33

[3] W.K. van Asselt et al., The Transition Jump System for the AGS, 1995 Particle Accel. Conf., Dallas, Texas, May 1995, p. 3022.

[4] J.E. Griffin et al., IEEE Trans. on Nucl. Sc. Vol. NS-30, No. 4, (1983) 3502

[5] J.M. Brennan and M.M. Blaskiewicz, these proceedings.

[6] R. Palmer, Progress on $\mu^+\mu^-$ colliders, Proc. of the 1997 Particle Accel. Conf., Vancouver, B.C., Canada, May 1997.

PROGRESS REPORT ON THE EUROPEAN SPALLATION SOURCE, MAY 1998

C.R. Prior
Rutherford Appleton Laboratory, Chilton, Didcot, U.K.

I INTRODUCTION

The European Spallation Source (ESS) is a 5 MW pulsed neutron source, study of which reflects growing demand from the world's leading neutron physicists for new means of exploring a diverse range of fields, from Polymers and Soft Matter to Fundamental Physics and Engineering. The design of the accelerator is the culmination of work carried out over a number of years by a group of European countries, and has benefited from advice from experts overseas. Phase 1 of the study was completed in 1996 and a report submitted to the European Commission [1]. The scientific case is accessible on the worldwide web at http://www.isis.rl.ac.uk/ESS and the report is also available in compact disc form. A summary of the accelerator design is given in [2]. An R&D programme is planned for the next phase and the various institutions are in the process of determining the level of their future commitment. Pending these decisions, work is being carried out on the ion source, the RF quadrupole linacs, improving the ring injection scheme and on beam diagnostic techniques. Some consideration has also been given to an alternative layout of the linac, which includes design of a 20 MeV funnelling system and a 2.5 MeV chopper line. More recently, ideas from Suzuki [3] have prompted investigation of a new ring in which the current method of charge-exchange injection using a graphite foil is replaced by stripping using a resonant laser optical system.

In this paper, after a brief description of the main design as in the ESS report, we cover the new developments and give the current position with regard to the R&D phase and likely progress in the future.

II THE ESS REFERENCE DESIGN

The ESS is a 5 MW pulsed neutron source with a repetition rate of 50 Hz and a pulse length of 1 μsec. Two targets are envisaged, one at 5 MW and 50 Hz for high intensity applications, and a second, using every fifth pulse, at 1 MW and 10 Hz for high resolution and long wavelength instruments. The reference design is dictated by requirements of operational reliability, low beam loss and relatively low cost.

It has the lowest ring space charge levels of the options considered and offers the greatest potential for development. Here we concentrate only on the accelerator aspects.

The layout of the system, depicted in Figure 1, comprises a 1.334 GeV linac producing pulses of H$^-$ ions at a mean current of 65 mA, which are transported and injected sequentially, via charge exchange foil stripping over 1000 turns, into two compressor rings. The full injection period is 1.2 msec and each ring accumulates 2.34×10^{14} protons per cycle in a 400 nsec bunch. Extraction is carried out successively to provide pulses of $\sim 1 \mu$sec giving 100 kJ per pulse at 50 Hz. The targets chosen for the reference design are liquid mercury, but solid targets are still being considered.

At the low energy end of the linac, the design requirements are for H$^-$ sources of the Penning or volume type producing 70 mA pulses with normalised rms emittances $0.1\,(\pi)\,\mu$rad m. These need to be matched into the first of two RF quadrupoles at 50 keV. The RFQs operate at 175 MHz and increase the beam energy to 5 MeV. Between the RFQs is a 2 MeV fast chopper to meet the demands of the compressor rings, chopping the pulses with a 60% duty factor at the ring revolution frequency (1.67 MHz). Two arms of this structure merge at a funnel, giving 107 mA of beam at 5 MeV. The proposed funnel is effectively a two-beam RFQ and a resonator-driven deflector. Beam is then accelerated, first to 70 MeV in a 350 MHz drift tube linac, and then to the design energy of 1.334 GeV by a cavity coupled linac at 700 MHz, 663 m long (Figure 2).

The two, vertically stacked, accumulator rings are designed for extremely low loss levels. With transverse and momentum collimators, the transfer line from the linac is designed to ensure the beam meets the precise requirements for ring injection. Immediately after the linac the momentum is ramped for horizontal phase space painting. A bunch rotation cavity reduces the momentum spread of the partly debunched beam by a factor 3. The beam then passes through a 180° achromatic bending section with mean radius 42.5 m. This consists of combined function magnets with low fields to avoid pre-stripping the H$^-$ beam and provides a high normalised dispersion for momentum collimation. Stripping foils at various points in the achromat are used for the collimation. Subsequent transport to the rings, with further debunching, is via two matching sections separated vertically by 2 m. The lower and upper accumulators are filled sequentially, with switching between the two carried out by a pulsed dipole. Dispersion in the transport line is zero at the stripping foils and betatron waists are created with $\beta_h = 2.4$, $\beta_v = 1.8$. For the ring, the values are much higher, at $\beta_h = 7.9$, $\beta_v = 4.9$ and a normalised dispersion $\alpha_p/\sqrt{\beta_h} = 2.0\,\text{m}^{1/2}$. The mismatching optimises the injection scheme and is used to reduce subsequent traversals of the foil by re-circulating protons and so minimise foil heating.

The arrangement of the rings is shown in Figure 3. They have a mean radius of 26 m and three superperiods each of 15 cells forming first order achromats around long low field dipoles, one of which is used for injection. The design bending field of 0.177 T is chosen to give negligible pre-stripping of H$^-$ ions ahead of the

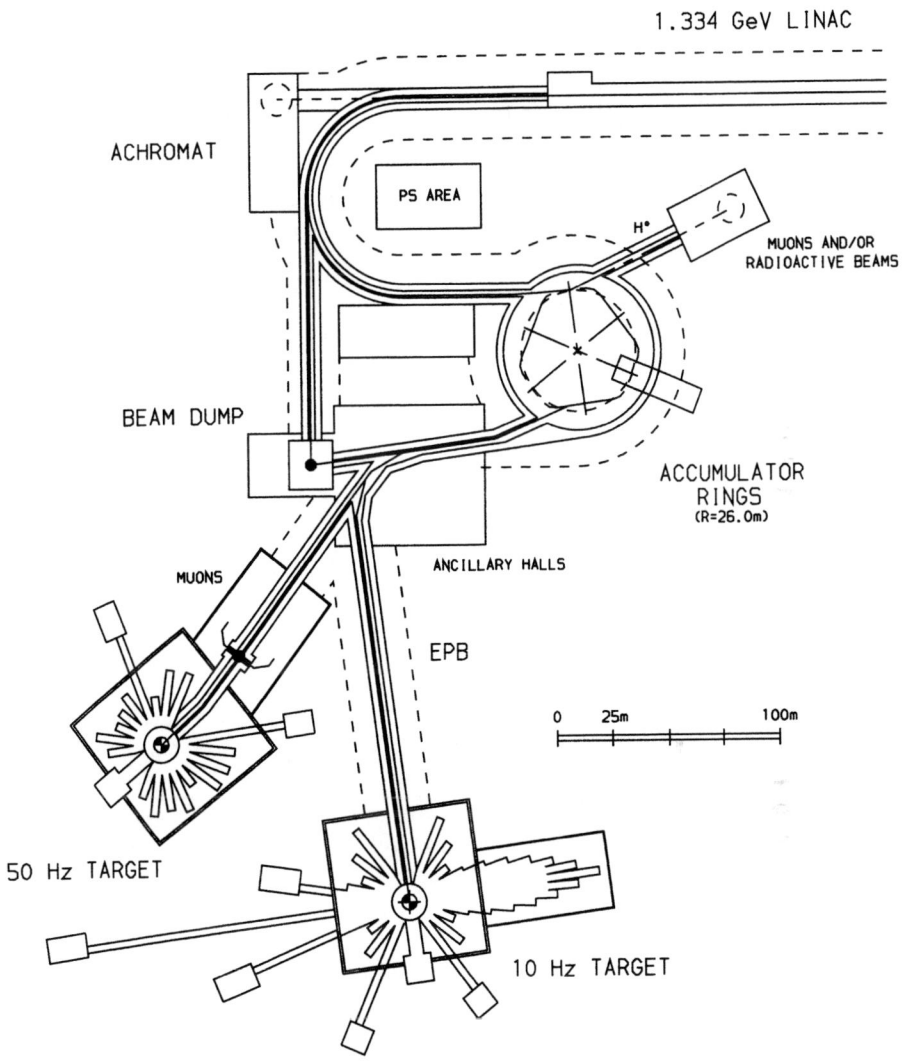

FIGURE 1. ESS Reference Design.

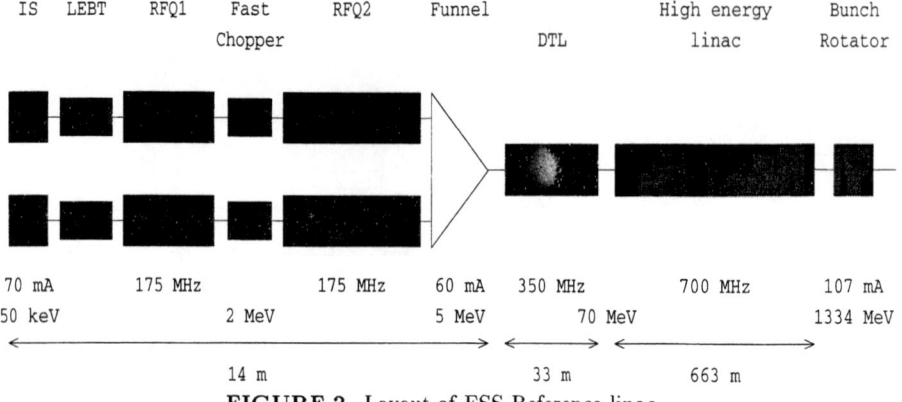

FIGURE 2. Layout of ESS Reference linac.

foil, to minimise delayed stripping of H^0 atoms in the ring and to bend stripped electrons directly to a collector. The foil is $545\,\mu g\ cm^{-2}$ graphite, and the chosen energy, $1.334\,GeV$, and dipole fields are consequences of the atomic physics of Stark lifetimes for the H^0 excited states.

Longitudinally, injection depends on phase space painting by ramping the beam momentum ($[0 \to 2] \times 10^{-3}$ to $[2 \to 4] \times 10^{-3}$ linearly over 1000 turns is chosen) and steering the RF bucket by varying the RF frequency [4]. Because of the non-zero ring dispersion, this correlates into the transverse plane giving initially large betatron oscillations which diminish as injection proceeds. Vertical painting is achieved by means of four vertical orbit bump magnets set for inversely correlated oscillations, small to large, as the ring fills. Computer simulations have determined the optimum RF voltages during injection and suggested the vertical orbit bump programme to keep subsequent foil traversals to a minimum. With a predicted average of seven foil interactions per particle, total ring losses from delayed stripping, inelastic collisions, scattering and momentum variations should be around 0.02%. Longitudinally, the bunches are well-confined, with no particles straying into the gaps and so being lost at extraction.

Fast extraction is single turn from a dispersion-free region in the horizontal plane, and kickers and septum extraction magnets are positioned so that the bunches from each ring are spaced closely together during transport to the targets. Bunch and pulse durations are 0.4 and $1.0\,\mu sec$ respectively.

Transport to the targets is via beam lines based on 90° and 60° FODO cells, allowing the use of several magnets of the same type and providing easy dispersion matching where appropriate [5]. In normal mode of operation, 1 MW of beam power passes from the rings to the 10 Hz target and 4 MW to the 50 Hz target. The full 5 MW can be exploited at the 50 Hz target when the other is out of service. A "muon" transmission target is envisaged in the 50 Hz line.

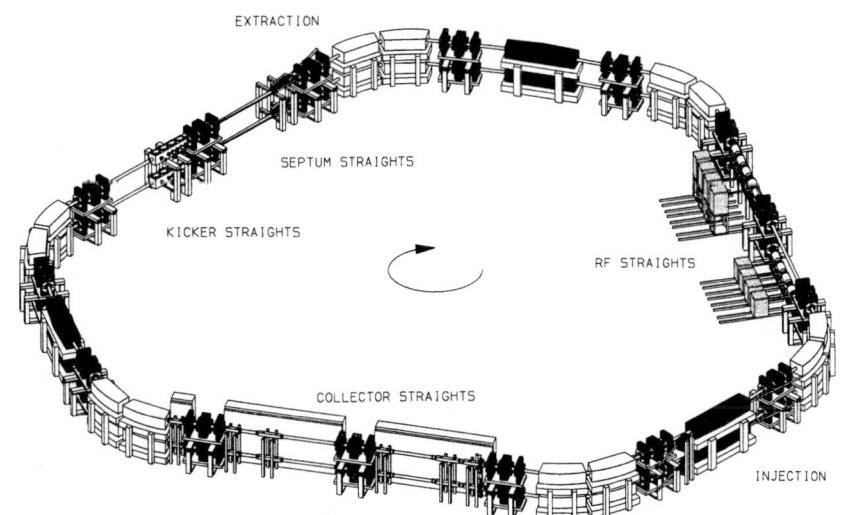

FIGURE 3. Layout of ESS accumulator rings.

III RECENT DEVELOPMENTS

Since publication of the ESS report, work has continued at the University of Frankfurt into the design of a volume source driven by arc discharge to produce the necessary 70 mA of H^- ions in 1.2 msec pulses. Such requirements are beyond the capabilities of current technology. 40 mA have been achieved through the use of pulsed caesium, with less than 3% of residual electrons, and continuous operation demonstrated over a period of eight days. The LEBT has already been constructed but has not yet been connected to the ion source. Other work at Frankfurt has been directed towards the RFQs, particularly in connection with a design to be built at RAL for ISIS, studies of which are hoped eventually to lead to full development of a unit for ESS.

A new RFQ computer code is being written to include more sophisticated geometries, realistic 3D space charge calculations and take account of neighbouring bunches [6]. When completed, this will allow full 3D simulation of the beam through the entire linac structure, RFQ, DTL and CCL. Simulated radial-axial coupling for halo particles has been observed in the CCL [7], and investigations are being carried out into halo production by mismatched bunched beams [8].

The RF system for the 700 MHz CCL has been cost optimised by connecting two klystrons with 2.4 MW peak RF power to a single 50 Hz bouncer-type modulator. The 8.5 MW AC peak power modulator is equipped with eight water-cooled, commercially available, IGBT switches, of a specification allowing switching $> 10^{10}$ times. A testbed for studying their thermal behaviour with 1.4 msec long pulses at 50 Hz repetition rate will be ready in June 1998.

The ESS report also confirmed that, during the R&D phase, other possible designs for the accelerating system would be considered.

A superconducting linac was an option mentioned, based on 700 MHz elliptical cavities with a constant 10 MV/m accelerating gradient along the linac. However, at low β, this corresponds to peak surface gradients of 45 MV/m, and recent experimental results from JAERI [9] of a 600 MHz, $\beta = 0.5$ single cell cavity indicate a peak surface gradient of only 25 MV/m at 2°K cooling temperature. The consequent implications for the ESS superconducting linac layout are the subject of current investigation.

Recently work has been carried out on an alternative (room temperature) linac. Discussions with commercial firms suggested there could be advantages, in terms of size and efficiency, in a scheme based on IOTs (inductive output tubes), rather than klystrons, operating at frequencies of 280 and 560 MHz. Figure 4 shows a possible layout with an RFQ-DTL system funnelled at 20 MeV, followed by CCDTL and CCL accelerating systems to the design energy. With lower space charge levels, funnelling at 20 MeV should be easier than at 5 MeV. Also, with this design the

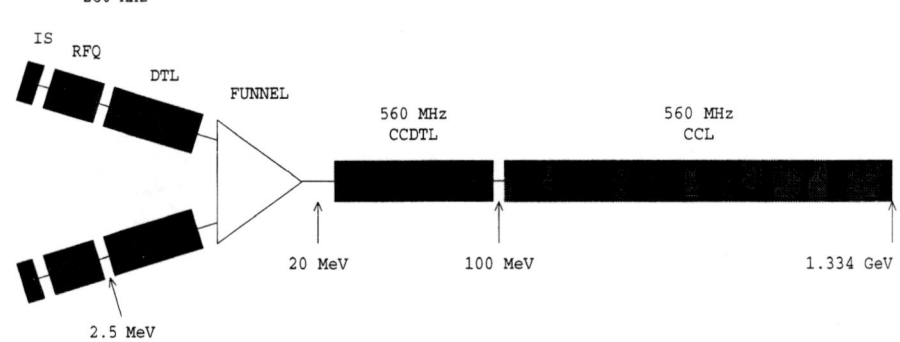

FIGURE 4. Alternative Linac for ESS.

sequence of focusing transitions ($2\beta\lambda$ (RFQ) $\rightarrow 4\beta\lambda$ (DTL) $\rightarrow 9\beta\lambda$ (Funnel) $\rightarrow 11\beta\lambda$ (CCDTL) $\rightarrow 11\beta\lambda \rightarrow 13\beta\lambda$ (CCL) per cell) should be smoother than in the reference design.

Study of the new proposals has been directed at three main areas: RFQ, chopper and funnel. A long 2.5 MeV RFQ has been designed for 280 MHz which gives high transmission efficiency and low emittance growth as well as minimal longitudinal filamentation [6]. Despite the relatively low energy, the beam can be well matched into the regular focusing of the DTL.

A possible funnel design is depicted in Figure 5. It consists of six FODO cells, each containing RF buncher cavities, mainly at harmonic number 3 (840 MHz). Arms of the funnel are merged together via an arrangement of two 10° septum magnets, a 2° septum cavity and a 2° electrostatic deflector. One of the buncher cavities is a specially-designed unit shared by both sets of beams and one of the quadrupoles is similarly shared. The third and fourth cells contain no bends but

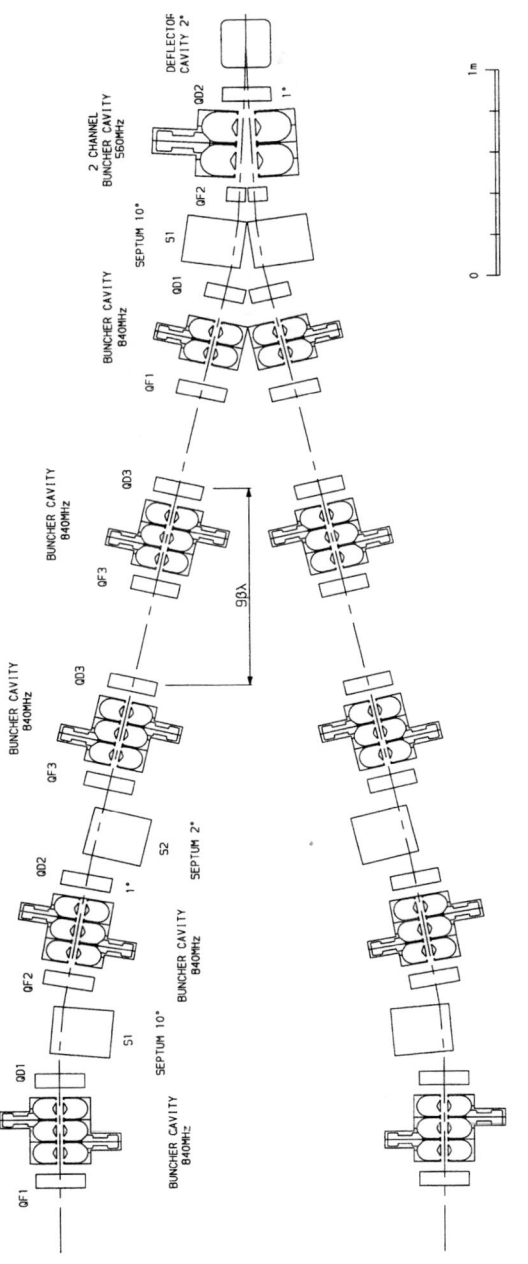

FIGURE 5. 20 MeV Funnel Design for Alternative ESS Linac.

provide space for diagnostics. Each cell has a phase advance under linear space charge of approximately 45° (depressed from a zero-current value of ∼ 80°) and the whole is part of a (first order) achromat in which the final two cells (without bends) are not required and are omitted. Longitudinally the bunch motion consists of a sequence of shearing and energy kicks in the cavities, and the applied voltages are adjusted so that the bunch exactly inverts in phase space for every π transverse phase shift. This gives strong and uniform longitudinal focusing. Transversely, quadrupole strengths are chosen to produce as uniform oscillations as possible, which keeps the ratio of non-linear to linear space charge forces constant. Care is taken to match the dispersion and its angle to zero, since residual dispersion is expected to lead to unwelcome effects in subsequent parts of the linac, emittance growth in particular. Simulations show well controlled beam dynamics through the structure with rms emittance growth for an initial Gaussian beam of less than 2%.

The chopper is also likely to be a source of emittance growth: 30% is predicted for the two-beam RFQ scheme in the ESS Report, and similar figures are suggested for designs for other spallation sources. An alternative scheme which has been studied is shown in Figure 6 and consists of a symmetrical arrangement of 13 quadrupoles, two buncher cavities and two sets of electrostatic voltage plates to deflect the beam to the collector, the split system making it possible to limit the overall cell length. Longitudinally the motion is as for the funnel and strong,

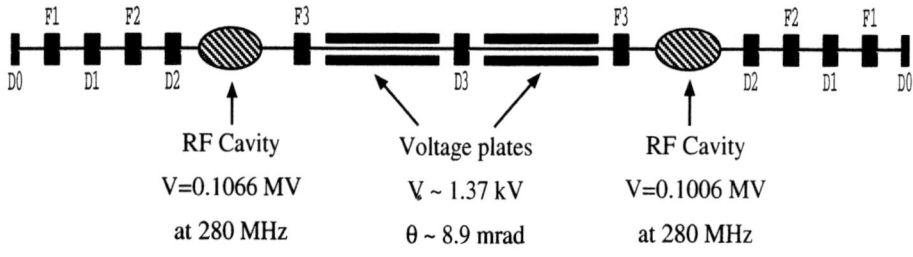

FIGURE 6. 2.5 MeV Chopper line for Alternative ESS Linac

uniform focusing is imposed. Transversely the requirement is for regular betatron oscillation amplitudes, as equal as possible in both planes. A figure of merit can be defined by

$$\mathcal{F} = \frac{2Vl}{\mathcal{A}}$$

where $\pm V$ is the voltage on the plates, l is the length of the plates and $\mathcal{A}$ is the acceptance. Quadrupole gradients can be determined to minimise this figure, meet the oscillation requirements and identify the optimum position for the collector. The result compares very favourably with other proposed designs, and, for a transverse Gaussian beam with self-consistent, longitudinal Hofmann-Pedersen distribution [10], shows an rms emittance growth of less than 1%.

Other work has been directed towards modifying the reference injection scheme in order to reduce the peak operating temperature of the foil, which at $\sim 2500°C$ is thought to be too high [11]. One idea is to chop at 70%, another is to increase the ring radius, both of which would alleviate pressure on the linac design and, by reducing the number of injection turns to 850 or 560 respectively, provide an opportunity to lower the number of foil traversals. There is further scope for this by re-optimising the vertical orbit bump programme in conjunction with RF steering and a slightly modified method of momentum ramping. However the two requirements of good beam distribution and a low mean number of foil interactions impose conflicting demands on the injection parameters, and the inevitable compromise is thought unlikely to reduce the temperatures by a great deal.

The main new idea in the theoretical development of the project is the proposal from Suzuki [3] of replacing H^- charge exchange using a foil with stripping by means of an intense laser and optical resonator. The scheme would avoid the overheating problems, there would be less chance of particle loss and emittance growth at injection. A superconducting option could be adopted with a lower peak beam intensity from the linac and a longer period for injection. A possible layout for ESS is shown in Figure 7. The ring would now have superperiod 2 and a larger mean radius (~ 40 m). Foil stripping could still be included as an option. Preliminary work does indeed appear to suggest that parameters at the foil are such that heating might be reduced.

The mechanism for laser charge-exchange is a two-stage process of neutralisation and ionisation. The incoming H^- beam is first converted to H^0 in the strong electric field generated by relativistic interaction with the magnetic field of an undulator. Excitation of the H^0 ground state to states of quantum number $n = 3p$ is effected by a laser and an optical ring resonator. The H^0 excited states are then ionised into H^+ in the magnetic field of a long tapered undulator. Analysis of the transition probabilities and the rates of excitation shows that the ionisation process needs to be repeated eight times to achieve a sufficiently high percentage of H^+. For this the ionising undulator needs to be approximately 5 m long; each period is 1 m, the field consisting of half sine-waves 0.25 m long interspaced with free spaces of 0.25 m. The free space provided helps avoid broadening of the atomic absorption spectrum by the Lorentz electric field during the relativistic interaction. Typically the peak magnetic fields would be of the order of 1 T. The efficiency of the process is thought to be high, but care is needed to minimise dispersive effects within the undulators.

IV FUTURE PLANS

With the completion of Phase 1 of the ESS project, application was made to the European Union for funding for an R&D programme of work. To date, support has only been forthcoming for further target studies and it remains for individual institutions to determine their own level of involvement in the next stage of accelerator development. FZJ (Jülich) is going ahead with an accelerator and tar-

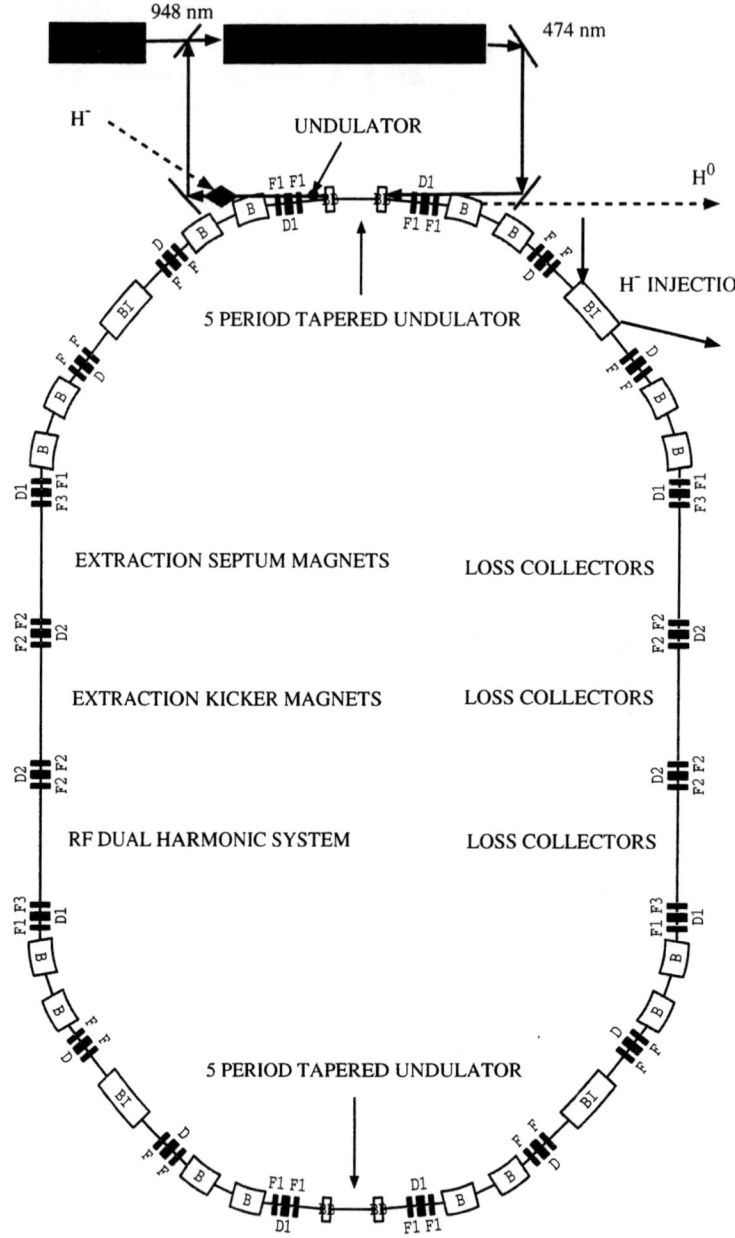

FIGURE 7. Alternative ESS Lattice for accumulator rings using laser stripping technique for injection.

get programme, which will involve a considerable amount of code development; a 500 MHz superconducting cavity with two input coupling ports has been ordered for studying pulsed behaviour under power splitting. At the University of Frankfurt, H$^-$ ion source and RFQ development will continue in close collaboration with RAL. A test stand will be built at RAL to explore the possibilities of the ion sources and an RFQ for ISIS, with the aim of following this with the first ESS RFQ design. Then the beam chopper will be built at RAL in the year 2000, followed by part of the funnel. If results are encouraging, the second leg and the complete funnel will be built and tested by 2003. Of the other European countries, Saclay (France) is prepared to cost the superconducting linac option and look into beam diagnostics; the University of Eindhoven (Netherlands) will work on detectors for neutron instruments; Denmark, Italy and Sweden are keen to become involved in an R&D programme; Hungary and Belgium would like to join in; and Russia and Switzerland are expected to show interest. It is hoped that the strong neutron users' community in Europe can exert some influence to ensure the R&D programme receives adequate central financial support.

REFERENCES

1. G. Bauer, T. Broome, D. Filges, H. Jones, H. Lengeler, A. Letchford, G. Rees, H. Stechemesser, G. Thomas (eds): *ESS: A next Generation Neutron Source for Europe. Vol III: The ESS Technical Study.*, November 1996.
2. I.S.K. Gardner, H. Lengeler, K. Bongardt, H. Klein, G.H. Rees, C.M. Warsop: *Status of the European Spallation Source Design Study.* Proceedings of the Particle Accelerator Conference PAC97, Vancouver, Canada.
3. Y. Sazuki: *A New Concept of Charge-Exchange Injection for the High Intensity Proton Storage Ring of JAERI.* To be published in the Proceedings of the Asian Particle Accelerator Conference APAC '98.
4. C.R. Prior: *Longitudinal Injection for the ESS.* ESS-94-7-R, November 1994.
5. V. Ziemann: *New Optics of the ESS Target Beam lines.* ESS-96-56-R, December 1996.
6. A.P. Letchford: Minutes of ESS Meeting 1, Rutherford Appleton Laboratory, April 1998.
7. K. Bongardt, M. Pabst and A.P. Letchford: *Core and Halo Particle Dynamics of High Intensity Proton Beams.* Proceedings of the Particle Accelerator Conference PAC97, Vancouver, Canada.
8. M. Pabst and K. Bongardt: *Analytical Approximation of the Three Mismatch Modes for Bunched Beams.* ESS 97-85-L, August 1997.
9. N. Ouchi et al. *Proton Linac Activities in JAERI.* Proceedings of the Superconducting Workshop, Legnaro, Italy, October 1997.
10. A. Hofmann and F. Pedersen: *Bunches with Local Elliptic Energy Distributions.* IEEE Trans. Nucl. Sci. Vol NS-26, No. 3, 1979.
11. J.P. Duke: *Foil Stripping Temperatures in the European Spallation Source.* Proceed-

SNS Accumulator Ring Design and Space Charge Considerations [1]

W.T. Weng

AGS Department, Brookhaven National Laboratory, Upton, NY 11973

Abstract. The goal of the proposed Spallation Neutron Source (SNS) is to provide a short pulse proton beam of about 0.5μs with average beam power of 1MW. To achieve such purpose, a proton storage ring operated at 60Hz with 1×10^{14} protons per pulse at 1GeV is required. The Accumulator Ring (AR) receives 1msec long H^- beam bunches of 28mA from a 1GeV linac. Scope and design performance goals of the AR are presented. About 1200 turns of charge exchange injection is needed to accumulate 1mA in the ring. After a brief description of the lattice design and machine performance parameters, space charge related issues, such as: tune shifts, stopband corrections, halo generation and beam collimation etc. will be discussed.

I INTRODUCTION

The Oak Ridge National Laboratory is leading a conceptual design for a next generation pulsed spallation neutron source, the Spallation Neutron Source (SNS). There are three major accelerator systems included in Brookhaven's area of responsibility. First is the High Energy Beam Transport (HEBT) system. Secondly, the Accumulator Ring (AR) system and thirdly, the Ring to Target Beam Transport (RTBT) system. This paper describes the design of the AR itself whose magnet and tunnel layout is shown in Fig. 1.

The proton Accumulator Ring is one of the major systems in the design of the SNS. The primary function of the AR is to take the 1GeV H^- beam of about 1msec length from the linac and convert it into a 0.5μs beam through a stripping foil in about one thousand turns. The final beam should have 1.0×10^{14} protons per pulse, resulting in 1MW design average beam power at 60Hz repetition rate. Provisions have been reserved for a future upgrade to 2MW beam power by doubling the stored current to 2.0×10^{14} proton per pulse without changes in both the magnet and vacuum system [1].

---

[1] Research on the SNS is sponsored by the Division of Material Sciences, U.S. Department of Energy, under contract number DE-AC05-96OR22464 with Lockheed Martin Energy Research Corporation for Oak Ridge National Laboratory.

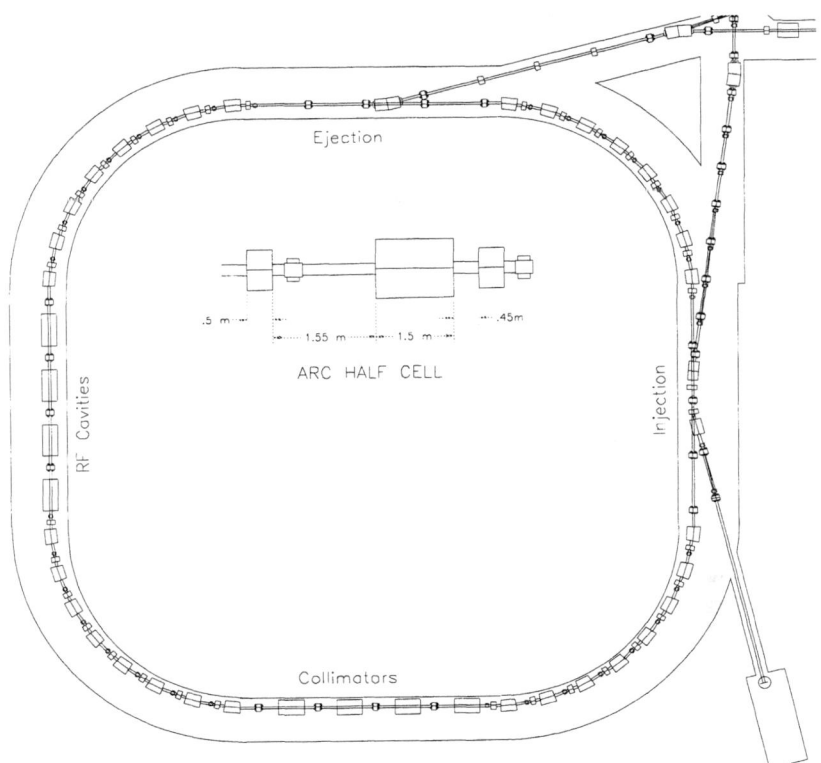

FIGURE 1. Layout of the accumulator ring.

One of the major performance requirements is to keep the average uncontrolled particle loss during the accumulation time to less than 2.0×10^{-4} per pulse. The reason of this stringent requirement is to keep the residual radiation to such a level that the hands-on maintenance is possible except for a few localized areas, such as: injection, extraction and collimation. To achieve this goal, special care have been exercised in the H^- stripping, the RF stacking, and the collimator design.

In Section 2, the lattice design, H^- injection and RF stacking process of the SNS accumulator ring will be described. Space charge tune shift and resonance stopband corrections is presented in Section 3. The study in halo formation and its implications on the collimator placement and design is given in Section 4.

II LATTICE, H^- INJECTION AND RF STACKING

The accumulator ring of the Spallation Neutron Source (SNS) will have a four-fold symmetric lattice. Its lattice function is shown in Fig. 2 [2]. The lattice will accommodate the long straight sections required for the injection system, the

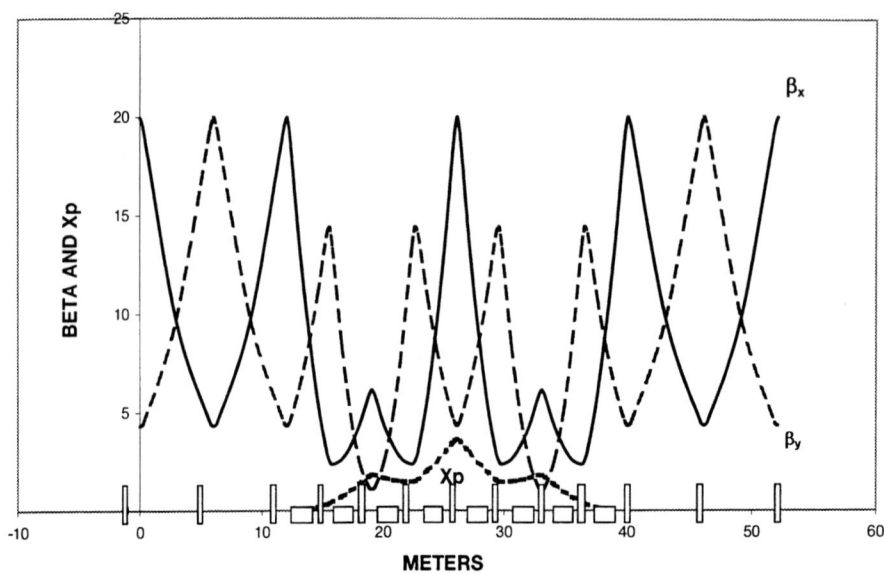

FIGURE 2. Accumulator lattice functions.

extraction system, the RF cavities, and the beam scraping system. The straight sections will be dispersion free, which is desirable, especially for the RF cavities and the injection system. The lattice will provide ease of betatron tuning and flexibility of operation. Unlike lattices of lower symmetry, a lattice of four-fold symmetry will assure that there are no dangerous betatron structure resonances other than for the integer tune. Lattice functions and other salient performance and design parameters of the accumulator ring are summarized in Table 1.

The most demanding system in the design of the 1MW short pulse spallation neutron source is the H^- multi-turn injection into the storage ring [3]. For the SNS accumulator ring, a carbon foil of $400\mu g/cm^2$ is assumed. The stripping efficiency for 1GeV incident H^- beam is about 99.8%. The temperature rise for the 1MW design is estimated to be about $3200°C$. In addition, it has been found that

a. Stripping losses in passage of the H^- beam through the B=3kG field in a DC bump dipole magnet upstream of the stripper foil are negligible.

b. A fraction $f(H^o) = 8.19 \times 10^{-3}$ of the incident H^- beam will emerge as H^o from a $400\mu g/cm^2$ carbon stripper foil and must be disposed of in an external dump.

c. Field ionization of the H^o component in the B=2.41kG field of a quadrupole downstream of the foil will lead to negligible uncontrolled losses.

d. Fractional losses from nuclear non-elastic interactions in the foil as low as

TABLE 1. SNS Accumulator ring parameters

Beam Average Power	1.0 MW
Kinetic Energy	1.0 GeV
Average Current	1.0 mA
Repetition Rate	60 Hz
Ion Source Peak Current	35 mA
Linac Peak Current	27.7 mA
Beam Duty Cycle	6.18 %
Linac Pulse Length	1.03 msec
Beam Loss (Controlled/Uncontrolled)	< 4 / 0.02 %
Number of Turns Injected	1158
Revolution Period	0.8413 μsec
Revolution Frequency	1.1886 MHz
Circumference	220.688 m
Space-Charge Tune-Shift	< 0.1
Bunching Factor (Dual RF Systems)	0.405
Number of Protons/Ring	1.04×10^{14}
Beam Emittance (Transverse, norm.)	217 πmm-mr
Tunes ν_x/ν_y	5.82 / 5.80
β_{max}(x/y)	19.2 / 19.2 m
Dispersion X_p(max/min)	4.1/0.0 m
Transition Energy γ_t	4.93
RF Voltage (1st Harmonic)	40 kV
RF Voltage (2nd Harmonic)	20 kV
RF bucket	17 eV-sec
Beam Emittance (Longitudinal)	10 eV-sec
Beam Gap (injection)	295 nsec
Injected Pulse Length	546 nsec
Extracted Pulse Length	591 nsec
Vacuum	10^{-9} Torr

1.26×10^{-5} can be realized with rapidly (exponentially) collapsing injection bumps and "smoke-ring" injection scheme which result in very small multiple foil traversals of $< N_t > = 2.43$ traversals/injected proton.

e. Multipole Coulomb and nuclear elastic scattering fractional losses out of a ring acceptance of $A_{x,y} = 330\pi$mm-mrad for the above injection conditions are 1.35×10^{-5}, well below our loss criterion of $<10^{-4}$. With the small area stripper foil (8mm H × 4mm V) used for the above estimates, 2.2% of the incident H^- beam misses the foil and will be deflected by a magnet to an external dump.

To accumulate all the particles needed and keep them in proper azimuthal distribution, a dual harmonic RF system is employed. The fundamental RF system will provide 40kV and the second harmonic RF system will provide 20kV to form a flattened RF bucket for particle trapping with resulting bunching factor of about 0.4. Such a RF system will reduce the incoherent tune shift by 25%. The 4-dimensional multi-particle tracking program *Accsim* is needed to follow the 1200 turns of particles in the ring. The resultant beam distributions in real space and in RF phase space are shown in Fig. 3 and Fig. 4 respectively [4]. Much more work has to be done in this exercise to determine the optimal combination of RF waveform and injection strategy.

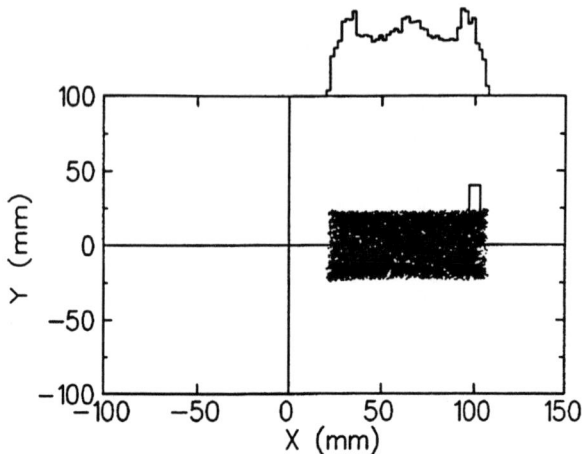

FIGURE 3. The resultant beam distributions in real space. *Accsim* snapshot at 1200 turns. Pseudo barrier RF.

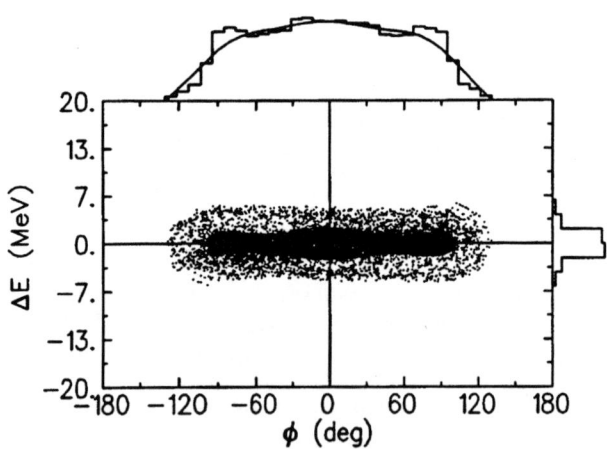

FIGURE 4. The resultant beam distributions in RF phase space. *Accsim* snapshot at 1200 turns. Pseudo barrier RF.

III SPACE CHARGE TUNE SHIFT AND RESONANCE CORRECTIONS

A Experiences at the AGS

We will briefly review the intensity evolution of the AGS from a few 10^{10} ppp to 6.3×10^{13} ppp to get some idea of the effect of space charge tune shift at injection.

Shown in Figure 2 is the intensity evolution of the AGS since its completion in 1961 [5,6]. Two major upgrades to the AGS complex were carried out since its operation for physics research in 1961. The first one was the "Conversion Project" from 1967 to 1972 to replace a 50 MeV linac by a 200 MeV linac and to raise the AGS rep-rate from 0.15 Hz to 0.5 Hz. The second one was the "Booster Project" from 1987 to 1991 to build a Booster to raise the injection energy from 200 MeV to 1.5 GeV and to provide high intensity high mass heavy ions for RHIC. Major improvements in the intensity record are summarized briefly in the following chronology of Table 2.

TABLE 2. Major improvements in the intensity record

Year	Parameters or Improvements	AGS Intensity
1970	50 MeV linac injector Space charge limited at injection. $\Delta\nu_y = 0.3$	3×10^{12}
1976	200 MeV linac H+ injector Resonance stopband correctors; Transverse damper	10^{13}
1990	200 MeV linac H- injection RF feedforward compensation. $\Delta\nu_y = 0.55$	1.6×10^{13}
1995	1.5 GeV Booster injection; Resonance stopband correctors Direct RF feedback γ-transition jump $\Delta\nu_y = 0.35$	6.3×10^{13}

It is clear that raising the injection energy is the most effective way to increase the achievable intensity for a space charge limited low energy proton synchrotron. It is also clear that many accelerator physics manipulations have to come into play to keep all those particles inside the synchrotron. The reason that raising injection energy helps to store more particles is due to the reduction of the incoherent space charge tune shift during injection. A good indication of the space charge effect at injection is given by the expression of incoherent space charge tune shift [7]:

$$\Delta\nu = \frac{Nr_p}{2\beta\gamma^2\varepsilon_N B_f} \quad (1)$$

where $r_p = 1.54\times10^{-18}$ m is the classical proton radius, N is the total number of protons, ε_N is the normalized beam emittance, B_f is the bunching factor, and β and γ are the beam velocity and energy. At the tune shift level about 0.35 unit, the acceptable proton intensity changed from 3×10^{12}, to 10^{13} and to 6×10^{13} ppp at three different injection energies respectively.

As shown in Table 2, the incoherent space charge tune shift at injection can reach as high as 0.55 unit. That means many particles in the beam can move cross half integer and third integer resonance lines. Properly placed quadrupoles and sextupoles are used to correct the stopband width of those resonances to minimize the amplitude growth. Experiences showed that each correction could account for from few percent to tens of percent of reduction of particle losses.

Due to the strong energy dependence of the amount of tune shift caused by the space charge force, raising injection energy can alleviate this limiting effect. This is also the reason that the US Spallation Neutron Source design chooses to have full energy injection at 1GeV.

B Design for the SNS Accumulator Ring

Now let's look at the space charge tune shift for the SNS accumulator ring. According to eq.(1), with $N=1.0\times10^{14}$ ppp, $\beta=0.875$, $\gamma=2.066$, $B_f=0.44$, $\varepsilon_N=217$, the expected tune shift is

$$\Delta\nu_{S.C.} \simeq 0.1 \tag{2}$$

Such a tune shift is relatively small compared to a value of about 0.3 at ISIS and 0.25 at PSR.

Because the ring will operate at very high intensities for which stringent limits on losses will be imposed, the possibility of beam loss due to resonance excitation must be considered. The second, third and fourth-order resonances between tunes of 5 and 6 are shown in Figure 5 [2]. Although the space-charge tune spread of the beam is expected to be small (~ 0.1), there are a number of resonance lines sufficiently close to the working point to be of concern. Moreover, even though lines such as those excited by normal and skew quadrupoles are far from the working point, they can cause unfavorable distortions of the betatron functions. The four-fold symmetric lattice allows placement of correction elements so that either even or odd harmonics (in azimuth θ) can be produced for the correction of the following resonances. Detailed calculations are underway to determine the required magnet strengths.

a. Second-Order Resonances

The $2\nu_x = 11$ and $2\nu_y = 11$ resonances can be corrected by exciting the trim windings on the quadrupoles with harmonic 11 in azimuth θ. (To produce an odd harmonic, magnets separated by $\theta = 180°$ are excited with equal but opposite currents.) Similarly, the sum resonance $\nu_x+\nu_y = 11$ can be corrected by exciting the skew quadrupole correctors with harmonic 11; the difference resonance $\nu_x - \nu_y = 0$ can be corrected (or enhanced if coupling between the two planes is desired) by exciting these correctors with harmonic 0.

b. Third-Order Resonances

The $3\nu_x = 17$ and $\nu_x + 2\nu_y = 17$ resonances can be corrected by exciting sextupoles with harmonic 17. The eight sextupole correctors allow for independent

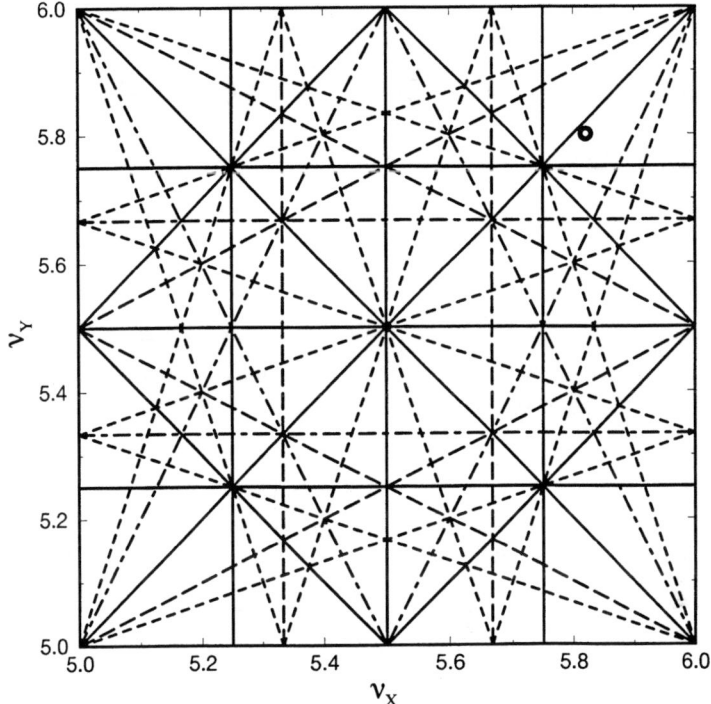

FIGURE 5. Accumulator Tune Chart. The circle shows the working point. Solid and short-dashed lines indicate second and fourth-order resonances; long-dashed and dot-dashed lines indicate third-order.

correction of the two lines. Similarly, the $3\nu_y = 17$ and $\nu_y + 2\nu_x = 17$ resonances can be corrected with the eight skew sextupole correctors.

  c. *Fourth-Order Resonances*

The fourth-order resonance lines are driven by octupoles (both normal and skew) and those closest to the working point pass through the point $\nu_x = \nu_y = 5.75$ in Figure 5. The effects of these lines can be reduced by exciting the octupole correctors with harmonic 0. Schemes for correcting specific lines are under investigation. The octupole correctors also allow for Landau damping of transverse instabilities.

  d. *Structure Resonances*

One consideration in our choice of a lattice was the avoidance of structure resonances. The 4-fold symmetric lattice has fourth-order structure resonances at betatron tunes of 5 and 6; third-order at tunes of 5.333 and 6.666; and second-order at a tune of 6. However, none of these are near the proposed working point. A 3-fold symmetric lattice that has been considered in previous design studies, on the other hand, has fourth-order structure resonances at a betatron tune of 3.75 which is right on top of the proposed working point for this lattice. Although these resonances may not hurt the ring performance, they are best avoided.

IV HALO FORMATION AND COLLIMATION

One of the major performance requirements of the SNS ring is to keep the average uncontrolled particle loss during the accumulation time to less than 2.0×10^{-4} per pulse. The reason of this stringent requirement is to keep the residual radiation to such a level that the hands-on maintenance is possible except for a few localized areas, such as: injection, extraction and collimation.

Typical beam losses of existing low power proton synchrotron are in the order of a few percent. Let us use the AGS and the proposed SNS ring as example. The relevant beam parameters are summarized in Table 3. It can be clearly seen that 1% loss of the SNS ring is equivalent to the entire flux of the AGS beam. Such a situation is totally unacceptable.

TABLE 3. AGS and SNS parameters

	AGS	SNS
Proton Intensity	6×10^{13} ppp	10^{14} ppp
Rep-Rate	0.5	60
Flux	3×10^{13} pps	6×10^{15} pps
Loss of 1%	3×10^{11} pps	6×10^{13}

All existing low energy high intensity proton synchrotrons have operated at a condition that the available physical apertures are fully occupied. Quantitatively, it can be characterized as

$$\varepsilon_A/\varepsilon_B \simeq 1.0 \qquad (3)$$

where ε_A signifies the equivalent emittance of the aperture and ε_B is the beam emittance. For the SNS accumulator ring design,

$$(\varepsilon_A/\varepsilon_B)_{SNS} \geq 3.0 \qquad (4)$$

In other words, there is about factor of 1.6 of space allowed in both vertical and horizontal dimension. The question now is that "is it enough?"

It has been found that the large amplitude particle can interact with the core particles to move either closer to the center or away from the center [8–10]. This process can be understood by an envelope oscillation created by the mismatch between the beam shape and the lattice of the focusing channels. A particle in the halo region tends to be driven away in such a mismatched focusing channel. Although the smaller amplitude particles stay close to the stable fixed point in the center, the larger amplitude particles can drift away following the multiple islands as show in Fig. 6 [9]. The crucial questions now, are how far the islands can extend away from the center, what are the dynamical nature of the islands, and when the chaotic motion will set in. Those are all important questions to be answered by any new high power accelerators. It can happen both in the linac and in the circular rings.

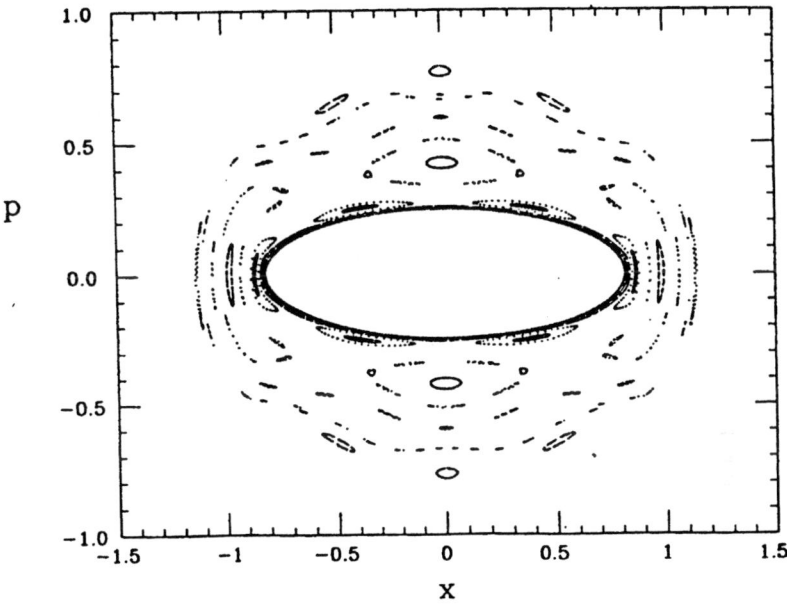

FIGURE 6. The Poincare surface of section in particle phase space from particle-core interaction [9].

A thorough understanding of the halo dynamics as function of mismatch, power supply ripple, space charge tune shift, and the lattice structure, etc., is necessary to be able to estimate the degree of beam losses and placement of collimators with confidence.

Assuming Gaussian distributions shown in Fig. 7, the development of proton synchrotron and main attention of accelerator physics in the past 40 years can be roughly classified into three period.

a. Period 1 located roughly between 1960 to 1975 when the total intensity for fixed target research was a major concern. The figure of merit in this period is the total intensity,

$$N = \int_{-2.5\sigma}^{2.5\sigma} f(z) dz.$$

b. Period 2 located roughly between 1970 to 1990 when the brightness for colliding beam research was a major concern. The corresponding figure of merit is the brightness, $B = N/\sigma$.

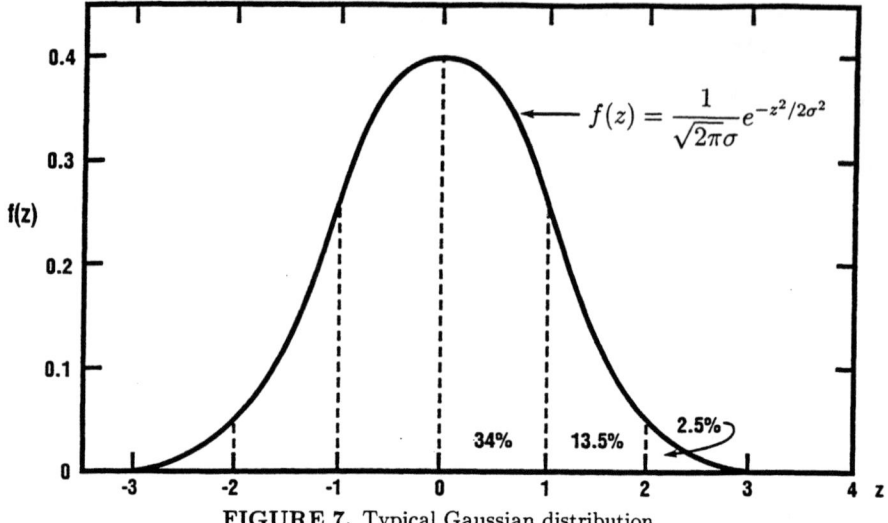

FIGURE 7. Typical Gaussian distribution.

c. Period 3 started from 1990 and could well extend to 2010 when the loss of the halo particles is a major concern. This concerns all high power proton accelerators from 1MW to 10MW range. The corresponding figure of merit will be the population and dynamics of the halo particles,

$$H = 2 \int_{4\sigma}^{\infty} f(z) dy.$$

We heard from T. Wangler that for a tune-depressed and mismatched linac beam, the amplitude growth can reach 5 to 6 times of the beam root-mean-square size. In a synchrotron, both the tune depression and mismatch is less than that of a linac, the maximum amplitude growth should be lower. However, the actual performance can only be understood after careful tracking studies.

The SNS collaboration initiated two programs to meet the need of reliable and flexible 6-dimensional tracking code for injection and halo studies. Such a code should include space charge effect in both transverse and longitudinal planes following the injection process of turn-by-turn build up from the linac with an aim to push the reliability limit far beyond what is available today for the design of the injection process and the collimation strategy of the SNS accumulator ring. To meet the design requirements, the code should be able to give reliable answer to particle dynamics at large amplitude to better than 1%.

Our first effort is to generate a new code which is called "Sensible Analysis Model for Beams in Accelerators" (SAMBA). Although there are similar codes, such as ACCSIM of TRIUMF, Simpson of KEK and Track2 of ESS, we feel that the SNS project needs a specialized code, that could be developed in parallel by all the members of the joint team, and is more comprehensive than existing codes, yet devoid of things that are not so important for our project. In doing so, we have freely borrowed from other codes we know, trying to collect the best in each one of them. We report the development of simulation codes by the joint team. The salient feature of this new code is summarized in the following [11].

SAMBA is written in C++. We did this for two main reasons:

- development of programming modules in parallel is more natural with C++ than with Fortran. This is why C++ has become the standard in the Industry, where a large group of programmers are often working on the same code, to bring it as early as possible to the market.

- the code is run under the supervision of a SuperCode that allows the coexistence of compiled and interpreted modules, making development and debugging very natural, as it will be explained later in detail.

the machine optics code that produces the main accelerator descriptor is MAD. This is a well developed and maintained code able to describe all of the features of the ring lattice, including higher order transformation matrices and the treatment of lattice errors.

SAMBA is independent from any specific graphic packages. Output to graphics can be done through files or by direct pipe from the code to graphics, in order to allow for animation. An intriguing possibility is to use the graphic firmware tools of the workstation to perform some of the calculations.

The real beam is represented by an ensemble of macro-particles in fully 6-dimensional phase space. The calculations include:

a. transport of the particles in the lattice using the first order and second order transformations calculated by MAD.

b. in the transverse space, space charge forces are calculated that apply an angle kick to the macros. Betatron tune distribution in the beam is calculated ("necktie"). Transverse impedances are used to calculate beam to wall coupling in a realistic vacuum chamber geometry.

c. in the longitudinal space, we have cavities with RF voltage plus a realistic longitudinal impedance budget, that allows the representation of longitudinal space charge kicks and the description of the bunch and the bucket.

Most of our design work on SNS injection and RF stacking has been performed using SAMBA. Work is continuing to complete this code development.

Since the most crucial physics input in this tracking code is the treatment of space charge force, to save time, some kind of approximations are always made. Such a simplified approach is necessary in comparing one scenario from another. However, its accuracy has to be checked by more reliable approach. To establish a benchmark for simplified treatment and establish accuracy of the calculation we create a particle in cell (PIC) module for space charge calculation.

Previous injection calculations for the SNS have employed the Schonauer model [12] for transverse space charge effects, which are limited to shifting the the macro-particle tunes. No other space charge effects on the transverse spatial charge distributions are calculated in this model. As such, to date the final ring emittance is determined by the choice of the linac beam distribution, by the closed orbit bumps, and by scattering of the beam by the foil. In this model only foil interactions produce beam halo. Here, preliminary results are reported using a full Particle-In-Cell (PIC), space charge calculation [13] for SNS injection. The impact of transverse space charge on emittance growth and halo are examined by repeating the previous Schonauer model optimized injection scenarios, here using the PIC model. Although the initial operation of the SNS will be at 1.0MW of beam power, these calculations are done for 2.0MW of beam power, since the ring is designed to allow for the 2.0MW upgrade.

We have two specific results to show following the space charge treatment of particle-core model using PIC. Due to the preliminary nature of our code development, some of the results shown may be improved in the future. The first result is the comparison of lattice dependence of the halo generation. The results of the halo growth of uniform, FODO, and doublet lattice are shown in Fig. 8 after 1250 turns of storage inside the ring [14]. These results show that there is no halo growth for uniform lattice, less than 10^{-3} halo growth for FODO lattice, and there is 1-2% halo growth for doublet. This may have implication of choosing lattice design for

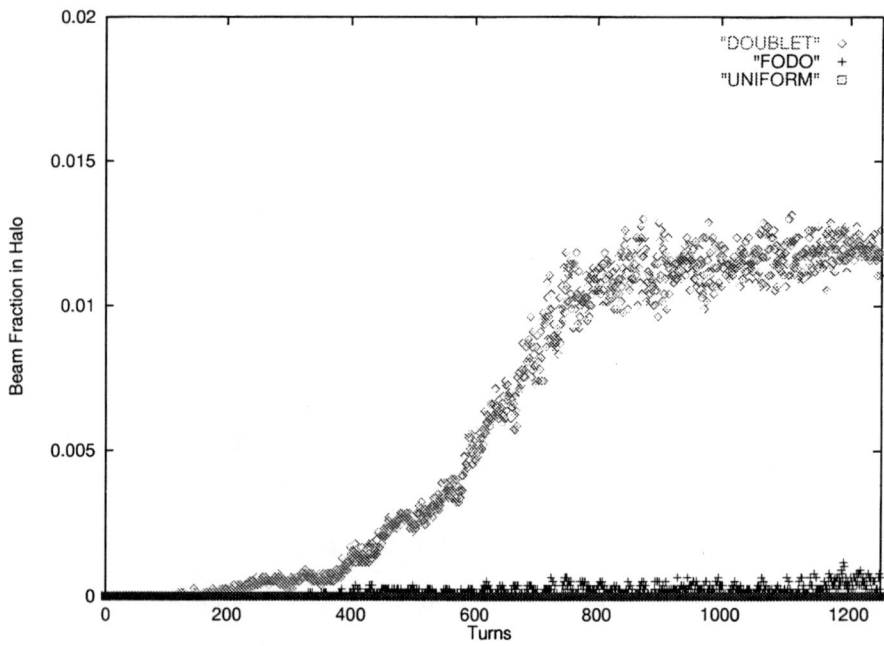

FIGURE 8. Halo growth for uniform, FODO and doublet lattice.

the accumulator ring to minimize the halo growth during injection.

Applying the PIC and particle-core model to the realistic injection study for the SNS ring, the particle distribution as a function of emittance is plotted in Fig. 9 for one of the injection scenario in vertical plane [13]. It can be seen that the result indicates that there is no particle outside of the 120π emittance using old tracking code, but there is a few percent of particles fall outside of 180π boundary. We would like to emphasize again that those results obtained here are very preliminary. There is a tendency for a PIC to overestimate the space charge effect if there are not sufficient number of particles in each bin. We will continue this study with more particles and more bins to search for convergence and reliability.

To contain those particles inadvertently migrating toward the wall, after all careful considerations and provisions, a collimator system has to be designed to catch the bulk of them before hitting the wall. For example, for the SNS four collimators, 3.2m each, enclosing a 4π solid angle around the source point and stuffed with

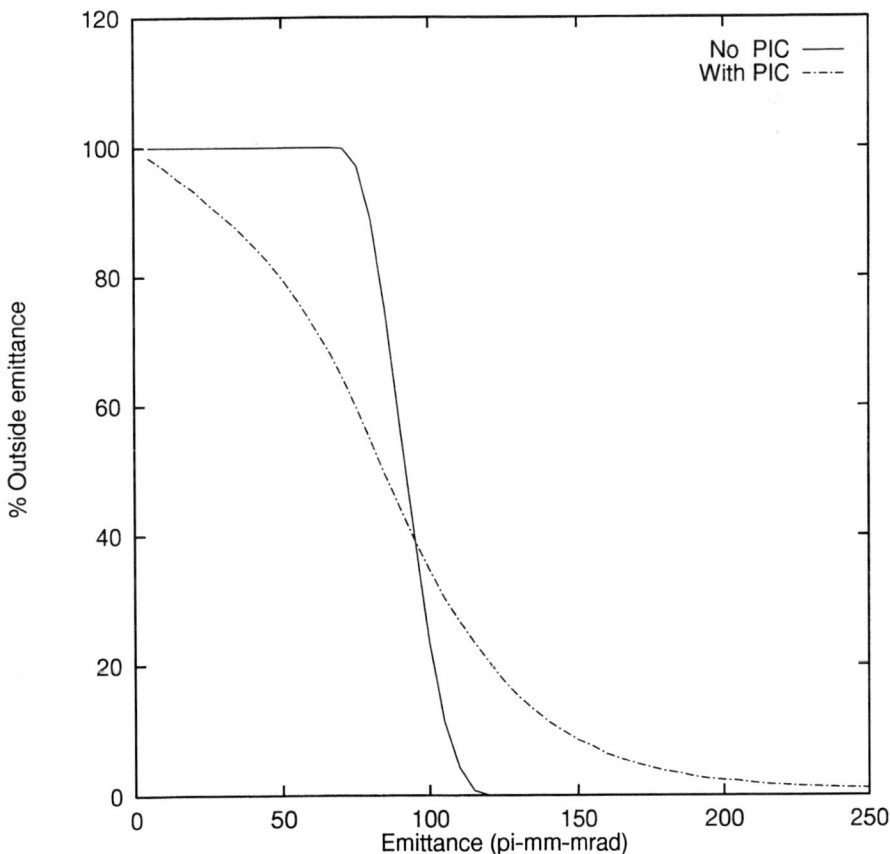

FIGURE 9. Particle distribution calculated by PIC and non-PIC methods.

segmented material to capture all secondary particles generated by the incident protons will be provided to reduce the radiation effects by a factor of 100. This way, most of uncontrolled losses will occur at the collimator, leaving ring components relatively intact for reliable operation [15].

For the SNS ring, the beam emittance is 120π and the physical acceptance is 360π. The hard decision is where to put the collimator? The possibility could be around 220π to 260π. The good answer can only be provided by the results from a reliable tracking code incorporating full space charge effects. We will continue to work in this direction to find satisfactory solution for SNS. One way to minimize the particle interception by the collimator is to introduce bend crystal upstream of the collimator to bring large amplitude particles outside of the accumulator ring. Depending on the final strategy adopted, the collimator will take a few percent to a few tenth of a percent beam losses and should be able to sustain direct hit of full beam for few pulses. With the help from this workshop, we will incorporate all the good ideas suggested to improve on the SNS design.

V ACKNOWLEGEMENT

The report given here is the results of both AGS and SNS accelerator physics staff. Contributions by J. Beebe-Wang, M. Brennan, J. Galambos, C. Gardner, J. Holmes, S.Y. Lee, Y.Y. Lee, A. Luccio, and D. Olsen are appreciated. The author would also like to thank J. Beebe-Wang for preparing this manuscript for publication.

REFERENCES

1. W.T. Weng, *et al.*, "Accumulator Ring Design for the NSNS Project", contributed paper to PAC97, May 12-16, 1997, Vancover, Canada.
2. Y.Y. Lee, C.J. Gardner and A.U. Luccio: "Accelerator Ring Lattice for the National Spallation Neutron Source", contributed paper to PAC97, May 12-16, 1997, Vancover, Canada.
3. L.N. Blumberg and Y.Y. Lee: "H^- Charge Exchange Injection Into the 1GeV NSNS Accumulator", BNL/NSNS Tech. Note #3, November, 1996.
4. A.U. Luccio, J. Beebe-Wang and D. Maletic: "Proton Injection and RF Capture in the National Spallation Neutron Source", contributed paper to PAC97, May 12-16, 1997, Vancover, Canada.
5. W.T. Weng, "Performance and Measurements of the AGS and Booster Beams", AIP Conf. Proc. No.377, p.145, Bloomington, IN, Oct. 10-13, 1995.
6. T. Roser: "High Intensity Performance and Upgrades at the Brookhaven AGS", in these workshop proceedings.
7. L.J. Laslett: "On Intensity Limitations Imposed by Transverse Space-Charge Effects in Circular Particle Accumulators", BNL Report 7534, p.324-67, 1963.

8. R.A. Jameson, "Self-Consistent Beam Halo Studies and Halo Diagnostic Development", *Frontiers of Accelerator Technology*, World Scientific, p.530, November, 1994.
9. S.Y. Lee and A. Riabko, "Envelop Hamiltonian of an Intense Charged-Particle Beam in Periodic Solenoidal Fields", Phys. Rev. E, Vol.51, Feb., 1995.
10. T. Wangler, "Space Charge in Proton Linacs", in these workshop proceedings.
11. A.U. Luccio, J. Beebe-Wang, M. Blaskiewicz, J. Galambos, J. Holmes and D. Olsen, "A Particle Simulation Code for a High Intensity Accelerator Ring", in these workshop proceedings.
12. H. Schonauer, "Addition of Transverse Space Charge to ACCSIM Code", Triumf Design Note TRI-DN-89-K50.
13. J. Galambos, J. Holmes, D. Olsen, J. Whealton, M. Blaskiewicz, A. Luccio and J. Beebe-Wang, "Progress Towards Understanding Transverse Space Charge Effects on SNS Ring Injection", in these workshop proceedings.
14. J.A. Holmes, J.D. Galambos, J.H. Whealton, D.K. Olsen, M. Blaskiewicz, A. Luccio, J. Beebe-Wang and S.Y. Lee, "Space Charge Calculations in Rings for Uniform, FODO, and Doublet Lattice", in these workshop proceedings.
15. H. Ludewig, S. Mughabghab and M. Todosow, "NSNS Ring System Design Study, Collimation and Shielding", BNL/SNS Tech. Note No.5, 1996.

CONTRIBUTED PAPERS: OBSERVATIONS

Longitudinal Space Charge Compensation at PSR

Filippo Neri

Los Alamos National Laboratory, Los Alamos, NM 87545 US

The longitudinal space-charge force in neutron spallation source compressor ring or other high-intensity proton storage rings can be compensated by introducing an inductive insert in the ring. The effect of the inductor is to cancel part or all of the space-charge potential, because the space charge impedance is capacitive. The longitudinal voltage, including the inductive effects has the form

$$V_s = \frac{\partial \lambda(s)}{\partial s}\left[\frac{g_0 Z_0}{2\beta\gamma^2} - \Omega_0 L\right] e\beta c R. \qquad (1)$$

Here V_s is the total voltage a particle at position s sees per turn, λ is the longitudinal charge distribution, s is the length along the bunch, Ω_0 is 2π times the revolution frequency, Z_0 is the vacuum impedance (377 Ω), R is the average machine radius. $g_0 = 2\log b/a + 1$, where b/a is the ratio of beam pipe to beam radius.

The Proton Storage Ring (PSR) at Los Alamos National Laboratory is a compressor ring used to produce short pulses of spallation neutrons. The PSR is about 90.2 m long; the proton energy is 797 MeV. Using Eq. 1 one can estimate that the inductance needed to completely compensate the longitudinal space-charge potential is about 12 µH. This corresponds to an impedance of about $j210$ Ω at the revolution frequency of 2.8 MHz.

In order to test this compensation scheme, two inductive inserts built at FNAL were installed in the PSR. Each insert was a vacuum tight stainless steel cylinder containing 30 ferrite cores. The ferrite toroids are 5 inch I.D., 8 inch O.D. and 1 inch thick. The relative permeability of the ferrite is about 50, with a Q of about 100, flat to about 50 MHz. To provide a method to vary the inserted inductance, a solenoid made of 88 turns of welder cable was wrapped around the outside of each tank. Because of the heating of the cables the bias could only turned on for a minute or so. This made taking data more difficult. The solenoid applied a biasing field orthogonal to the direction of the polarization induced by the beam. Fig. 1 is a plot of the measured inductance of both inserts versus the solenoid current. The inductance can be varied from about 7.5 µH to 2.4 µH. This is less than the 12 µH that are necessary to completely compensate the longitudinal space-charge voltage of the beam.

In preparation for the experiment, simulations were made of the longitudinal beam shapes at the end of injection for injected beams of various lengths. The injected bunch train length (pattern width) was 50 ns or 150 ns. A modified version of ACCSIM that include the effect of longitudinal impedances in the ring was used for the simulations. Since the inductor was measured only after the experiment, the calculations assumed that the applied bias (900 A) would reduce the inductance from 8 µH to 2 µH, not far from the measured values of 7.5 and 2.4 µH.

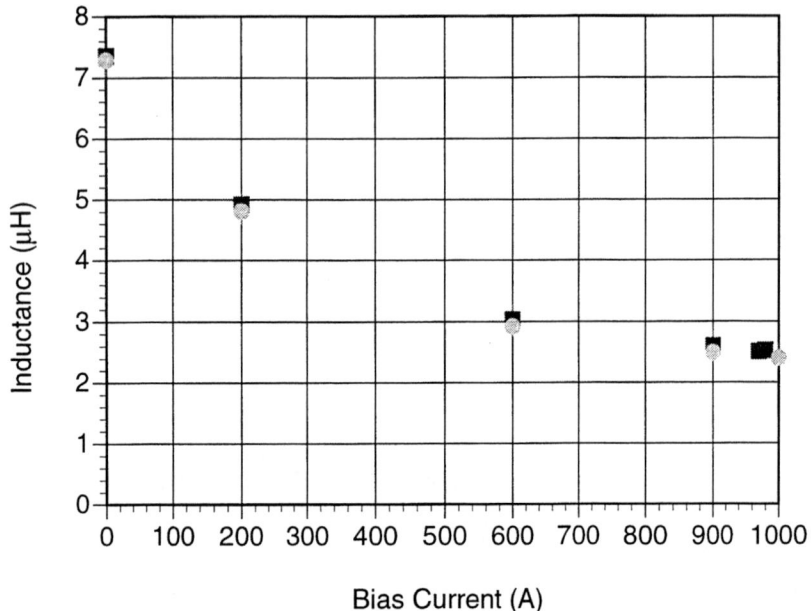

Fig. 1 Measured inductance of both modules for different bias currents.

A problem in this experiment is to accurately estimate the absolute number of injected protons. The absolute normalization of the current monitor is not better than about 10%. From the current monitor curves we conclude that the number of protons was about 4.0×10^{12} at end of injection for the 50 ns pattern-width, 625 µs injection experiments and 1.2×10^{13} for the 150 ns pattern-width, 625 µs injection experiments.

Fig. 2 (top) shows the digitized longitudinal bunch shape at 625 µsec after injection start, with a pattern width of 50 ns. The plotted beam shapes are obtained at end of injection, not at extraction. The buncher voltage was 7.5 kV. The black curve is the bias-off case, the gray curve is the case with a 900 A bias on the inductor solenoid. Fig. 2, bottom is the result of the simulations for the same case. The simulation curves are normalized so that the areas are the same as the areas of the experimental curves. The normalization is the same, up to a 1-2% error. The simulations used an injected beam with a momentum spread of 0.08% dp/p (RMS). This corresponds to an energy width of 0.985 MeV.

By taking the bias-on and bias-off curves in rapid succession, we deduced that the total number of particles was the same for the two cases, with at most a 2% difference.

Since the shift of the peak of the distributions is about 16% (50 ns pattern width,) we conclude that the difference of the effects with bias on and off is not due to fewer particles in the bias on case. This would be in any case very improbable, as the direction of the effect was shown to be repeatable. Also one can clearly see that the bias-on experimental curve (gray curve in Fig. 2, top) is wider then the bias-off curve (black curve in Fig. 2, top).

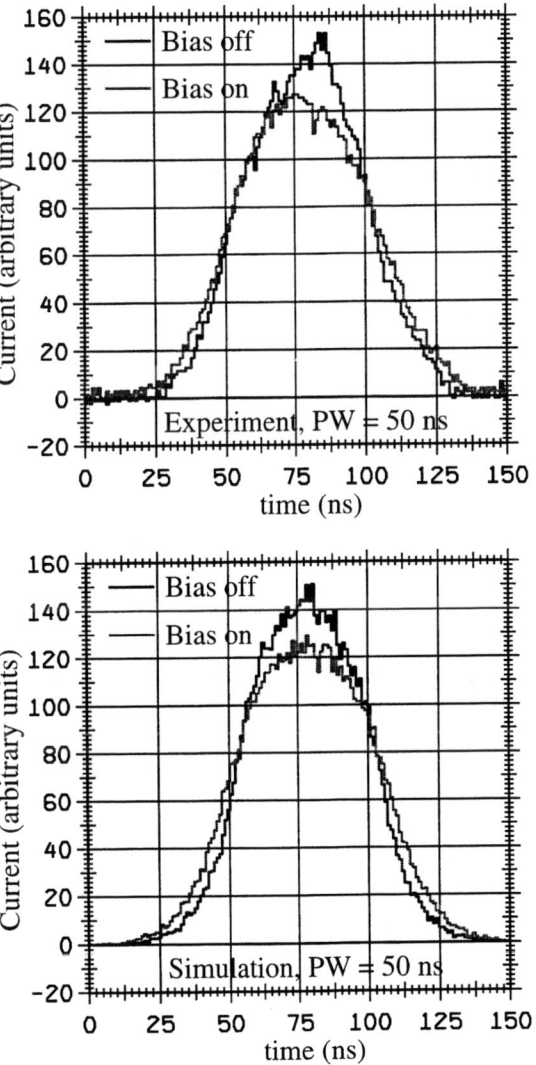

Fig. 2 Pulse shape after 625 μsec, with an injected pattern width of 50 ns. The simulations used 4.0×10^{12} protons. Solid curves represent the shape with no bias on the inductor (8 μH in the simulation). Dotted curves represent the shapes with a 900 A bias applied to the inductor. This is assumed to reduce the inductance to 2 μH in the simulations. Buncher voltage was 7500 V.

The agreement is reasonably good and the peak to peak ratio (about 16%) between the bias on and bias off cases agrees with the experiment with an error of 2-3%. In the simulations, the inductance with the 900 A bias on the solenoid was assumed to be 2 µH. Fig. 3 shows the results for the 150 ns injected beam width. The effect is smaller, but still consistent with the simulated predictions.

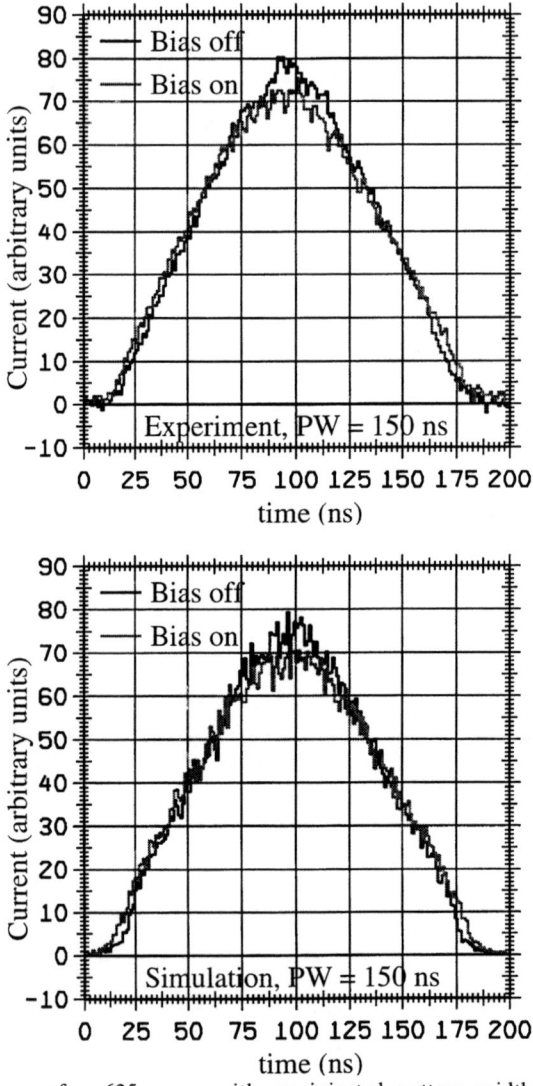

Fig. 3 Pulse shape after 625 µsec, with an injected pattern width of 150 ns. The simulations used 1.2×10^{13} protons. Solid curves represent the shape with no bias on the inductor (8 µH in the simulation). Dotted curves represent the shapes with a 900 A bias applied to the inductor. This is assumed to reduce the inductance to 2 µH in the simulations. Buncher voltage was 7500 V.

The simulations are in good agreement with the experimental results in the 50 ns pattern width case and not inconsistent in the 150 ns case. Because the inductor only compensates the space-charge potentially partially, the effects observed are not very large, but they support the results of the simulations. Simulations using the same assumptions show that an inductor of about 8 µH could significantly reduce the buncher voltages needed to confine 6.5×10^{13} protons per pulse (200 µA at 20 Hz). This is shown in Fig. 4 and 5.

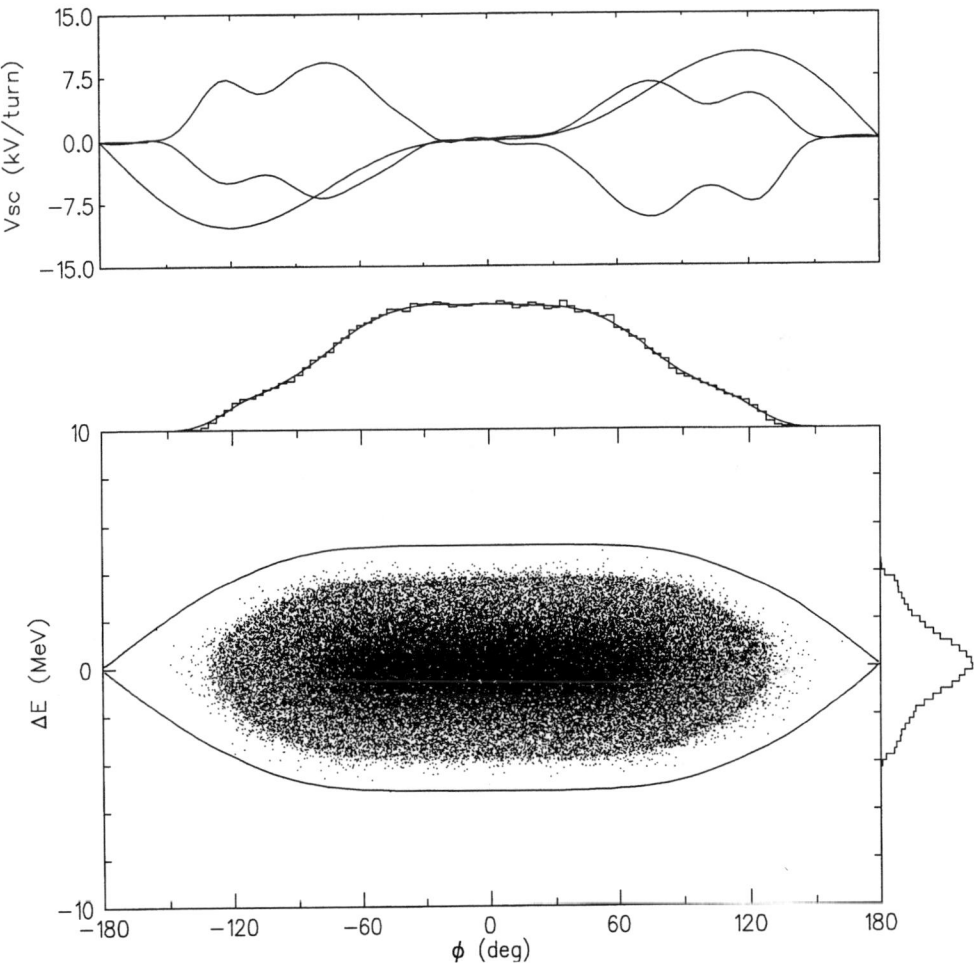

Fig. 4 Simulation of PSR after SPSS upgrade. 200 µA @20 Hz. 6.4×10^{13} protons are injected into the ring in 1800 turns. Buncher voltage is ramped from 6 to 8 kV. (-3 to -4 kV second harmonic). An 8 µH inductor is included in the simulation. The combination of RF voltage and inductor voltage keeps the gap in the beam clear. The bunching factor is about 0.49.

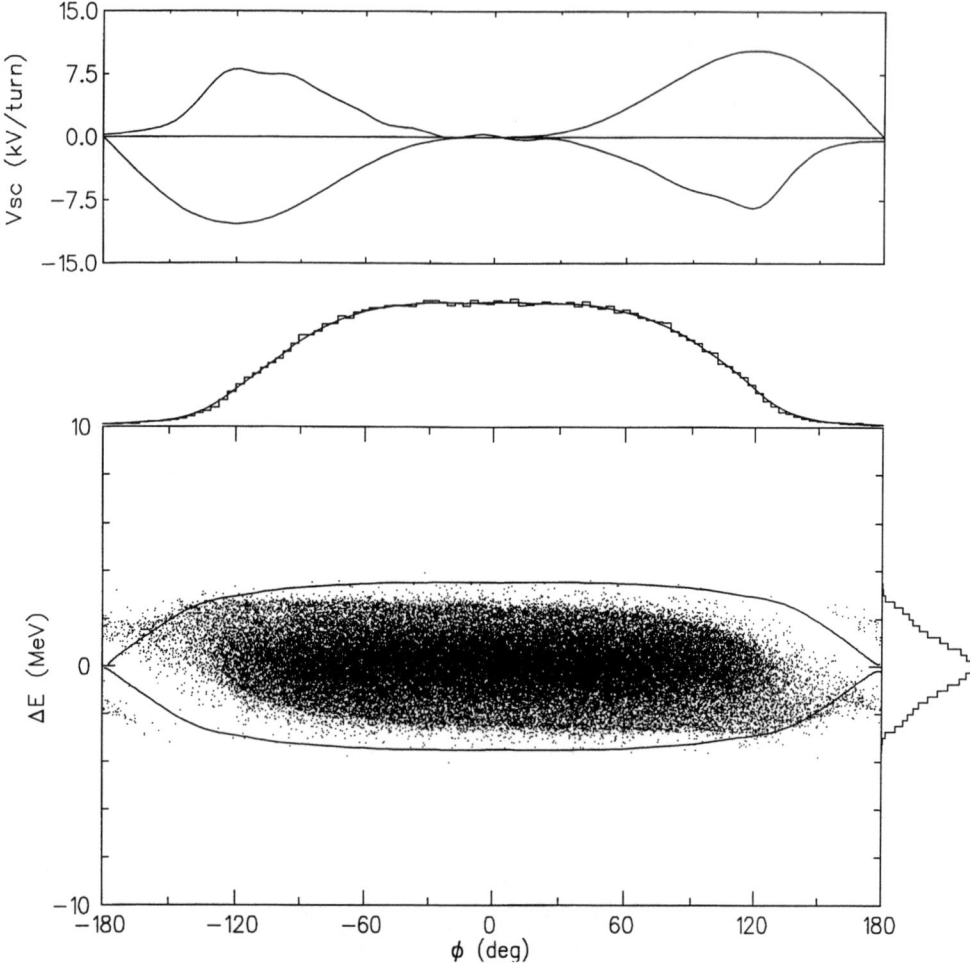

Fig. 5 Simulation of PSR after SPSS upgrade. 200 μA @20 Hz. 6.4×10^{13} protons are injected into the ring in 1800 turns. Same parameters as in Fig. 4, except for the absence of the inductor. The RF voltage is not sufficient to keep the gap clear. About 18 kV first harmonic plus 9 kV second harmonic would be required to keep the gap clear in this case

The PSR experiment confirmed the theoretical results for the effects of an inductive insert in an high-intensity storage ring. No instability was produced by the insert. The effect of the inductor on the PSR fast transverse instability was more ambiguous, but there were some indications that the inductor produced a current-dependent decrease in the buncher voltage needed to prevent the instability. Also, with the inductor, the best results for the instability were obtained with a flat-top rf profile. This is contrary to the observation without the inductor, where an increasing rf profile is needed to keep the beam longitudinally matched in presence of an increasing space-charge voltage.

We want to thank our colleagues at FNAL, J.E. Griffin, K.Y. Ng and Z.B. Qian for originally suggesting the experiment and participating in it. D. Wildman, also from FNAL built the inductors used in the experiment. P. Walstrom measured the inductors modules, after the experiment at Los Alamos.

Some Issues on the RF System in the 3 GeV Fermilab Pre-Booster

K.Y. Ng

Fermi National Accelerator Laboratory,[1] P.O. Box 500, Batavia, IL 60510

Abstract. Some issues are presented on the rf system in the future Fermilab pre-booster, which accelerates 4 bunches each containing 0.25×10^{14} protons from 1 to 3 GeV kinetic energy. The problem of beam loading is discussed. The proposal of having a non-tunable fixed-frequency rf system is investigated. Robinson's criteria for phase stability are checked and possible Robinson instability growth is computed.

I INTRODUCTION

The proposed future Fermilab pre-booster has a circumference of 158.07 m. It accelerates 4 bunches each containing 0.25×10^{14} protons from kinetic energy 1 to 3 GeV [1]. Because of the high intensity of the beam, the problems of space charge and beam loading must be addressed. A preliminary rf system has been proposed by Griffin [2]. Here, we wish to examine the issues of beam loading and Robinson instabilities. We first present a possible resonant ramp curve with space-charge distortion of the rf waveform compensated. Then the rf system which will eliminate most of the transient beam loading is reviewed. We next consider the possibility of non-tunable rf cavities and the possible Robinson instabilities that follow. We find that the Robinson growth in synchrotron amplitude is small and the high-intensity Robinson phase-stability criterion is well satisfied.

II THE RAMP CURVE

Because of the high beam intensity, the longitudinal space-charge impedance per harmonic is $Z_\|/n|_{\rm spch} \sim -j100 \; \Omega$. But the beam pipe discontinuity will contribute only about $Z_\|/n|_{\rm ind} \sim j20 \; \Omega$ of inductive impedance. The space-charge force will be a large fraction of the rf-cavity gap voltage that intends to focus the bunch. A proposal is to insert ferrite rings into the vacuum chamber to counteract this space-charge force [3]. An experiment of ferrite insertion was performed at the Los Alamos Proton Storage Ring and the result has been promising [4]. Here we assume such an insertion will over-compensate all the space-charge force leaving behind about $Z_\|/n|_{\rm ind} \approx j25 \; \Omega$ of inductive impedance. It may be a good idea to over-compensate the space charge, because an inductive impedance will help bunching so that the required rf voltage needed will be smaller.

---

[1] Operated by the Universities Research Association, Inc., under contract with the U.S. Department of Energy.

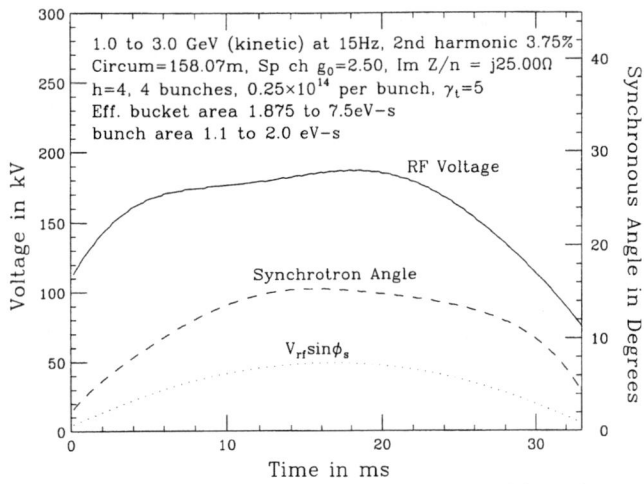

FIGURE 1. A typical ramp curve for the future Fermilab pre-booster.

The acceleration from kinetic energy 1 to 3 GeV in 4 buckets at a repetition rate of 15 Hz is to be performed by resonant ramping. In order to reduce the maximum rf voltage required, about 3.75% of second harmonic is added. A typical ramp curve is shown in Fig. 1, which will be used as a reference for the analysis below. This fraction of second harmonic was chosen because, in the present choice of initial and final bucket areas and bunch areas, raising this fraction beyond 3.75% will only flatten the rf gap voltage in the ramp but will not decrease the maximum significantly.

III THE RF SYSTEM

According to the ramp curve in Fig. 1, the peak voltage of the rf system is $V_{\rm rf} \approx$ 185 kV. Griffin proposed 10 cavities [2], each delivering a maximum of 18.5 kV. Each cavity contains 30 cm of ferrite rings with inner and outer radii 20 and 35 cm, respectively. The ferrite has a relative magnetic permeability of $\mu_r = 21$. The inductance and capacitance of the cavity are $L \sim 0.61$ μH and $C \sim 820$ pF. Assuming an average ferrite loss of 134 kW/m^3, the dissipation in the ferrite and wall of the cavity will be $P \sim 15$ kW. The mean energy stored is $W \sim 0.15$ J. Therefore each cavity has a quality factor $Q \sim 459$ and a shunt impedance $R_s \sim$ 12.5 kΩ. One such cavity is shown in Fig. 2.

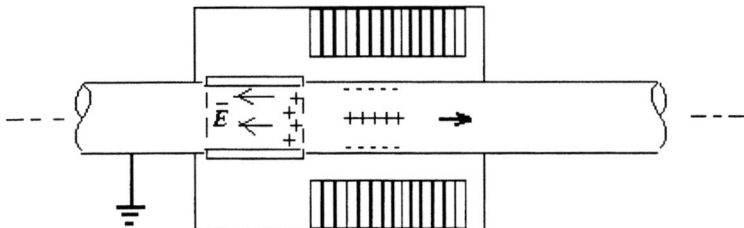

FIGURE 2. A ferrite-loaded cavity with a dielectric gap in the beam pipe. Protons move to the right.

Because each bunch contains $q = 4.005~\mu\mathrm{C}$, the transient beam loading is large. For the passage of one bunch, 4.005 μC of negative charge will be left at upstream end of the cavity gap. Since another negative image current will start from the downstream end of the cavity gap following the bunch, an equal amount of positive charge will accumulate there, as illustrated in Fig. 2. Thus a voltage V_{t0} will be created at the gap opposing the beam current. For a short bunch, this transient beam loading voltage can have a maximum of

$$V_{t0} \sim \frac{q}{C} = 5.0~\mathrm{kV}~, \qquad (3.1)$$

where $C = 820$ pF is the gap capacitance. We note from Fig. 1 that the accelerating gap voltages at both ends of the ramp are only about or less than 10 kV in each cavity. If the wakes due to the bunches ahead do not die out, we need to add up the contribution due to all previous bunch passages. Assuming a loaded quality factor of $Q_L = 45$, we find from Appendix C that the accumulated beam-loading voltage can reach a magnitude of $|V_t| = 37$ kV when the detuning angle is zero.

Griffin suggested to use a feed-forward system [2], which will deliver via a tetrode the same amount of negative charge to the downstream end of the gap so as to cancel the transient beam loading. This is illustrated in Fig. 3.

A feed-forward system is not perfect and we assume that the cancellation is 85 %. For a δ-function beam, the component at the fundamental rf frequency is 56.0 A. Therefore, the remaining image current across the gap is $i_{\mathrm{im}} = 8.4$ A. To counter this remaining 15% of beam loading in the steady state, the cavity must be detuned by the angle (see Appendix A)

$$\psi = \tan^{-1}\left(\frac{i_{\mathrm{im}}\cos\varphi_s}{i_0}\right)~, \qquad (3.2)$$

where φ_s is the synchronous angle and $i_0 = V_{\mathrm{rf}}/R_s$ is the cavity current *in phase* with the cavity gap voltage V_{rf}. For high quality factor of $Q = 459$ which is accompanied by a large shunt impedance, the detuning angle will be large. Corresponding

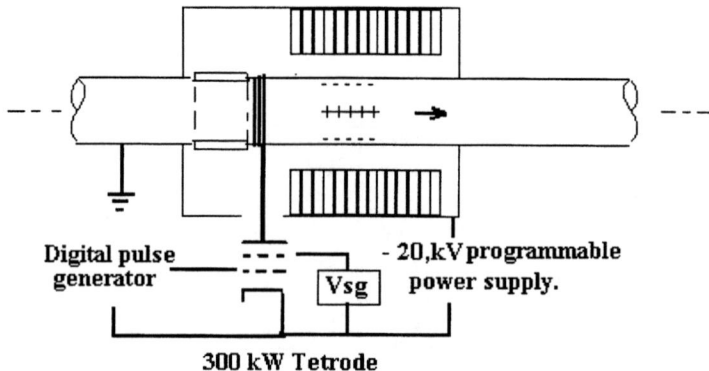

FIGURE 3. Transient beam-loading power tetrode connected directly to a rf cavity gap.

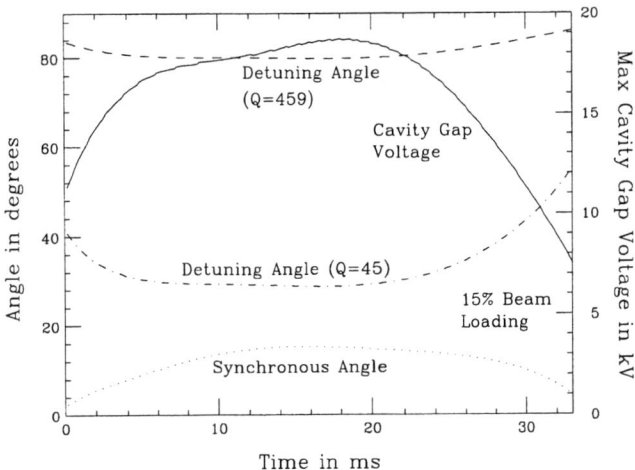

FIGURE 4. Detuning angle for the high $Q = 459$ and low $Q_L = 45$ situations.

to the ramp curve of Fig. 1, the detuning angle is plotted as dashes in Fig. 4 along with the synchronous angle and maximum cavity gap voltage. We see that the detuning angle is between 80° and 86°, which is too large. If a large driving tube is installed with anode (or cathode follower) dissipation at ~ 140 kW, the quality factor will be reduced to the loaded value of $Q_L \sim 45$ and the shunt impedance to $R_s \sim 1.22$ kΩ. The detuning angle then reduces to $\psi \sim 29°$ at the center of the ramp and to $\sim 40°$ or $\sim 56°$ at either end. This angle is also plotted in Fig. 4 as a dot-dashed curve for comparison. Then, this rf system becomes workable.

IV FIXED-FREQUENCY RF CAVITY

Now we want to raise the question whether it is possible to have a fixed resonant frequency for the cavity. A fixed-frequency cavity can be a very much simpler device because it may not need any biasing current at all. Thus the amount of cooling can be very much reduced and even unnecessary. It appears that the resonant frequency of the cavity should be chosen as the rf frequency at the *end* of the ramp, or $f_R = 7.37$ MHz so that the whole ramp will be immune to Robinson's phase-oscillation instability [5]. However, the detuning will be large. For example, at the beginning of the ramp where $f_{\rm rf} = 6.39$ MHz, the detuning angle becomes $\psi = 85.2°$. Since the beam-loading voltage $V_{\rm im}$ is small, the generator voltage phasor $\tilde{V}_g$ will be very close to the gap voltage phasor $\tilde{V}_{\rm rf}$. As a result, the angle θ between the gap voltage $\tilde{V}_{\rm rf}$ and the generator current phasor $\tilde{i}_g$ will be close to the detuning angle, as demonstrated in Fig. 5. For example, Fig. 6 shows that, at the beginning of the ramp, the detuning angle is $\psi = 85.2°$. Although the total power delivered by the generator

$$\tfrac{1}{2}\tilde{i}_g \cdot \tilde{V}_{\rm rf} = \frac{V_{\rm rf}^2}{2R_s} + \tfrac{1}{2}i_{\rm im}V_{\rm rf}\cos\varphi_s \tag{4.1}$$

is independent of θ, the energy capacity of the driving tube has to be very large.

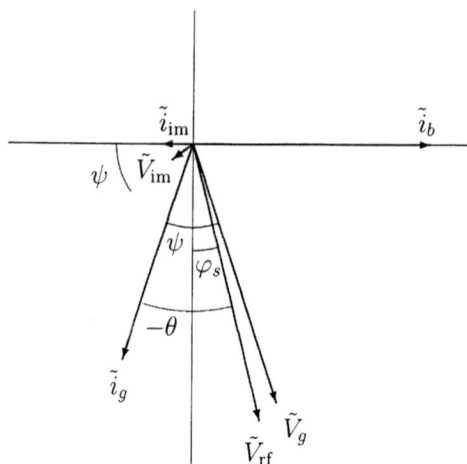

FIGURE 5. For a fixed cavity resonant frequency, the detuning angle ψ is fixed at each ramp time. When beam-loading is small, the angle θ between the gap voltage $\tilde{V}_{\rm rf}$ and the generator current $\tilde{i}_g$ will be close to ψ and will be large.

Another alternative is to choose the resonant frequency of the cavity to be the rf frequency near the *middle* of the ramp. Then the detuning angle ψ and therefore the angle θ between $\tilde{V}_{\rm rf}$ and $\tilde{i}_g$ will be much smaller at the middle of the ramp when the gap voltage is large. Although θ will remain large at both ends of the ramp, however, this is not so important because the gap voltages are relatively smaller there. Figure 7 shows the scenario of setting the cavity resonating frequency f_R equal to $f_{\rm rf}$ at the ramp time of 13.33 ms.

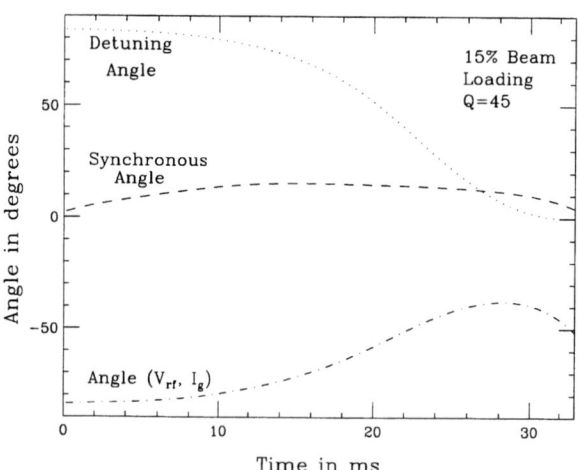

FIGURE 6. When the cavity resonant frequency is chosen as the rf frequency at the end of the ramp, both the detuning angle as well as the angle between the cavity gap voltage $\tilde{V}_{\rm rf}$ and the generator current $\tilde{I}_g$ are large.

There is a price to pay for this choice of f_R; namely, there will be Robinson phase instability for the second half of the ramp when the rf frequency is larger than f_R. The instability comes from the fact that, below transition, the particles with larger energy have higher revolution frequency and see a smaller real impedance of the cavity, thus losing less energy than particles with smaller energy. Therefore, the synchrotron amplitude will grow. In other words, the upper synchrotron sideband of the image current interacts with a smaller real impedance of the cavity resonant peak than the lower synchrotron sideband. However, since the loaded quality factor Q_L is not small, the difference in real impedance at the two sidebands is only significant when the rf frequency is very close to the cavity resonant frequency. Thus, we expect the instability will last for only a very short time during the second half of the ramp. The growth rate of the synchrotron oscillation amplitude has been computed and is equal to [6]

$$\frac{1}{\tau} = -\frac{i_{\text{im}} \beta \omega_s (R_+ - R_-)}{2 V_{\text{rf}} \cos \varphi_s}, \qquad (4.2)$$

where

$$R_+ - R_- = \mathcal{R}e\left[Z_{\text{cav}}(\omega_{\text{rf}} + \omega_s) - Z_{\text{cav}}(\omega_{\text{rf}} - \omega_s)\right], \qquad (4.3)$$

i_{im} is the image current, β is the velocity with respect to light velocity, $\omega_s/(2\pi)$ is the synchrotron frequency, and Z_{cav} is the longitudinal impedance of the cavity. We see from Fig. 7 that the growth occurs for only a few ms and the growth time is at least ~ 25 ms. The total integrated growth increment from ramp time 13.33 ms is $\Delta G = \int \tau^{-1} dt = 0.131$ and the total growth is $e^{\Delta G} - 1 = 14.0\%$ which is acceptable.

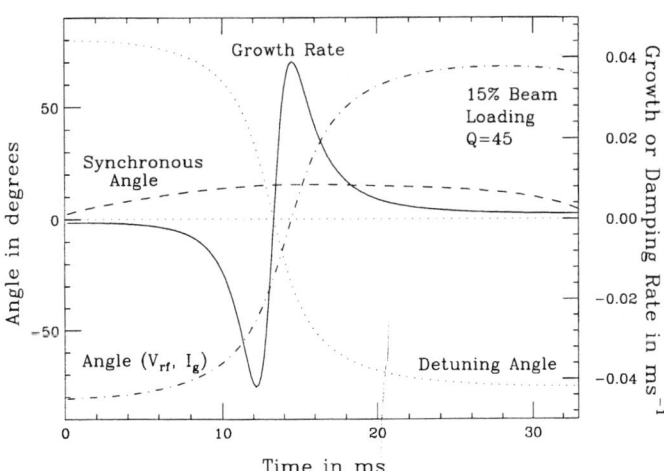

FIGURE 7. When the cavity resonant frequency is chosen as the rf frequency at the middle of the ramp at 13.33 ms, although the detuning angle as well as the angle between the cavity gap voltage $\tilde{V}_{\text{rf}}$ and the generator current $\tilde{I}_g$ are large at both ends of the ramp, they are relatively smaller at the middle of the ramp where the gap voltage is large.

We also want to see whether Robinson's criterion for stable phase oscillation is satisfied for this rf consideration. For the second half of the ramp where the detuning angle $\psi < 0$, the phase is stable because we are below transition and the synchronous angle φ_s is between 0 and $\frac{1}{2}\pi$ (see Appendix B). For the first half of the ramp where $\psi > 0$, the sufficient condition for stability is the high-intensity Robinson's criterion:

$$\frac{i_{\text{im}}}{i_0} < \frac{\cos\varphi_s}{\sin\psi \cos\psi} . \tag{4.4}$$

Figure 8 plots both sides of the criterion and shows that the criterion is well satisfied.

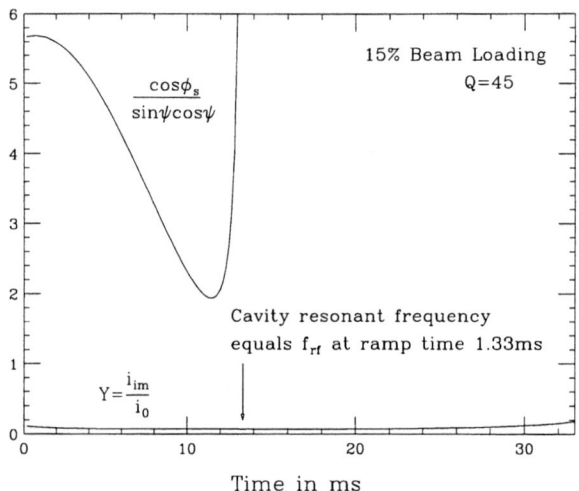

FIGURE 8. Plot showing the high-intensity Robinson's phase-stability criterion is satisfied.

V CONCLUSION

We started from a rf system designed by Griffin for the future Fermilab prebooster with space charge slightly over-compensated and transient beam loading 85% compensated. Based on a resonant ramp curve with 3.75% second harmonic, we studied the possibility of having fixed-frequency non-tunable rf cavities. In order to reduce the phase angle between the rf voltage and the generator current, we proposed to set the resonant frequency of the cavity equal to the rf frequency at the middle of the ramp. Robinson phase instability would result in the second half of the ramp. The total integrated growth in synchrotron oscillation amplitude was found to be only $\sim 14\%$ which is small enough to be acceptable. We also checked that the whole ramp satisfies Robinson's high-intensity criterion for stable phase oscillation.

APPENDIX

In this appendix, we try to gather together the derivations of some of the formulas used in the paper. Most of them are well known. They are included here just for completeness.

A RF DETUNING

The rf cavity has a loaded shunt impedance R_s, a loaded quality factor Q_L, and resonates at frequency $f_R = \omega_R/(2\pi)$. Corresponding to a beam particle revolving with frequency $f_0 = \omega_0/(2\pi)$, the rf frequency is $f_{\rm rf} = \omega_{\rm rf}/(2\pi) = h\omega_0$, where h is the rf harmonic. The impedance of the cavity seen by the particle at $f_{\rm rf}$ can be written approximately as

$$Z_{\rm cav} = \frac{R_s}{1 - jQ_L\left(\frac{\omega_R}{\omega_{\rm rf}} - \frac{\omega_{\rm rf}}{\omega_R}\right)} \approx R_s \cos\psi\, e^{j\psi}, \tag{A.1}$$

where ψ is the rf tuning angle, which is defined as

$$\tan\psi = 2Q_L \frac{\omega_R - \omega_{\rm rf}}{\omega_R}. \tag{A.2}$$

This detuning is necessary because (1) we want the load to appear real to the generator (the generator current i_g in phase with the cavity gap voltage $V_{\rm rf}$) so that there will not be any power reflection to the generator, and (2) both the generator voltage V_g and the beam-loading voltage $V_{\rm im}$ contribute to the cavity gap voltage. This is illustrated in the phasor diagram in Fig. 9, where the tilde represents a phasor rotating counter-clockwise with angular frequency $\omega_{\rm rf}$. Here, we assume most of the transient beam-loading has been cancelled; therefore, the image current phasor $\tilde{i}_{\rm im}$ has a magnitude much smaller than that of the beam current phasor $\tilde{i}_b$. According to Eq. (A.1), we see from Fig. 9 that both the beam-loading voltage phasor $\tilde{V}_{\rm im}$ and the generator voltage phasor $\tilde{V}_g$ are at a phase ψ ahead of their respective current phasors $\tilde{i}_{\rm im}$ and $\tilde{i}_g$. Since these two voltage phasors add up to give the gap voltage phasor $\tilde{V}_{\rm rf}$ which has a synchronous angle φ_s, we must have after dividing by $R_s \cos\phi$,

$$i_g \sin\psi = i_{\rm im} \sin(\tfrac{\pi}{2} - \varphi_s + \psi). \tag{A.3}$$

Resolving the current contributions along $\tilde{i}_g$, we have

$$i_g = i_0 + i_{\rm im}\sin\varphi_s, \tag{A.4}$$

where $i_0 = V_{\rm rf}/R_s$ is the total current in phase with the cavity gap voltage. Eliminating i_g, we arrive at

$$\tan\psi = \frac{i_m \cos\varphi_s}{i_0}. \tag{A.5}$$

185

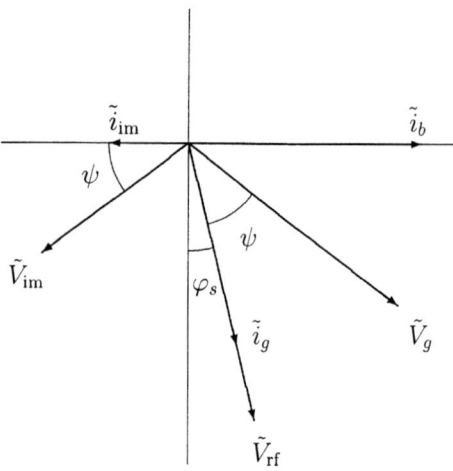

FIGURE 9. Phasor plot showing the vector addition of the generator voltage phasor $\tilde{V}_g$ and the beam-loading voltage phasor $\tilde{V}_{im}$ to give the gap voltage phasor $\tilde{V}_{rf}$ in a rf cavity. Note the detuning angle ψ which put the gap current phasor $\tilde{i}_g$ in phase with the gap voltage phasor.

B ROBINSON'S STABILITY CRITERIA

Now let us study the conditions for phase stability. Suppose that the beam particle has a slightly larger energy than the synchronous particle. After a revolution or h rf periods, $\tilde{i}_b$ in Fig. 9 will be ahead of the x-axis by a small angle $\epsilon > 0$ if it is below transition. Then the accelerating voltage it sees will be $V_{rf} \sin(\varphi_s - \epsilon)$ instead of $V_{rf} \sin \varphi_s$, or an extra decelerating voltage of $\epsilon V_{rf} \cos \varphi_s$, and it receives less energy from the cavity than the synchronous particle. The motion is therefore stable. This is Robinson's criterion for establishing stable phase oscillation when beam loading can be neglected [5]. In other words, one requires

$$\begin{cases} 0 < \varphi_s < \frac{\pi}{2} & \text{below transition,} \\ \frac{\pi}{2} < \varphi_s < \pi & \text{above transition.} \end{cases} \quad \text{(B.1)}$$

When beam loading is included, the gap voltage phasor $\tilde{V}_{rf}$ will be modified also, because the image current phasor $\tilde{i}_{im}$ and hence the beam-loading voltage phasor $\tilde{V}_{im}$ also advance by the small angle ϵ after h rf periods. The extra beam-loading voltage phasor is $\epsilon i_{im} R_s \cos \psi \, e^{j(\psi + 3\pi/2)}$. If $\psi < 0$, this phasor will point into the 3rd quadrant and decelerate the particle in concert with $\epsilon V_{rf} \cos \varphi_s$, causing no instability. On the other hand, if $\psi > 0$, this phasor will point into the 4th quadrant and accelerate the particle instead. To be stable, the extra accelerating voltage on the beam must be less than the amount of decelerating voltage $\epsilon V_{rf} \cos \varphi_s$, or

$$\frac{i_{im}}{i_0} < \frac{\cos \varphi_s}{\sin \psi \cos \psi} \qquad \begin{cases} \psi > 0 \text{ below transition,} \\ \psi < 0 \text{ above transition,} \end{cases} \quad \text{(B.2)}$$

which is called Robinson's high-intensity criterion for phase stability. Satisfying this criterion just enables stable oscillating like sitting inside a stable potential well and there will not be any damping. Violating this criterion the particle will be in an unstable potential well so that phase oscillation will not be possible.

C TRANSIENT BEAM LOADING INCLUDING PREVIOUS PASSAGES

We follow closely the approach by Boussard [7]. Let the bunch spacing be h_b rf buckets or T_b in time. The cavity time constant or filling time is $T_f = 2Q_L/\omega_R$ and the e-folding voltage decay decrement between two successive bunch passages is $\delta_L = T_b/T_f$. During this time period, the phase of the rf fields changes by $\Psi = \omega_R T_b - 2\pi h_b$, which can also be written in terms of the detuning angle,

$$\Psi = (\omega_R - \omega_{\rm rf})T_b = \delta_L \tan\psi , \tag{C.1}$$

where Eq. (A.2) has been used. The transient beam-loading voltage left by the first passage of a short bunch carrying charge q is $V_{t0} = q/C$. The total beam-loading voltage V_t seen by a short bunch is obtained by adding up vectorially the beam-loading voltages for all previous bunch passages. The result is

$$V_t = \tfrac{1}{2}V_{t0} + V_{t0}(e^{-\delta_L}e^{j\Psi} + e^{-2\delta_L}e^{j2\Psi} + \cdots) , \tag{C.2}$$

where the $\tfrac{1}{2}$ in the first term on the right side is the result of Wilson's fundamental theorem of beam-loading, which states that a particle sees only one-half of its own induced voltage. It is worth pointing out that these voltages are not phasors or components at the rf frequency. They are vectors that contain components of all frequencies. The summation can be performed exactly giving the result

$$V_t = \frac{q}{C}\left[F_1(\delta_L,\psi) + jF_2(\delta_L,\psi)\right] , \tag{C.3}$$

with

$$F_1 = \frac{1 - e^{-\delta_L}}{2D} , \quad F_2 = \frac{e^{-\delta_L}\sin(\delta_L\tan\psi)}{D} ,$$

$$D = 1 - 2e^{-\delta_L}\cos(\delta_L\tan\psi) + e^{-2\delta_L} . \tag{C.4}$$

Notice that $\delta_L \approx \pi h_b/Q_L$, which is 0.0698 for $h_b = 1$ and $Q_L = 45$. When the detuning angle $\psi = 0$, $|V_t| \approx V_{t0}/(2\delta_L)$. The functions F_1 and F_2 are computed at some other values of ψ, which are listed in Table 1 and plotted in Fig. 10.

We see that the total transient beam loading V_t falls rapidly as the detuning angle ψ increases. It vanishes approximately $\sim 88.7°$ and oscillates rapidly after that. However, the choice of a large ψ is not a method to eliminate beam loading, because the steady-state beam loading will not be reduced.

TABLE 1. F_1 and F_2 for some values of the detuning angle ψ.

ψ	$\Psi = \delta_L \tan\psi$	F_1	F_2
0°	0°	$\sim \dfrac{1}{2\delta_L}$	0
84.9°	45°	0.061	1.197
87.5°	90°	$\sim \dfrac{\delta_L}{4}$	$\sim \dfrac{1}{2}$
88.7°	180°	$\sim \dfrac{\delta_L}{8}$	0

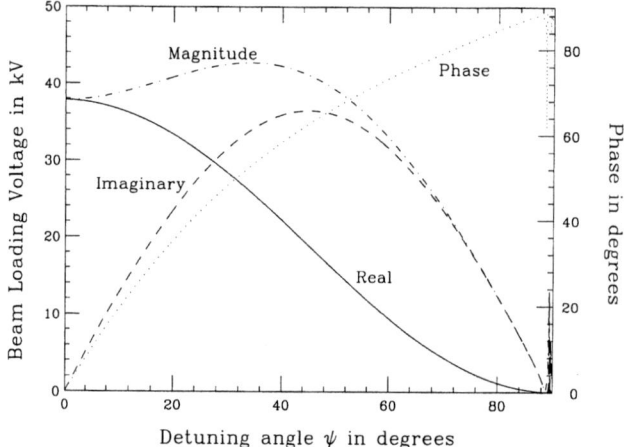

FIGURE 10. Plot of transient beam-loading voltage including all previous bunch passages, $\dfrac{q}{C}(F_1 + jF_2)$, versus detuning angle ψ.

REFERENCES

1. C. Ankenbrandt, these proceedings.
2. J.E. Griffin, *RF System Considerations for a Muon Collider Proton Driver Synchrotrons*, Fermilab report FN-669, 1998.
3. K.Y. Ng and Z.B. Qian, Proc. Phys. at the First Muon Collider and at Front End of a Muon Collider, Fermilab, Batavia, Nov. 6-9, 1997.
4. J.E. Griffin, K.Y. Ng, Z.B. Qian, and D. Wildman, *Experimental Study of Passive Compensation of Space Charge Potential Well Distortion at the Los Alamos National Laboratory Proton Storage Ring*, Fermilab Report FN-661, 1997.
5. P.B. Robinson, *Stability of Beam in Radiofrequency System*, Cambridge Electron Accel. Report CEAL-1010, 1964.
6. See for example, H. Wiedemann, *Particle Accelerator Physics II*, Springer, 1995, p.203.
7. D. Boussard, *Beam Loading*, Fifth Advanced CERN Accelerator Physics Course, Rhodes, Greece, Sep. 20-Oct. 1, 1993, CERN 95-06, p.415.

Studies of Space-Charge Physics in Beams for Advanced Accelerator Applications

J. G. Wang, S. Bernal, P. Chin, R. A. Kishek, Y. Li, M. Reiser, M. Venturini, W. W. Zhang, Y. Zou,
Institute for Plasma Research, University of Maryland, College Park, MD 20742

T. Godlove and D. Kehne, FM Technologies, Fairfax, VA 22032

I. Haber, Naval Research Laboratory, Washington DC 20375

R. C. York, NSCL, Michigan State University, East Lansing, MI 48824

Abstract. We review experimental observations of space-charge effects and collective phenomena in charged particle beams for accelerator applications. These include halo formation and emittance growth, bunch profile evolution, space-charge waves, and longitudinal instabilities. We also report on the development of the University of Maryland Electron Ring for the study of space-charge physics in a circular lattice.

Introduction

Advanced accelerator applications, such as high-intensity linacs injecting synchrotrons (or pulse-compressor rings) for spallation neutron sources, drivers for heavy ion inertial fusion, high energy boosters, free electron lasers, high-power microwave generators, etc., require ever-increasing beam intensity. Important beam dynamics issues for these types of applications are space-charge effects and the collective behavior of charged particles. The interaction of charged particles with external and self-fields limits the maximum transportable current, degrades beam quality through emittance dilution and increased energy spread. Hence, it is crucial to fully understand these phenomena in order to develop advanced accelerators for high intensity applications.

At the University of Maryland we have been conducting research programs to study space-charge physics in charged particle beams for advanced accelerator applications. Both transverse and longitudinal dynamics have been experimentally investigated by employing low-energy, high-current, non-relativistic electron beams. The main objective of our measurements is to provide a testbed to check theory and simulation codes against experimental observations. We will briefly review some of the important results from our experiments.

An electron recirculator is being developed for studying the space-charge physics of charged particle beams in circular systems. This facility will extend our ongoing beam physics experiments to a much longer distance than is available today. More importantly, the crucial new beam physics issues in a circular lattice will be investigated. These include space-charge tune shift, resonance traversal and maximum transportable current, thermal equilibrium and equipartitioning, bending and dispersion, space-charge waves and instabilities. We will report on the design, progress, and status of the E-ring project.

CP448, *Workshop on Space Charge Physics in High Intensity Hadron Rings*
edited by A. U. Luccio and W. T. Weng
© 1998 The American Institute of Physics 1-56396-824-X/98/$15.00

Experimental Observations

1. Halo formation and emittance growth [1-3]

A highly space-charge dominated electron beam distributed in the form of five beamlets is matched by two solenoid lenses into a 5-m long transport channel focused by 36 short solenoids. For well matched conditions, no halo particles are observed in the experiments and computer simulation. Halo appears at a distance of one betatron oscillation if the beam is rms-mismatched. The experimental observation of the beam transverse images on an axially-movable fluorescent screen is shown in Fig.1, where computer simulation plots are also depicted. Both theory and simulation indicate an emittance growth due to halo particles, which are not detected by an emittance meter due to its low sensitivity.

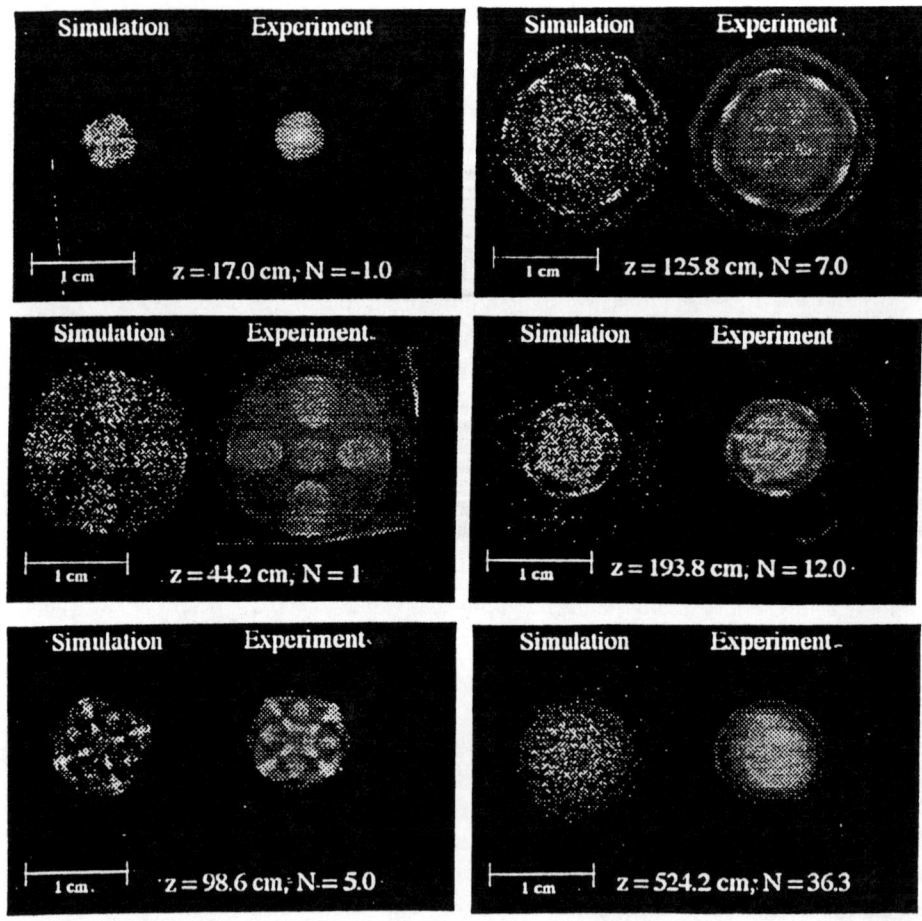

Fig. 1 Simulation plots and fluorescent-screen pictures of the beam images at six different locations along the transport channel.

2. Bunch profile evolution [4-6]

Parabolic bunches are a natural Boltzmann distribution in the space-charge dominated case. A parabolic bunch features a linear space-charge force and linear velocity distribution along the bunch. Its parabolic profile is preserved during drift expansion and linear compression. The dynamics of parabolic bunches is governed by the longitudinal envelope equation. In contrast, a rectangular bunch is associated with highly nonlinear space-charge forces along the bunch. It suffers rapid edge erosion during drift expansion due to strong space-charge forces on its edges, which generate rarefaction waves. Hence, linear compression of a rectangular bunch is not a good scheme to handle the space-charge forces on the edges of the bunch. The dynamics of rectangular bunches is governed by a one-dimensional nonlinear fluid equation with a solution by the method of characteristics. The experimental observation of an initially rectangular bunch during drift expansion is shown in Fig. 2, where the evolution of both the bunch current profile and energy distribution is depicted.

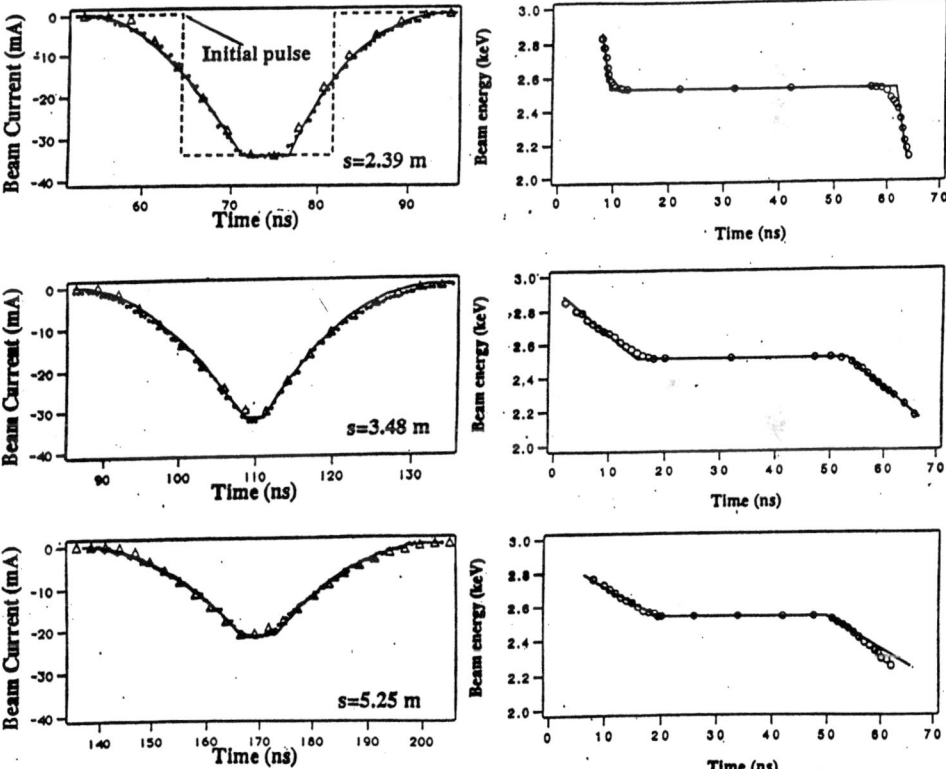

Fig. 2 Current profiles and energy spread of a rectangular bunch at different locations along the transport channel, where the solid lines are from theory, circles and triangles are from experiment, while dots from simulation.

3. Space-charge waves [7-9]

Longitudinal space-charge waves can be produced on beams due to external perturbations. Experiments have been performed to generate localized space-charge waves and to observe their propagation on beams. With an active device such as a gridded electron gun, a single slow wave or a single fast wave can be generated. A typical picture of slow and fast space-charge waves propagating on beams is shown in Fig. 3. This measurement yields the speed of the space-charge waves, a measure of the strength of longitudinal space-charge effect, and leads to a determination of the geometry factor due to perturbations. The reflection and transmission of space-charge waves at bunch ends are also observed. The technique with localized space-charge waves can also be employed to diagnose important beam parameters, such as space-charge impedance, etc.

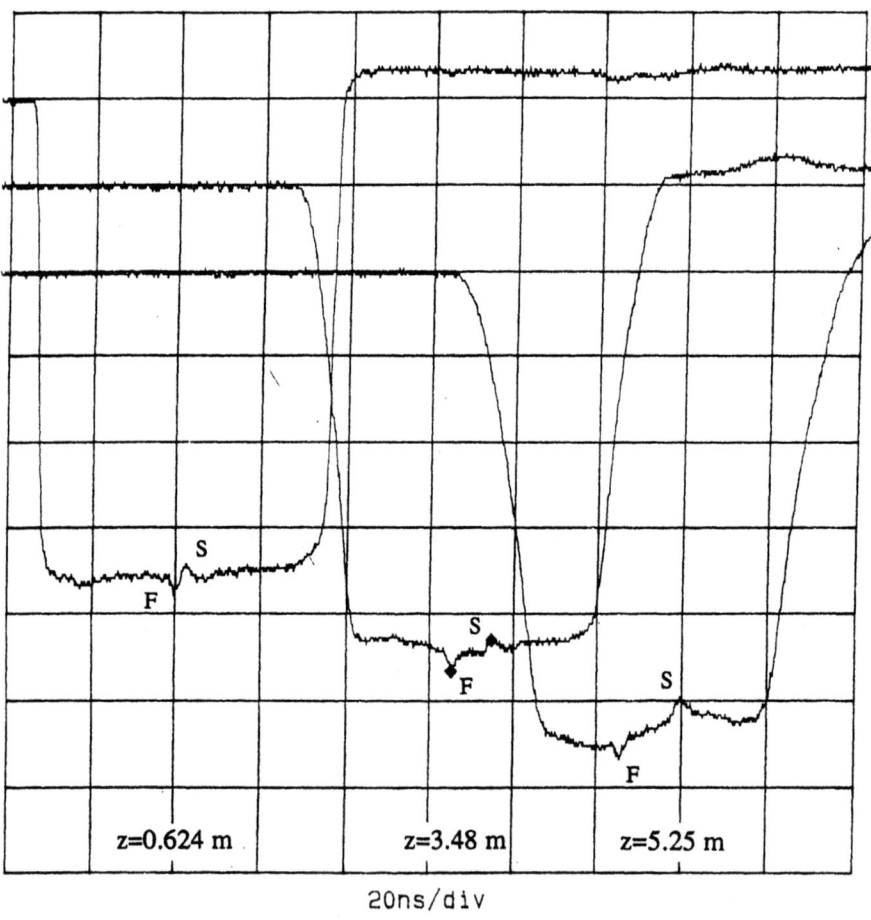

Fig. 3 Beam current signals with slow (S) and fast (F) space-charge waves at three different location along a transport channel.

4. Longitudinal instabilities [10-11]

Space-charge waves suffer longitudinal instability in a dissipative environment. An experiment is performed in a resistive-wall channel focused by a uniform, long solenoid. The channel length is about 1 meter and has a total resistance of 5 - 10 kΩ. Localized space-charge waves are generated and transported through the channel. The diagnostics includes beam current monitors and energy analyzers. The properties of the instability are studied in detail in the long wavelength range. Both the growth of a slow wave, as shown in Fig. 4, and the decay of a fast wave are observed. The growth/decay rates are measured and compared with theory.

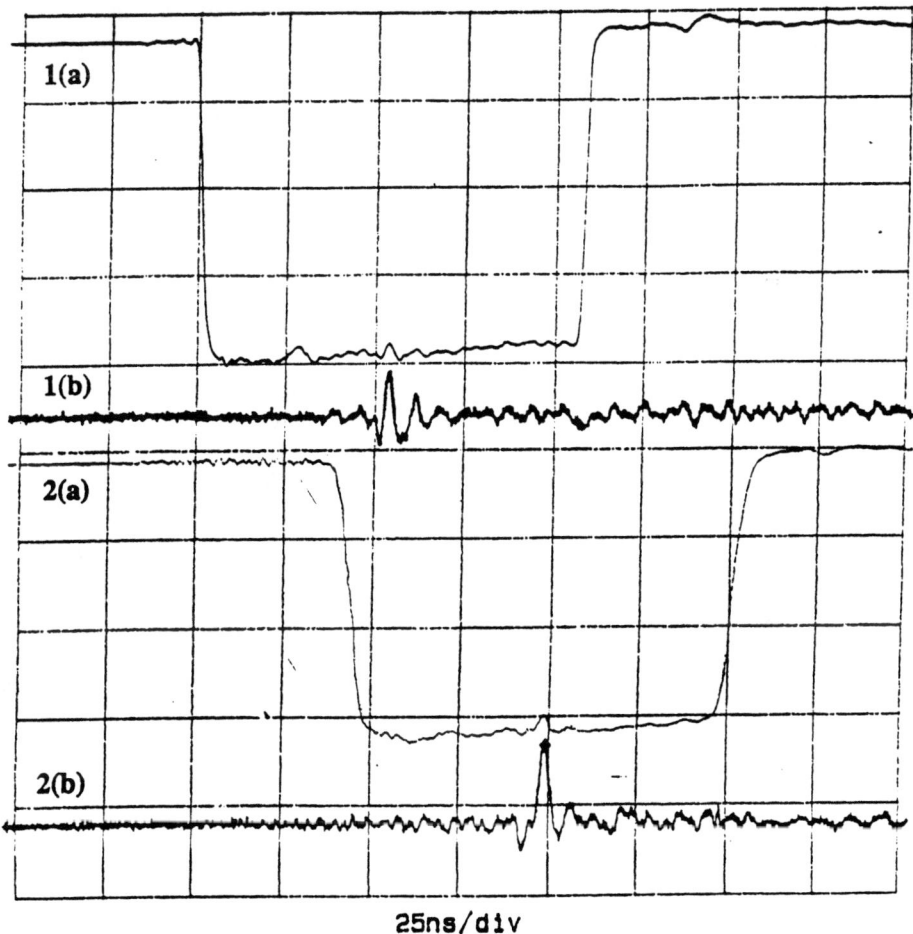

Fig. 4 Growth of a single slow wave in a resistive channel, where 1(a) is the beam current with a slow wave before the resistive channel, 1(b) is the slow wave only, 2(a) is the beam current with the slow wave after the resistive channel, and 2(b) is the slow wave only.

Development of an Electron Ring

The design and development of a small electron ring to study the beam dynamics in a recirculator has been reported elsewhere [12]. The schematic of the E-ring layout is shown in Fig. 5. The E-ring will transport an electron beam of 10 keV, 100 mA, 50 ns in a smooth circular lattice of 11.5 m in circumference. The focusing elements are 70 printed-circuit quadrupoles, plus two Panofsky-type magnets of special design for injection and extraction. The beam is deflected by 36 printed-circuit dipoles along the ring. Three induction modules provide longitudinal focusing and, in Phase II of the project, acceleration up to 40 keV. The diagnostics include BPMs, energy analyzers, and phosphor screens. A pulsed extraction system similar to the injector is included for beam analysis in an extraction chamber housing an emittance meter, an energy analyzer and a phosphor screen viewer. The ring main parameters are summarized in Table I.

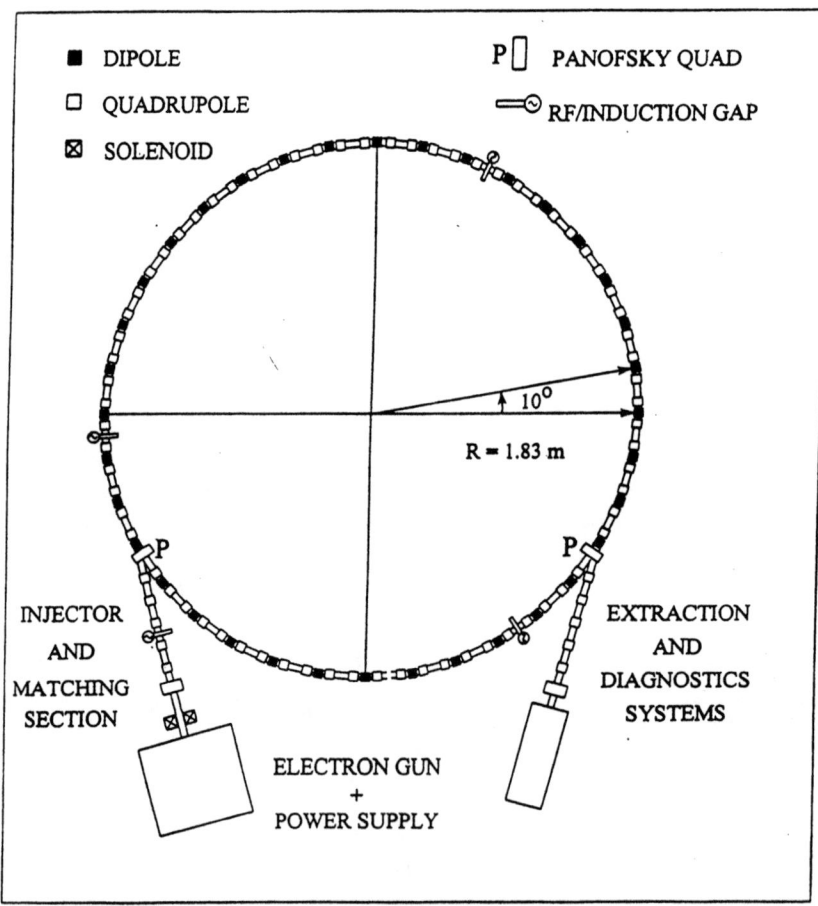

Fig. 5 Schematic of electron ring.

Table I. E-ring design parameters

Injection Energy	10 keV
Injection Current	100 mA
Generalized Perveance	0.0015
Initial Emittance	10 mm mrad*
Mean Beam Radius	1.04 cm
Tube Bore Diameter	4.90 cm
Circumference	11.5 m
Lap Time	197 ns
Lattice Periods	36
Half-Lattice Length	16 cm
Quadrupole Length	4.4 cm
Quadrupole Diameter	5.3 cm
Gradient	8.1 G/cm
Tune υ_o	7.6
Phase advance σ_o	76°
Phase advance σ	9° - 25°

* Normalized effective emittance (4 x rms)

The magnetic quadrupoles and dipoles for the ring lattice are made of flexible printed circuit board to form a pattern of current loops designed to minimize aberrations [13]. A major concern of the magnet performance was the rather large fringe fields caused by the short aspect ratio (<1) of the magnets. The field distribution and multipole contents of the dipoles and quadrupoles have been measured extensively [14]. The performance of the quads has been studied in a scaled injector experiment [15]. A prototype of the induction module (and its modulator) has also been developed. The device can provide 3 kV, 3 ns rise time "ear fields" at a repetition rate of 5 MHz [16]. For the diagnostics, a resistive-wall current/position monitor and a capacitive current/position monitor have been built and tested. The monitors have achieved fast rise time, high resolution and accuracy of beam current and centroid positions [17].

The mechanical design of the E-ring has been completed in collaboration with the National Superconducting Cyclotron Lab at Michigan State University. The ring physically consists of 18 mechanical sections. Each section contains two FODO lattice periods, two 10° bends. A diagnostic chamber serving as a pumping port and housing a capacitive BPM and a phosphor screen viewer is included in every other sections. A prototype section has been completed and tested for mechanical, vacuum, and beam transport properties. The injector consists of a gridded electron gun, a short solenoid and five printed quadrupoles for matching, a fast Panofsky quadrupole and a fast deflecting dipole [18]. The extraction structure is a mirror image of the injector, except that the electron gun is replaced by a diagnostic chamber. The injector is being developed at FM Technologies, VA, under a DOE small business project. The new electron gun has already been built.

Extensive computations with single particle codes such as MARYLIE [19], DIMAD, and TRACE have been performed for error analyses, earth-field compensation, and correction schemes. Computer simulation with the WARP code has shown very promising performance of the E-ring for ten turns [20]. Theoretical studies on the beam envelope including space charge and dispersion reveal interesting beam physics [21].

Summary

The space-charge effects and collective phenomena in space-charge dominated electron beams are studied at the University of Maryland for advanced accelerator applications. The experimental observations presented in the paper include the formation of halo particles due to mismatch, evolution of an initially rectangular bunch during drift expansion, propagation of localized space-charge waves on beams, and growth of a slow wave in a resistive-wall channel. The development of a small electron ring is in progress. We have completed the design and prototyped the major components of the ring, such as magnets, induction modules, resistive and capacitive BPMs, mechanical sections, etc. The construction of the E-ring will start soon.

Acknowledgments

We would like to acknowledge the contributions to the experimental observations by Drs. D.X. Wang and H. Suk. This research is supported by the U.S. Department of Energy.

References

[1] M. Reiser, Theory and Design of Charged Particle Beams (John Wiley, New York, 1994).
[2] D. Kehne, M. Reiser, and H. Rudd, AIP Conf. Proc. **253**, 47 (1991).
[3] I. Haber, D. Kehne, M. Reiser and H. Rudd, Phys. Rev. A **44**, 5194 (1991).
[4] J. G. Wang, D. X. Wang, and M. Reiser, *Appl. Phys. Lett.* **62**(6), 645 (1993).
[5] D. X. Wang, J. G. Wang, D. Kehne, and M. Reiser, Appl. Phys. Lett. **62** (25), 3232 (1993).
[6] D. X. Wang, J. G. Wang, and M. Reiser, Phys. Rev. Lett. **73** (1), 66 (1994).
[7] J. G. Wang, D. X. Wang, and M. Reiser, Phys. Rev. Lett. **71** (12), 1836 (1993).
[8] J. G. Wang, H. Suk, D. X. Wang, and M. Reiser, Phys. Rev. Lett. **72** (13), 2029 (1994).
[9] J. G. Wang, D. X. Wang, H, Suk, and M. Reiser, Phys. Rev. Lett. **74** (16), 3153 (1995).
[10] J. G. Wang, H. Suk, and M. Reiser, Phys. Rev. Lett. **79** (6), 1042 (1997).
[11] J. G. Wang and M. Reiser, *Physics of Plasmas*, **5**(5), 2064 (1998).
[12] M. Reiser, S. Bernal, A. Dragt, M. Venturini, J. G. Wang, H. Onishi, and T. Godlove, Fusion Engineering and Design **32-33**, 293 (1996).

[13] T.F. Godlove, S. Bernal, M. Reiser, IEEE CAT. 95CH35843, 2117 (1995).
[14] W. W. Zhang, S. Bernal, P. Chin, R. Kishek, M. Reiser, M. Venturini, J. G. Wang, in Bulletin of the American Physical Society, **43** (2) 1204 (1998).
[15] S. Bernal, R. Kishek, Y. Li, M. Reiser, J. G. Wang, and T. Godlove, in Bulletin of the American Physical Society, **43** (2) 1204 (1998).
[16] J. J. Deng, J. G. Wang, M. Reiser, and T. Godlove, Proc. 1997 Part. Accel. Conf., Vancouver, Canada, May 1997.
[17] H. Suk, J. G. Wang, S. Bernal, and M. Reiser, ibid.
[18] T. Godlove, S. Bernal, J. J. Deng, Y. Li, M. Reiser, J. G. Wang, and Y. Zou, ibid.
[19] M. Venturini, S. Bernal, A. Dragt, M. Reiser, J. G. Wang, and T. F. Godlove, *Fusion Engineering and Design* **32-33**, 283 (1996).
[20] R. A. Kishek, I. Haber, M. Venturini, and M. Reiser, "PIC Code Simulations of the Space-Charge-Dominated Beam in the University of Maryland Electron Ring", these Proceedings.
[21] M. Venturini, R. A. Kishek, and M. Reiser, "Dispersion and Space Charge", these Proceedings.

Space Charge Effect and the Booster to AGS Transfer

S.Y. Zhang
AGS Department, BNL

I. INTRODUCTION

The envelope equation [1] predicts that when the maximum tune shift is depressed beyond the half integer resonance, or integer resonance, by a quarter of the shift, the envelope resonance will happen.

Simulations [2,3] further show that under similar conditions, the beam emittance will blow-up.

Several experiments conducted on low energy proton synchrotrons [4-9] have confirmed the simulation.

For high intensity proton runs, emittance blow-up is often accompanied with nontrivial beam loss. It has been suggested [10] that the largest portion of the Booster to AGS (BTA) transfer loss is located in the AGS end. It was also found that the space charge tune spread, immediately after the bunched beam enters the AGS ring, is large enough for the envelope resonance and emittance blow-up [11].

It was decided, in 1997 high intensity proton run, to raise the BTA transfer kinetic energy from $1.54\ GeV$ to $1.9\ GeV$. The transfer efficiency was increased, as expected. The beam emittance in the AGS ring did not increase, even the AGS acceptance was enlarged [12], which is probably because of the reduced space charge tune spread.

II. SPACE CHARGE EFFECT AND THE EMITTANCE GROWTH

A. Envelope Oscillation

The defocusing space charge force can be shown as k in the usual equation

$$x'' + Kx = kx \tag{1}$$

where $K > 0$ is the external focusing force.

The effect of space charge is resemble to the quadrupole error, or the half-integer error. However, there are two fundamental differences. Firstly, the space charge force as an internal force affects each other among the particles. This force is distributed among the particles unequally. Secondly, a strong space charge effect will produce larger beam size. In turn, a larger beam size necessarily implies to a weaker space charge effect, as shown in,

$$k = \frac{\alpha}{\sigma^2} \quad (2)$$

where α is a constant, and σ is the beam size.

To deal with these problems, we let,

$$x = \sigma \cos \phi \quad (3)$$

where both the envelope σ and the derivative of phase advance, ϕ', are not constant. Substituting (3) into (1), we get,

$$\sigma'' - \frac{\epsilon_0^2}{\sigma^3} + K\sigma - k\sigma = 0 \quad (4)$$

where ϵ_0 is the beam emittance defined by the equilibrium beam size σ_0 as,

$$\epsilon_0 = \sqrt{K}\sigma_0^2 \quad (5)$$

Substituting (2) into (4), we get,

$$\sigma'' - \frac{\epsilon_0^2}{\sigma^3} + K\sigma - \frac{\alpha}{\sigma} = 0 \quad (6)$$

which is the envelope equation.

Assuming symmetric beam, the Laslett tune shift is [13,14],

$$\Delta \nu = \frac{-N r_0 R}{4\pi \nu_0 \beta^2 \gamma^3 \sigma^2} \quad (7)$$

where N is the total particle number, r_0 is the classical radius of proton, R is the machine radius, and ν_0 is the bare tune. Note that here σ is the rms beam size. The coefficient α is associated with the tune shift $\Delta \nu$ by

$$\alpha = \frac{-2\nu_0 \Delta \nu \sigma_0^2}{R^2} \quad (8)$$

Substituting (8) into (6), an equation describing the envelope σ with respect to the maximum tune shift $\Delta \nu$ is established. Analyzing this equation, Sacherer [1] found that the envelope resonance happens when the maximum tune shift $\Delta \nu$ is depressed beyond the half integer resonance by a quarter of the shift. For instance, let the AGS vertical tune at the injection be 8.85, then the envelope resonance will happen if the Laslett tune shift is larger than 0.47.

We now give some comments on this result.

1. Qualitatively, the envelope equation analysis is very suggestive. The Laslett tune shift can be depressed beyond the half integer resonance by a certain amount of the shift, without incurring envelope resonance.

2. One reason of this is that the envelope resonance is relevant to the resonant modes, rather than to individual particles. Note that the Laslett tune shift is the maximum tune shift, which is relevant to only a small portion of the particles [14].

3. The envelope resonance can be understood as follows. When certain amount of particles have tune shift larger than the half integer resonance, these particles perform betatron oscillation with large amplitude in terms of modes, due to the beta function distortion. Though this large amplitude oscillation is not the emittance blow-up in usual sense, it reduces the tune shift. In this way, envelope oscillation happens.

4. In the envelope equation analysis, the quadrupole (half integer) error is not essential, in terms of envelope oscillation. It does, however, affects the strength of the resonance, as seen from the envelope equation with the exciting force f,

$$\sigma'' - \frac{\epsilon_0^2}{\sigma^3} + K\sigma - \frac{\alpha}{\sigma} = f \qquad (9)$$

5. The emittance growth is not directly related to the envelope resonance. However, it is clear that the envelope resonance is a source of the emittance growth, given the machine imperfections.

6. Under weak envelope resonance, the beam loss can be understood by the beam scraping as the emittance approaches the machine acceptance. Under strong envelope oscillation, the turbulent resonance certainly implies large increase of the oscillation amplitude of some modes, which can cause massive beam loss within short period of time, accompanied by rapid emittance blow-up.

B. Emittance growth and simulation

The envelope oscillation implies a significant equilibrium emittance growth only for very large space charge force, which happens for the low energy end of Linac. For most synchrotrons, therefore, the envelope oscillation will not give rise to essential emittance growth.

In the simulation performed for LEB of SSC [2], it was found that the beam emittance grows when the bare tune becomes smaller than 11.66, for a fixed tune spread $\Delta\nu_{inc} = -0.33$. With half integer resonance, the particle distribution change has been observed, which is believed to cause the emittance growth. The simulation is also applied to some machine studied, and relatively consistent results were obtained [3].

In reality, other factors might contribute to the emittance growth if the envelope oscillation becomes strong. For instance, the nonlinearity of the quadrupole focusing, and the nonlinearity of the space charge force itself.

It is of interest, therefore, to confirm the relevance between the onset of the envelope oscillation and the emittance growth.

III. STUDY OF LOW ENERGY PROTON SYNCHROTRONS

In the past two decades, experiments have been performed on several low energy proton synchrotrons. The emittance growth is indeed observed, when the space charge incoherent tune spread is depressed beyond half integer, or integer, by about a quarter of the tune shift.

For convenience, let us define the variable χ to represent the percentage of the tune spread beyond the half integer, or integer.

A. Fermilab Booster

In an experiment performed at the Fermilab Booster, the beam intensity was increased, while the emittance was observed [4]. The Linac beam emittance was $\epsilon_{N,95\%} = 7\ \pi\mu m$. At $N = 1.2\ TP$, $(1TP = 10^{12})$, emittance growth was observed. The tune spread is calculated as $\Delta\nu_{inc} = -0.38$, which is kept constant for higher intensities, because of the emittance growth. For example, at $N = 3\ TP$, the emittance is proportionally enlarged to $\epsilon_{N,95\%} = 17\ \pi\mu m$.

In this case the bare tune is $\nu_y = 6.8$, and we have $\chi = 21$.

B. CERN PS

At the CERN Proton Synchrotron (PS), for a fixed beam intensity, the RF voltage variation was used for the space charge tune shift adjustment, instead of using the variable intensity. The growth of the emittance was observed [5,6]. For the vertical bare tune of $\nu_y = 6.22$, at $\Delta\nu_{inc} = -0.28$, emittance growth was observed. For shorter bunches, i.e. stronger space charge effects, the emittance grows such that the real incoherent tune spread is kept constant. We calculate that $\chi = 21$.

Another set-up shows that at the vertical bare tune of $\nu_y = 6.28$, at $\Delta\nu_{inc} = -0.37$, emittance grows. In this case, we have $\chi = 24$.

C. PS Booster

No report for dedicated study of the emittance growth was available. Instead, some information can be learned in the intensity push, reported for the PS Booster [7].

The space charge formulation in [7] uses a form factor for the transverse particle distribution, which has been, in fact, accounted in the equation (7). Therefore, this factor is double counted. Moreover, other complications involved in the intensity push, such as that a debuncher in Linac is used to change the particle distribution, the vertical beam mis-steering and x-y coupling are used in the injection, double RF is used, the beam horizontal emittance is 3 times of the vertical one, and the ring is filled up, with heavy beam losses. All these made the use of the original data difficult.

It is known that the PS Booster vertical limiting physical half aperture is 32 mm, similar to the AGS Booster. The limiting average beam rms size is, therefore, about 10 mm. Using $\sigma = 10$ mm and the typical high intensity $N = 6 \times 10^{12}$ particle per ring for PSB, we get $\Delta\nu_{inc} = -0.66$, meanwhile $\nu_y = 5.45$. That is $\chi = 32$.

D. PSR

Various set up for the bare tune and intensity were made to observe the emittance growth at the Proton Storage Ring (PSR) at the Los Alamos [8].

The space charge incoherent tune shift defined in [8] is smaller than (7) by a factor of 2, which for Gaussian distribution happens to be the rms tune spread. For consistency, we still use the tune spread calculated by (7). Let the original rms beam size be 4.25 mm, then the emittance growth with respect to the intensity increase and also the bare tune is shown in Table 1. We observe that only two cases, i.e. A1 and B1, have no emittance growth. In other cases, the emittance grows proportionally to the intensity increase. This happens for each bare tune set up. We may also see that this experiment failed to identify the emittance growth thresholds, for that we can only say $\chi < 34$, which comes from the cases A2 and C1.

Case	ν_y	N TP	σ_y rms size, mm	$\epsilon_{N,95\%}$ emittance, $\pi\mu m$	χ %
A1	2.193	6	4.25	11.75	-30
A2	2.193	11.8	4.95	15.94	34
A3	2.193	23	6.65	28.76	66
B1	2.142	6	4.2	11.47	7
B2	2.142	11.8	5.75	21.51	53
B3	2.142	23	7.75	39.07	76
C1	2.100	6	4.75	14.68	35
C2	2.100	11.8	7.25	34.19	67
C3	2.100	23	10.75	75.17	83
D1	2.059	6	6.3	25.82	63
D2	2.059	11.8	10.3	69.01	81

Table 1

E. AGS Booster

At the AGS Booster multiturn injection, the vertical mis-steering is used to define the vertical emittance, which we assume to be equivalent to $\epsilon_{N,95\%} = 30\ \pi\mu m$. For '95 and '96 HEP run, the beam vertical emittance observed at the BTA multiwire 006 clearly indicates the emittance growth along with the intensity increase [9], which is shown in Fig.1. It is known that the Booster extraction septum has a vertical aperture at about $\epsilon_{N,95\%} = 65\ \pi\mu m$. This limitation is shown in Fig.1. Since there is emittance growth, the Booster vertical acceptance is not determined at 200 MeV, but at about 10 ms after the injection, 340 MeV, where the bare tune has been reduced from $\nu_y = 4.95$ at the injection to $\nu_y = 4.80$. If we take $\epsilon_{N,95\%} = 40\ \pi\mu m$ for $N = 40\ TP$, and $\epsilon_{N,95\%} = 60\ \pi\mu m$ for $N = 60\ TP$, shown by a solid line in Fig.1, then we can calculate that $\Delta\nu_{inc} = -0.36$. Therefore, we get $\chi = 17$.

The relation between the space charge tune spread and the beam loss is shown in Fig.2, where the bare tune tuning along the cycle was used to calculate the maximum tune shift beyond the half integer. It is interesting to observe that when the space charge tune spread is below the onset of the envelope oscillation, the beam loss becomes trivial.

F. Summary

The studies performed on the low energy proton synchrotrons have shown the emittance growth at the threshold predicted as the onset of the envelope oscillation. Therefore, it is convinced that the envelope oscillation can cause beam emittance growth. The results are summarized as follows.

	FermiB	CPS	CPS	PSB	PSR	AGSB
Tune, ν_y	6.80	6.22	6.28	5.45	2.19	4.80
Shift, $-\Delta\nu_{inc}$	0.38	0.28	0.37	0.66	<0.29	0.36
χ	21	21	24	32	<34	17

Table 2

IV. AGS INJECTION LOSS

Extensive studies of 1996 high intensity proton run have shown the following results.

1. Mainly using the beam loss monitors, it was found that the loss at the Booster extraction and the BTA transfer line accounts about 5 TP for 90 TP Booster late, per AGS cycle [15,16].

2. Meanwhile, the AGS injection loss accounts for about 3/4 of the total BTA loss, i.e. at the Booster Late 90 TP, the AGS injection loss is about 18 TP [10].

3. Due to horizontal scraping at the Booster extraction septum F6, and the vertical aperture limitation, also at F6, the beam in the BTA transfer line has both horizontal and vertical emittance about 60 $\pi\mu m$ [17]. In AGS, the emittance becomes 100 $\pi\mu m$.

Take the beam intensity arriving the AGS ring $N = 84\ TP$, the bunching factor $B_f = 0.2$, (the bunch length about 110 ns), and the normalized beam emittance $\epsilon_{N,95\%} = 60\ \pi\mu m$. We get the incoherent space charge tune spread $\Delta\nu_{inc} = -0.80$, at the BTA transfer kinetic energy 1.54 GeV. This shows that the space charge effect at AGS injection might be responsible for the beam loss and emittance blow-up in the AGS ring.

Other experience in the machine operation seems agree with this argument.

1. The BTA loss was strongly intensity-dependent.

2. The AGS vertical tune at injection porch was very sensitive to the beam loss.

3. The BTA optics matching, even the steering to the AGS, was not very sensitive to the beam loss.

4. Since the BTA line has limited the beam emittance, only the bunch length at the Booster extraction was sensitive, which affects the space charge effect in terms of beam peak current arriving the AGS ring.

5. Since the space charge type loss is distributed around the AGS ring, the loss pattern does not change much in different machine tuning.

For high intensity proton run, the AGS vertical aperture was filled, the normalized beam emittance is $\epsilon_{N,95\%}^V = 100\ \pi\mu m$. The implied vertical aperture limit is the same as the one reported for pre-Booster era high intensity proton run [18]. The increase of the beam emittance in the AGS ring from the one in the BTA line, 60 $\pi\mu m$, also indicates the possibility of the turbulent envelope oscillation and the accompanied emittance blow-up.

V. HIGH MOMENTUM BTA TRANSFER

It was decided, in 1997 SEB run, to raise the BTA transfer kinetic energy to 1.9 GeV. The main parameters at the transfer line are as follows.

E_k	1.54	1.90	GeV
$\epsilon_{N,95\%}$	60	70	$\pi\mu m$
N	84	84	TP
B_f	0.2	0.2	
$\Delta\nu_{inc}$	0.80	0.51	

Table 3

The beam normalized emittance arriving at the AGS ring is limited by the BTA transfer line acceptance, which is enlarged as the physical aperture unchanged. The bunching factor should be smaller at higher transfer momentum. However, the beam was expected to extract from the Booster at the zero $\dot{B}$, thus the RF voltage could be lowered. Therefore, similar bunching factor was expected. In overall, the space charge incoherent tune spread can be reduced by about 40%.

A. Booster

The highest Booster late intensity was less than 81 TP per AGS cycle in 1997, compared with larger than 92 TP in 1996 SEB run. Very limited running period of only 8 weeks might be the major reason of this.

Some difficulties were encountered at the zero $\dot{B}$ extraction, probably the beam loading effect. A little higher RF voltage was applied, and the extraction was moved 1.5 ms before the zero $\dot{B}$.

The Booster magnetic cycle was extended from 133 ms to 150 ms, keeping the Booster injection change to the minimum, in order to concentrate the efforts on the BTA transfer.

B. Transfer Efficiency

The BTA transfer efficiency was improved as expected. Five morning numbers in the Daily Log for 1996 and 1997 operations, with largest Booster late and AGS CBM (Early) intensities, are shown in the following table.

1996	Booster Late	AGS CBM	Unit
(1)	90	67.0	TP
(2)	92	64.5	TP
(3)	89	67.2	TP
(4)	90	66.4	TP
(5)	90	66.5	TP
Average	90.2	66.3	TP
1997	Booster Late	AGS CBM	
(1)	80.3	65.8	TP
(2)	78.2	65.2	TP
(3)	78.8	65.2	TP
(4)	76.2	65.2	TP
(5)	74.5	63.5	TP
Average	77.6	65.0	TP

Table 4

C. Beam Emittance

From the observation at the BTA transfer line multiwire 006, we found that the horizontal beam size did not change, which implies that the emittance was increased from 60 $\pi\mu m$ of 1996 to 70 $\pi\mu m$ in 1997. The vertical beam emittance was increased from 65 $\pi\mu m$ of 1996 to 70 $\pi\mu m$ in 1997.

A serious concern of the high momentum transfer was the emittance blow up in the AGS ring. The AGS extraction encountered problems in 1996 run, due to the limited acceptance of extraction lines. At 1.9 GeV, the AGS acceptance is enlarged by 17%. If the space charge effect was still strong enough to let the AGS ring be filled, then the beam could be unable to extract from the AGS. The measurement of the beam emittance at the AGS extraction line, however, had shown that the beam size was not enlarged [19].

The bunch length was a little less than 100 ns, i.e., the bunching factor is 0.18, smaller than 0.2 in 1996. Since the beam intensity arriving at the AGS was never get close to 84 TP anyway, so the space charge incoherent tune spread was probably about 0.5. A naive conclusion, therefore, is that at $\Delta\nu_{inc} = -0.5$ and the tune 8.85, the space charge induced emittance growth is probably not strong enough to fill up the AGS ring.

D. Transfer Line Beam Loss

If the calibration of the loss monitor readings for higher momentum at $(1.9\ GeV/1.54\ GeV)^2 = 1.5$ is used, then the beam loss at the Booster extraction and the AGS injection section was a little larger than 1996, but comparable, as shown in Table 5.

	1996	1997	
Booster Late	82 TP	80 TP	
AGS CBM	60 TP	67 TP	
Loss Monitor	Sum	Sum	'97/'96
Booster F8	15,890	28,280	1.78
Booster A1	2,753	7,955	2.89
BTA (7)	75,697	16,460	0.22
AGS L20	23,350	54,126	2.32
AGS A5	2,250	6,770	3.01

Table 5

Due to the loss monitor saturation of Booster F6 (BF6), BF7, and BTA023 in 1996, the comparison was made by the loss monitors located at immediate downstream, for instance BF8 and BA1 for BF6 and BF7. It seems that some improvements could be made at the Booster extraction and the AGS injection sections.

Meanwhile, the loss in the transfer line was reduced significantly, but this is probably due to improvements independent of the momentum shift.

VI. Research Bibliography

[1] F. Sacherer, Ph.D Thesis, UCRL-18454, Lawrence Radiation Laboratory, 1968.

[2] S. Machida, Nucl. Inst. Meth, A-309, p.43, 1991.

[3] S. Machida, Proc. Particle Accelerator Conf. p.3224, 1993.

[4] C. Ankenbrandt and S.D. Holms, Proc. 1987 IEEE Particle Accelerator Conf. p.1066, 1987.

[5] R. Cappi et al. Proc. Particle Accelerator Conf. p.171, 1991.

[6] R. Cappi et al. Proc. Particle Accelerator Conf. p.3570, 1993.

[7] J.P. Delahaye et al. Proc. 11th Int. Conf. High Energy Acc. p.299, 1980.

[8] D. Neuffer et al. Proc. Particle Accelerator Conf. p.1893, 1991.

[9] S.Y. Zhang, SNS Tech. Note, No.32, Apr. 1997.

[10] L.A. Ahrens, AGS Studies Report, No.349, Aug. 1996.

[11] S.Y. Zhang, AGS Tech. Note, No.451, Oct. 1996.

[12] S.Y. Zhang, AGS Tech. Note, No.474, Jan. 1998.

[13] L.J. Laslett, Proc. of 1963 summer Study of Storage Rings, BNL-Report 7534, p. 324, 1963.

[14] S.Y. Zhang, T. Roser, and W.T. Weng, AGS Tech. Notes, No.449, Oct. 1996.

[15] L.A. Ahrens, AGS Studies Report, No.347, Aug. 1996.

[16] S.Y. Zhang, AGS Studies Report, No.351, Sep. 1996.

[17] S.Y. Zhang, AGS Studies Report, No.350, Aug. 1996.

[18] E. Raka et al. IEEE Trans. Nucl. Sci. NS-32, No.5, p.3110, 1985.

[19] J.W. Glenn and K. Brown, private communication.

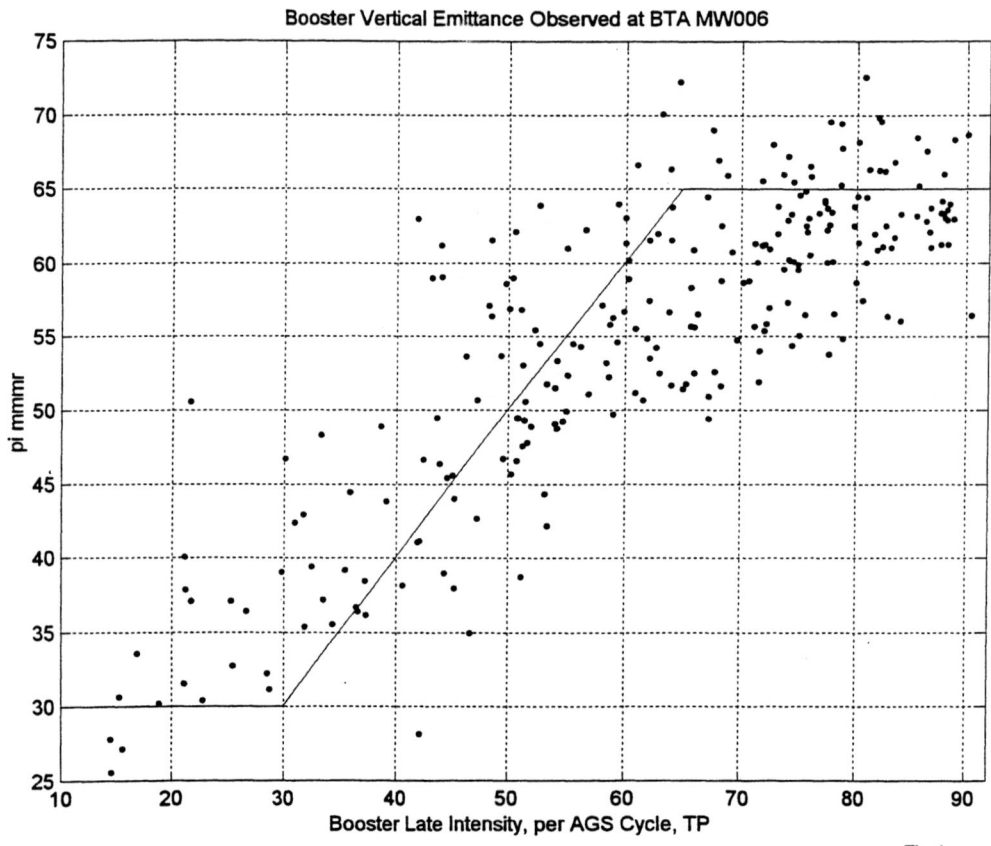
Fig.1

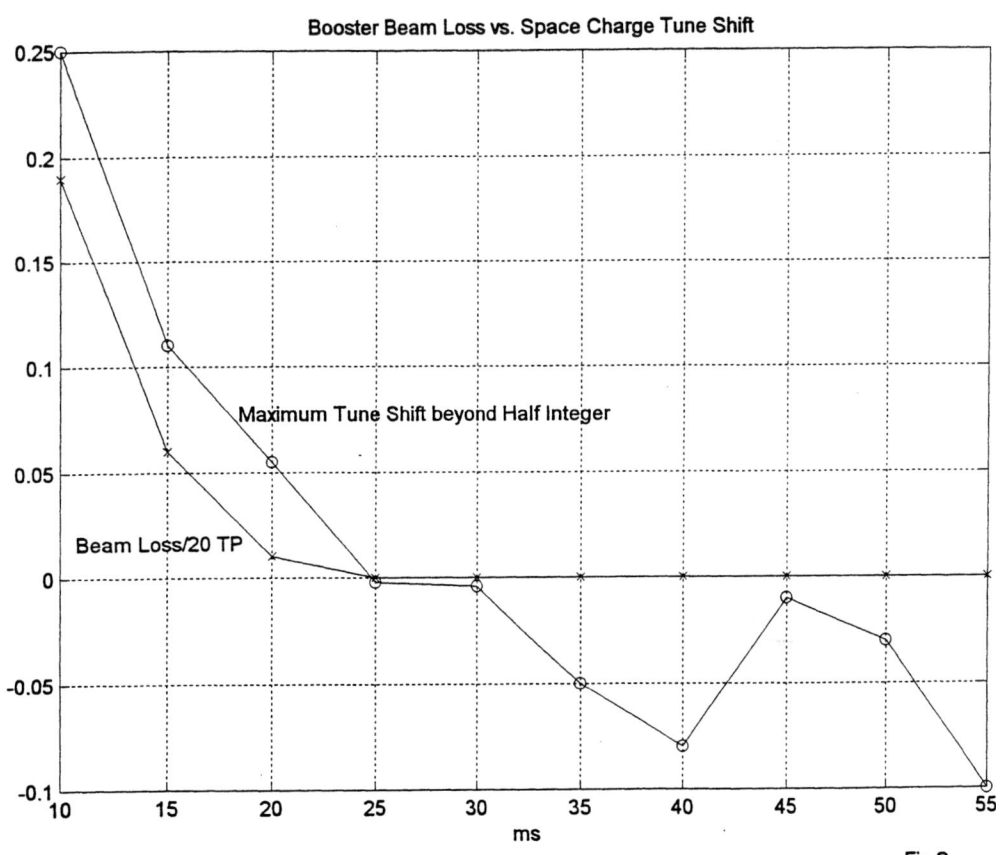

Fig.2

CONTRIBUTED PAPERS: THEORY

An Exactly Solvable Model of Transverse Stability in Bunched Beams

Michael Blaskiewicz*
AGS Department Brookhaven National Laboratory

Abstract

An exactly solvable model of transverse stability in bunched beams is presented. The model assumes a square well longitudinal potential with a hollow longitudinal phase space distribution. The effects of space charge in conjunction with both short and long range wake fields as well as damping systems are investigated.

INTRODUCTION

Transverse instabilities have been known since the early days of accelerators [1]. For coasting beams where the betatron tune Q_x is independent of betatron amplitude the eigenvalue problem is relatively straightforward and reduces to a one dimensional dispersion integral. For bunched beams the problem is more difficult. Various approximate techniques have been developed [2, 3, 4, 5] but they fail in parameter regimes commonly used in hadron accelerators [6]. The main difficulty appears to be the space charge force which can lead to tune shifts ΔQ_{sc} which are large compared to the synchrotron tune Q_s. Particle tracking techniques are reliable over a larger region of parameter space than the approximate techniques, but there are serious convergence problems for values of $\Delta Q_{sc}/Q_s \gtrsim 10$ [6]. In the Brookhaven AGS and AGS Booster $\Delta Q_{sc}/Q_s \approx 100$ at injection energies. For RHIC injection $\Delta Q_{sc}/Q_s \approx 20$ is expected.

In this paper a model which is precisely solvable for large space charge tune shifts is discussed. The model involves fairly serious approximations of both the longitudinal dynamics and the longitudinal phase space distribution. However, for zero chromaticity this model predicts growth rates and thresholds which are close (within 10%) to those obtained by particle tracking using more realistic longitudinal dynamics [6].

---

*Work supported by the United States Department of Energy

The square well model is reviewed and a method of incorporating a simple damping system is introduced. The stability of the system is evaluated for a range of accelerator parameters.

THE SQUARE WELL MODEL

Take the machine azimuth, θ, to be the time-like variable which increases by 2π each turn. The longitudinal coordinate is the relative arrival phase within the bunch, $\tau = \omega_0 t - \theta$ where ω_0 is the angular revolution frequency and t is real time. The bunch occupies $0 \leq \tau \leq \tau_b$, where τ_b is the bunch length. Assume that particles experience no longitudinal forces except at the ends of the bunch, where they are reflected. Take the longitudinal phase space distribution where all particles have $|d\tau/d\theta| = v_0$ so that the line density is a boxcar function and there is a well defined synchrotron tune

$$Q_s = v_0 \pi / \tau_b.$$

Let $X_+(\theta, \tau)$ be the transverse offset of particles with $d\tau/d\theta = -v_0$ (positive momentum offset below transition) at azimuth θ and arrival phase τ. Define $X_-(\theta, \tau)$ in similar fashion for particles with $d\tau/d\theta = +v_0$. Assume that all betatron tune shifts are small compared to the betatron tune so that

$$X_+(\theta, \tau) = e^{-iQ_x \theta} x_+(\theta, \tau)$$

where the variation of $x_+(\theta, \tau)$ is slow in θ, and similarly for X_-. Then

$$\begin{aligned}
\frac{dx_+}{d\theta} &= \frac{\partial x_+}{\partial \theta} - v_0 \frac{\partial x_+}{\partial \tau} \\
&= i\Delta Q_{sc}\{x_+(\theta,\tau) - \bar{x}(\theta,\tau)\} - i\xi v_0 x_+ \\
&\quad + i\kappa \int_0^\tau ds\, W(\tau - s)\bar{x}(\theta,s) \\
&\quad + \text{contributions from previous turns} \quad (1)\\
\frac{dx_-}{d\theta} &= \frac{\partial x_-}{\partial \theta} + v_0 \frac{\partial x_-}{\partial \tau} \\
&= i\Delta Q_{sc}\{x_-(\theta,\tau) - \bar{x}(\theta,\tau)\} + i\xi v_0 x_- \\
&\quad + i\kappa \int_0^\tau ds\, W(\tau - s)\bar{x}(\theta,s) \\
&\quad + \text{contributions from previous turns.} \quad (2)
\end{aligned}$$

In these equations $W(\tau)$ is the transverse wake potential, ξ is proportial to the chromaticity, and $\bar{x} = (x_+ + x_-)/2$ is the transverse centroid of the beam. The scaling factor is

$$\kappa = \frac{I_0 c}{4\pi\beta(E_T/q)\omega_0^2 Q_x}.$$

Consider the eigenvalue problem where $\partial/\partial\theta \to -i\Delta Q_x$. Assume the wake potential is given by

$$W(\tau) = \sum_{j=1}^{N} W_j \exp(-\alpha_j \tau), \tag{3}$$

for $\tau > 0$ and $W(\tau) = 0$ for $\tau \leq 0$. In addition suppose there is a damper which produces a rigid mode tune shift $-i\Delta Q_d$. An ideal damper would add

$$\left.\frac{dx_\pm(\theta,\tau)}{d\theta}\right|_d = -\frac{\Delta Q_d}{\tau_b}\int_0^{\tau_b} d\tau' \bar{x}(\tau'), \tag{4}$$

to the second part of equations (1) and (2). This would require a fairly major revision of the code and is not necessary. Instead, a decay length β_d satisfying $\beta_d \tau_b \ll 1$ is introduced. The equations become

$$\frac{dx_+(\tau)}{d\tau} = -i\left\{\frac{\Delta Q_{sc}/2 + \Delta Q_x - \xi v_0}{v_0}\right\} x_+ - (i\sum_j F_j + G)/v_0 + i(\Delta Q_{sc}/2v_0)x_-$$

$$\frac{dx_-(\tau)}{d\tau} = +i\left\{\frac{\Delta Q_{sc}/2 + \Delta Q_x + \xi v_0}{v_0}\right\} x_- + (i\sum_j F_j + G)/v_0 - i(\Delta Q_{sc}/2v_0)x_+$$

$$\frac{dF_j(\tau)}{d\tau} = -\alpha_j F + (\kappa W_j/2)\{x_- + x_+\}$$

$$\frac{dG(\tau)}{d\tau} = -\beta_d G - (\beta_d \Delta Q_d/2)\{x_- + x_+\} \tag{5}$$

In these units the head-tail phase shift is $\chi = -\xi\tau_b/2$. The boundary conditions are $x_+(0) = x_-(0)$, and $x_+(\tau_b) = x_-(\tau_b)$, since an instantaneous change in $d\tau/d\theta$ leaves x and p_x unchanged. The boundary condition on the damping term is $G(0) = G(\tau_b)$ which yields a nearly constant $G(\tau)$ for small β_d. Also, the wake force is continuous

$$F_j(0) = e^{-2\pi i(Q_x + \Delta Q_x) - [2\pi - \tau_b]\alpha_j} F_j(\tau_b). \tag{6}$$

The coupling between bunches can be ignored by setting the right hand side of the last equation to zero. The ΔQ_x in the exponential is usually too small to be interesting and will be ignored. The differential equations are of the form

$$\frac{d\mathbf{U}}{d\tau} = \mathbf{MU}$$

where **M** is a constant matrix and, except for special cases,

$$\mathbf{U} = \sum_k \mathbf{U}_k e^{\lambda_k \tau}.$$

The $\mathbf{U}_k$s and λ_ks can be obtained to machine precision with standard routines. The computational strategy is:
1) guess a value for ΔQ_x
2) find the $\mathbf{U}_k$s and λ_ks for ΔQ_x
3) find an error term due to not satisfying the boundary conditions
4) modify ΔQ_x to reduce the error term
5) go to 2 if error is too big, otherwise end

For small coherent forces the eigenvalues are $\Delta Q_x = $ integer $\times Q_s$ so one has some idea of the range the tune shift will occupy. By using many starting points one can be fairly certain that the growth rate of the most unstable mode has been found. From a computational point of view steps 3 and 4 are the most problematic. The error term in step 3 is obtained by ignoring the boundary condition $x_+(\tau_b) = x_-(\tau_b)$ and replacing this with the constraint $x_+(0) = 1$. This yields as many constraints as unknowns, and $\Delta x(\Delta Q_x) \equiv x_+(\tau_b) - x_-(\tau_b)$ is taken as the error term. To improve the tune shift value calculate $\Delta x(\Delta Q_x + \epsilon)$ and $\Delta x(\Delta Q_x + i\epsilon)$ where ϵ is a small real quantity. This yields numerical estimates of $\partial Re(\Delta x)/\partial Re(\Delta Q_x)$, $\partial Re(\Delta x)/\partial Im(\Delta Q_x)$, $\partial Im(\Delta x)/\partial Re(\Delta Q_x)$, and $\partial Im(\Delta x)/\partial Im(\Delta Q_x)$. An updated value of ΔQ_x is obtained in the linear approximation. To reduce the likelyhood of jumping over weak zeroes there is an upper limit on the magnitude of the change in ΔQ_x, which is of order the distance between the inital starting points.

An example of the tune search is shown in Figure 1. A small upper limit on the ΔQ_x step was used to produce the graph but a value of 0.1 found all the eigenmodes. The input parameters are $\Delta Q_{sc}/Q_s = 2$, $\Delta Q_d = 0$, $N = 1$, $\alpha_1 \tau_b = 0$, and $\kappa W_1 \tau_b / Q_s = 6$. Additionally, the boundary condition $F_1(0) = 0$ was used. For space charge alone the tune shifts are given by $\Delta Q_x = -\Delta Q_{sc}/2 \pm \sqrt{(\Delta Q_{sc}/2)^2 + (kQ_s)^2}$, where k is an integer and the $+$ is used for $k \geq 0$ [6]. The rightmost tune shift in Figure 1 is $\Delta Q_x/Q_s = 5.07$ while $\Delta Q_x/Q_s = 5.08$ for space charge alone and $k = 6$. The leftmost tune shift in Figure 1 is $\Delta Q_x/Q_s = -5.11$ while $\Delta Q_x/Q_s = -5.12$ for space charge alone and $k = -4$.

Since there are $11 = 4+6+1$ unique tune values in the figure there are no missed eigenvalues in the interval $-5 \leq \Delta Q_x/Q_s \leq 5$. It is reasonable to believe that large growth rates will not occur for larger values of ΔQ_x and was verified by computations using a larger search interval.

The eigenmode corresponding to the largest growth rate of Figure 1 is shown in Figure 2. The magnitude of $x_+ - x_-$ at the endpoints is less than 10^{-10}. Identifying $\Delta Q_x/Q_s \approx 5$ with $k = 6$ and $\Delta Q_x/Q_s \approx -5$ with $k = -4$ suggests that the eigenmode results from coupling between the $k = 0$ and $k = -1$ sidebands.

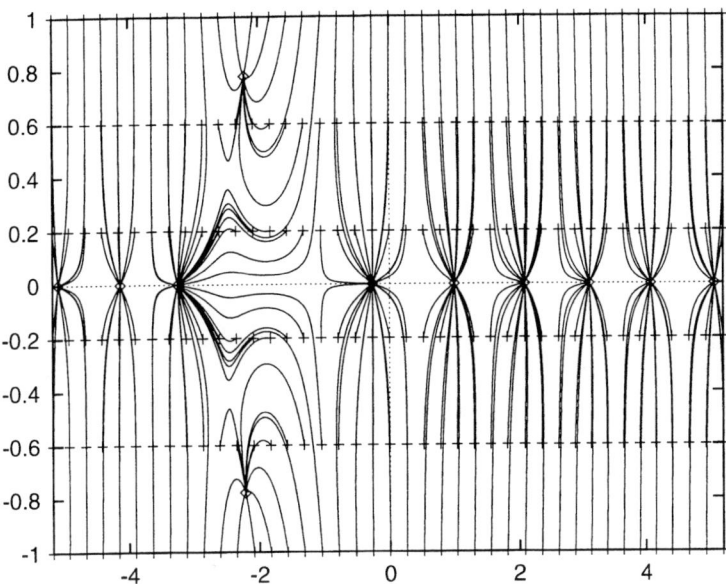

Figure 1: $Im(\Delta Q_x/Q_s)$ .vs. $Re(\Delta Q_x/Q_s)$ during the tune search. The crosses are the initial values, the diamonds are the final values, and the solid lines show the evolution of the search.

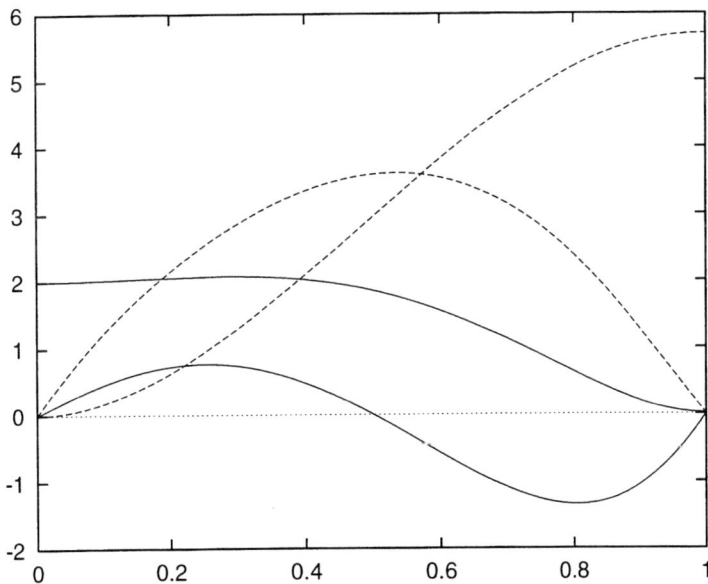

Figure 2: $x_+ + x_-$ and $x_+ - x_-$ .vs. τ/τ_b for the most unstable eigenmode of Figure 1. The real parts are solid lines and the imaginary parts are dashed.

MODEL PREDICTIONS

The model described in the previous section is easily applied to a range of accelerator parameters. The most striking prediction to date involves the effect of space charge on the fast head tail instability due to short range wakes. Consider a one-pole wakefield ($N = 1$) and ignore coupling between bunches ($F_1(0) = 0$). Then,

$$\frac{\Delta Q_x}{Q_s} = \frac{\Delta Q_x}{Q_s}\left(\chi, \alpha_1 \tau_b, \frac{\Delta Q_{sc}}{Q_s}, \frac{\kappa W_1 \tau_b}{Q_s}\right) \quad (7)$$

For zero chromaticity the threshold value of the wake force $\kappa W_1 \tau_b / Q_s$ depends only on $\alpha_1 \tau_b$ and $\Delta Q_{sc}/Q_s$. This is show in Figure 3. For $\Delta Q_{sc} = 0$ and $\alpha_1 \tau_e \gg 1$ the threshold wake satisfies $\kappa W_1 \approx 0.5 * v_0 \alpha^2$. For $\Delta Q_{sc}/Q_s > 10$ the curves in Figure 3 are nearly straight lines.

$$\left\{\frac{\kappa W_1 \tau_b}{Q_s}\right\}_{thresh} = a(\alpha_1 \tau_b) + b(\alpha_1 \tau_b)\frac{\Delta Q_{sc}}{Q_s}. \quad (8)$$

Using a least square fit, the coefficients are $a(0) = 2.16$, $b(0) = 2.63$; $a(5) = 5.48$, $b(5) = 6.18$; $a(10) = 11.80$, $b(10) = 10.22$; $a(20) = 37.4$, $b(20) = 15.6$. The rms difference between the fit and the plotted curves was less that 0.2% of the minimum value in the fit interval.

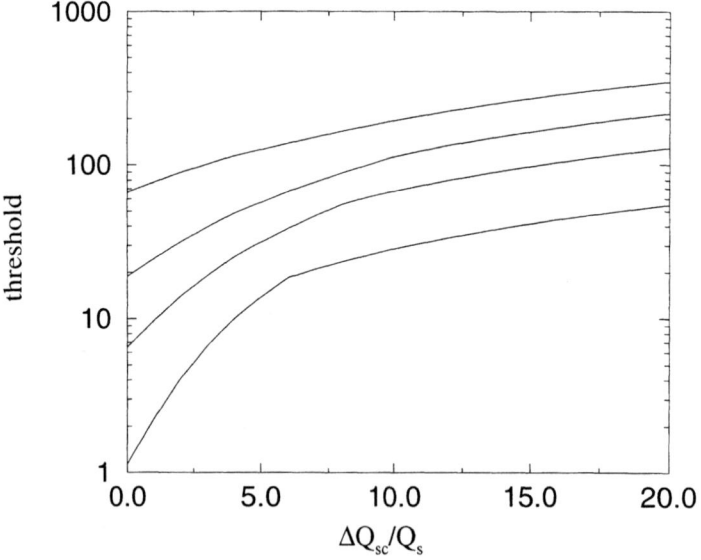

Figure 3: Threshold value of $\kappa W_1 \tau_b / Q_s$ versus $\Delta Q_{sc}/Q_s$ for the analytic square well with $\chi = 0$ and $\alpha_1 \tau_e = 0, 5, 10, 20$ from bottom to top.

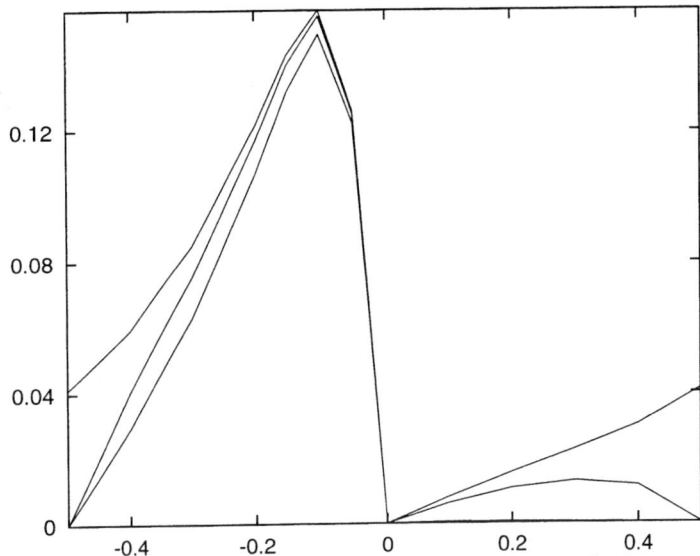

Figure 4: $Im(\Delta Q_x/Q_s)$ for the most unstable mode .vs. Q_x for $\alpha_1 = 0.1$, $\kappa W_1 \tau_b/Q_s = 0.2$, $\tau_b = 1$. For $\Delta Q_{sc} = 0$ the fast head tail threshold is $\kappa W_1 \tau_b/Q_s = 0.51$ for $Q_x = 0$ and $\kappa W_1 \tau_b/Q_s = 2.1$ for $Q_x = 0.5$. $\Delta Q_{sc}/Q_s = 0, 10, 20$ from bottom to top

When the range of the wakefield is long and equation (6) is used, space charge is less benign. Figure 4 illustrates the effect. For $\Delta Q_{sc} = 0$, $Im(\Delta Q_x/Q_s) < 10^{-3}$ for $0 \leq Q_x \leq 0.5$. This is a well known low intensity effect[2]. Conversely, the growth rates for larger values of ΔQ_{sc} could be very serious in a real machine. Also, for $\Delta Q_{sc}/Q_s = 20$ and $Q_x = 0.5$ there was an unstable eigenmode. From symmetry considerations this mode must be the result of mode coupling, since the weak coupling growth rates vanish identically.

The effectiveness of dampers can be reduced by space charge. For $\Delta Q_{sc}/Q_s = 100$ and optimal gain, Figure 5 shows that the growth rate is nearly 100 times larger than for $\Delta Q_{sc} = 0$.

CONCLUSIONS

The square well model and its computer implementation have been described. The model yields precise answers in parameter regimes where other techniques are unreliable. The model predictions are mixed. While instabilities due to short range wake fields are damped by space charge it can decrease stability when long range wake fields are present.

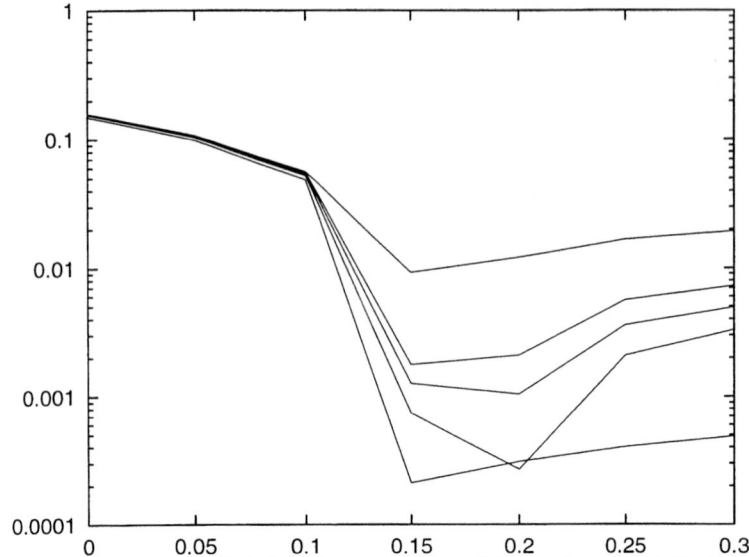

Figure 5: $Im(\Delta Q_x/Q_s)$ for the most unstable mode .vs. $\Delta Q_d/Q_s$ for $Q_x = -0.1$ and other parameters as in Figure 4. $\Delta Q_{sc}/Q_s = 0, 10, 20, 50, 100$ from bottom to top

References

[1] L.J. Laslett, V.K. Neil, A.M. Sessler, *The Review of Scientific Instruments* **36**, # 4, p. 436, (1965).

[2] F. Sacherer, CERN Report 77-13 (1977) p. 198.

[3] G. Besnier, D. Brandt, B. Zotter, *Particle Accelerators* **17**, 51-77, (1985).

[4] Y. Chin, K. Satoh & K. Yokoya, *Particle Accelerators* **13**, p45 (1983).

[5] K. Satoh & Y. Chin, *Nuclear Instruments and Methods* **207**, p309 (1983).

[6] M. Blaskiewicz, submitted to *Phys Rev ST Accel Beams* .

Emittance Growth in Heavy Ion Rings Due to the Effects of Space Charge and Dispersion*

John J. Barnard, George D. Craig, Alex Friedman,
David P. Grote, Bojan Losic, and Steven M. Lund

Lawrence Livermore National Laboratory, L-645, Livermore, CA 94550

Abstract

We review the derivation of moment equations which include the effects of space charge and dispersion in bends first presented in ref. [1]. These equations generalize the familiar envelope equations to include the dispersive effects of bends. We review the application of these equations to the calculation of the change in emittance resulting from a sharp transition from a straight section to a bend section, using an energy conservation constraint. Comparisons of detailed 2D and 3D simulations of intense beams in rings using the WARP code (refs. [2,3]) are made with results obtained from the moment equations. We also compare the analysis carried out in ref. [1], to more recent analyses, refs. [4,5]. We further examine self-consistent distributions of beams in bends and discuss the relevance of these distributions to the moment equation formulation.

Introduction

There are many applications in which beams having non-negligible space charge forces are transported through bends. In heavy ion fusion (HIF), recirculating induction accelerators (recirculators), with large tune depressions, and with rapid acceleration through resonances, are being considered to ignite inertial confinement fusion targets. Even in linac approaches to HIF, designs of the final transport to the target usually include transport through 180 degrees or more of bend section. In some Acclerator Production of Tritium designs, a final bent transport section is being considered as part of an upgrade option. For the application of studying high energy density in matter, a beam pulse in a storage ring will be longitudinally compressed, reaching tune shifts for short periods much larger than allowed by the Laslett-tune shift limit. Even in traditional synchrotrons and storage rings obeying the Laslett limit, it is useful to have a framework in which space charge and dispersion are both included.

In the HIF application, the normalized emittance of the beam must remain small to be able to focus the beam on a small spot. The growth of the normalized emittance of an accelerated beam is also of interest for many other applications in which high brightness is required. The concept of transverse energy conservation was used in ref. [1] to study emittance growth in bends. This built upon earlier studies which have calculated changes in emittance also using a transverse energy constraint. For example, emittance growth associated with non-uniform space-charge distributions was examined in refs. [8]- [10]. Emittance growth due to initial beam displacements and mismatches with and without space-charge and momentum spread has been studied in, refs. [11-13,17], and references therein.

In the work reviewed here the beams propagate in continuous or alternating gradient focusing channel, with phase advances that are depressed due to space

---

* Work performed under the auspices of the U. S. Department of Energy by LLNL under contract W-7405-ENG-48.

charge. In addition, bends are present, which provide a displacement in the center of oscillation for ions which are off of the design momentum. Moment equations are employed to estimate emittance growth arising from the transition from straight sections to bends. (See also ref. [14] for an estimate of emittance growth due to the transitions in the absence of space charge.) On a transition from a bend to a straight section, or from a straight section to a bend, if the transition is sufficiently sharp, the beam becomes mismatched. We assume that small non-linear forces act to phase mix particles, and we find the asymptotic emittance of such a beam. Further, if we assume that the process of phase mixing is completed before the beam goes through another straight/bend transition, we may calculate the emittance growth through a "racetrack" configuration consisting of two 180° bends and two straight sections, even without a detailed knowledge of the rate at which the phase mixing occurs.

Model Equations of Motion

The force equation in the radial (bend) direction using cylindrical coordinates, (ρ, θ, y) is:

$$\ddot{\rho} - \frac{v_\theta^2}{\rho} = F_{bend} - v_\theta^2 k_{\beta 0x}^2 (\rho - \rho_0) + v_\theta^2 k_{sx}^2 (\rho - <\rho>) \qquad (1)$$

Here, ρ is the radial coordinate of a particle in a bend, θ is the azimuthal coordinate, y is the vertical coordinate and $v_\theta \equiv \rho\dot{\theta}$ is the azimuthal velocity, and k_{sx}^2 is a defocusing constant of the assumed linear space charge force in the radial direction (defined below). For simplicity, non-relativistic kinematics are assumed. Also, ρ_0 is the nominal radius of a particle with the azimuthal component of the design momentum p_0 and design velocity $v_0 \equiv \beta c$.
The component of the bending force F_{bend} in the radial direction is given by:

$$F_{bend} = \left(\frac{qe}{Am_a}\right) \begin{cases} v_\theta B_y & \text{(magnetic bends)} \\ E_\rho & \text{(electric bends)} \end{cases} \qquad (2)$$

Here q is the ion charge state (+1 for protons), e is the proton charge in Coulombs, A is the ion mass in amu, m_a is the atomic mass unit in kg. B_y is the vertical bending field (for magnetic bends) or E_ρ is the radial electric field (for electric bends).
We let $x \equiv \rho - \rho_0$ and define the increment in path length along the design orbit $ds \equiv \rho_0 d\theta$. The equations of motion are then given by,

$$x'' = -k_{\beta 0x}^2 (x - x_m) + k_{sx}^2 (x - x_c) - \frac{\partial h_{nl}(x,y)}{\partial x}. \qquad (3)$$

$$y'' = -k_{\beta 0y}^2 y + k_{sy}^2 (y - y_c) - \frac{\partial h_{nl}(x,y)}{\partial y}. \qquad (4)$$

$$k_{sx}^2 \equiv \frac{K}{2(\Delta x^2 + (\Delta x^2 \Delta y^2)^{1/2})}; \qquad k_{sy}^2 \equiv \frac{K}{2(\Delta y^2 + (\Delta x^2 \Delta y^2)^{1/2})}. \qquad (5)$$

Here, x is the in-plane deviation from the design orbit and y is the vertical coordinate in a particular transverse slice of the beam. The beam travels in the $+s$ direction, and prime (') indicates derivative with respect to s; $k_{\beta 0x}$ and $k_{\beta 0y}$ can represent either alternating gradient focusing (if they are s dependent) or they can represent the focusing effects in the smooth approximation, in which case, $k_{\beta 0x} = k_{\beta 0y} = \sigma_0/2L$ where σ_0 is the undepressed phase advance, and L is the half-lattice period. Dispersion effects enter through the term x_m, where

$x_m \equiv (1/k_{\beta 0x}^2 \rho_0)(\delta p/p_0)$ for magnetic focusing and $x_m \equiv (2/k_{\beta 0x}^2 \rho_0)(\delta p/p_0)$ for electric focusing. The quantity $\delta p/p_0$ is the fractional difference between the longitudinal momentum of a particle and the design momentum p_0, and $K \equiv 2qI/(\beta^3 A I_o)$ is the perveance. Here $I_o \equiv 4\pi\epsilon_0 m_a c^3/e$ is the characteristic proton current ($\cong 31$ MA). Finally, for generality, we have included an unspecified external non-linear potential h_{nl} that is a function of x,y, and possibly s.

We adopt the notation of ref. [1], throughout this paper in which the quantity Δ is reserved for the two argument operator in which centroid quantities are subtracted off: $\Delta ab \equiv <ab> - <a><b>$ (e.g. $\Delta x^2 \equiv <x^2> - <x>^2$), where $<>$ indicates average over all particles in a slice; $x_c \equiv <x>$, and $y_c \equiv <y>$.

These equations are identical to the equations found in ref. [1], except here we no longer assume $k_{\beta 0x}$, $k_{\beta 0y}$, and ρ_0 to be independent of s, nor do we require $k_{\beta 0x} = k_{\beta 0y}$. In deriving the moment equations in ref. [1], no use was made of the assumed constancy or equality of $k_{\beta 0x}$ and $k_{\beta 0y}$ nor the constancy of ρ_0, so the generalization simply amounts to a relabeling of the focusing constants.

Eqs. (3) and (4) represent, in an approximate way, the effects of: linear focusing, linear space charge defocusing, dispersion in a bend, and external non-linearities in the focusing field. The physical approximations that have been made include the following: (1) Eqs. (3) and (4) have been linearized in the small quantities $k_{\beta 0x}$, $k_{\beta 0y}$, and $\delta p/p_0$. (The non-linear term h_{nl} has also been included in some of the derivations). (2) The non-linearity is small: ($|h_{nl}| << |k_{\beta 0x}^2 x^2|, |k_{\beta 0y}^2 y^2|$). (Terms which are non-linear in $\delta p/p_0$, such as $k_{\beta 0x} \delta p/p_0$, have been neglected.) (3) Space charge forces depend only on lowest order moments. (We have used the KV formula for the electrostatic potential, which is equivalent to assuming uniform density elliptical beam. Centroid position and semi-major axes are, however, allowed to vary with s). (4) The beam is coasting: (p_0, β, and δp are constants). (5) The beam is non-relativistic: ($\beta << 1$).

Let $f(x, x', y, y', \frac{\delta p}{p_0}, s) = dN/dxdx'dydy'd\frac{\delta p}{p_0}$ where dN is the number of particles within incremental phase volume $dxdx'dydy'\, d\frac{\delta p}{p_0}$. For the model equations (3)-(5) the Vlasov equation becomes:

$$\frac{\partial f}{\partial s} + x'\frac{\partial f}{\partial x} + \left(-k_{\beta 0x}^2(x-x_m) + k_{sx}^2(x-x_c) - \frac{\partial h_{nl}}{\partial x}\right)\frac{\partial f}{\partial x'} + y'\frac{\partial f}{\partial y} + \left(-k_{\beta 0y}^2 y + k_{sy}^2(y-y_c) - \frac{\partial h_{nl}}{\partial y}\right)\frac{\partial f}{\partial y'} = 0. \tag{6}$$

The average of a variable ξ over the continuous distribution is given by:
$$<\xi>(s) \equiv \int dx \int dx' \int dy \int dy' \int d\frac{\delta p}{p_0} \xi f(x,x',y,y',\frac{\delta p}{p_0},s)/N,$$
where
$$N \equiv \int dx \int dx' \int dy \int dy' \int d\frac{\delta p}{p_0} f(x,x',y,y',\frac{\delta p}{p_0},s).$$

Following ref. [7], we take all second order moments of the Vlasov eq. (6), yielding eight (first-order with respect to s) coupled moment equations:

$\frac{d}{ds}\Delta x^2 = 2\Delta xx'$

$\frac{d}{ds}\Delta x'^2 = (-2k_{\beta 0x}^2 + 2k_{sx}^2)\Delta xx' + 2k_{\beta 0x}^2 \Delta x' x_m - 2\Delta(x'\frac{\partial h_{nl}}{\partial x})$

$\frac{d}{ds}\Delta xx' = \Delta x'^2 - k_{\beta 0x}^2 \Delta x^2 + k_{sx}^2 \Delta x^2 + k_{\beta 0x}^2 \Delta xx_m - \Delta(x\frac{\partial h_{nl}}{\partial x})$

$\frac{d}{ds}\Delta y^2 = 2\Delta yy'$

$\frac{d}{ds}\Delta y'^2 = (-2k_{\beta 0y}^2 + 2k_{sy}^2)\Delta yy' - 2\Delta(y'\frac{\partial h_{nl}}{\partial y})$

$\frac{d}{ds}\Delta yy' = \Delta y'^2 - k_{\beta 0y}^2 \Delta y^2 + k_{sy}^2 \Delta y^2 - \Delta(y\frac{\partial h_{nl}}{\partial y})$

$\frac{d}{ds}\Delta xx_m = \Delta x' x_m$

$\frac{d}{ds}\Delta x' x_m = -k_{\beta 0x}^2 \Delta xx_m + k_{sx}^2 \Delta xx_m + k_{\beta 0x}^2 \Delta x_m^2 - \Delta(x_m \frac{\partial h_{nl}}{\partial x})$ \hfill (7)

Similarly, the first order moments of eq. (6) yield the following:

$\frac{d}{ds} x_c = x'_c$

$\frac{d}{ds} x'_c = -k^2_{\beta 0x} x_c + k^2_{\beta 0x} <x_m> - <\frac{\partial h_{nl}}{\partial x}>$

$\frac{d}{ds} y_c = y'_c$

$\frac{d}{ds} y'_c = -k^2_{\beta 0y} y_c - <\frac{\partial h_{nl}}{\partial y}>$ (8)

Note that if $h_{nl} = 0$, eq. (7) forms a closed sets of 8 equations, and eq. (8) forms two sets of two closed equations. If $h_{nl} \neq 0$, eqs. (7) and (8) form the beginning of an infinite hierarchy of moment equations.

Transverse Energy Conservation

For the case of alternating gradient focusing, and when the bends occupy only a fraction of the lattice, the focusing constants $k^2_{\beta 0x}$, $k^2_{\beta 0y}$, and the bend radius of curvature ρ_0 are dependent on s. This s-dependence of the external forces implies that there will not be a constant energy-like quantity. However, as in ref. [1], if $k_{\beta 0x}, k_{\beta 0y}$, and ρ_0 are constants representing average quantities, we may define a transverse energy H:

$$2H = k^2_{\beta 0x} \Delta x^2 + k^2_{\beta 0y} \Delta y^2 + \Delta x'^2 + \Delta y'^2 - 2k^2_{\beta 0x} \Delta x x_m - K \ln((\Delta x^2)^{1/2} + (\Delta y^2)^{1/2})$$

$$+2 <h_{nl}> +k^2_{\beta 0x} x_c^2 + k^2_{\beta 0y} y_c^2 + x_c'^2 + y_c'^2 \qquad (9)$$

Use of eqs. (7) and (8) shows that:

$$\frac{d}{ds} H = \frac{d}{ds} <h_{nl}> \qquad (10)$$

Thus if h_{nl} is not a function of s, H is an invariant.

Emittance Growth

We define separate x and y emittances:

$$\epsilon_x^2 \equiv 16(\Delta x^2 \Delta x'^2 - \Delta x x'^2); \qquad \epsilon_y^2 \equiv 16(\Delta y^2 \Delta y'^2 - \Delta y y'^2). \qquad (11)$$

Using eqs. (7), the following emittance evolution equations can be derived:

$$\frac{d}{ds} \epsilon_x^2 = 32k^2_{\beta 0x}(\Delta x^2 \Delta x' x_m - \Delta x x' \Delta x x_m) + 32(\Delta(x\frac{\partial h_{nl}}{\partial x})\Delta x x' - \Delta x^2 \Delta(x'\frac{\partial h_{nl}}{\partial x})) \qquad (12)$$

$$\frac{d}{ds} \epsilon_y^2 = 32(\Delta(y\frac{\partial h_{nl}}{\partial y})\Delta y y' - \Delta y^2 \Delta(y'\frac{\partial h_{nl}}{\partial y})) \qquad (13)$$

Thus the emittance would be constant if non-linearities were not present ($h_{nl} = 0$) and the momentum spread were absent ($x_m = 0$ for all particles.) Eqs. (12) and (13) are valid for both continuous and alternating gradient focusing.

Equilibrium Beam

In the continuous focusing approximation, the left-hand-sides of eqs. (7) and (8) can be set to zero to obtain the following equilibrium (constant moment) conditions:

$$\Delta x'^2 = (k_{\beta 0x}^2 - k_{sx}^2)\Delta x^2 - k_{\beta 0x}^2 \Delta xx_m + \Delta(x\frac{\partial h_{nl}}{\partial x})$$

$$\Delta y'^2 = (k_{\beta 0y}^2 - k_{sy}^2)\Delta y^2 + \Delta(y\frac{\partial h_{nl}}{\partial y})$$

$$\Delta xx_m = \frac{k_{\beta 0x}^2 \Delta x_m^2 - \Delta(x_m \frac{\partial h_{nl}}{\partial x})}{k_{\beta 0x}^2 - k_{sx}^2}$$

$$x_c = <x_m> - \frac{1}{k_{\beta 0x}^2} <\frac{\partial h_{nl}}{\partial x}>$$

$$y_c = -\frac{1}{k_{\beta 0y}^2} <\frac{\partial h_{nl}}{\partial y}>$$

$$\Delta xx' = \Delta yy' = \Delta x'x_m = \Delta(x'\frac{\partial h_{nl}}{\partial x}) = \Delta(y'\frac{\partial h_{nl}}{\partial y}) = 0$$

$$x_c' = y_c' = 0 \qquad (14)$$

Assuming that $h_{nl} = x_c = y_c = 0$ the transverse energy (9) in equilibrium ($H = H_{eq}$) reduces to:

$$2H_{eq} = (2k_{\beta 0x}^2 - k_{sx}^2)\Delta x^2 + (2k_{\beta 0y}^2 - k_{sy}^2)\Delta y^2 - \frac{3k_{\beta x0}^4 \Delta x_m^2}{k_{\beta 0x}^2 - k_{sx}^2} - K\ln((\Delta x^2)^{1/2} + (\Delta y^2)^{1/2}) \quad (15)$$

Note that for a given H_{eq}, the ratio of Δx^2 to Δy^2 is still unspecified. A further assumption is required to specify the final state of the beam. It is often reasonable to assume that transverse energy equipartition results in a beam in which the two transverse temperatures are equal, i.e. $\Delta x'^2 = \Delta y'^2$. (Note that we have implicitly assumed that the timescale for complete equipartition $[\Delta x'^2 = \Delta y'^2 = \Delta(\delta p/p_0)^2]$ is much larger than timescales of interest.) The condition that $\Delta x'^2 = \Delta y'^2$ can be expressed as a relation between Δy^2 and Δx^2:

$$\Delta y^2 = \frac{(k_{\beta 0x}^2 - k_{sx}^2)}{(k_{\beta 0y}^2 - k_{sy}^2)}\Delta x^2 - \frac{k_{\beta 0x}^4}{(k_{\beta 0x}^2 - k_{sx}^2)}\Delta x_m^2 \qquad (16)$$

When $k_{\beta 0x} = k_{\beta 0y} = k_{\beta 0}$ and $\Delta x_m^2 << \Delta x^2$ this result reduces to:

$$\Delta y^2 \cong \Delta x^2 - 2k_{\beta 0}^4 \Delta x_m^2 / \left(k^2(k^2 + k_{\beta 0}^2)\right), \qquad (17)$$

where $k^2 \equiv k_{\beta 0}^2 - K/(4\Delta x^2)$.

Rings

Suppose a beam is in equilibrium in a straight section, and then enters a continuous bend (i.e. a ring). If the lattice parameters (such as the bend radius of curvature) abruptly change to new values, the beam becomes mismatched to the bend. Physically, particles that are not on the design momentum for the bend initially become spatially separated, creating non-linear space-charge forces, allowing phase mixing of the coherent mismatch oscillations until a new equilibrium is reached. For concreteness, we consider a lattice in which $k_{\beta 0x} = k_{\beta 0y} \equiv k_{\beta 0}$, and also assume that $k_{\beta 0}$ is the same both in the straight section and in the bend section. Thus, in this example, we assume the contribution to focusing from the bends is included in $k_{\beta 0}$. We assume that the initial beam (subscript 0) is matched to the straight section. Thus $\Delta y_0^2 = \Delta x_0^2$, and $\Delta x_0'^2 = \Delta y_0'^2 = k_0^2 \Delta x_0^2$, and all other moments are equal to zero. The initial transverse energy satisfies $2H_0 = (2k_{\beta 0}^2 + 2k^2)\Delta x_0^2 - K\ln[2(\Delta x_0^2)^{1/2}]$. To calculate the final equilibrium beam parameters, we set the final transverse energy equal to the initial transverse energy, and simultaneously solve this constraint with the equalized temperature constraint, eq. (16). Thus, there are two equations in two unknowns (Δx^2 and Δy^2), yielding the final equilibrium values of Δx^2 and Δy^2 in the bend. From the equilibrium values (eq. 14), all other second order moments may be calculated, including the emittance.

In ref. [3], comparisons of the results of the continuous focusing theory were made to 2D WARP PIC simulations of the transition from straight to bend for parameters of a small recirculator experiment being built at Lawrence Livermore National Laboratory. The relevant parameters of the simulation were: the ion species was singly charged Potassium, (mass 39), at an energy of 80 kV, and a current of 2 mA, leading to a perveance K of 3.54×10^{-4}. The average focusing constants are $k_{\beta 0x} = k_{\beta 0y} = 1.89$ m^{-1}, corresponding to a phase advance of 78 degrees and half-lattice period $L = 0.36$ m. The average bend radius of curvature is $\rho_0 = 2.29$ m. The normalized emittances $\epsilon_{nx} \equiv \gamma \beta \epsilon_x$ and $\epsilon_{ny} = \gamma \beta \epsilon_y$ (where γ is the Lorentz factor of the beam) were both set to 0.03 mm-mrad at injection into the ring. $\delta p/p_{0rms} = 7.2 \times 10^{-4}$ was assumed.

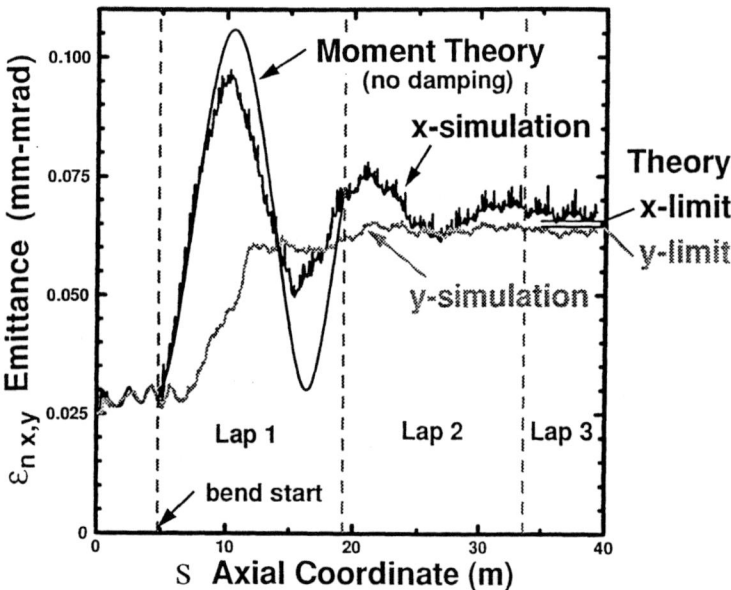

Figure 1. Comparison of WARP simulations with model results (from ref. [3]). x and y emittance evolution in a ring geometry, after initialization in a straight section, as calculated by WARP, by direct integration of the moment equations (Moment Theory, no damping) and the asymptotic value using energy conservation.

Reprinted from *Nuclear Instruments and Methods in Physics Research A*, J. J. Barnard, ©1998 with permission from Elsevier Science.

In Figure 1, (from ref. [3]), we plot the WARP simulations of the normalized x and y emittances over 3 laps of the small recirculator. In addition, we plot the initial evolution of the emittance as predicted by direct integration of the moment equations, (indicated as Moment Theory (no damping)) in the figure. The simulations include all of the details of alternating gradient lattice including fringe fields and image effects, as well as the non-linear space charge fields. The theory calculations use only the uniform focusing and bending approximation. Also, because the moment equations do not include non-linearities and the associated non-linear phase mixing, the amplitude of the x-emittance oscillations remain constant and the y-emittance does not grow. In the simulations, small non-linearities cause the oscillations to damp and the y-emittance to gradually grow closer to the x-emittance. Although direct integration of the moment equations does not capture the damping of the oscillations in the x-emittance or the growth in the y-emittance, the moment equations accurately predict the initial amplitude and frequency of the oscillations. Also shown on the figure are the final equilibrium results (indicated as x-limit theory, and y-limit theory) calculated using the prescription indicated above, which is based on using the moment equations to calculate the transverse energy and assuming equality of the final velocity spreads in the x and y directions. As can be seen, the theory closely predicts the asymptotic values of the x and y emittances as found by the fully 3D simulations, and also captures the simulation result that the y-emittance equilibrates to a value less than the x-emittance. Simulations of the University of Maryland Ring (ref. [15]) shows a similar increase in emittance with an ultimate saturation.

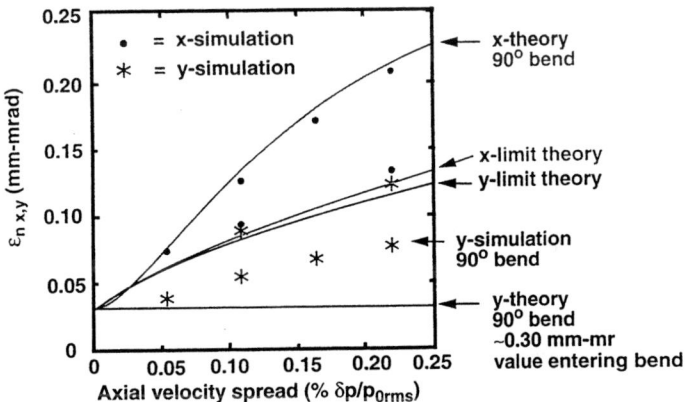

Figure 2. Comparison of WARP simulations with model results (from ref. [3]). x and y emittance after the 90 degree bend and asymptotic results as a function of the fractional momentum spread (in percentage), after initialization in a straight section, as calculated by WARP, by direct integration of the moment equations (x- and y- theory 90° bend) and the asymptotic value using energy conservation (x- and y- limit theory).

In Figure 2, (from ref. [3]), we plot the emittance at the end of ninety degrees of bend and also the asymptotic values of the emittance, all as a function of $(\Delta(\delta p/p_0)^2)^{1/2} \equiv \delta p/p_{0rms}$. This quantity is largely unknown in the experiment, and it could range anywhere from the value expected from accelerative cooling (see e.g. ref. [16]) of the longitudinal momentum spread induced by the ~ 0.1 eV ion source, which results in a spread of order 10^{-6}, or if the fractional error in the injector diode voltage errors is as large as 0.005 (at high enough frequency), the resulting induced fractional momentum spread, $\delta p/p_{0rms}$ would be as large as 2.5×10^{-3}. A third possible source of momentum spread comes from instabilities associated with an anisotropic velocity distribution (ref. [22]). If this instability heats the longitudinal component until it is of the same temperature as the transverse, the resulting moment spread would be $\delta p/p_{0rms} \cong \epsilon_x/4(\Delta x^2)^{1/2} \cong 7.2 \times 10^{-4}$. (For a more complete discussion cf. ref. [3]). As can be seen from the plot, direct integration of the moment equations closely captures the simulation value of the emittance after 90 degrees (during the initial emittance oscillation) and closely matches the mean rise in emittance and difference between the x and y emittances.

When $k_{\beta 0x} = k_{\beta 0y}$, and when the change in emittance is much less than the original emittance, one may analytically estimate the change in emittance squared in an abrupt transition from straight to bend:

$$\epsilon_x^2 - \epsilon_{x0}^2 \cong 16 \frac{k_{\beta 0}^4(2k_{\beta 0}^2 + k^2)}{k^2(k_{\beta 0}^2 + k^2)} \Delta x_0^2 \Delta x_m^2; \qquad \epsilon_y^2 - \epsilon_{y0}^2 \cong 16 \frac{k_{\beta 0}^6}{k^2(k_{\beta 0}^2 + k^2)} \Delta x_0^2 \Delta x_m^2 \qquad (18)$$

Eqs. (18) are valid only for small changes in emittance, and so are not applicable to the parameters of figure 1.

Racetracks

In a recirculator that is composed of two 180° bends connected by two straight sections in the shape of a racetrack, if phase mixing is rapid enough the equilibrium can be reached before each transition. Transverse energy is conserved as a beam enters a bend from a straight, but since the beam acquires a finite $\Delta x x_m$ as it finds equilibrium in the bend, the transverse energy will be discontinuous entering a straight from a bend, (where ρ_0 becomes infinite, and hence $\Delta x x_m$ abruptly changes to zero.) The quantities Δx^2, $\Delta x'^2$, Δy^2, $\Delta y'^2$ are, of course, continuous at all transitions. A new value of H is calculated which is again constant throughout the straight section. At the beginning of the bend the process repeats. In Fig. (3), we have applied this formulation to a small scale racetrack recirculator, which is not undergoing acceleration. This prescription for calculation of the emittance was carried out numerically, and compared with the 3-D version of the WARP code. As can be seen, the emittance growth is tracked closely although the higher frequency oscillatory behavior associated with lattice and mismatch oscillations are, of course, not seen. (For small values of $\Delta(\delta p/p_0)^2$, or large values of σ/σ_0 the prescription overestimates the emittance growth, since the assumption of complete phase-mixing between transitions is not achieved.)

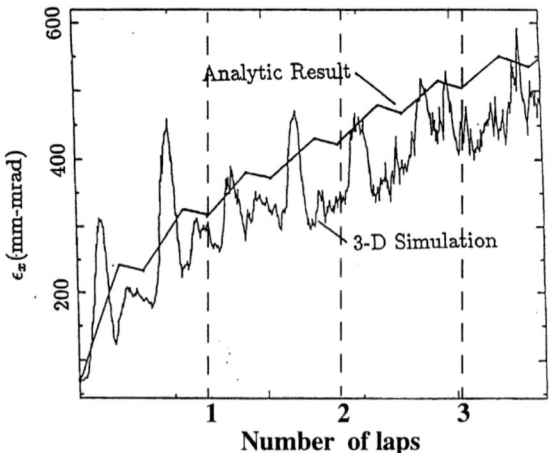

Figure 3. Emittance growth in a racetrack. The parameters are $\sigma_0 = 72°$, $\sigma = 8°$,

$\rho = 3.6$m, $Am_a\beta^2c^2/2 = 10$MeV, $(\Delta(\delta p/p_0)^2)^{1/2} = 0.0079$. The length of each straight section was 7.2 m.

Comparison to Other Recent Work

Recently, in ref. [4, 18], envelope equations which include the same physics (i.e. dispersion linear in $\delta p/p_0$, and space charge with an assumed elliptical symmetry) as the moment equations (eqs. 7,8) were derived. Venturini and Reiser found a generalized emittance ϵ_{dx} which, when expressed in terms of the notation of this paper can be written as:

$$\epsilon_{dx}^2 = (\Delta x_m^2 \Delta x^2 - \Delta x x_m^2)(\Delta x_m^2 \Delta x'^2 - \Delta x' x_m^2) - (\Delta x_m^2 \Delta x x' - \Delta x x_m \Delta x' x_m)^2 \quad (19)$$

By taking the derivative of ϵ_{dx}^2 and using eqs. (7), it is straightforward to show that ϵ_{dx}^2 is constant. By using the constancy of ϵ_{dx} and ϵ_y, two of the eight first order moment equations could be eliminated, leaving six first order moment equations, equivalent to the three second order equations of ref. [4]. Thus the envelope equations in ref. [4] contain the same physical content as the previously derived (ref. [1]) moment equations. In ref. [4], it is suggested that it would be possible to eliminate much of the growth in emittance by matching the envelope and the dispersion function $D(s) = (\Delta x \delta p/p_0)/\Delta(\delta p/p_0)^2$ (as defined in ref. [4]). This is equivalent to finding the matched periodic solution of the moment equations in the ring, and then constructing the section which injects the beam into the ring such that the values of the moments match those of the matched periodic solution within the ring. In order to prevent mismatch oscillations of the centroid, the centroid equations (8) must also be matched on the transition from straight to bend. Another method (ref. [1]) of preventing emittance growth is by slowly varying the radius of curvature, allowing an adiabatic transition into the bend.

Also, recently in ref. [5], vertical and horizontal dispersion functions are derived. The horizontal dispersion function derived by Lee is identical to that of ref. [4], except that horizontal/vertical coupling is allowed such as can occur if there are quadrupole rotation errors (see e.g. ref. [19]). The vertical dispersion function is identical to that of a straight lattice, but again with the inclusion of horizontal/vertical coupling. The envelope equations derived are not consistent with the moment eqs. (7,8) or the envelope equations of ref. [4], however, due to additional approximations.

Self-consistent distributions

Recently, in ref. [20], a self-consistent KV solution to the Vlassov-Poisson system in a bend was obtained. The solutions in the non-relativistic case are of the form:

$$f(x, x', y, y', \delta p/p_0) = f_\perp(H_\perp) \exp[-(\delta p/p_0)^2/\delta_0^2] \quad (20)$$

Here $2H_\perp = x'^2 + y'^2 + k_{\beta 0}^2 x^2 + k_{\beta 0}^2 y^2 + 2q\phi/mv_0^2 - 2(x/\rho_0)\delta p/p_0$. In ref. [20], generalizations to the KV distribution were investigated of the form $f_\perp(H_\perp) = f_0 \delta(H_\perp - H_0)$. In ref. [21] thermal equilibrium distributions of the form $f_\perp(H_\perp) = f_0 \exp(-H_\perp/k_b T)$ have been examined. Figure 4 illustrates the two distributions for the parameters of the University of Maryland electron ring experiment (ref. [20]) ($k_{\beta 0}^2 = 17.437$ m^{-2}, current = 105 mA, energy = 10 keV, $\rho_0 = 1.82$ m, with $k/k_{\beta 0} = 0.33$, and $\delta p/p_{0rms} = .01$.)

The moment eqs. require averages over xE_x and yE_y where E_x and E_y are the electric field components due to space charge. Although in ref. [1], an ellipse with uniform charge density was used to calculate E_x and E_y, as pointed out in ref. [7], the results also apply if the density is a function only of $x^2/\Delta x^2 +$

$y^2/\Delta y^2$, i.e. constant on nested elliptical surfaces (ref. [4]). As can be seen for the generalized KV distribution, the assumption of a density distribution that is constant on nested ellipses is poor for the KV distribution, but appears to be a better approximation for thermal equilibrium beams, which underlies the calculation of asymptotic emittance growth above. This may, in part, explain why the WARP simulation results agree well with the moment model.

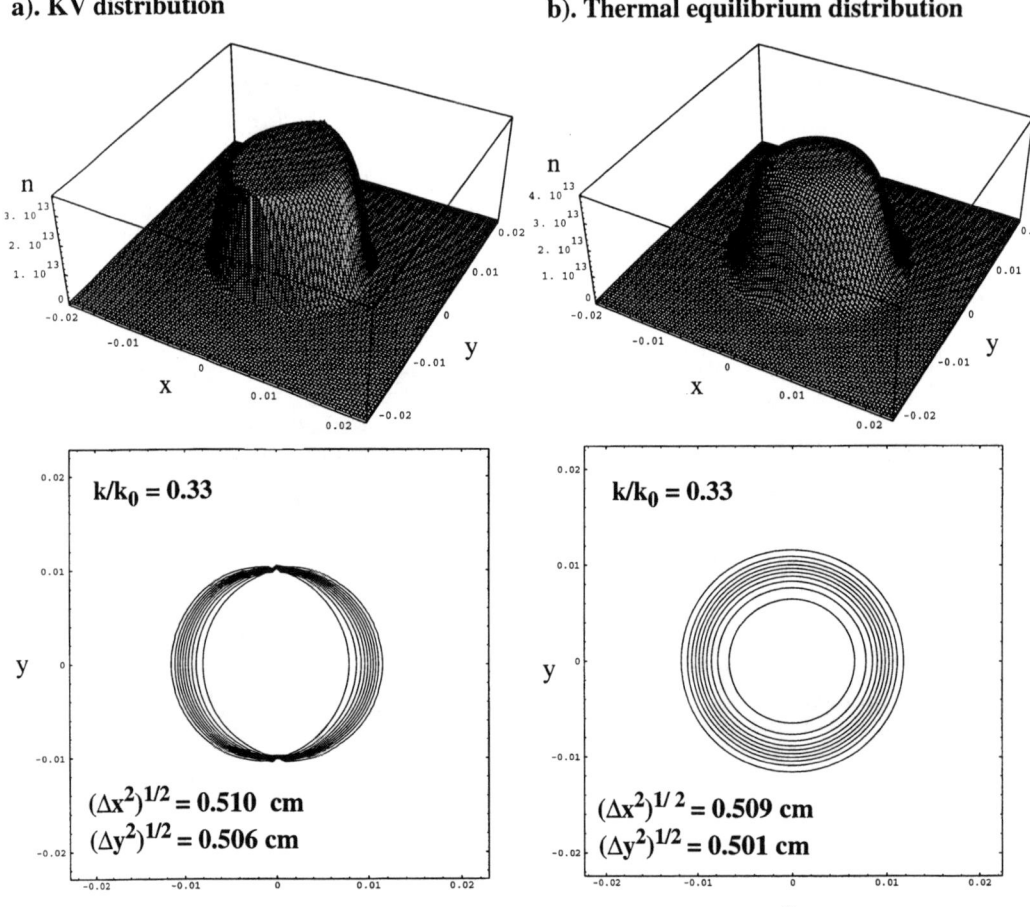

Figure 4. Self consistent beam density distributions in bends. a) Surface plot (upper) and contour plot (lower) of generalized KV distribution (ref. [20]); and b) Thermal distribution (ref. [21]).

Discussion

Emittance growth from sharp transitions, as discussed above, provides one source of emittance growth. Others, such as misalignments of quadrupoles, field strength errors, non-linear applied fields, etc., provide additional mechanisms to degrade the emittance. In the recirculator design of ref. [6], an insertion and extraction region occurs over a 100 m long straight section, which gives the machine some of the features of the racetrack in that equilibration can occur on passing from bend to straight and from straight to bend. Since the energy is increasing on each lap it would be difficult to design the insertion/ extraction section which is "matched" at all energies. Assuming abrupt transitions, use of the moment equations together with the parameters of the beam at the exit of the High Energy Ring of ref. [6] lead to an estimated emittance growth by a factor of about 2. Since the entrance beam parameters lead to a much smaller emittance growth, the normalized emittance will grow by less than a factor of 2. This is within the emittance "budget" in the design of ref. [6]. It is also possible that the transitions between bends and straights can be made gradual enough so that equilibria are reached adiabatically, with little associated growth in the normalized emittance.

Conclusions

We have reviewed the derivation of moment equations in which focusing, space charge, and dispersion in a bend, are included. We have shown that the moment equations derived in ref. [1], using the average bend and continuous focusing approximation, accurately predicts the initial amplitude and frequency of emittance oscillations which occur at a sharp transition from a straight section to a bend (ref. [3]). We have also reviewed the method of estimating the asymptotic value of the emittance growth due to straight/bend mismatches from considerations of transverse energy conservation as the beam equilibrates. By assuming the transverse energy of the beam is conserved during the equilibration, and assuming that the beam reaches equilibrium, and also that the equilibrium transverse velocity spread is the same in x as it is in y we can calculate all moments and thus the change in emittance. In racetracks, in which four such transitions are made per lap, we have calculated the emittance growth under the assumption that the equilibrium state is reached between each transition. In small scale rings the analytic result agreed well with 2-D and 3-D WARP simulations when σ/σ_0 was small and the velocity spread was sufficiently large (so that the assumption of phase mixing between transitions was realized). In the High Energy Ring of ref. [6], this prescription yielded an emittance growth of less than a factor of 2.

References

1. J.J. Barnard, H.D.Shay, S.S. Yu, A. Friedman, and D.P. Grote, "Emittance Growth in Heavy-Ion Recirculators," 1992 Linear Accelerator Conference Proceedings 1992 August 24-28, Ottawa, Ontario, Canada, C.R. Hoffman, ed. (1992). AECL 10728 (AECL Research, Chalk River, Canada) p. 229.
2. A. Friedman, D. P. Grote, D. A. Callahan, A. B. Langdon, I. Haber, "3D Particle Simulations of Axially Confined Heavy Ion Beams Using the WARP code: Transport Around Bends," Particle Accelerators, **37**, 131, (1992).
3. S.M. Lund, J.J. Barnard, G.D. Craig, A. Friedman, D.P. Grote, H.S. Hopkins, T.C. Sangster, W.M. Sharp, S. Eylon, T.J. Fessenden, E. Henestroza, S. Yu, and I. Haber, "Numerical Simulation of Intense-Beam Experiments at LLNL and LBNL," Nuclear Instruments and Methods A (in press).
4. Marco Venturini and Martin Reiser, "RMS Envelope Equations in the Presence of Space Charge and Dispersion," submitted to Phys. Rev. Letters (1998).
5. S.Y. Lee and H. Okamoto, "Space Charge Dominated Beams in Synchrotrons," these proceedings (1998).
6. J.J. Barnard, F. Deadrick, A. Friedman, D.P. Grote, L.V. Griffith, H.C. Kirbie, V.K. Neil, M.A. Newton, A.C. Paul, W.M. Sharp, H.D. Shay, R.O. Bangerter, A. Faltens, C.G. Fong, D.L. Judd, E.P. Lee, L.L. Reginato, S.S.

Yu, and T.F. Godlove, "Recirculating Induction Accelerators as Drivers for Heavy Ion Fusion," Physics of Fluids B: Plasma Physics, **5**, 2698 (1993). Also, "Study of Recirculating Induction Accelerators as Drivers for Heavy Ion Fusion," Lawrence Livermore National Laboratory UCRL-LR-108095 (1992).
7. F. J. Sacherer, "RMS Envelope Equations with Space Charge," IEEE Transactions on Nuclear Science **NS-18**, 1105, (1971).
8. P. M. Lapostolle, "Possible Emittance Increase through Filamentation Due to Space Charge in Continuous Beams," IEEE Transactions on Nuclear Science, **NS-18**, 1101, (1971).
9. T. P. Wangler, K. R. Crandall, R. S. Mills, and M. Reiser, "Relation Between Field Energy and RMS Emittance in Intense Particle Beams," IEEE Transactions on Nuclear Science, **NS-32**, (1985).
10. O. A. Anderson, "Internal Dynamics and Emittance Growth in Space-Charge-Dominated Beams," Particle Acclerators, **21**, 197, (1987).
11. J. J. Barnard, "Anharmonic Betatron Motion in Free Electron Lasers" Nuclear Instruments and Methods in Physics Research **A296** (1990).
12. M. Reiser "Free Energy and Emittance Growth in Nonstationary Charged Particle Beams," Journal of Applied Physics **70**, 1919 (1991).
13. O. A. Anderson, "Emittance Growth Rates for Displaced or Mismatched High Current Beams in Nonlinear Channels," Proc. of the Fourth NPB Techn. Symp., Argonne National Laboratory, (1992).
14. K. T. Nguyen "Emittance Growth and Energy Bandwidth in the IFRR," Proceedings of the 1990 DARPA/ SDIO/Services Annual Charged Part. Beam Review, p. 71, Nav. Res. Lab., Washington D. C. (1991).
15. R. A. Kishek, I. Haber, M. Venturini, and M. Reiser, "PIC Code Simulations of the Space-Charge-Dominated Beam in the University of Maryland Electron Ring," these proceedings.
16. M. Reiser, "Theory and Design of Charged Particle Beams," (Wiley & Sons, New York, 1994).
17. J.J. Barnard, J. Miller, I. Haber, "Emittance Growth in Displaced Space Charge Dominated Beams with Energy Spread," Proceedings of the 1993 Particle Accelerator Conference, Washington, D.C., May 1993, **5**, 3612. (1993)
18. M. Venturini, R.A. Kishek, and M. Reiser, "Dispersion and Space Charge," these proceedings (1998).
19. John J. Barnard, "Emittance Growth from Rotated Quadrupoles in Heavy Ion Accelerators," Proceedings of the 1995 Particle Accelerator Conference, Dallas TX, held May 1-5, 1995 (**5**), 3241 (1996).
20. M. Venturini and M. Reiser, "Self-consistent beam distributions with space charge and dispersion in a circular ring lattice," Phys. Rev. E, **57** 4725, (1998).
21. John J. Barnard, and Bojan Losic, "Emittance Growth from Bend/Straight transitions for Beams approaching Thermal Equilibrium," Proceedings of the 1998 Linear Accelerator Conference," In Preparation (1998).
22. I. Haber, D.A. Callahan, A. Friedman, D.P. Grote, S.M. Lund, T.F. Wang, "Characteristics of an Electrostatic Instability Driven by Transverse-Longitudinal Temperature Anisotropy", Nuclear Instruments and Methods A, in press (1998).

Micromap Approach to Space Charge in a Synchrotron

G. Franchetti*†‡, I. Hofmann‡ and G. Turchetti*†

‡ *GSI, Plankstr. 1, D-64291, Darmstadt, Germany*
* *Dipartimento di Fisica, Università di Bologna, Via Irnerio 46, 40126, Bologna, Italy*
† *INFN, Sezione di Bologna, Via Irnerio 46, 40126, Bologna, Italy*

Abstract. A micromap approach to space charge is presented as a concept easily implemented in a computer code. A map of arbitrary length is found to be a suitable tool to include space charge by means of a kick approximation. We adopt space charge description where a transverse elliptic particle symmetry is assumed, and all analysis is carried out in a coasting beam approximation. This approach is particularly useful to investigate the single particle dynamics when the working point of a machine lies in a resonance driven by a skew quadrupole. Since the improvement of the multiturn injection efficiency due to linear coupling is based mainly on the exchange of emittance between x- and y- planes, we have investigated the emittance exchange dynamics when space charge is not negligible. We found that a coherent effect of space charge improves the efficiency of the exchange with respect to the case of single particle emittance exchange with space charge neglected.

I MICROMAP

We will describe the transverse dynamics of a single particle in a magnetic focusing lattice in a coordinate system where x, y denotes the horizontal and vertical axes perpendicular to the reference orbit ($x = y = 0$) and s is the axial (curvilinear) coordinate. The particle evolution in the horizontal $x - y$ plane is calculated as a function of s. In these coordinates the transverse equations of motion are

$$x'' - \left(k(s) - \frac{1}{\rho(s)^2}\right)x = \frac{1}{\rho(s)}\delta p + f_x(x,y,s) \tag{1a}$$

$$y'' + k(s)y = f_y(x,y,s) \tag{1b}$$

where $' = d/ds$, $k(s)$ is the quadrupole gradient, $\rho(s)$ is the x−plane radius of curvature of the reference orbit, δp is the off momentum, and $f_x(x,y,s)$ and $f_y(x,y,s)$ describe the nonlinearities of the applied focusing lattice [1].

In a linear lattice ($f_x = f_y = 0$) the solution of the Eq. (1) for the initial condition $\mathbf{x}(s) = (x, x', y, y')_s$ can be expressed in terms of the transfer map

$$\mathbf{x}(s + \Delta s) = L_{s,s+\Delta s}\mathbf{x}(s) + \mathbf{D}_{s,s+\Delta s}\delta p \qquad (2)$$

where $L_{s,s+\Delta s}$ is a block-diagonal matrix [1] and $\mathbf{D}_{s,s+\Delta s} = (D_{s,s+\Delta s}, D'_{s,s+\Delta s})$ is a particular solution of Eq. (1a) corresponding to the initial condition $x(s) = 0$ and $x'(s) = 0$.

If the lattice is not linear we can represent the effects of f_x, f_y in the dynamics while preserving the simplecticity of the transfer map Eq. (2) by using a single kick approximation [1]. The nonlinear effects on the particle moving in the interval $[s, s + \Delta s]$ are compressed in a kick $\mathbf{K}$ as follows

$$\mathbf{x}(s) \xmapsto{\mathbf{K}} \mathbf{x}(s)' \xmapsto{\mathbf{L}} \mathbf{x}(s + \Delta s) \qquad (3)$$

with $\mathbf{K}(\mathbf{x}) = (x, x' + \Delta s f_x, y, y' + \Delta s f_y)_s$. Therefore, the transfer map with kick approximation nonlinear forces is

$$\mathbf{x}(s + \Delta s) = L_{s,s+\Delta s} \begin{pmatrix} x \\ x' + \Delta s f_x \\ y \\ y' + \Delta s f_y \end{pmatrix}_s + \mathbf{D}_{s,s+\Delta s}\delta p \qquad (4)$$

Validity of this method requires that the space step Δs be chosen consistently with the strengths of f_x and f_y and how they change with s.

In order to include space charge effects in the transverse map given by Eq. (4) using the single kick approximation, we have to evaluate the space charge contribution $f_{\zeta,sc}$ to the functions f_x and f_y (here $\zeta = x, y$). If we consider a particle moving in a drift, its equation of motion is

$$\frac{dp_\zeta}{dt} = F_{\zeta,sc} \qquad (5)$$

It is straightforward to show that the relative error ΔA between dp_ζ/dt and $m\gamma_0(dv_\zeta/dt)$ is less then $(\gamma_0^2 - 1)(x'^2 + y'^2)$ where $\gamma_0 = (1 - \beta_0^2)^{-1/2}$. Using the parameters of the SIS ring [2], $\epsilon_x = 200$ mm mrad, $\epsilon_y = 20$ mm mrad, $\gamma_{xmax} = 0.86$, $\gamma_{ymax} = 1.98$, $\beta_{xmax} = 16.46$, $\beta_{ymax} = 28.52$, and $\beta_0 = 0.155$, we find that $\Delta A < 5.2 \cdot 10^{-3}$. Here we have denoted the $x-$ and $y-$plane emittances as ϵ_x and ϵ_y, $\gamma_{x,y\,max}$ and $\beta_{x,y\,max}$ denote the maximum values of the $x-$ and $y-$plane γ and beta functions describing the particle orbits, and β_0 is the relativistic factor of the design particle.

For such parameters Eq. (5) is well approximated as

$$\zeta'' = f_{\zeta,sc} \qquad (6)$$

where $f_{\zeta,sc} = qE_{\zeta,sc}/v_0 p_0 \gamma_0^2$, q is the electric charge of the particle, $E_{\zeta,sc}$ is the space charge electric field, v_0 and p_0 are the axial velocity and momentum of the design particle, and γ_0^2 represents the leading order correction to the electric force law due to self-magnetic fields.

The final form of the transverse micromap with space charge kicks is

$$\mathbf{x}(s+\Delta s) = L_{s,s+\Delta s} \begin{pmatrix} x \\ x' + \Delta s(f_{x,nl} + \frac{qE_x}{v_0 p_0 \gamma_0^2}) \\ y \\ y' + \Delta s(f_{y,nl} + \frac{qE_y}{v_0 p_0 \gamma_0^2}) \end{pmatrix}_s + \mathbf{D}_{s,s+\Delta s} \delta p \qquad (7)$$

The evolution of an off-momentum particle in the step Δs is obtained by using the micromap in Eq. (7)

In order to include the effect of the bend frame, the axial space step Δs is related to the time step Δt by means of the formula

$$\Delta s = \frac{1}{1+\frac{\mathcal{D}}{\rho}\delta p} v \Delta t$$

where v is the axial velocity of the particle, ρ the radius of curvature of the bend, and $\mathcal{D}$ is the dispersion function [3]. Note that each particle moves a different axial Δs for a fixed timestep.

II COASTING BEAM MODEL

A Macroparticle Description

In a coasting beam model with elliptic symmetry the particle density has the form

$$n(x, y, s) = n(\phi)$$

where ϕ is the isodensity parameter $\phi = x^2/a_x^2 + y^2/a_y^2$, and a_x, a_y are the rms transverse radii of the beam. Fig. 1 shows the scheme followed to introduce a macroparticle description of the coasting beam. We fix a series of transverse planes separated a distance dL from each other (Fig. 1a). Then we concentrate the particles between two successive planes as a macroparticle distribution on a single transverse plane as shown in Fig. 1b. We choose, arbitrarily, to put each of the macroparticles accumulation planes at the end of the accumulation interval. Each of the macroparticles has mass M^* and charge q^*. If n is the particle density of the coasting beam, the two dimensional particle density on a macroparticle plane α is $\delta = n \cdot dL$.

We can fill the transverse macroparticles planes of the beam with a sufficient number N^* of macroparticles to preserve the initial transverse symmetry of the beam. Since we have chosen to keep the macroparticle distribution on planes separated by a distance dL, it is also reasonable to keep the same average inter-particle distance dL_α in the transverse direction. From this assumption we find a relation between N^* and dL

$$dL = \sqrt{\frac{\pi AB}{N^*}}$$

where $A = 2a_x$ and $B = 2a_y$. These assumptions determine the charge q^* and mass M^* of the macroparticles.

B Space Charge Calculation

The symmetry of the beam allows us to calculate the transverse component of the elettrostatic space charge electric field and the longitudinal component is zero by assumption. We assume free boundary condition. The transverse electric field components are given by the formulas [4]

$$E_x = \frac{q^*}{\epsilon_0} M_x x; \quad E_y = \frac{q^*}{\epsilon_0} M_y y$$

where

$$M_u = \frac{a_x a_y}{2} \int_0^\infty \frac{n(\lambda)ds}{(a_u^2 + s)\sqrt{(a_x^2 + s)(a_y^2 + s)}} \qquad (8)$$

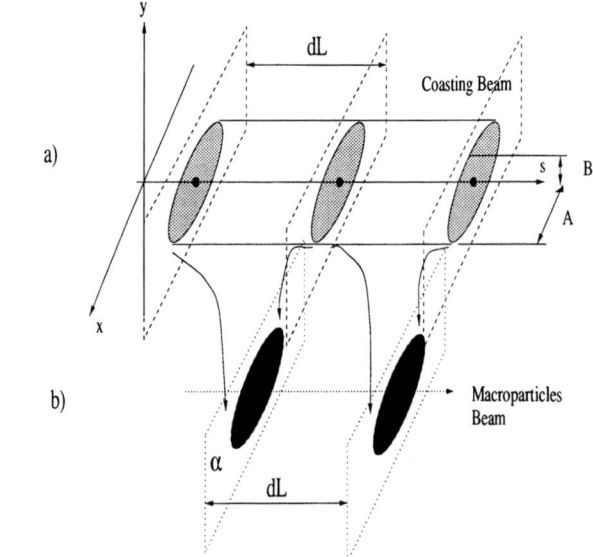

FIGURE 1. Scheme used to introduce the macroparticles description of a coasting beam: the real beam is represented by a macroparticle distribution on a series of transverse planes separated by an axial distance dL.

and $\lambda = x^2/(a_x^2 + s) + y^2/(a_y^2 + s)$ with $u = x, y$. To evaluate the integral in Eq. (8) we, analogously to a previous bunched beam analysis [5], employ the change of variables $s = \alpha^2(1/u^2 - 1)$, where $\alpha = (a_x a_y)^{1/3}$. Then E_x, for instance, becomes

$$E_x = \frac{q^*}{2\epsilon_0} x \int_0^1 \frac{n(\phi) 2u^2 du}{[(\frac{a_x^2}{\alpha^2} - 1)u^2 + 1]^{3/2}[(\frac{a_y^2}{\alpha^2} - 1)u^2 + 1]^{1/2}} \quad (9)$$

Following [5,6], we expand $n(\phi)$ in a Fourier series as

$$n(\phi) = \frac{c_0}{2} + \sum_{\ell=1}^{\infty} c_\ell \cos\left(\frac{\ell\pi\phi}{\Phi}\right)$$

where

$$c_\ell = \frac{2}{\Phi} \int_0^\Phi n(\phi) \cos\left(\frac{\ell\pi\phi}{\Phi}\right) d\phi \quad (10)$$

and Φ is the maximum value of ϕ (i.e. $n(\Phi) = 0$ defines the beam edge). The relation between the volume of the radial shell where $\phi \in [\phi_i, \phi_i + \Delta\phi]$ and the number of macroparticles ΔN contained within the shell is

$$\frac{\Delta N}{\pi a_x a_y dL} = n(\phi_i)\Delta\phi$$

It is straightforward to see that c_ℓ can be expressed as

$$c_\ell = \frac{2}{\Phi \pi a_x a_y dL} \sum_{i=1}^{N^*} \cos\left(\frac{\ell\pi\phi_i}{\Phi}\right)$$

If we substitute the expansion for $n(\phi)$ in Eq. (9), we can express the electric field, for instance E_x, as

$$E_x = \frac{c_0}{2} F_0^x(x, y, z) + \sum_{\ell=1}^{\infty} c_\ell F_\ell^x(x, y, z) \quad (11)$$

where $F_\ell^x(x, y, z) = x M_x q_x^*/\epsilon_0$ and M_x is given by Eq. (9) with the density n replaced by $\cos(\ell\pi\phi/\Phi)$.

This method of describing the particle density as smooth distribution can represent the electrostatic field produced an arbitrary transverse density profile that is constant on elliptical surfaces.

C Space Charge and Difference Resonances

When there is coupling between the transverse equantions of motion, the space-charge ellipse of the coasting beam can be rotated about the longitudinal axis.

Such coupling can be produced by skew quardupoles and enhance various resonance effects. The rotation angle α of the space-charge ellipse is function of the longitudinal coordinate s. In this case, we can employ the previous technique to calculate the space charge through an appropriate symmetry transformation. The coordinate of a particle $\mathbf{x} = (x, y)$ in the laboratory frame is rotated back by an angle α to $\tilde{\mathbf{x}} = (\tilde{x}, \tilde{y}) = R(-\alpha)\mathbf{x}$, where $\tilde{x}, \tilde{y}$ are the coordinates in the rotated frame. The space charge is then computed for the untilted ellipse in the rotated frame and then transformed back to the laboratory frame. If $\tilde{\mathbf{E}}(\mathbf{x})$ is the electric field for the untilted ellipse in the rotated frame, then the laboratory frame electric field is given by

$$\mathbf{E}(\mathbf{x}) = R(\alpha)\tilde{\mathbf{E}}(R(-\alpha)\mathbf{x})$$

where $R(\alpha)$ denotes a counter-clockwise rotation by angle α. To determine the tilting angle α from the spatial particle distribution, we use the fact that the quantity $\overline{xy} = 0$ in the rotated frame. The condition $\overline{\tilde{x}\tilde{y}} = \overline{(x\cos\alpha - y\sin\alpha)(x\sin\alpha + y\cos\alpha)} = 0$ leads to the equation

$$-m^2 \overline{xy} + m(\overline{x^2} - \overline{y^2}) - \overline{xy} = 0$$

where $m = \tan\alpha$. Solving this quadratic equation for m then determines α. The precision of this method in evaluating m depends of the number of macroparticles, the tilting angle α, and the ratio a_x/a_y. If $a_x/a_y \simeq 1$, the relative error of m is large. Hence we take $m = 0$ when the correlation coefficient $\sigma = \overline{xy}^2/(\overline{x^2} \cdot \overline{y^2})$ satisfies $\sigma < 0.2$.

III EMITTANCE EXCHANGE WITH SPACE CHARGE

One limitation of multiturn injection schemes [7,8] is the loss of particles hitting the vertical septum. A skew quadrupole couples the transverse planes exciting a difference resonance. When the bare tune is on such a resonance, horizontal oscillation energy is transfered to the vertical plane and the horizontal amplitude of the injected particle will diminish during the few revolutions of the multiturn injection. This effect can be exploited to move particles away from the septum, thereby improving the injection efficiency.

The single particle dynamics nearby a skew quadrupole driven resonance will be analyzed in sections B and C. When a coasting beam is considered, space charge can be important for the single particle dynamics since it changes the single particle tunes. Since the bandwidth of the resonance can be small, space charge can be a significant effect: the amplitude of exchange and the period of exchange is strongly dependent on the location of the single particle tunes with respect the resonance. As a consequence, the evolution of the emittance exchange is not similar to that of a single particle evolving in the external focusing fields. If the current is strong enough, space charge can push the single particle tunes out of the resonance and suppress the exchange process.

A Lattice Used

The SIS (SchwerIonen-Synchrotron) is a 18 Tesla-meter fast cycling synchrotron that can be used to accelerate a variety of ions. The SIS [2] has a strictly periodic lattice with 12 identical cells. The ion beam is injected at 11.4 MeV/u and a triplet focusing system is employed in each cell. A skew quadrupole scheme is under consideration in the SIS to improve the injection efficiency. We will consider a beam of U^{+28} with $\epsilon_x = 200$ mm mrad and $\epsilon_y = 20$ mm mrad. The emittance exchange is expected to take ~ 30 turns. We have introduced a thin skew quadrupole in the SIS lattice in order to simulate a real skew quadrupole. The skew quadrupole is located at the beginning of the second cell.

B Single Particle Emittance Exchange

As reported in [9], if the bare tunes are on a difference resonance ($q_x - q_y = k - \delta$ with k integer and $\delta \ll 1$) and if the initial vertical emittance is zero, then the emittances exchange is described by the formula

$$\epsilon_x = \epsilon_{x0} - \frac{4Q_c^2 \sin^2 \Theta}{\delta^2 + 4Q_c^2} \epsilon_{x0} \qquad (12)$$

Here,

$$\Theta = 2\pi q_x N_t \sqrt{\frac{\delta^2}{4} + Q_c^2} \qquad (13)$$

and

$$\epsilon_x + \epsilon_y = const$$

with initial values $\epsilon_{x,0}$ and $\epsilon_{y,0} = 0$, where ϵ_x, ϵ_y are the instantaneous Courant Snyder invariants, Q_c is proportional to the strength of the skew, and N_t is the number of turns.

Eq. (12) shows the mechanism for the emittance exchange: when the initial vertical emittance is zero, ϵ_x is oscillating and the exchange is due to the preservation of $\epsilon_x + \epsilon_y$. On the other hand, Eq. (12) was derived by neglecting second-order derivatives in the equation of motion and assumes only slowly varying amplitudes. These assumptions are strongly related to the strength of the skew: if we require a fast exchange they are not satisfied and the evolution of the Courant Snyder emittances can differ from Eq. (12). For instance, single particle simulations showed that $\epsilon_x + \epsilon_y$ can exhibit a dependence on the longitudinal position (Fig. 2a). However in spite of this difference, simulations indicate that even for a strong skew strength, the $\sin^2$-like evolution of the Courant Snyder emittance predicted by Eq. (12) is still present besides a modulation. Thus it makes sense to consider the amplitude and the period of emittance exchange. Some simulations indicate that an integrated gradient of $j = 0.01$ m^{-1} leads to an emittance exchange in 25 turns. Such a simulated exchange is plotted in Fig. 2b.

C Resonance Bandwidth

In this section we present how the evolution of the emittances can change as function of the tunes. with space-charge neglected. Fig. 2b shows the definition of the amplitude and period of the emittance exchange for a single particle. The discontinuities in the second-derivative of the curve are due to the strong skew kick experienced each turn. Fig. 3 summarizes the simulated behaviour of a single particle near the resonance. In Fig. 3 we have taken 40 working points on a line orthogonal to the resonance $q_x - q_y = 1$ with $q_x = 4.29$ and $q_y = 3.29$. For each working point we determined $\epsilon_{x,max}$ and $\epsilon_{x,min}$ by tracking a particle with initial coordinates $\hat{x} = \sqrt{\epsilon_x}, \hat{p}_x = 0$, and $\hat{y} = \sqrt{\epsilon_y}, \hat{p}_y = 0$ ($\hat{}$ = Courant Snyder coordinates) for 2000 turns. The exchange amplitude was calculated as $\Delta \epsilon_x = \epsilon_{x,max} - \epsilon_{x,min}$ (Fig. 3a,c). The period of the exchange T was computed as follows. 20000 turns were simulated using the same initial condition and measuring (in turns) the position when the instantaneous emittance ϵ_x crosses the average emittance $(\epsilon_{x,max} + \epsilon_{x,min})/2$. All these crossings were used to determined a series of periods from which the average period T was calculated (Fig. 3b,d). When the particle is exactly on the resonance, the linear coupling allows a periodic exchange of energy between the two planes. As the particle moves from the resonance at $q_x = 4.29$, the difference in phase between the horizontal and vertical oscillations reduces the time in which the exchange acts in one direction. This change reduce both the amplitude and period of exchange. The peak near $q_x = 4.5$ in Fig. 3a,b shows a higher order effect.

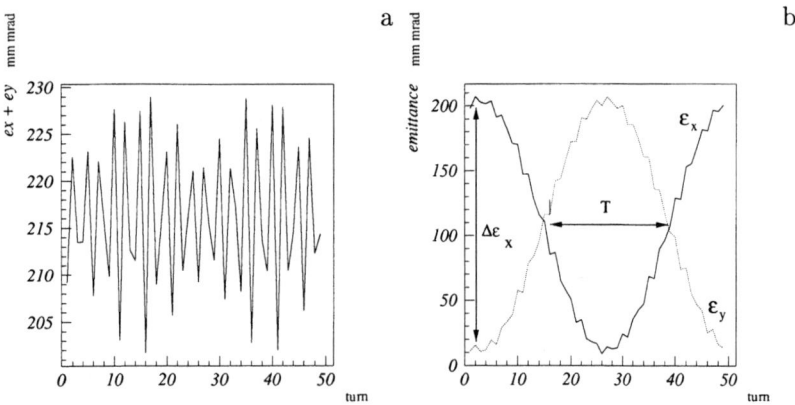

FIGURE 2. Simulated single-particle emittance exchange. a) the sum of the $x-$ and $y-$ emittances is not conserved; b) definition of amplitude and frequency of emittance exchange.

D Simulations of a Coasting Beam on Resonance

We employed the micromap technique to simulate the evolution of a coasting beam in the presence of space charge with an initial transverse K-V distribution. Simulations were carried out with $\delta p = 0$ and for currents of $I = 20$ mA and $I = 140$ mA. 1000 macroparticles were used and the space charge was calculated up to the 5th order (i.e., in Eq. (11) the sum is cut off for $\ell > 5$). The bare tunes of the machine were chosen to lie exactly on a $q_x - q_y = 1$ resonance with $q_x = 4.29$ and $q_y = 3.29$. The time step was chosen to correspond to a longitudinal increment of 1/20 of an SIS cell, i.e. $\Delta t = 0.02$ μs. In Figs. 4a,b,c,d are plotted the horizontal and vertical rms emittances ($\epsilon_{x\ rms}$ and $\epsilon_{y\ rms}$) and beam sizes (x_{rms} and y_{rms}). The discontinuity between the short horizontal lines in Figs. 4c,d show the effect of the skew quadrupole kick, which reaches its maximum strength when $\epsilon_x \simeq \epsilon_y$. The evolution in the rms beam sizes is driven by the emittance exchange. In Figs.

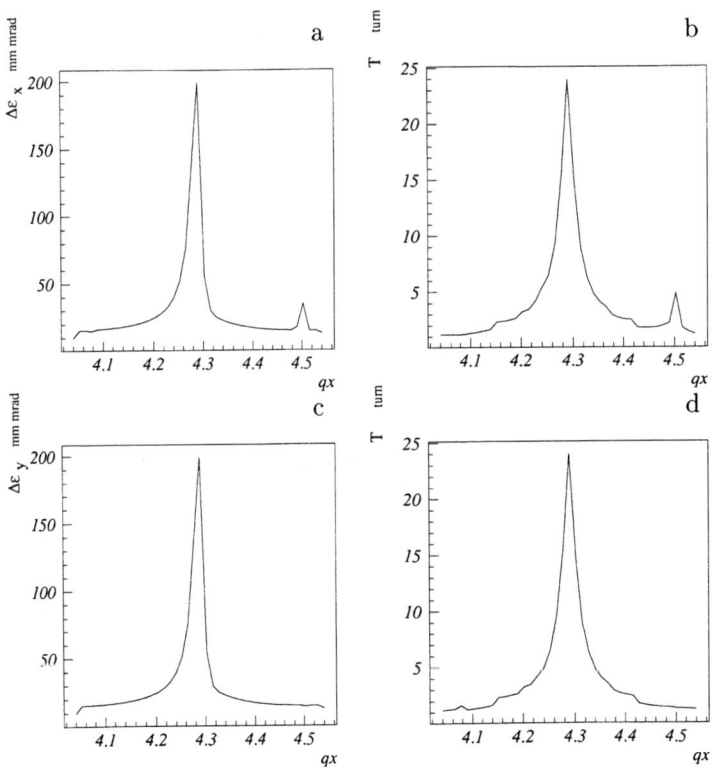

FIGURE 3. a) Amplitude and period of emittance oscillations as a function of q_x. b) Period of ϵ_x oscillation; c) ϵ_y oscillation amplitude; d) Period ϵ_y oscillation.

4a,b the thickness of the band of x_{rms} and y_{rms} values is due to the time step: since Δt is small with respect the time needed to go through one cell, the rms size shows envelope oscillations which when plotted in discrete time intervals look like a band rather than a continuous line. For $I = 20$ mA the initial tuneshifts due to the space charge are $\Delta q_x = -0.0049$ and $\Delta q_y = -0.0178$, so that the tunes of each particle are still near resonance (see Fig. 3) and then a complete rms emittance exchange occurs in 25 turns. Each particle "exchange" starts from different initial conditions, but the evolution of the rms emittances preserve the quantity $\epsilon_{x,rms} + \epsilon_{y,rms}$ better than for the single particle emittances (Fig. 2a). Figs. 5 shows results of a simulation repeated for a current of $I = 140$ mA, with initial tuneshifts of $\Delta q_x = -0.034$ and $\Delta q_y = -0.12$, which brings the single particle tunes to the border of the resonance. In this case the "exchange" is weaker but not as much as indicated by the single particle results in Fig. 3: The exchange period T is reduced to ~ 22 turns and $\Delta \epsilon_x = \Delta \epsilon_y = 80$ mm mrad, while from Fig. 3 we would expect that $\Delta \epsilon_x = \Delta \epsilon_y = 40$ mm mrad and $T = 4$ turns. This decreased exchange rate stems from the collective motion driven by the skew and a coherent effect of the space charge. In fact the skew quadrupole causes the emittance exchange that in turn induces a variation in the single particle tunes through the resonance, causing a partial sampling of the resonance. Figs. 3b,d show that the wave number Θ/N_t of the exchange function in Eq. (13) depends on the single particle tune. Therefore, the phase advance results from the integration of the wave number over the number of turns. In order to have half emittance oscillation, the (phase advance $= \pi$) we need more turns since T is not constant. An analogous argument can be used to explain the larger amplitude of emittance exchange obtained in Fig. 5 with respect the 40 mm mrad predicted in Figs. 3a,c. In these simulations were also included the coherent tuneshift which could be relevant in the exchange process.

IV CONCLUSION

We presented a micromap technique as an extension of a standard transfer map to include space charge defocusing forces for a coasting beam. A macroparticle model was developed to calculate the space charge forces under the assumption of elliptic beam symmetry. A code employing this procedure was written and used to study emittance exchange driven by a skew quadrupole. Results provide evidence of a coherent effect of space charge which improves the exchange with respect the single particle exchange in the absence of space charge. This improvement stems from the coherent motion of the beam. More detailed simulations are needed to investigate this process when the working point of the machine doesn't lie near a resonance and when non-K-V distributions are used to model the initial particle distribution of the beam.

ACKNOWLEDGMENTS

The authors have benefitted from critical comments by Steve Lund.

REFERENCES

1. A. Bazzani, E. Todesco, G. Turchetti and G. Servizi, *A Normal Form Approach to the Theory of Nonlinear Betatronic Motion*, CERN **94-02**, (1994).
2. B. Franczak, GSI-SIS-TN / 87-13, 10. Sept. 1987
3. J. Rossbach and P. Schmüser, *Basic Course on Accelerator optics*, Cern Accelerator School **94 - 01** (1994);
4. F. J. Sacherer, *IEEE Trans. Nucl. Sci.* **NS-18**, 1105 (1971);

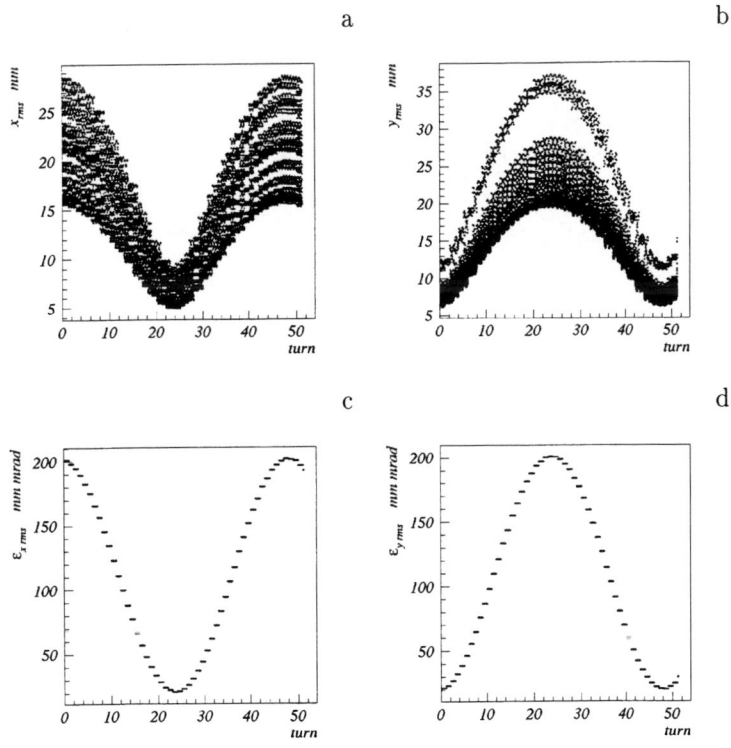

FIGURE 4. Simulated multi-particle evolutions as a function of turn number for I=20 mA a) and b) rms sizes; c) and d) rms emittances.

5. G. Franchetti, I. Hofmann and G. Turchetti *Six-dimensional approach to the beam dynamics in HIDIF scenario.*, Proc. of the 12th Int. Symp. on Heavy Ion Inertial Fusion, Heidelberg (Germany), Sept. 24-27, 1997, to be pubblished on Nuclear Instruments and Methods A;
6. R.W. Garnett and T.P. Wangler, IEEE Particle Accelerator Conference, San Francisco, CA, May 6-9, 1991, IEEE, New York, p. 330.
7. R.W. Hasse and I. Hofmann, *Space charge limit of multiturn injection in HIDIF by 2D and 3D PIC simulation*, Proc. of the 12th Int. Symp. on Heavy Ion Inertial Fusion, Heidelberg (Germany), Sept. 24-27, 1997, to be published on Nuclear Instruments and Methods A;
8. C. Prior and G. Rees, *Multiturn Injection and Lattice Design for HIDIF*, Proc. of the 12th Int. Symp. on Heavy Ion Inertial Fusion, Heidelberg (Germany), Sept. 24-27, 1997, to be published on Nuclear Instruments and Methods A;
9. K. Schindl and P. Van der Stock, *IEEE Trans. Nucl. Sci.* **NS-24**, 1390 (1977).

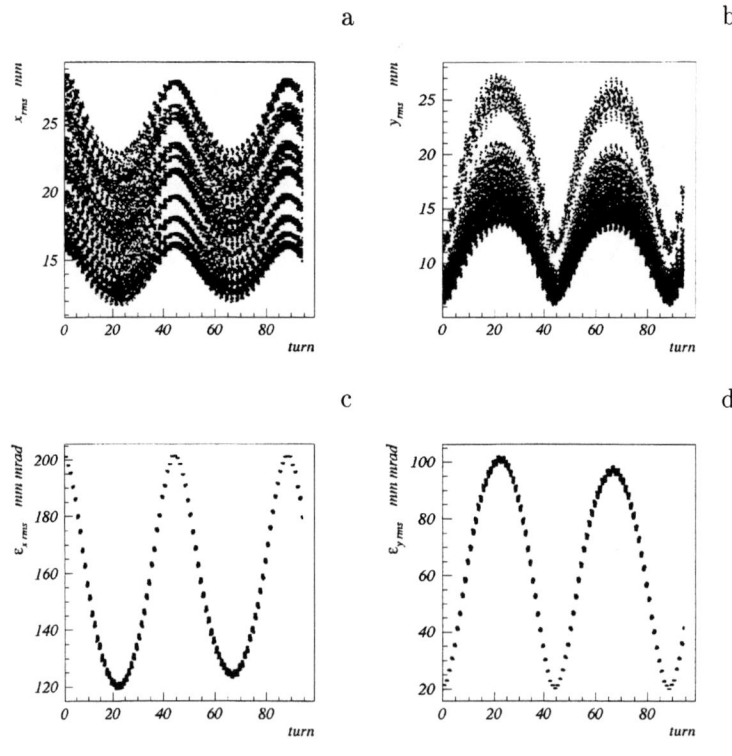

FIGURE 5. Simulated multi-particle evolutions as a function of turn number for I=140 mA a) and b) rms sizes; c) and d) rms emittances.

Longitudinal Halo in Beam Bunches with Self-Consistent 6-D Distributions

R.L. Gluckstern, A.V. Fedotov * and S.S. Kurennoy, R.D. Ryne [†]

* *Physics Department, University of Maryland, College Park, Maryland 20742* [†] *LANSCE Division, Los Alamos National Lab, Los Alamos, New Mexico 87545*

Abstract. We have explored the formation of longitudinal and transverse halos in 3-D axisymmetric beam bunches by starting with a self-consistent 6-D phase space distribution. Stationary distributions allow us to study the halo development mechanism without being obscured by beam redistribution and its effect on halo formation. The beam is then mismatched longitudinally and/or transversely, and we explore the rate, intensity and spatial extent of the halos which form, as a function of the beam charge and the mismatches. We find that the longitudinal halo forms first because the longitudinal tune depression is more severe than the transverse one for elongated bunches and conclude that it plays a major role in halo formation.

I INTRODUCTION

A great deal of attention has been focused recently on the design of ion linacs with currents one or two orders of magnitude greater than previously achieved. A major issue in such designs is the activation caused by the formation of halos around the beam, a phenomenon whose origin is only now being understood.

It now appears that rms mismatch of a high intensity beam in a focusing channel is a major source of halo formation. Specifically, the dominant effect seems to be a parametric-like resonance between the collective oscillation (most likely a breathing mode) of the beam bunch due to mismatches, and the non-linear oscillation of individual ions in the bunch. Analytic predictions [1] for a 2-D KV [2] beam are found to be in excellent agreement with numerical simulations. In fact, most attention has been given to the 2-D case [3–7]. Recently Barnard and Lund [8] have performed numerical studies with a 3-D beam bunch, drawing attention to the existence and importance of longitudinal halos for a spheroidal bunch.

With few exceptions, the simulations have been performed by artificially mismatching an rms matched beam which is *not* itself a stationary solution of the Vlasov equation. As a result, when halos appear it is not easy to identify their origin as due to the mismatch, or to the unavoidable relaxation and redistribution of the non-stationary beam in the 6-D phase space. For this reason, we have chosen to start with a stationary 6-D phase space distribution for our spheroidal beam bunch for which no such redistribution occurs. Our analysis assumes smoothed external transverse and longitudinal restoring force gradients, k_x, k_y, k_z. In general, the distribution can be chosen to have an approximately ellipsoidal boundary. However, for simplicity, we treat the azimuthally symmetric case ($k_x = k_y$) for which the beam bunch is approximately spheroidal.

II STATIONARY 6-D PHASE SPACE DISTRIBUTION

We take for the azimuthally symmetric 6-D phase space distribution

$$f(\boldsymbol{x},\boldsymbol{p}) = \frac{N}{(H_0 - H)^{1/2}} \equiv \frac{N}{(G(\boldsymbol{x}) - mv^2/2)^{1/2}}, \qquad (1)$$

where N and H_0 are constants, H is the Hamiltonian, and

$$G(\boldsymbol{x}) = H_0 - k_x r^2/2 - k_z z^2/2 - e\Phi_{\text{sc}}(\boldsymbol{x}). \qquad (2)$$

Here $\boldsymbol{p} = m\boldsymbol{v}$, $r^2 = x^2 + y^2$ and k_x, k_z are the smoothed transverse and longitudinal restoring force gradients. The quantity $\Phi_{\text{sc}}(\boldsymbol{x})$ is the electrostatic potential due to the space charge of the bunch. The distribution is normalized so that $\int d\boldsymbol{x} \int d\boldsymbol{p} f(\boldsymbol{x},\boldsymbol{p}) = 1$.

Integrating f over $d\boldsymbol{p}$ leads to the charge density

$$\rho(\boldsymbol{x}) = Qg(\boldsymbol{x})/\int d\boldsymbol{x} g(\boldsymbol{x}), \qquad (3)$$

where Q is the total bunch charge and $G(\boldsymbol{x}) = (k_s/\kappa^2)g(\boldsymbol{x})$, with $\kappa^2 = (eQ/\epsilon_0)/\int G(\boldsymbol{x})d\boldsymbol{x}$ and $k_s = 2k_x + k_z$. The Poisson equation then becomes

$$\nabla^2 g(\boldsymbol{x}) = -\kappa^2 + \kappa^2 g(\boldsymbol{x}). \qquad (4)$$

The solution of Eq. (4) can then be written as

$$g(\boldsymbol{x}) = 1 + \sum_{\ell=0}^{\infty} \alpha_\ell P_{2\ell}(\cos\theta) i_{2\ell}(\kappa R), \qquad (5)$$

where R, θ are the spherical counterparts to r, z, and where our focus on a spheroidal beam bunch implies the need for only even order Legendre polynomials $P_{2\ell}(\cos\theta)$. The spherical Bessel functions of imaginary argument, regular at the origin, are denoted by $i_{2\ell}(\kappa R)$. The edge of the bunch is defined as the border $g(\boldsymbol{x}) = 0$, where the density first vanishes. We therefore choose the α_ℓ in Eq. (5) to reproduce the spheroidal surface

$$\frac{r^2}{a^2} + \frac{z^2}{c^2} = 1 \qquad (6)$$

as closely as possible, where $2a$, $2c$, are the minor, major axes of the (prolate) spheroid, with $c \geq a$.

Because f is only a function of the Hamiltonian H, the distribution is stationary. Starting with the x^2, $x\dot{x}$, $\dot{x}^2$ moments of the stationary Vlasov equation and the equation of motion for x, we obtain the envelope-like equation

$$k_x \langle x^2 \rangle = e \langle x E_x \rangle + m \langle \dot{x}^2 \rangle \tag{7}$$

where $\langle \ \rangle$ denotes an average over the 6-D phase space of the beam bunch. A similar equation holds for z. Using Eq. (7) to define the rms tune depressions η_x and η_z leads to

$$\frac{1-\eta_x^2}{\eta_x^2} = \frac{e \langle x E_x \rangle}{m \langle \dot{x}^2 \rangle}, \quad \frac{1-\eta_z^2}{\eta_z^2} = \frac{e \langle z E_z \rangle}{m \langle \dot{z}^2 \rangle}. \tag{8}$$

We can perform the averages in Eq. (8) in terms of dimensionless parameters. Specifically we write

$$m \langle \dot{x}^2 \rangle = m \langle \dot{y}^2 \rangle = m \langle \dot{z}^2 \rangle = (k_s/2\kappa^2)(g_2/g_1), \tag{9}$$

$$\phi_x \equiv (a\epsilon_0/Q)\langle x E_x \rangle, \quad \phi_z \equiv (a\epsilon_0/Q)\langle z E_z \rangle, \tag{10}$$

where $g_n \equiv \int d\boldsymbol{x} g^n(\boldsymbol{x})/a^3$, and ϵ_0 is the permittivity of free space. This leads to

$$\frac{1-\eta_x^2}{\eta_x^2} = \frac{2\kappa^2 a^2 \phi_x g_1^2}{g_2}, \quad \frac{1-\eta_z^2}{\eta_z^2} = \frac{2\kappa^2 a^2 \phi_z g_1^2}{g_2}. \tag{11}$$

Starting with Eqs. (5) and (6), one can see that α_ℓ, g_n, ϕ_x, ϕ_z, η_x, η_z are each only functions of the two variables κa, c/a. Futhermore, our choice of a stationary distribution of the form $f(H)$ automatically corresponds to equipartition, as illustrated in Eq. (9). Finally, the bunch charge can be written in the dimensionless form

$$Q/e = B \cdot q(\kappa a, c/a) \equiv 2B\kappa^2 a^2 g_1^2 a^2 / g_2 \langle x^2 \rangle, \tag{12}$$

where $B = mc_0^2 \epsilon_0 \epsilon_x^2 / e^2 a$ is a dimensionless scaling constant. Here mc_0^2 is the rest energy of the ion and $\epsilon_x^2 = \langle \dot{x}^2 \rangle \langle x^2 \rangle / c_0^2$ is the square of the normalized rms transverse emittance. The parameter $q(\kappa a, c/a)$ contains the dependence of the bunch charge on κa and c/a. From another perspective κa is determined by the total charge and the eccentricity of the spheroid.

III NUMERICAL INVESTIGATION OF PARAMETERS

We have explored the range of parameters from $c/a = 1$ to 4 and $\eta_z = 1$ to 0.3. This range of parameters corresponds to the proposed bunches for the Accelerator

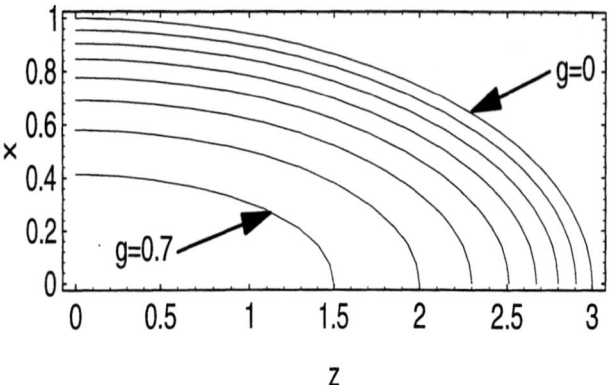

FIGURE 1. Contours $g(x) =$ constant for $c/a = 3$, $\eta_x = 0.65$, $\eta_z = 0.49$.

Production of Tritium (APT) project [9]. In Fig. 1 we plot the contours $g(x) =$ const for $c/a = 3$ and $\kappa a = 3$, showing the spheroidal shape of the bunch and its charge distribution. We find that $(1 - \eta_x^2)/\eta_x^2$ and $(1 - \eta_z^2)/\eta_z^2$ are proportional to $\kappa^2 a^2$ for fixed c/a to surprisingly great precision and we are able to construct the simple universal formulas

$$\eta_x \cong [1 + (\kappa a/2.82)^2 (c/a)^{0.20}]^{-1/2}, \tag{13}$$

$$\eta_z \cong [1 + (\kappa a/2.82)^2 (c/a)^{0.91}]^{-1/2}, \tag{14}$$

each accurate to a few percent. We also find a similar universal fit for the total charge parameter:

$$q(\kappa a, c/a) \cong 44\kappa^2 ac. \tag{15}$$

In Figs. 2 and 3, we then plot η_x and η_z against q for different c/a.

Clearly, the longitudinal tune depression, η_z, is more severe than the transverse one for $c > a$. Thus we expect, and find, that a longitudinal halo is more likely to form first for comparable horizontal and transverse mismatches. Furthermore, if the rate of halo formation is controlled by the tune depression, Fig. 3 suggests that we can accelerate slightly more current for higher c/a if the longitudinal halo is of primary concern, and Fig. 2 shows that we can accelerate significantly more current for higher c/a if the transverse halo is of primary concern.

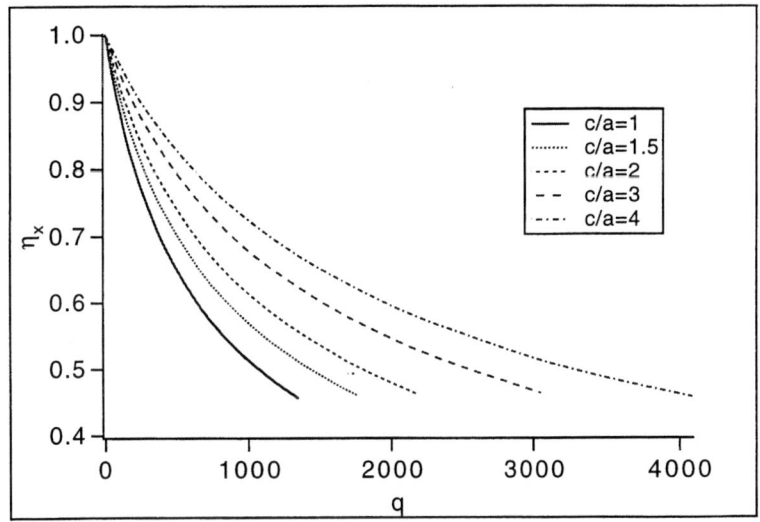

FIGURE 2. Transverse tune depression η_x vs bunch charge.

IV ORBIT SIMULATIONS

For axisymmetric beam bunches, there are two independent mismatch parameters, one associated with the transverse bunch size, the other with the longitudinal bunch size. Unfortunately, an analytic treatment of the mismatch oscillation is possible only for a uniformly charged spheroid, which we now outline. We then assume that the results of a correct self-consistent 6-D treatment resemble our model with uniform charge density.

The breathing modes of the uniformly charged spheroid are governed, for small envelope oscillations, by the coupled equations

$$\delta\ddot{a} + \nu_x^2 \delta a = -\nu_{xz}^2 \delta c, \quad \delta\ddot{c} + \nu_z^2 \delta c = -2\nu_{xz}^2 \delta a, \qquad (16)$$

where, for $c/a \gg 1$, $\nu_x^2 \gg \nu_{xz}^2 \gg \nu_z^2$. The high frequency mode ($\nu_h = \nu_x$) is primarily transverse, resembling the 2-D breathing mode, with δa and δc in phase and

$$\frac{(\delta c)_{\max}}{(\delta a)_{\max}} = 2\frac{\nu_{xz}^2}{\nu_x^2} \simeq \frac{a}{c} \cdot \frac{1 - \eta_x^2}{1 + \eta_x^2}. \qquad (17)$$

The low frequency mode ($\nu_\ell = \nu_z$) is primarily longitudinal, with δa and δc out of phase by π, and

$$\frac{(\delta a)_{\max}}{(\delta c)_{\max}} = \frac{-\nu_{xz}^2}{\nu_x^2} \simeq \frac{-a}{2c} \cdot \frac{1 - \eta_x^2}{1 + \eta_x^2}. \qquad (18)$$

We make the conjecture that, for the beam bunch of Eqs. (3) and (5), the form of Eqs. (16) remains unchanged.

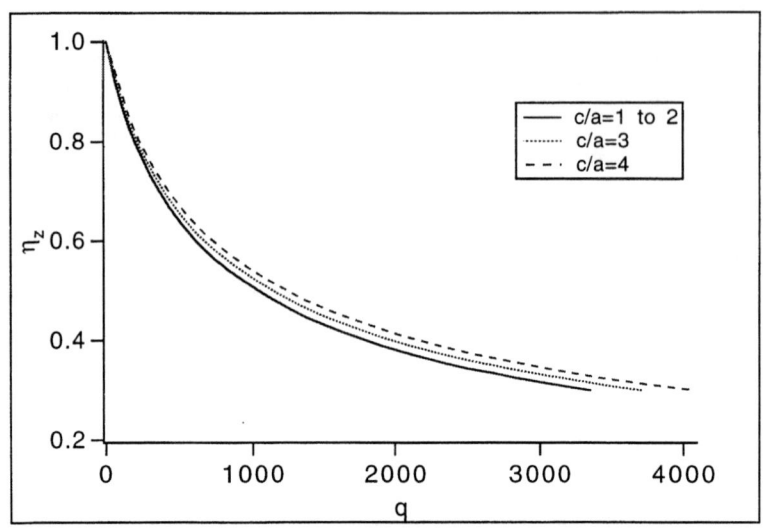

FIGURE 3. Longitudinal tune depression η_z vs bunch charge.

As an illustrative example, we take the parameters $c/a = 3$, $\kappa a = 3.0$, corresponding to $\eta_z = .49$, $\eta_x = .65$. Using these values in Eqs. (17)-(18), we obtain $\delta c/\delta a \approx 1/7$ for the transverse mode and $\delta a/\delta c \approx -1/15$ for the longitudinal mode. These numbers are confirmed to reasonable accuracy by our numerical simulations.

We have developed a 3-D particle-in-cell (PIC) code to test the analytic model described above, and to explore halo formation. The single-particle equations of motion are integrated using a symplectic, split-operator technique [11]. The space charge calculation uses area weighting ("Cloud-in-Cell") and implements open boundary conditions with the Hockney convolution algorithm [12]. The code runs on parallel computers, and in particular, the space charge calculation has been optimized for parallel platforms using the Ferrell-Bertschinger method [13].

We initially populate the 6-D phase space according to Eq. (1), and then mismatch the x, y, z coordinates by factors $\mu_x = \mu_y = 1 + \delta a/a$, $\mu_z = 1 + \delta c/c$ and the corresponding momenta by $1/\mu_x = 1/\mu_y$, $1/\mu_z$. In Fig. 4, we start with $\delta c/c = 0.4$, $\delta a/a = -0.06$ (primarily longitudinal mode), plotting the maximum x and z among the million particles in our run as a function of time (in the Lorentz frame of the bunch). The development of a longitudinal halo is clearly visible. In Fig. 5, the longitudinal phase space clearly shows the typical peanut diagram similar to the one observed in 2-D calculations. Similar results are obtained for different tune depressions and for other mismatches and will be presented later [10].

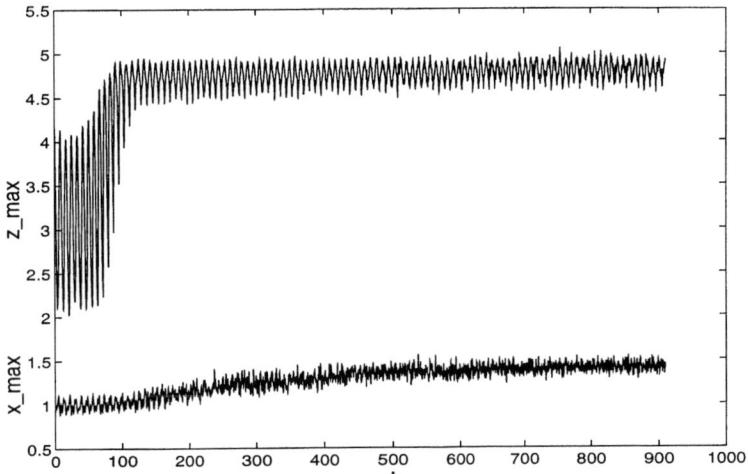

FIGURE 4. Maximum x and z as a function of time in arbitrary units. In this figure one longitudinal breathing oscillation takes about 8.3 such units. ($c/a = 3$, $\eta_x = .65$, $\eta_z = .49$, $\mu_z = 1.4$, $\mu_x = 0.94$).

V DISCUSSIONS AND CONCLUSIONS

Most of the previous studies were concerned with transverse halos in long beams. Present simulations with a 6-D stationary distribution show the importance of the longitudinal halo in a beam bunch, as is evident from the following observations:

1. Due to the coupling between r and z, a transverse or longitudinal halo is observed even for a very small mismatch as long as there is a significant mismatch in the other plane.

2. The longitudinal halo develops earlier than the transverse halo for elongated bunches with comparable mismatch in r and z when the mismatches are not very large and tune depressions are not severe simply because we have $\eta_z < \eta_x$ for elongated bunches with fixed charge. For severe mismatches and/or tune depressions both halos develop very quickly.

3. We find that halo intensity is governed primarily by the mismatch.

4. The maximum halo extent in units of the matched radius depends on the mismatch, but is almost independent of tune depression if it is modest. Our new result is that the longitudinal halo extent starts to grow when the tune depression becomes more severe, exhibiting a significant rise when η_z drops below 0.4 for $c/a = 3$. (Note that no strong rise is seen for a transverse halo, both in the present and previous [7] simulations.)

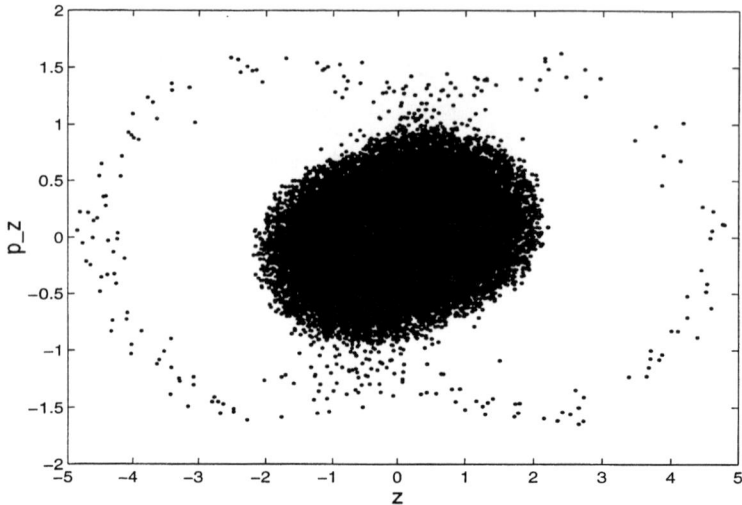

FIGURE 5. Phase space diagram of a longitudinal halo ($c/a = 3$, $\eta_x = .65$, $\eta_z = .49$, $\mu_z = 1.4$, $\mu_x = 0.94$).

5. The rate at which the halo develops also depends on both the mismatch and tune depression. For more severe tune depression the halo starts to develop earlier.

The above conclusions were based on simulations for different c/a in the range of interest [9]. Details supporting the above conclusions will be found in [10].

VI SUMMARY

We have constructed, analytically and numerically, a new class of 6-D phase space stationary distributions for an azimuthally symmetric beam bunch of arbitrary charge in the shape of a prolate spheroid (see Fig. 1), and have determined the rms tune depressions η_x, η_z as a function of the bunch charge and eccentricity (see Figs. 2 and 3). Our choice of parameters automatically assures equipartition. Such an approach allows us to study the fundamental mechanism of halo formation associated with the beam mismatch. This method proved to be very efficient in previous 2-D calculations [5,7]. The fact that "real" beam is non-stationary and not in thermal equilibrium will lead to additional effects beyond those associated with the mismatch.

Our main conclusion is that the longitudinal halo is of great importance because it develops earlier than the transverse halo for elongated bunches with comparable longitudinal and transverse mismatches, and because it occurs even for mismatches of order 10%. The main characteristics of the longitudinal halo are briefly discussed in Section V.

VII ACKNOWLEDGMENT

We wish to acknowledge the support of DOE, Division of High Energy Physics, and Division of Mathematical, Information, and Computational Sciences. This research used resources of the National Energy Research Scientific Computing Center, which is supported by the DOE Office of Energy Research. We thank Tom Wangler for helpful conversations. In addition RLG and AVF wish to thank Andy Jason and the LANSCE1 group for its hospitality during part of these studies.

REFERENCES

1. R.L. Gluckstern, Phys. Rev. Letters **73**, 1247 (1994).
2. See I.M. Kapchinsky, Theory of Resonance Linear Accelerators (Harford Academic Press, New York, 1985), p. 247 ff.
3. M. Reiser, in Proceedings of the 1991 Particle Accelerator Conference, edited by L.Lizama et al. (IEEE, San Francisco, CA, 1991) p. 2497; A. Cucchetti, M. Reiser and T. Wangler, ibid., p. 251; J.S. O'Connell, T.P. Wangler, R.S. Mills, and K.R. Crandall, in Proceedings of the 1993 Particle Accelerator Conference, edited by S.T. Corneliussen (IEEE, Washington, DC, 1993), p. 3657.
4. R.L. Gluckstern, W-H. Cheng, and H. Ye, Phys. Rev. Letters **75**, 2835 (1995).
5. R.L. Gluckstern, W-H. Cheng, S.S. Kurennoy and H. Ye, Phys Rev. E **54**, 6788 (1996).
6. H. Okamoto and M. Ikegami, Phys. Rev. E **55**, 4694 (1997).
7. R.L. Gluckstern and S.S. Kurennoy, in Proceedings of the 1997 Particle Accelerator Conference, Vancouver, Canada (unpublished).
8. J.J. Barnard and S.M. Lund (I), and S.M. Lund and J.J. Barnard (II), in Proceedings of the 1997 Particle Accelerator Conference, Vancouver, Canada (unpublished).
9. APT Conceptual Design Report, Los Alamos Report No. LA-UR-97-1329, 1997.
10. Submitted for publication in Phys. Rev. E.
11. E. Forest et al., Phys. Lett. A **158**, 99 (1991).
12. R.W. Hockney and J.W. Eastwood, Computer Simulation Using Particles, Adam Hilger, NY (1988).
13. R. Ferrell and E. Bertschinger, Int. J. Mod. Phys. C **5**, 933 (1994).

An RMS Particle Core Model for Rings

J. A. Holmes*, J. D. Galambos*, D. K. Olsen*, and S. Y. Lee**

*Oak Ridge National Laboratory, Oak Ridge, TN 37831
**Indiana University, Bloomington, IN 47408

Abstract. A self-consistent set of equations for the azimuthal variation of rms betatron oscillation amplitudes, including the effects of dispersion and space charge, is derived. These effective envelope equations can be integrated over the beam energy distribution to provide space charge forces in a particle core model for rings. The derivation of the envelope equations involves an accelerator ordering scheme for the beam dynamics and a statistical moments analysis of the canonical distribution function in the six-dimensional phase space of the beam Hamiltonian. The azimuthal variation of the second moments of the transverse canonical coordinates, x^β and z^β, integrated over the kinetic distribution function of the beam, provides the rms equations. These equations, at fixed beam energy, are integrated over the beam energy distribution to provide the overall space charge distribution and force. Because the envelope equations and dispersion function both depend upon and determine the space charge forces, the consistency of the particle core model requires either analytic closure assumptions or numerical iteration.

INTRODUCTION

The analysis and understanding of space charge effects for particle beams in linear accelerators has been greatly facilitated by the use of particle core models (1). Particle core models represent the dynamics of the beam by envelope equations that contain the effects of the lattice focusing forces, the beam emittances, and the space charge. In addition to providing a collective model for beam dynamics, the envelope equations are also used to calculate space charge forces in particle tracking. Particle core models are employed in conjunction with particle-tracking calculations to simultaneously study both collective and individual particle dynamics.

The alternatives to particle core models for the computational study of space charge effects are the Particle in Cell (PIC) models (2). In comparison with PIC models, particle core models have both advantages and disadvantages. One advantage of particle core models involves computing time. The computational work involved in advancing the envelope equations is comparable to that in tracking a single particle, and the evaluation of the resulting space charge forces is fast. Another advantage of particle core models is their simplicity, making them amenable to analytic calculations. A third advantage is that individual particle orbits can be studied one at a time using space charge fields given by the particle core model. However, a major limitation of the particle core model is that it provides a simplified representation of the space charge distribution, both in the envelope equations and in the particle-tracking equations. Even so, particle core models provide a practical middle ground between analytic theory and large PIC code calculations.

Up until now, particle core models have been applied primarily to the study of beam dynamics and halo generation in straight channels with strong space charge forces. Now, with the advent of a number of applications involving rings with high beam intensities and small beam loss requirements (3), the application of particle core models to rings is necessary. However, the representation of the beam via the envelope equations must be generalized to include the effects of dispersion. One treatment of this dispersion problem has recently been carried out at University of Maryland (4), and we present a somewhat different approach here.

The purpose of this paper is to extend the particle core model to rings by including dispersion. The approach involves a moments analysis of the betatron oscillations averaged over the distribution function in the canonical phase space of a beam Hamiltonian which is derived using an ordering scheme that separates the betatron motion from the longitudinal motion and the dispersion. Effective envelope equations are derived for the azimuthal variation of the rms values of the canonical transverse phase space coordinates, with the statistical averaging at each energy over the beam distribution function in phase space. These envelope equations are independent of dispersion and valid, in principle, for arbitrary space charge distributions. The effects of dispersion are incorporated in a straightforward fashion by integrating the envelope equations over beam energy to obtain the overall space charge distribution. For self-consistency, this space charge distribution must be the same as that appearing in the dispersion and envelope equations. In subsequent sections we present the accelerator ordering scheme and the beam Hamiltonian derivation; the behavior of the distribution function and integrated quantities under the accelerator ordering scheme; the derivation of the rms envelope equations; and the incorporation of dispersion to obtain the overall charge distribution.

THE BEAM HAMILTONIAN

This section presents a derivation of the beam Hamiltonian in a form that decouples the transverse betatron motion from the longitudinal motion and from explicit dependence on the dispersion. Although a derivation was recently published by one of the authors (5), the present version uses an accelerator ordering scheme, which is later applied to derive the envelope equations. The main purposes of presenting this section here are 1) to introduce the ordering scheme, and 2) to obtain the canonical variables of the phase space used later in deriving envelope equations.

Begin with the standard Hamiltonian for a charged particle in an electromagnetic field:

$$H = q\Phi + [m^2c^4 + (c\vec{p} - q\vec{A})^2]^{\frac{1}{2}} \qquad (1)$$
$$= E_0 + \Delta E,$$

where $E_0 \equiv \gamma mc^2$ is the reference energy of the beam; ΔE is the energy deviation; the canonical coordinates and momenta are (x, p_x, z, p_z, s, p_s); and the time, t, is the independent parameter. In order to make the azimuthal coordinate, s, the independent parameter, it is customary (6) to define a new Hamiltonian by $H_1 = -p_s$. First, separate the vector potential $\vec{A}$ into contributions $\vec{A}^{ext}$ from the lattice and $\vec{A}^{sc}$ from the beam space charge. We assume the lattice contributions to the magnetic fields are transverse to the azimuthal direction, so that the external vector potential can be chosen to satisfy $A_x^{ext} = A_z^{ext} = 0$. Define a reference momentum that satisfies the equation $p_0^2 c^2 \equiv E_0^2 - m^2c^4 = \beta^2 E_0^2$. With these assumptions and definitions, Eq. (1) can be manipulated to obtain:

$$H_1 = -\frac{qA_s^{ext}}{c} - \frac{qA_s^{sc}}{c} - (1 + \frac{x}{\rho}) \times$$
$$[p_0^2 + 2\frac{E_0}{c^2}(\Delta E - q\Phi) + \frac{(\Delta E - q\Phi)^2}{c^2} - (p_x - \frac{qA_x^{sc}}{c})^2 - (p_z - \frac{qA_z^{sc}}{c})^2]^{\frac{1}{2}}. \qquad (2)$$

The canonical variables are now $(x, p_x, z, p_z, t, -\Delta E)$. To adapt Eq. (2) to the analysis of space charge in rings, we adopt the following accelerator ordering scheme:

$$O(\varepsilon^{-1}) \quad - \frac{\partial}{\partial x}, \frac{\partial}{\partial z}, \frac{\partial}{\partial p_x}, \frac{\partial}{\partial p_z}, \frac{\partial}{\partial(\Delta E)}$$

$$O(1) \quad - H_1, \rho, H, mc^2, E_0, t, s, \frac{\partial}{\partial t}, \frac{\partial}{\partial s}$$

$$O(\varepsilon) \quad - \vec{A}^{ext}, x, p_x, z, p_z, \Delta E$$

$$O(\varepsilon^2) \quad - \vec{A}^{sc}, \Phi.$$

The application of this ordering to Eq. (2) yields, valid through $O(\varepsilon^2)$:

$$H_1 = -p_0 \frac{x}{\rho} p_0 - \frac{qA_s^{ext}}{c} + \frac{p_0}{2}[(\frac{p_x}{p_0})^2 + (\frac{p_z}{p_0})^2]$$
$$-\frac{E_0}{p_0 c^2}\Delta E - \frac{x}{\rho}\frac{E_0}{p_0 c^2}\Delta E + \frac{1}{2p_0 c^2}\frac{1}{\gamma^2 \beta^2}(\Delta E)^2 + q(\frac{E_0}{p_0 c^2}\Phi - \frac{A_s^{sc}}{c}). \quad (3)$$

The external vector potential is next expressed in terms of the magnet fields and the space charge vector potential in terms of the electrostatic potential. We will not consider accelerating channels here, but rather confine our attention to nonaccelerated beams, so that A_s^{ext} is independent of time. An expression for the external potential, valid through second order, for bending and quadrupole magnets is $A_s^{ext} = -B_0 x - \frac{1}{2}(\frac{B_0}{\rho} + \frac{\partial B_0}{\partial x})x^2 + \frac{1}{2}\frac{\partial B_0}{\partial x}z^2$, where B_0 is the bending field on the reference orbit. The vector potential due to the beam is related to the electrostatic potential by $A_s^{sc} = \beta\Phi$. Together with these substitutions, we make the following canonical scaling transformation:

$$H_2 = \frac{H_1}{p_0}$$
$$\tilde{p}_x = \frac{p_x}{p_0} \quad \tilde{p}_z = \frac{p_z}{p_0} \quad \tilde{\delta} = \frac{\Delta E}{\beta^2 E_0} = \frac{\delta p}{p_0} \quad (4)$$
$$\tilde{x} = x \quad \tilde{z} = z \quad \tilde{t} = \beta c t.$$

With the above transformation and substitutions, and using $\frac{qB_0}{p_0 c} = \frac{1}{\rho}$, we obtain:

$$H_2(\tilde{x},\tilde{p}_x,\tilde{z},\tilde{p}_z,\tilde{t},-\tilde{\delta},s) = \frac{1}{2}[(\tilde{p}_x)^2 + K_x(s)(\tilde{x})^2 + (\tilde{p}_z)^2 + K_z(s)(\tilde{z})^2]$$
$$-\tilde{\delta} - \frac{x}{\rho}\tilde{\delta} + \frac{1}{2\gamma^2}(\tilde{\delta})^2 + \frac{q}{\gamma^2\beta^2 E_0}\Phi(\tilde{x},\tilde{z},s), \quad (5)$$

where $K_x(s) = \frac{1}{\rho^2}(1 + \frac{\rho}{B_0}\frac{\partial B_0}{\partial x})$; $K_z(s) = -\frac{1}{\rho B_0}\frac{\partial B_0}{\partial x}$; and a term of constant value -1 was dropped. The canonical variables are now $(\tilde{x},\tilde{p}_x,\tilde{z},\tilde{p}_z,\tilde{t},-\tilde{\delta})$ and the independent parameter is s.

Equation (5) contains the term $-\frac{x}{\rho}\tilde{\delta}$, which couples the betatron motion to the dispersion. The definitive step in Ref. (5) removes this term through the following canonical transformation of the second type:

$$F_2(\tilde{x}, p_x^\beta, \tilde{z}, p_z^\beta, \tilde{t}, -\delta, s) = (\tilde{x} - D_x\delta)p_x^\beta - \tilde{t}\delta + \tilde{z}p_z^\beta + D_x'\tilde{x}\delta - \frac{1}{2}D_xD_x'\delta^2, \quad (6)$$

where $D_x = D_x(s)$ is a function of s that is yet to be defined. The quantities D_x' and D_x'' are the first and second derivatives of D_x with respect to s. With this transformation the new coordinates and momenta are

$$\begin{aligned} x^\beta &= \tilde{x} - D_x\delta & z^\beta &= \tilde{z} & \delta &= \delta \\ p_x^\beta &= \tilde{p}_x - D_x'\delta & p_z^\beta &= \tilde{p}_z & \tau &= \tilde{t} + D_x\tilde{p}_x - D_x'\tilde{x}, \end{aligned} \quad (7)$$

and the new Hamiltonian is

$$\begin{aligned} H(x^\beta, p_x^\beta, z^\beta, p_z^\beta, \tau, -\delta, s) &= \frac{1}{2}[(p_x^\beta)^2 + K_x(s)(x^\beta)^2 + (p_z^\beta)^2 + K_z(s)(z^\beta)^2] \\ &\quad -\delta - \frac{1}{2}\delta^2[\frac{D_x(s)}{\rho} - \frac{1}{\gamma^2}] \\ &\quad + x^\beta\delta[D_x''(s) + K_x(s)D_x(s) - \frac{1}{\rho}] \\ &\quad + \frac{1}{2}\delta^2 D_x(s)[D_x''(s) + K_x(s)D_x(s) - \frac{1}{\rho}] \\ &\quad + \frac{q}{\gamma^2\beta^2 E_0}\Phi(x^\beta + D_x(s)\delta, z^\beta, s). \end{aligned} \quad (8)$$

Clearly, from Eqs. (7) and (8), D_x is closely related to the dispersion function, and the new transverse coordinates and momenta, $x^\beta, p_x^\beta, y^\beta, p_y^\beta$, are pure betatron oscillations about the closed orbit at the specific corresponding energy. Although δ is the relative momentum deviation, we will use the terms momentum and energy interchangeably in referring to δ here.

To complete the separation of the Hamiltonian, we consider the dependence of the space charge term on D_x. To do this, we expand $\Phi(x^\beta + D_x\delta, z^\beta, s)$ about the reference orbit:

$$\Phi(x^\beta + D_x\delta, z^\beta, s) = \sum_{m,n=0}^{\infty} \frac{\Phi^{mn}(s)}{m!n!}[x^\beta + D_x(s)\delta]^m (z^\beta)^n. \tag{9}$$

We assume that the beam is centered at the origin, so that $\Phi^{01} = \Phi^{10} = 0$, and is sufficiently symmetric that $\Phi^{11} = 0$. This is certainly true for most commonly used analytic beam distributions. Then, setting the constant term $\Phi^{00} = 0$, Eq. (9) can be written as

$$\begin{aligned}\Phi(x^\beta + D_x\delta, z^\beta, s) &= \frac{\Phi^{20}}{2}(x^\beta + D_x\delta)^2 + \frac{\Phi^{02}}{2}(z^\beta)^2 + \Delta\Phi \\ &= \frac{\Phi^{20}}{2}(x^\beta)^2 + \frac{\Phi^{02}}{2}(z^\beta)^2 + \Phi^{20}D_x x^\beta\delta + \frac{\Phi^{20}}{2}(D_x\delta)^2 + \Delta\Phi,\end{aligned} \tag{10}$$

where the term $\Delta\Phi = O(\varepsilon^3)$. Substituting this into Eq. (8) yields

$$\begin{aligned}H(x^\beta, p_x^\beta, z^\beta, p_z^\beta, \tau, -\delta, s) &= \frac{1}{2}\{(p_x^\beta)^2 + [K_x(s) + \frac{q\Phi^{20}(s)}{\gamma^2\beta^2 E_0}](x^\beta)^2\} \\ &+ \frac{1}{2}\{(p_z^\beta)^2 + [K_z(s) + \frac{q\Phi^{02}(s)}{\gamma^2\beta^2 E_0}](z^\beta)^2\} \\ &- \delta - \frac{1}{2}\delta^2[\frac{D_x(s)}{\rho} - \frac{1}{\gamma^2}] \\ &+ x^\beta\delta\{D_x''(s) + [K_x(s) + \frac{q\Phi^{20}(s)}{\gamma^2\beta^2 E_0}]D_x(s) - \frac{1}{\rho}\} \\ &+ \frac{1}{2}\delta^2 D_x(s)\{D_x''(s) + [K_x(s) + \frac{q\Phi^{20}(s)}{\gamma^2\beta^2 E_0}]D_x(s) - \frac{1}{\rho}\} \\ &+ \frac{q}{\gamma^2\beta^2 E_0}\Delta\Phi(x^\beta + D_x\delta, z^\beta, s).\end{aligned} \tag{11}$$

We now choose D_x to be the dispersion function with lowest order space charge correction, meaning that D_x satisfies the equation

$$D_x''(s) + [K_x(s) + \frac{q\Phi^{20}(s)}{\gamma^2\beta^2 E_0}]D_x(s) - \frac{1}{\rho} = 0. \tag{12}$$

With this choice of dispersion function, the Hamiltonian finally simplifies to

$$H(x^\beta, p_x^\beta, z^\beta, p_z^\beta, \tau, -\delta, s) = \frac{1}{2}\{(p_x^\beta)^2 + [K_x(s) + \frac{q\Phi^{20}(s)}{\gamma^2\beta^2 E_0}](x^\beta)^2\}$$
$$+ \frac{1}{2}\{(p_z^\beta)^2 + [K_z(s) + \frac{q\Phi^{02}(s)}{\gamma^2\beta^2 E_0}](z^\beta)^2\} \quad (13)$$
$$-\delta - \frac{1}{2}\delta^2[\frac{D_x(s)}{\rho} - \frac{1}{\gamma^2}]$$
$$+O(\varepsilon^3).$$

The canonical coordinates and momenta are $(x^\beta, p_x^\beta, z^\beta, p_z^\beta, \tau, -\delta)$. The motion for the Hamiltonian in Eq. (13) is completely separated into independent contributions for betatron oscillations, (x^β, p_x^β) and (z^β, p_z^β), and longitudinal dynamics, (δ, τ). The betatron terms are independent of dispersion and of the momentum shift, δ; while the coordinate, τ, is cyclic, so that δ is constant. The space charge terms provide tune shifts in the betatron oscillations (first two lines) and modify the dispersion function (Eq. (12)) appearing in the third line. In lowest order all space charge terms are expressed with coefficients obtained from expanding the potential about the reference orbit. The remaining effects of space charge are third order and higher, and will be neglected in determining the rms envelope equations. In terms of the present notation, the ordering scheme is

$$O(\varepsilon^{-1}) \quad - \quad \frac{\partial}{\partial x^\beta}, \frac{\partial}{\partial z^\beta}, \frac{\partial}{\partial p_x^\beta}, \frac{\partial}{\partial p_z^\beta}, \frac{\partial}{\partial(-\delta)}$$

$$O(1) \quad - \quad K_x, K_z, q\Phi^{20}, q\Phi^{02}, D_x, \rho, \gamma, \beta, E_0, \tau, s, \frac{\partial}{\partial\tau}, \frac{\partial}{\partial s}$$

$$O(\varepsilon) \quad - \quad x^\beta, p_x^\beta, z^\beta, p_z^\beta, -\delta$$

$$O(\varepsilon^2) \quad - \quad H + \delta.$$

DISTRIBUTION FUNCTION AND AVERAGING

Let $f(x^\beta, p_x^\beta, z^\beta, p_z^\beta, \tau, -\delta, s)$ be the distribution function of the beam in the phase space of canonical coordinates. Define the phase space volume elements $dV = dx^\beta dp_x^\beta dz^\beta dp_z^\beta d\tau d(-\delta)$ and $dV_\perp = dx^\beta dp_x^\beta dz^\beta dp_z^\beta$. We take the normalization of f to be given by $\int f dx^\beta dp_x^\beta dz^\beta dp_z^\beta d\tau d(-\delta) = \int f dV = N$, where N is the number of particles in the beam. The distribution function f satisfies a kinetic equation of the form

$$\frac{\partial f}{\partial s}+(x^\beta)'\frac{\partial f}{\partial x^\beta}+(p_x^\beta)'\frac{\partial f}{\partial p_x^\beta}+(z^\beta)'\frac{\partial f}{\partial z^\beta}+(p_z^\beta)'\frac{\partial f}{\partial p_z^\beta}+\tau'\frac{\partial f}{\partial \tau}+(-\delta)'\frac{\partial f}{\partial(-\delta)}=S, \quad (14)$$

where the source term S accounts for non-Hamiltonian processes, such as gains due to injection and losses due to collisions with the foil or with other beam particles. We extend the adopted ordering scheme by assuming that the distribution function is $O(1)$ and that the non-Hamiltonian sources and sinks are $O(\varepsilon)$:

$$O(1) - f$$
$$O(\varepsilon) - S.$$

Then, to lowest order the distribution function obeys the Vlasov Equation in canonical phase space.

Useful averages and moments are calculated by integrating the product of the quantity to be averaged times the distribution function over phase space at fixed values of azimuth, s, time, t, and momentum, δ. According to Eqs. (4) and (7), $\tau = \beta ct + D_x p_x^\beta - D_x' x^\beta$. Hence, to average at fixed time, multiply the function to be averaged by the Dirac delta function, $\delta(\tau - \beta ct - D_x p_x^\beta + D_x' x^\beta)$. Similarly, to average at fixed momentum, δ_0, multiply by $\delta(\delta - \delta_0)$. Using this prescription, we define the following quantities:

$$\lambda(s,t,\delta_0) = \int \delta(\tau - \beta ct - D_x p_x^\beta + D_x' x^\beta)\delta(\delta - \delta_0)fdV$$
$$= \int f(x^\beta, p_x^\beta, z^\beta, p_z^\beta, \beta ct + D_x p_x^\beta - D_x' x^\beta, -\delta_0, s)dV_\perp$$
$$\lambda_{tot}(s,t) = \int \delta(\tau - \beta ct - D_x p_x^\beta + D_x' x^\beta)fdV \quad (15)$$
$$= \int f(x^\beta, p_x^\beta, z^\beta, p_z^\beta, \beta ct + D_x p_x^\beta - D_x' x^\beta, -\delta, s)dV_\perp d(-\delta)$$
$$P(\delta) = \frac{\lambda(s,t,\delta)}{\lambda_{tot}(s,t)}.$$

In Eq. (15), $\lambda(s,t,\delta_0)$ is the azimuthal particle density per unit momentum at δ_0, $\lambda_{tot}(s,t)$ is the overall azimuthal particle density, and $P(\delta)$ is the momentum probability distribution: note that $\int P(\delta)d(-\delta) = 1$. Furthermore, the averages of any function, $g(x^\beta, p_x^\beta, z^\beta, p_z^\beta, \tau, -\delta, s)$, are given by:

$$\langle g(s,t,\delta)\rangle = \frac{1}{\lambda(s,t,\delta)}\int [g\times f](x^\beta,p_x^\beta,z^\beta,p_z^\beta,\beta ct+D_x p_x^\beta - D_x' x^\beta,-\delta,s)dV_\perp$$
$$\langle g(s,t)\rangle_{tot} = \int \langle g(s,t,\delta)\rangle P(\delta)d(-\delta).$$
(16)

In order to derive rms envelope equations, it is necessary to differentiate averaged quantities in the azimuthal direction. Because these are dynamic equations, the derivatives in s are accompanied by changes in time. From Eqs. (4), (5), (7), and (13) it is seen that

$$\beta ct' = \frac{\partial H_2}{\partial(-\delta)} = 1+\delta(\frac{D_x}{\rho}-\frac{1}{\gamma^2})+\frac{x^\beta}{\rho}$$
$$\tau' = \frac{\partial H}{\partial(-\delta)} = 1+\delta(\frac{D_x}{\rho}-\frac{1}{\gamma^2}).$$
(17)

In the s differentiation of quantities averaged over transverse phase space, it is appropriate to evaluate $\beta ct'$ on the closed orbit, $x^\beta = 0$. In moving from s to $s^+ = s+\Delta s$, the quantity $\tau = \beta ct + D_x(s)p_x^\beta - D_x'(s)x^\beta$ becomes
$\tau^+ = \beta ct +[1+\delta(\frac{D_x}{\rho}-\frac{1}{\gamma^2})]\Delta s + D_x(s+\Delta s)p_x^\beta - D_x'(s+\Delta s)x^\beta$. Accordingly, define
$$\frac{\partial \tau}{\partial s} = 1+\delta(\frac{D_x}{\rho}-\frac{1}{\gamma^2})+D_x' p_x^\beta - D_x'' x^\beta = \tau' + D_x' p_x^\beta - D_x'' x^\beta.$$
Then

$$\frac{d}{ds}[\lambda\langle g(s,t,\delta)\rangle] = \lim_{\Delta s\to 0}\frac{1}{\Delta s}\left[\begin{array}{c}\int [g\times f](x^\beta,p_x^\beta,z^\beta,p_z^\beta,\tau^+,-\delta,s^+)dV_\perp \\ -\int [g\times f](x^\beta,p_x^\beta,z^\beta,p_z^\beta,\tau,-\delta,s)dV_\perp\end{array}\right],$$
(18)

and differentiation inside the integral and application of the chain rule yields
$\frac{d}{ds}[\lambda\langle g(s,t,\delta)\rangle] = \int[(\frac{\partial g}{\partial \tau}\frac{\partial \tau}{\partial s}+\frac{\partial g}{\partial s})\times f + g\times(\frac{\partial f}{\partial \tau}\frac{\partial \tau}{\partial s}+\frac{\partial f}{\partial s})]dV_\perp$. Using the Vlasov Equation to substitute for $\frac{\partial f}{\partial s}$, we get

$$\frac{d}{ds}[\lambda\langle g(s,t,\delta)\rangle] = \lambda\left\langle\frac{\partial g}{\partial \tau}\frac{\partial \tau}{\partial s} + \frac{\partial g}{\partial s}\right\rangle - \int\left(\begin{array}{c}g\times[(x^\beta)'\frac{\partial f}{\partial x^\beta} + (p_x^\beta)'\frac{\partial f}{\partial p_x^\beta} + (z^\beta)'\frac{\partial f}{\partial z^\beta} + (p_z^\beta)'\frac{\partial f}{\partial p_z^\beta} \\ +(\tau' - \frac{\partial \tau}{\partial s})\frac{\partial f}{\partial \tau} + (-\delta)'\frac{\partial f}{\partial(-\delta)}]\end{array}\right)dV_\perp$$

Finally, noting that $\delta' = 0$ and $\tau' = \frac{\partial \tau}{\partial s} - D_x' p_x^\beta + D_x'' x^\beta = \frac{\partial \tau}{\partial s} + O(\varepsilon)$, we integrate the transverse derivatives by parts, also noting that for Hamiltonian motion $\frac{\partial (x^\beta)'}{\partial x^\beta} + \frac{\partial (p_x^\beta)'}{\partial p_x^\beta} = \frac{\partial (z^\beta)'}{\partial z^\beta} + \frac{\partial (p_z^\beta)'}{\partial p_z^\beta} = 0$, and average to obtain

$$\frac{d}{ds}[\lambda\langle g(s,t,\delta)\rangle] = \lambda\left\langle\begin{array}{c}(x^\beta)'\frac{\partial g}{\partial x^\beta} + (p_x^\beta)'\frac{\partial g}{\partial p_x^\beta} + (z^\beta)'\frac{\partial g}{\partial z^\beta} \\ +(p_z^\beta)'\frac{\partial g}{\partial p_z^\beta} + \tau'\frac{\partial g}{\partial \tau} + \frac{\partial g}{\partial s}\end{array}\right\rangle + O(\varepsilon). \quad (19)$$

By choosing $g = 1$ we obtain the result that $\frac{d\lambda}{ds} = O(\varepsilon)$, so that to lowest order

$$\frac{d}{ds}\langle g(s,t,\delta)\rangle = \left\langle(x^\beta)'\frac{\partial g}{\partial x^\beta} + (p_x^\beta)'\frac{\partial g}{\partial p_x^\beta} + (z^\beta)'\frac{\partial g}{\partial z^\beta} + (p_z^\beta)'\frac{\partial g}{\partial p_z^\beta} + \tau'\frac{\partial g}{\partial \tau} + \frac{\partial g}{\partial s}\right\rangle. \quad (20)$$

Before proceeding with the derivation of rms envelope equations, consider two functions $g(x^\beta, p_x^\beta, z^\beta, p_z^\beta, \tau, -\delta, s)$ and $h(x^\beta, p_x^\beta, z^\beta, p_z^\beta, \tau, -\delta, s)$. Defining variations $\Delta g = g - \langle g\rangle$ and $\Delta h = h - \langle h\rangle$, we find that

$$\langle \Delta g \Delta h\rangle = \langle gh\rangle - \langle g\rangle\langle h\rangle. \quad (21)$$

Setting $h = g$ leads to the standard deviation of g

$$\sigma_g^2 = \langle(\Delta g)^2\rangle = \langle g^2\rangle - \langle g\rangle^2. \quad (22)$$

Equations (20), (21), and (22) will be used extensively in obtaining the rms envelope equations.

THE RMS ENVELOPE EQUATIONS

Let $q = x^\beta$ or $q = z^\beta$ be one of the transverse canonical coordinates and let $p = q'$ be the momentum canonical to q. Then, according to Eq. (21), to lowest order in ε

$$\frac{d}{ds}\sigma_q^2 = \frac{d}{ds}\{\langle q^2 \rangle - \langle q \rangle^2\}$$
$$= 2\{\langle q'q \rangle - \langle q' \rangle\langle q \rangle\}$$
$$= 2\{\langle pq \rangle - \langle p \rangle\langle q \rangle\}$$
$$= 2\langle \Delta p \Delta q \rangle$$

$$\frac{d^2}{ds^2}\sigma_q^2 = \frac{d}{ds}\{\frac{d}{ds}\sigma_q^2\}$$
$$= 2\frac{d}{ds}\{\langle pq \rangle - \langle p \rangle\langle q \rangle\}$$
$$= 2\{\langle p^2 \rangle - \langle p \rangle^2 + \langle qp' \rangle - \langle q \rangle\langle p' \rangle\}$$
$$= 2\{\langle \sigma_p^2 \rangle - \langle \Delta q \Delta \frac{\partial H}{\partial q} \rangle\}$$

$$\frac{d}{ds}\sigma_p^2 = \frac{d}{ds}\{\langle p^2 \rangle - \langle p \rangle^2\}$$
$$= 2\{\langle pp' \rangle - \langle p \rangle\langle p' \rangle\}$$
$$= -2\{\langle p\frac{\partial H}{\partial q} \rangle - \langle p \rangle\langle \frac{\partial H}{\partial q} \rangle\}$$
$$= -2\langle \Delta p \Delta \frac{\partial H}{\partial q} \rangle, \tag{23}$$

where $p' = -\dfrac{\partial H}{\partial q}$.

Let us now define an effective emittance in terms of the rms statistical emittance by

$$\varepsilon_q = \eta_\varepsilon \{\sigma_p^2 \sigma_q^2 - \langle \Delta p \Delta q \rangle^2\}^{\frac{1}{2}}, \tag{24}$$

where the factor η_ε is a constant normalization factor. For a K-V distribution the factor $\eta_\varepsilon = 4$ makes the effective emittance equal to the area of the maximal (q, p) phase space ellipse. Using Eqs. (23) and (24), it is straightforward to show that

$$\frac{d}{ds}(\frac{\varepsilon_q}{\eta_\varepsilon})^2 = -2\{\langle(\Delta q)^2\rangle\langle\Delta p\Delta\frac{\partial H}{\partial q}\rangle - \langle\Delta q\Delta\frac{\partial H}{\partial q}\rangle\langle\Delta p\Delta q\rangle\}. \tag{25}$$

If the dependence of $\frac{\partial H}{\partial q}$ on q is linear, i.e. $\frac{\partial H}{\partial q} = K_q q$, then Eq. (25) shows that the effective emittance is constant. For the Hamiltonian of Eq. (13), $\frac{\partial H}{\partial q} = K_q q + O(\varepsilon^2)$ for both x and z, so the effective emittance is constant to leading order.

The relations $\langle \Delta p \Delta q \rangle = \frac{1}{2}\frac{d}{ds}\sigma_q^2$ and $\langle \sigma_p^2 \rangle = \frac{1}{2}\frac{d^2}{ds^2}\sigma_q^2 + \langle \Delta q \Delta\frac{\partial H}{\partial q}\rangle$, Eq. (23), can be combined with the definition, Eq. (25), of the effective emittance to obtain, using the identity $(\sigma_q^2)'' = \frac{[(\sigma_q^2)']^2}{2\sigma_q^2} + 2\sigma_q \sigma_x''$, the rms envelope equation for q:

$$(\eta_\varepsilon^{\frac{1}{2}}\sigma_q)'' - \frac{\varepsilon_q^2}{(\eta_\varepsilon^{\frac{1}{2}}\sigma_q)^3} + \frac{\eta_\varepsilon}{(\eta_\varepsilon^{\frac{1}{2}}\sigma_q)}\langle\Delta q\Delta\frac{\partial H}{\partial q}\rangle = 0. \tag{26}$$

Equation (26) shows that the normalization constant η_ε provides a scale factor relating the rms value of Δq to an envelope radius $\eta_\varepsilon^{\frac{1}{2}}\sigma_q$.

Using the adopted beam Hamiltonian, we obtain

$$a'' - \frac{\varepsilon_x^2}{a^3} + [K_x(s) + \frac{q\Phi^{20}(s)}{\gamma^2\beta^2 E_0}]a = 0$$

$$b'' - \frac{\varepsilon_x^2}{b^3} + [K_y(s) + \frac{q\Phi^{02}(s)}{\gamma^2\beta^2 E_0}]b = 0, \tag{27}$$

where $a = \eta_\varepsilon^{\frac{1}{2}}\sigma_x$ and $b = \eta_\varepsilon^{\frac{1}{2}}\sigma_z$. Equation (27) presents the rms envelope equations, correct to lowest order in ε, for betatron oscillations at momentum δ. These equations are independent of dispersion and momentum. For a monoenergetic beam, no momentum spread, and an elliptical K-V distribution of radii a and b, the potential is given by $\Phi_{K-V} = -\frac{1}{2}[\frac{4q\lambda_{tot}}{a(a+b)}]x^2 - \frac{1}{2}[\frac{4q\lambda_{tot}}{b(a+b)}]z^2$, and Eq. (27) takes the usual form

$$a'' - \frac{\varepsilon_x^2}{a^3} + K_x(s)a = [\frac{4q^2\lambda_{tot}}{\gamma^2\beta^2 E_0}]\frac{1}{a+b} = \frac{2K}{a+b}$$

$$b'' - \frac{\varepsilon_x^2}{b^3} + K_y(s)b = [\frac{4q^2\lambda_{tot}}{\gamma^2\beta^2 E_0}]\frac{1}{a+b} = \frac{2K}{a+b},$$

(28)

where $K = \frac{2q^2\lambda_{tot}}{\gamma^2\beta^2 E_0}$ is the generalized perveance (7).

THE SPACE CHARGE DISTRIBUTION

The rms envelope equations derived above provide only part of the formulation of a particle core model for rings. It remains to obtain the space charge forces to close the model. This can be done by superposing the particle distributions at each energy, which can be inferred from the envelope equations, over the momentum distribution, while taking into account the effects of dispersion in spreading the beam. To do this it is necessary to make some assumptions about the particle distribution function. This is a standard problem when dealing with moments expansions, and we state below where this is required.

For a given momentum, δ, the overall displacement relative to the center of the reference beam is seen from Eq. (7) to be

$$\tilde{x} = x^\beta + D_x\delta$$
$$\tilde{z} = z^\beta.$$

(29)

The total particle density at each value of $\tilde{x}$ and $\tilde{z}$ is given by the distribution function as follows

$$n(\tilde{x},\tilde{z}) = \int \begin{pmatrix} f(x^\beta, p_x^\beta, z^\beta, p_z^\beta, \tau, -\delta, s)\delta(\tau - \beta ct - D_x p_x^\beta + D_x' x^\beta) \\ \times \delta(x^\beta + D_x\delta - \tilde{x})\delta(z^\beta - \tilde{z})dV \end{pmatrix}$$

$$= \int f(\tilde{x}- D_x\delta, p_x^\beta, \tilde{z}, p_z^\beta, \beta ct + D_x p_x^\beta - D_x' x^\beta, -\delta, s)dp_x^\beta dp_z^\beta d(-\delta)$$

(30)

$$= \int n(\tilde{x},\tilde{z},-\delta)d(-\delta).$$

Equation (30) requires detailed knowledge of the distribution function, but it is of interest to extend the moments approach of the previous sections to obtain rms values of the overall charge distribution.

We now calculate the overall average and rms values of $\tilde{x}$ and $\tilde{z}$:

$$\langle \tilde{x} \rangle_{tot} = \int (\langle x \rangle + D_x \delta) P(\delta) d(-\delta)$$
$$= \langle x \rangle_{tot} + D_x \langle \delta \rangle_{tot}$$
$$\langle \tilde{z} \rangle_{tot} = \int \langle z \rangle P(\delta) d(-\delta) \qquad (31)$$
$$= \langle z \rangle_{tot}.$$

If the particle distribution at each energy is centered about the closed orbit, then $\langle x \rangle = \langle z \rangle = \langle x \rangle_{tot} = \langle z \rangle_{tot} = 0$; and if the reference orbit is centered in the momentum distribution, then $\langle \delta \rangle_{tot} = 0$. In this case, $\langle \tilde{x} \rangle_{tot} = \langle \tilde{z} \rangle_{tot} = 0$. To evaluate the total rms size of the beam, we calculate

$$\langle (\Delta \tilde{x})^2 \rangle_{tot} = \langle [(x^\beta + D_x \delta) - (\langle x^\beta \rangle_{tot} + D_x \langle \delta \rangle_{tot})]^2 \rangle_{tot}$$
$$= \left\langle \begin{array}{c} (x^\beta - \langle x^\beta \rangle_{tot})^2 \\ +2D_x(x^\beta - \langle x^\beta \rangle_{tot})(\delta - \langle \delta \rangle_{tot}) + D_x^2(\delta - \langle \delta \rangle_{tot})^2 \end{array} \right\rangle_{tot}$$
$$= [\langle (x^\beta)^2 \rangle_{tot} - (\langle x^\beta \rangle_{tot})^2] \qquad (32)$$
$$+ 2D_x[\langle x^\beta \delta \rangle_{tot} - \langle x^\beta \rangle_{tot} \langle \delta \rangle_{tot}] + D_x^2[\langle \delta^2 \rangle_{tot} - (\langle \delta \rangle_{tot})^2]$$

$$\langle (\Delta \tilde{z})^2 \rangle_{tot} = \langle (z - \langle z \rangle_{tot})^2 \rangle_{tot}$$
$$= \langle z^2 \rangle_{tot} - (\langle z \rangle_{tot})^2.$$

If $\langle x \rangle_{tot} = \langle z \rangle_{tot} = 0$, if the betatron oscillation coordinates are uncorrelated with δ, so that $\langle x\delta \rangle_{tot} = 0$, and if $\langle \delta \rangle_{tot} = 0$, then the effective emittances are independent of momentum, and Eq. (32) simplifies to

$$\sigma_{\tilde{x}}^2 = \langle (\Delta \tilde{x})^2 \rangle_{tot} \qquad\qquad \sigma_{\tilde{z}}^2 = \langle (\Delta \tilde{z})^2 \rangle_{tot}$$
$$= \langle x^2 \rangle_{tot} + D_x^2 \langle \delta^2 \rangle_{tot} \qquad = \langle z^2 \rangle_{tot} \qquad (33)$$
$$= \frac{a^2}{4} + D_x^2 \langle \delta^2 \rangle_{tot} \qquad\qquad = \frac{b^2}{4}.$$

Consequently the rms standard deviations of the overall displacement contain separate contributions from betatron oscillations and from dispersion in the case of $\tilde{x}$ and from betatron oscillations only in the case of $\tilde{z}$.

The above results can be used to construct simple overall particle distributions for the purpose of calculating the space charge fields for particle core model dynamics and particle tracking. The use of Eq. (30) to define the particle density requires detailed information about the distribution function. A simpler computational approach involves assuming some standard form for the distribution function in transverse phase space. Then the envelope equations can be solved to provide the rms radii of the assumed distribution, and hence its contribution to the space charge density, at each energy. Discretizing momentum space into a number of bins, each of size $\Delta\delta$, and assuming a fraction of the beam, $P(\delta)\Delta\delta$, to be inside each bin, the associated charge distributions can be added in a weighted sum with weights proportional to $P(\delta)\Delta\delta$. The total space charge field is thus obtained by superposition of the fields given by the transverse distribution parameterized by the envelope solutions. For self-consistency, the total space charge potential obtained in this manner, expanded about the reference closed orbit, must yield the coefficients used in Eqs. (10) and (13). To enforce this, iteration of the process may be required.

Even more simply, the overall rms beam parameters in Eq. (33) can be associated with an assumed transverse distribution, and the summation over beam momentum bypassed. For example, for an elliptical K-V distribution with semi-axes

$$A = 2\sigma_{\tilde{x}} \qquad\qquad B = 2\sigma_{\tilde{z}}$$
$$= (a^2 + 4D_x^2 \langle \delta^2 \rangle_{tot})^{\frac{1}{2}} \qquad = b, \qquad (34)$$

the space charge potential is

$$\Phi_{K-V} = -\frac{1}{2} [\frac{4q\lambda_{tot}}{A(A+B)}] \tilde{x}^2 - \frac{1}{2} [\frac{4q\lambda_{tot}}{B(A+B)}] \tilde{z}^2$$
$$= -\frac{1}{2} [\frac{4q\lambda_{tot}}{A(A+B)}] (x^\beta + D_x \delta)^2 - \frac{1}{2} [\frac{4q\lambda_{tot}}{B(A+B)}] (z^\beta)^2, \qquad (35)$$

so that $\Phi^{20} = -\dfrac{4q\lambda_{tot}}{A(A+B)}$ and $\Phi^{02} = -\dfrac{4q\lambda_{tot}}{B(A+B)}$. With this approximation, the envelope equations must be solved in the form of Eq. (27), rather than Eq. (28). Also, through the space charge terms, the radii A and B appear in Eq. (12) for the dispersion, D_x. An iterative approach to solving for D_x for a matched beam prior to the dynamic calculations can be developed starting with the usual dispersion function without space charge. The approach requires solving Eqs. (27), (34), and (35) simultaneously for matched core boundary conditions, using the available dispersion function. With an explicit symplectic integration scheme of independent kicks and linear transport steps, all information would be available when required. After solving for the matched envelope, the resulting space charge potential could be used to solve for the space-charge-corrected closed orbit dispersion function in Eq. (12). This would complete one iteration, and the approach could be repeated to convergence. The converged space-charge-corrected dispersion function for a matched beam would then be used as the dispersion function for the dynamic calculations. This simple scheme provides a self-consistent particle core model using a space-charge-corrected dispersion function, a pair of rms envelope equations, and K-V beam distribution.

CONCLUSIONS

A self-consistent particle core model for transverse beam dynamics in rings, including the effects of space charge and dispersion, was derived using a moments approach. The model includes rms envelope equations for betatron oscillations, a space-charge-corrected dispersion function, and a prescription for the evaluation of the space charge potential, all coupled together self-consistently. In addition to describing the collective dynamics of the beam, this model can provide space charge forces for particle tracking calculations. The derivation was carried out using an accelerator ordering scheme and a statistical moments analysis based on the canonical distribution function in the six-dimensional phase space of the beam Hamiltonian. The azimuthal variation of the second moments of the transverse canonical (betatron oscillation) coordinates, x^β and z^β, averaged at fixed beam energy over the kinetic distribution function of the beam, leads to the rms envelope equations. These envelope equations are found to be independent of dispersion. A subsequent integration over beam energies provides the rms values of the overall displacements, $\tilde{x}$ and $\tilde{z}$, and the spatial beam distribution and space charge force. This integration can be carried out with varying degrees of complexity to obtain particular models of the charge distribution. Self-consistency must be assured by using the resulting space charge force to calculate the dispersion and envelope equations until the system is converged, thus closing the loop.

* Research on the Spallation Neutron Source is sponsored by the Division of Materials Science, U.S. Department of Energy, under contract number DE-AC05-96OR22464 with Lockheed Martin Energy Research Corporation for Oak Ridge National Laboratory.

REFERENCES

1. Riabko, A., Ellison, M., Kang, X., Lee, S.Y., Liu, J.Y., Li, D., Pei, A., and Wang, L., "Hamiltonian Formalism for Space Charge Dominated Beams in a Uniform Focusing Channel," in *Proc. of the 1995 Particle Accelerator Conf. and International Conf. on High-Energy Accelerators*, Dallas, TX, May 1-5, 1995, pp. 3182-4; O'Connell, J.S., Wangler, T.P., Mills, R.S., and Crandall, K.R., "Beam Halo Formation from Space-Charge Dominated Beams in Uniform Focusing Channels," in *Proc. of the 1993 Particle Accelerator Conf.*, Washington, D.C., May 17-20, 1993, pp. 3657-9.
2. Hockney, R.W. and Eastwood, J.W., *Computer Simulation Using Particles*, New York: Adam Hilger, 1988.
3. National Spallation Neutron Source Conceptual Design Report, Volumes 1 and 2, NSNS/CDR-2/V1,2, (May, 1997).
4. Venturini, M. and Reiser, M., *Phys. Rev. E* **57**, 4725-32 (1998).
5. Lee, S.Y. and Okamoto, H., *Phys. Rev. Lett.* **80**, accepted for publication (1998).
6. Guignard, G., "A General Treatment of Resonances in Accelerators," *Lectures Given in the Academic Training Program of CERN 1977-1978*, CERN 78-11, Geneva, Nov. 10, 1978.
7. Wangler, T.P., "Emittance Concept and Growth Mechanisms," in Space Charge Dominated Beams and Applications of High Brightness Beams, AIP Press, 1996, pp. 3-22.

Achromat with Linear Space Charge for Bunched Beams

D. Raparia, J.G. Alessi, Y. Y. Lee and W. T. Weng

Brookhaven National Laboratory
Upton, NY 11973, USA

Abstract. The standard definition for an achromat is a transport line having zero values for the spatial dispersion (R16) and the angular dispersion (R26). In the presence of linear space charge this definition of achromaticity does not hold. The linear space charge in the presence of a bend provides coupling between (a) bunch spatial width and bunch length (R15) and (b) bunch angular spread and bunch length (R25). Therefore, achromaticity should be redefined as a line having zero values of the spatial dispersion (R16), the angular dispersion (R26), and matrix elements R15 & R25. These additional conditions (R15=R25=0) can be achieved, for example, with two small RF cavities at appropriate locations in the achromat, to cancel space charge effects. An example of the application of this technique to the Spallation Neutron Source (SNS) high energy beam transport line will be presented.

I INTRODUCTION

A transport line is said to be achromatic if spatial & angular widths of the beam are independent of its momentum spread. In other words, the spatial dispersion (R16) and the angular dispersion (R26) have zero values. The spatial & angular widths are also independent of the bunch length in absence of linear space charge. The beam ellipse is upright in the x-z plane. A property of an achromatic bend is that the total phase advance should be $n\pi (n = 1, 2 \cdots)$ if all dipoles in achromat bend the beam in the same direction, or $2n\pi$ if some of the dipoles in achromat bend the beam in the opposite direction. In the presence of linear space charge the transport line no longer remains achromatic for two reasons. (1) The total phase advance is no longer $n\pi$ or $2n\pi$ due to tune depression. The total phase advance can be changed back to $n\pi$ or $2n\pi$ by readjusting quadrupoles in achromat. (2) When beam passes through the bend it is no longer upright in x-z plane because space charge provides coupling between (a) bunch spatial width and bunch length (R15) and (b) bunch angular width and bunch length (R25). The beam ellipse can be made upright in x-z plane by introducing two rf cavities in the achromat.

CP448, *Workshop on Space Charge Physics in High Intensity Hadron Rings*
edited by A. U. Luccio and W. T. Weng
© 1998 The American Institute of Physics 1-56396-824-X/98/$15.00

II ACHROMAT WITH LINEAR SPACE CHARGE FOR BUNCHED BEAM

A Linear Space Charge for Bunched Beams

Following the formulism of TRACE3D [3], the electric field components due to a uniformly charged ellipsoid are given by [1]

$$E_x = \frac{1}{4\pi\epsilon_0} \frac{3I\lambda}{c\gamma^2} \frac{(1-f)}{r_x(r_x+r_y)r_z} x,$$

$$E_y = \frac{1}{4\pi\epsilon_0} \frac{3I\lambda}{c\gamma^2} \frac{(1-f)}{r_y(r_x+r_y)r_z} y,$$

$$E_z = \frac{1}{4\pi\epsilon_0} \frac{3I\lambda}{c} \frac{f}{r_x r_y r_z} z,$$

where r_x, r_y and r_z are the semi-axes of a ellipsoid, I is the average electrical current over one RF period, λ is the free-space wavelength of the RF, c is the velocity of light, and ϵ_0 is the permittivity of the free space. The form factor f is a function of $p \equiv \frac{\gamma r_z}{\sqrt{r_x r_y}}$ given by

$$f(p) = \begin{cases} \frac{1}{1-p^2} - \frac{p}{(1-p^2)^{3/2}} \cos^{-1} p & \text{if } p < 1 \\ \frac{p \cosh^{-1} p}{(p^2-1)^{3/2}} - \frac{1}{(p^2-1)} & \text{if } p > 1 \\ \frac{1}{3} & \text{if } p = 1 \end{cases}$$

where $\cosh^{-1} p = \ln\left(p + \sqrt{p^2-1}\right)$.

The space charge is applied in the kick approximation as a change in the normalized momentum components as the beam traverses Δs, and is given by

$$\Delta(\beta\gamma)_u = \frac{qE_u \Delta s}{m_0 c^2 \beta}$$

where u represents x, y, or z. This kick formulation is correct as long as the ellipsoid is upright in local x-y, y-z, and z-x planes. When beam traverses the bend the ellipsoid is tilted in the local x-z plane. To calculate the space charge kick first one has to transform to the coordinate system in which the ellipsoid is upright. The ellipsoid is transformed back to the local coordinate system after applying the space charge kick. The transfer matrix for the space-charge kick is

$$R_S = \begin{pmatrix} 1 & 0 & 0 & 0 & 0 & 0 \\ \frac{qE_x \Delta s}{m_0 c^2 \beta} & 1 & 0 & 0 & 0 & 0 \\ 0 & 0 & 1 & 0 & 0 & 0 \\ 0 & 0 & \frac{qE_y \Delta s}{m_0 c^2 \beta} & 1 & 0 & 0 \\ 0 & 0 & 0 & 0 & 1 & 0 \\ 0 & 0 & 0 & 0 & \frac{qE_z \Delta s}{m_0 c^2 \beta} & 1 \end{pmatrix}.$$

B Coupling Between Bunch Length and Bunch Spatial & Angular Widths

This coupling is induced by linear space charge, which can be shown as follows. The transfer matrix of a bending magnet for bend angle α and length L is

$$R_B = \begin{pmatrix} \cos(k_x L) & \frac{1}{k_x}\sin(k_x L) & 0 & 0 & 0 & \frac{h(1-\cos(k_x L))}{k_x^2} \\ -k_x \sin(k_x L) & \cos(k_x L) & 0 & 0 & 0 & \frac{h\sin(k_x L)}{k_x} \\ 0 & 0 & \cos(k_y L) & \frac{1}{k_y}\sin(k_y L) & 0 & 0 \\ 0 & 0 & -k_y \sin(k_y L) & \cos(k_y L) & 0 & 0 \\ \frac{-h\sin(k_x L)}{k_x} & \frac{-h(1-\cos(k_x L))}{k_x^2} & 0 & 0 & 1 & \frac{-1}{\rho^2 k_x^3}(k_x L \beta^2 - \sin(k_x L)) + \frac{L}{\gamma^2}(1 - \frac{1}{\rho^2 k_x^2}) \\ 0 & 0 & 0 & 0 & 0 & 1 \end{pmatrix}$$

where

$$h = \frac{1}{|\rho|}\frac{\alpha}{|\alpha|}$$

$$k_x = \sqrt{(1-n)h^2}$$

$$k_y = \sqrt{nh^2}$$

$$L = |\rho|\alpha$$

$$\rho = \frac{m_0 c \beta \gamma}{qB_y}$$

$$n = -\left(\frac{\rho}{B_y}\frac{\partial B_y}{\partial x}\right)_{x=0,y=0}$$

When space-charge is included in the calculation for the bending magnet in the kick approximation, the two matrices R_S and R_B are multiplied, and the resultant matrix R_{SB} will have $R_{SB}15$ and $R_{SB}25$ non-zero and given by

$$R_{SB}15 = R_B 16 * R_S 65 = \frac{h(1-\cos(k_x L))}{k_x^2} \times \frac{qE_z L}{m_0 c^2 \beta}$$

$$R_{SB}25 = R_B 26 * R_S 65 = \frac{h\sin(k_x)L}{k_x} \times \frac{qE_z L}{m_0 c^2 \beta}.$$

Even after applying the usual conditions for an achromat, R16=R26=0, the system is not achromatic. To completely remove coupling between x-x' and z-z' planes, not only the determinant of sub-matrix R[xz] has to be zero, but all of its elements have to be zero. If any element of this sub-matrix is non-zero, it means that beam is not upright in the x-z plane. The coupling provided by the space charge will effectively produce dispersion (R16), angular dispersion (R26) and non-zero matrix elements R15 & R25. The beam ellipse can be made upright in x-z plane by introducing two rf cavities in the achromat.

III EXAMPLE

As shown, all four elements of the R[xz] sub-matrix should have zero values for a transport line to be achromatic. We will use as an example the SNS high energy transport line [2]. This transport line is about 180 meters long and has an achromat which is six cells long, with a total phase advance of 360 degrees. The lay out of HEBT is shown in the figure 1, and the corresponding amplitude and dispersion functions are shown in figure 2.

The transfer matrix of the achromat in the HEBT for the zero current (using TRACE3D) is

$$R_{acro} = \begin{pmatrix} 1 & 0 & 0 & 0 & 0 & 0 \\ 0 & 1 & 0 & 0 & 0 & 0 \\ 0 & 0 & 1 & -1.305 & 0 & 0 \\ 0 & 0 & 0.01778 & 0.98846 & 0 & 0 \\ 0 & 0 & 0 & 0 & 1 & 14.15989 \\ 0 & 0 & 0 & 0 & 0 & 1 \end{pmatrix}$$

with the following units

$$\begin{pmatrix} 1 & m & 1 & m & 1 & m \\ m^{-1} & 1 & m^{-1} & 1 & m^{-1} & 1 \\ 1 & m & 1 & m & 1 & m \\ m^{-1} & 1 & m^{-1} & 1 & m^{-1} & 1 \\ 1 & m & 1 & m & 1 & m \\ m^{-1} & 1 & m^{-1} & 1 & m^{-1} & 1 \end{pmatrix}$$

The R[xz] sub-matrix of the achromat plus the following two cells, for zero current, is

$$R[xz]_{acro+2c} = \begin{pmatrix} 0 & 0 \\ 0 & 0 \end{pmatrix}$$

The R[xz] sub-matrix for the achromat, with 28 mA current, is

$$R[xz]_{acro} = \begin{pmatrix} 0.133349 & -0.18495 \\ 0.00469 & 0.8130 \end{pmatrix}$$

Now one can apply the conditions R16 = R26 = 0, by adjusting two families of quadrupoles in the achromat, and the resultant R[xz] sub-matrix of the achromat is

$$R[xz]_{acro} = \begin{pmatrix} 0.14278 & 0 \\ 0.00339 & 0 \end{pmatrix}$$

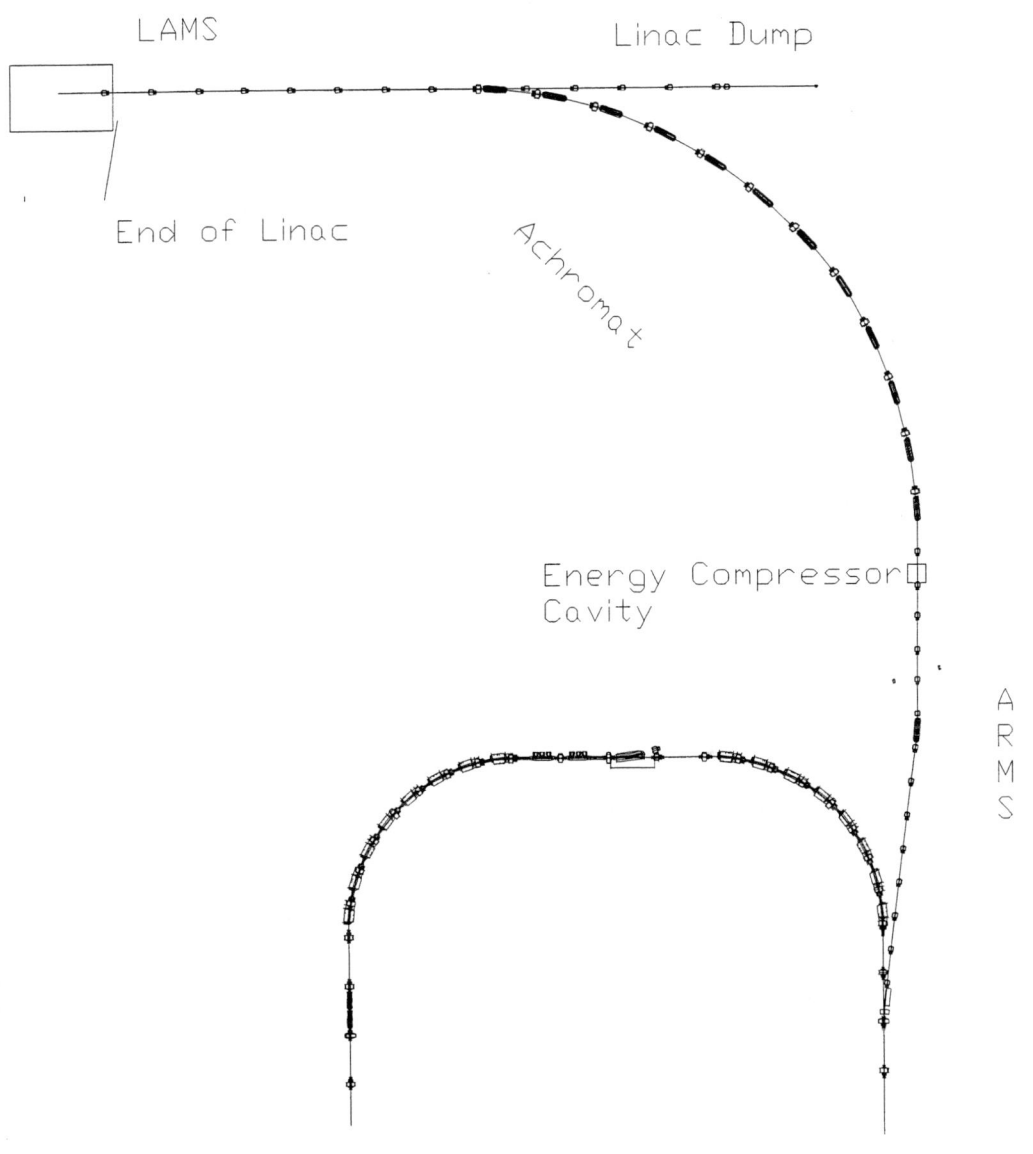

FIGURE 1. Layout of HEBT line.

The R[xz] sub-matrix of the achromat plus the following two cells, for the 28 mA, is then

$$R[xz]_{acro+2c} = \begin{pmatrix} -0.00608 & -0.68454 \\ -0.01605 & 0.01919 \end{pmatrix}$$

This shows the condition R16 = R26 = 0, will not yield a achromatic system if one includes space charge.

Next, one can apply the conditions R16 = R26 = R15 = R25 = 0 (instead of R16 = R26 = 0) by adjusting the two families of quadrupoles in the achromat, and adding two small 1 meter RF cavities in the 2nd and 5th cells of the achromat set at appropriate voltages. One can then get the following R[xz] sub-matrix, similar to the zero current case.

$$R[xz]_{acro} = \begin{pmatrix} 0 & 0 \\ 0 & 0 \end{pmatrix}$$

For this case, however, the R[xz] sub-matrix for the achromat plus the following two cells, for 28 mA, is

$$R[xz]_{acro+2c} = \begin{pmatrix} 0 & 0 \\ 0 & 0 \end{pmatrix}$$

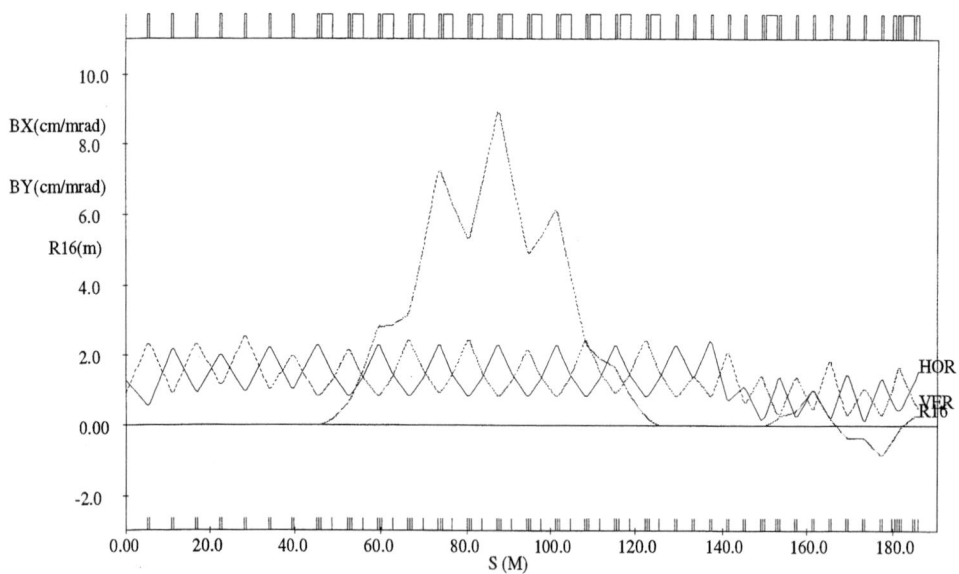

FIGURE 2. TRANSPORT output for the amplitude (β) functions and dispersion function (η) along the HEBT

Therefore, achromaticity should be redefined as a line having zero values of matrix elements R15, R25, R16 and R26. The additional condition can be achieved with two small rf cavities.

REFERENCES

1. K. R. Crandall and D. P. Rusthoi, Los Alamos Report LA-UR-97-886.
2. P. M. Lapostolle, CERN report AR/Int. SG/65-15, Geneva, Switzerland, July 1965.
3. D. Raparia, J. Alessi, Y.Y. Lee, W. T. Weng, to be published in Proceedings of 1997 Particle Accelerator Conference, Vancouver, B.C. Canada, May 12-16, 1997.

Dispersion and Space Charge

Marco Venturini*, Rami A. Kishek† and Martin Reiser†

*Department of Physics and Institute for Plasma Research
†Department of Electrical Engeneering and Institute for Plasma Research, University of Maryland, College Park MD 20742

Abstract. The presence of space charge affects the value of the dispersion function. On the other hand dispersion has a role in shaping the beam distribution and therefore in determining the resulting forces due to space charge. In this paper we present a framework where the interplay between space charge and dispersion for a continuous beam can be simultaneously treated. We revise the derivation of a new set of rms envelope-dispersion equations we have recently proposed in [1]. The new equations generalize the standard rms envelope equations currently used for matching to the case where bends and a longitudinal momentum spread are present. We report a comparison between the solutions of the rms envelope-dispersion equations and the results obtained using WARP, a Particle in Cell (PIC) code, in the modeling of the Maryland Electron Ring.

INTRODUCTION

The aim of this paper is to provide a suitable framework to describe the way space-charge forces and dispersion function affect each other in a beam line where both bends and a longitudinal momentum spread are present. Let us first review dispersion from a single particle perspective. At any point along a beam line a particle off-momentum experiences betatron oscillations in the transverse plane around a point shifted from the location of the design orbit. The amount of the shift is proportional to the relative deviation in the longitudinal momentum from the design value and the local value of the dispersion function. The average of the dispersion function is roughly inversely proportional to the focusing in the horizontal plane evaluated in the smooth approximation. As for the other lattice functions, the dispersion function has to be properly matched in order to avoid degradation of beam quality.

How does the scenario change if we allow space charge get into the picture? First of all the dispersion function ceases to be a pure lattice function. In fact, a problem one has to face is to define a proper meaning for the dispersion function. In some way we can guess it has to depend on the particle beam distribution. In particular, because an effect of the space charge is to shield the external focusing we should expect an increase in the average value of the dispersion function. On the other hand since the space-charge force depends on the beam distribution through its transverse size (a larger size, for example, makes the effect of space charge more diluted and the shielding of the external focusing less effective), the dispersion function has to be dependent on the beam envelopes. Moreover the converse is also to be expected: that is in some way the beam envelopes themselves also should depend on the local value of the dispersion function.

For a simple model of a ring lattice in the smooth approximation the problem of the interplay between space charge and dispersion can be investigated in a fully self-consistent way by solving the Vlasov-Poisson equation. This method was used in [2] to characterize the equilibrium distributions in the form of generalized KV beams, which have KV beam-like distributions in the transverse variables and gaussian distributions in the longitudinal momentum. Although insightful, an analysis based on solving the Vlasov-Poisson equation is not practical for a lattice with discrete magnet elements. In addition, having in mind the need to solve the matching problem we would like to be able to describe the beam envelopes and the dispersion in terms of a set of differential equations. The derivation of such a set of equations has been presented in [1]. Here we will review that derivation and provide some additional numerical tests. Our approach also makes use, in a different way, of the equations for the second moments of the beam distribution that were first introduced in [3] for the purpose of studying the effect of bends in space charge dominated beams.

The idea is to generalize the rms envelope equations first derived by Sacherer in [4] for the case of straight beam line. In Sacherer's derivation a key point is the assumption that the rms emittance, which is an exact invariant in the linear approximation, is roughly preserved even in the presence of nonlinearities due to the space charge. When bends and a longitudinal momentum spread are present the standard rms emittance is no longer a linear invariant. If we want to generalize Sacherer's equations we have to replace the standard rms emittance with an rms emittance-like quantity that displays the due linear invariance. At the same time we have to identify the proper form of the dispersion function. As we will see, such an identification arises naturally when one writes the equations for the moments of the beam distribution.

The result of the generalization of Sacherer's approach is a new set of three second order differential equations in the horizontal and vertical envelopes and the dispersion function. The equations for the horizontal envelope and the dispersion function are coupled, as expected. The new set of equations should make it possible in principle to achieve simultaneous matching of both the envelopes and the dispersion function for highly space charge dominated beams. In particular, we believe that proper matching of the dispersion function could significantly reduce the amount of emittance growth that has been observed in the numerical simulations of space charge dominated beams in a lattice with bends and a longitudinal momentum spread. We shall comment more on this point at the end of the paper.

THE RMS ENVELOPE-DISPERSION EQUATIONS

The analysis presented in this paper applies to continuous beams of particles of mass m charge q described by the Hamiltonian

$$H = \frac{1}{2}(p_x^2 + p_y^2) + \frac{k_x(z)}{2}x^2 + \frac{k_y(z)}{2}y^2 + \frac{q}{mv_z^2\gamma^3}\psi(x,y,z) - \frac{\delta}{\rho(z)}x + \frac{m^2c^4}{E_o^2}\delta^2. \tag{1}$$

Here v_z is the longitudinal velocity, γ the relativistic factor, ψ, the self-potential which includes contribution from both the magnetic and electric field (see e.g. [5]). Each particle has a longitudinal momentum $p_z = p_o(1+\delta)$ with a relative deviation δ from the design momentum p_o, with E_o being the corresponding energy. In writing (1) we assumed that the nonlinearities due to the external focusing and the transverse current due to the bending of the beam are negligible. Also, we are assuming there is no focusing in the longitudinal direction (i.e. δ is a constant of the motion). From the Vlasov equation associated with the Hamiltonian (1) one can derive the following equations for the second moments ($g_o = q/(mv_z^2\gamma^3)$):

$$\frac{d}{dz}\langle x^2 \rangle = 2\langle xp_x \rangle, \tag{2a}$$

$$\frac{d}{dz}\langle p_x^2 \rangle = -2k_x\langle xp_x \rangle - 2g_o\langle p_x \frac{\partial}{\partial x}\psi \rangle + \frac{2}{\rho}\langle p_x\delta \rangle, \tag{2b}$$

$$\frac{d}{dz}\langle xp_x \rangle = \langle p_x^2 \rangle - k_x\langle x^2 \rangle - g_o\langle x\frac{\partial}{\partial x}\psi \rangle + \frac{1}{\rho}\langle x\delta \rangle, \tag{2c}$$

$$\frac{d}{dz}\langle x\delta \rangle = \langle p_x\delta \rangle, \tag{3a}$$

$$\frac{d}{dz}\langle p_x\delta \rangle = -k_x\langle x\delta \rangle - g_o\langle \delta\frac{\partial}{\partial x}\psi \rangle + \frac{1}{\rho}\langle \delta^2 \rangle. \tag{3b}$$

The equations for $\langle y^2 \rangle$, $\langle p_y^2 \rangle$, $\langle yp_y \rangle$ are similar to those for $\langle x^2 \rangle$, $\langle p_x^2 \rangle$, $\langle xp_x \rangle$ when dispersion is absent ($\rho = \infty$). Here $\langle \cdot \rangle$ denotes averaging over the set of all the dynamical variables x, p_x, y, p_y, δ. In our model we assume for simplicity that the beam centroid remains located on the geometrical center of the beam pipe $\langle x \rangle = \langle y \rangle = 0$. The average of the transverse momenta is also assumed to vanish $\langle p_x \rangle = \langle p_y \rangle = 0$.

If there is no dispersion, let us say $\rho = \infty$, the equations (2) are the same as the ones considered by Sacherer [4]. Let us briefly review his analysis. In the general case the terms involving the self-potential will depend on higher order moments. Therefore self-consistency would require that one should consider all the equations for the higher order moments as well. Sacherer proved that under the assumption of a beam density with an elliptic symmetry and the invariance of the rms emittance one can disregard the equations for the higher order moments and combine eqs. (2a), (2c) into a single second differential equation that depends only on the envelopes $\sigma_x = \sqrt{\langle x^2 \rangle}$ and $\sigma_y = \sqrt{\langle y^2 \rangle}$. In particular, the first assumption allows one to write $\langle x \frac{\partial}{\partial x} \psi \rangle$ as a function of the envelopes only, while the invariance of the emittance makes it possible to eliminate $\langle p_x^2 \rangle$ in terms of σ_x and its first derivative.

As we pointed out, if we want to extend this reasoning to the case with dispersion we have to find an rms emittance-like quantity that is linearly preserved. But first we need to find a suitable representation for the dispersion function in this multiparticle context. This is done by observing that in the zero space charge limit the two equations (3) can be combined into a single second order equation that has the same form as the equation for the dispersion function $D(z)$ for a single particle. This allows us to identify $\langle x\delta \rangle = \langle \delta^2 \rangle D(z)$ and $\langle p_x \delta \rangle = \langle \delta^2 \rangle D'(z)$. [1] We will then take these as the defining relations for the dispersion function in the general case with space charge. One can easily check by direct calculation using (2), (3) in the zero space charge limit that the following quantity is an exact linear invariant

$$\epsilon_{dx}^2 = (\langle x^2 \rangle - D^2 \langle \delta^2 \rangle)(\langle p_x^2 \rangle - D'^2 \langle \delta^2 \rangle) - (\langle x p_x \rangle - DD' \langle \delta^2 \rangle)^2. \quad (4)$$

Notice ϵ_{dx} coincides with the rms emittance at those location where $D = D' = 0$. For a more insightful proof of the invariance of (4) we refer to [1].

As we allow space charge to get into the picture we can again use eqs. (3) to describe the dispersion function provided we resolve the term $\langle \delta \frac{\partial}{\partial x} \psi \rangle$. This can be done in the linear approximation by using the result proved in [4] that, under the assumption of a beam density with elliptic symmetry, the linear part of the self force defined by a least square method is given by $Kx/[2\sigma_x(\sigma_x + \sigma_y)]$. The quantity K is the generalized perveance [5]. Therefore the dispersion function obeys:

$$D'' + \left[k_x(z) - \frac{K}{2\sigma_x(\sigma_x + \sigma_y)} \right] D = \frac{1}{\rho(z)}. \quad (5)$$

The key assumption at this point is that the linear invariant defined in (4) is approximately conserved when space charge is present. This mirrors the assumption that the standard rms emittance be invariant in Sacherer's derivation of the usual rms envelope equations. The new invariant can be used to express $\langle p_x^2 \rangle$ in terms of the horizontal envelope and dispersion function. Consequently, we can write the equations for the envelopes as:

$$\sigma_x'' = \frac{\epsilon_{dx}^2 + (\sigma_x \sigma_x' - DD' \langle \delta^2 \rangle)^2}{\sigma_x(\sigma_x^2 - D^2 \langle \delta^2 \rangle)} - \frac{(\sigma_x')^2}{\sigma_x}$$

$$- k_x \sigma_x + \frac{K}{2(\sigma_x + \sigma_y)} + \frac{\langle \delta^2 \rangle}{\sigma_x} \left(\frac{D}{\rho} + D'^2 \right), \quad (6)$$

$$\sigma_y'' = \frac{\epsilon_y^2}{\sigma_y^3} - k_y \sigma_y + \frac{K}{2(\sigma_x + \sigma_y)}, \quad (7)$$

---

[1] The identification is unambiguous if the beam distribution is such that at the point where $D = D' = 0$ (e.g. at injection) it is $\langle x\delta \rangle = \langle p_x \delta \rangle = 0$.

The three equations (5),(6),(7), in the variables D, σ_x, σ_y, provide a consistent description for the evolution of the rms envelopes of a beam in a dispersive channel. We should mention that eq. (5) was first written by A. Garren [6] but used in combination with the standard rms-envelope equation. In that way the dependence of the envelopes on the dispersion function was not properly described.

NUMERICAL TESTS

The rms envelope-dispersion equations have already been tested against the fully self-consistent calculation discussed in [2] in the smooth approximation and the results reported in [1]. Here we present some additional checks where we compare the solutions of the new set of equations against the results obtained by using the PIC code WARP [9] in the modeling of the Maryland Electron Ring [7]. The numerical simulations using WARP are part of a larger effort to study various beam physics problems and design issues in the Maryland Electron Ring [8].

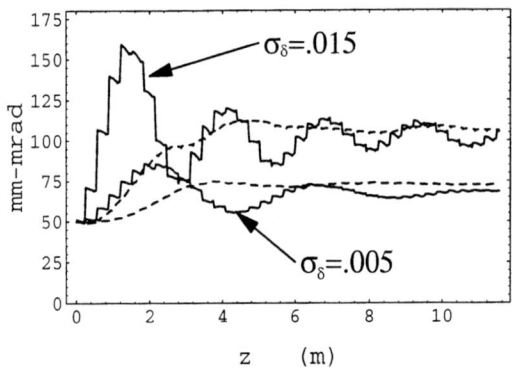

FIGURE 1. Effective emittances $4\epsilon_x$ (solid lines), $4\epsilon_y$ (dashed line) as calculated by WARP as a function of z for two beams with different relative rms longitudinal momentum spreads σ_δ.

In the study reported here we adopted a hard-edge model of the Ring design. The model of the Ring we use consists of 36 FODO-cells. Each FODO-cell is 32 cm long and includes a 10° bending dipole. We refer to [7] for a detailed description of the Ring lattice, magnet and beam parameters.

Let us first consider the evolution of the rms emittances after injection of a continuous beam of electrons with a longitudinal momentum spread as calculated by the PIC code. Fig. 1 shows the results of the simulations after injecting two beams with an initially uniform distribution in space and gaussian distribution in the transverse momenta. The relative rms longitudinal momentum spread, $\sigma_\delta = \sqrt{\langle \delta^2 \rangle}$, for the two beams is indicated in the figure. The beams are initially matched with regard of the envelopes only and the matching is done using the standard rms envelope equations. The initial effective emittance in both planes is $4\epsilon_x = 4\epsilon_y = 50$ mm-mrad; the beam energy is 10 keV and the current 100 mA.

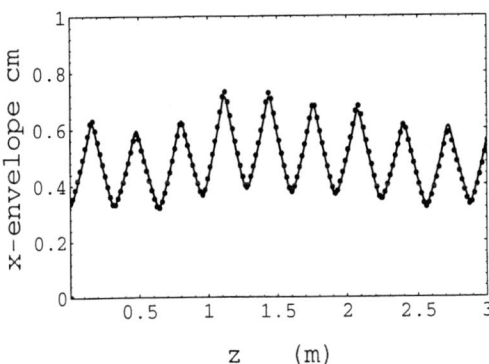

FIGURE 2. Beam envelope σ_x as calculated by WARP (solid line) and from the rms envelope-dispersion equations ($\sigma_\delta = .015$). The envelope is plotted over a distance corresponding to about one period of the betatron oscillation.

The picture shows large emittance oscillations that eventually dye off leaving a net emittance growth. Those oscillations can have a qualitative explanation in terms of a dispersion mismatch. Let us rewrite equation (4) in the form

$$\epsilon_x^2 = \epsilon_{dx}^2 + \langle \delta^2 \rangle \langle [p_x D(z) - x D'(z)]^2 \rangle. \qquad (8)$$

The expression above shows that if ϵ_{dx}, the new quantity we have defined, is preserved the standard rms emittance undergoes oscillations that depend on the dispersion function. The large amplitude of the oscillations is due to the large oscillations that an unmatched dispersion experiences.

It is easy to write in the smooth approximation a solution of eq. (5) with the initial conditions $D = D' = 0$. From this approximated solution one can observe the dispersion function undergoes sinusoidal oscillations around what, in the smooth approximation, would be its matched value. The frequency of oscillation is the same as the incoherent depressed tune. On the other hand, the depressed tune is sensitive to the beam envelopes which in turn depend on the longitudinal momentum spread. For a larger longitudinal momentum spread we expect larger envelopes and therefore a less depressed tune. This is consistent with the scaling of the period of the oscillations we observe in Fig. 1. That is, the period is longer in the case $\sigma_\delta = 0.005$ than in hte case $\sigma_\delta = 0.015$. There are however two patterns that an interpretation based on eq. (8) cannot evidently explain, i.e. the damping in the horizontal and the increase in the vertical emittance. Eq. (8) as well the rms envelope-dispersion equations presented in this paper do not account for the thermalization process we see acting here between the two degrees of freedom. This effect, triggered by the nonlinear coupling between x and y due to space charge is responsible for both the damping and the increase in the vertical rms emittance. We believe that the thermalization process is considerably enhanced by a dispersion mismatch of the beam. A properly matched injection is likely to reduce both the amplitude of the emittance oscillations and the final net emittance growth.

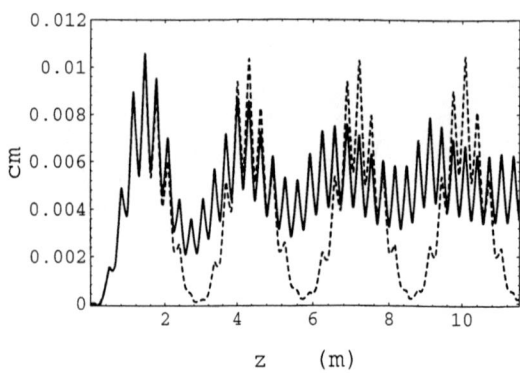

FIGURE 3. The second moment $\langle x\delta \rangle$ as calculated by WARP (solid line) and $\langle \delta^2 \rangle D(z)$ as calculated from the rms envelope-dispersion equations ($\sigma_\delta = .015$).

The remaining figures presented here show a comparison between the beam envelopes and the dispersion function as calculated from the rms envelope-dispersion equations and by the PIC code, for the same case of an initially dispersion-mismatched beam. Having in mind the discussion above we should expect that an agreement, if any, between the two calculations should take place only over a time scale shorter than the thermalization time. Indeed this is the case. We observe a very good agreement between theory and simulation over the first half of the (depressed) betatron oscillation period. This defines the length scale over which the matching of the dispersion has to be accomplished. In particular, Fig. 2 shows the horizontal envelope σ_x as calculated using equations (dots) and WARP (solid line) over the first betatron oscillation period. The envelope profile as calculated by WARP is well reproduced by the solutions of the rms envelope-dispersion equation. In Fig. 3 and Fig. 4 we show the second moment $\langle x\delta \rangle$ as calculated by WARP and the scaled dispersion function $\langle \delta^2 \rangle D(z)$. The two quantities, as indicated in the previous section, are expected to be equal. The two figures refer to two different values of the rms longitudinal momentum spread σ_δ as indicated in the captions.

Again, thermalization can be seen in action in the way the two curves (dashed curve from the theory, solid line from the simulation) separate after the first oscillation period. Notice however how the agreement is very good up to separation, and how the scaling of the period of the oscillation is faithfully reproduced by the solutions of the rms envelope-dispersion equations. Also the period of oscillation is consistent with that displayed by the horizontal rms emittances in Fig. 1 in the corresponding case of equal longitudinal momentum spread, as expected from (8).

In conclusion: we have presented a new set of rms envelope-dispersion equations that generalize the usual rms envelope equations to the case where a longitudinal momentum spread and bends are present. We have also shown some numerical tests that confirm the validity of the theory. At the same time the numerical tests show the domain of applicability of the equations is limited by thermalization. We

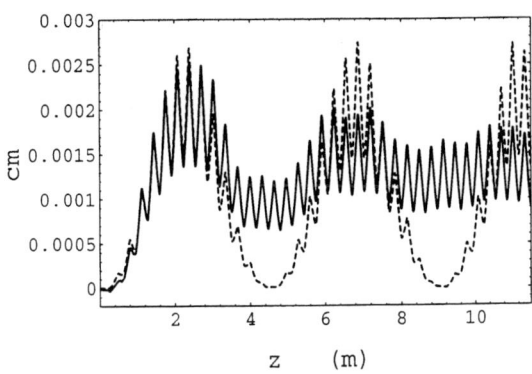

FIGURE 4. As in Fig. 2 but with a smaller rms longitudinal momentum spread $\sigma_\delta = .005$.

believe, however, that if proper matching conditions are met the effect of thermalization should be substantially reduced, as well as the net emittance growth observed in the simulations. Matching schemes to be applied to the Maryland Electron Ring are planned to be explored and tested.

We would like to thank A. Friedman, D. Grote, S. Lund, and I. Haber for support with the WARP code. This work was sponsored by the US Department of Energy.

REFERENCES

1. M. Venturini, M.Reiser, RMS Envelope Equations in the Presence of Space Charge and Dispersion, to appear in Phys. Rev. Lett.
2. M. Venturini, M.Reiser, Phys. Rev. E, **57**, 4 (1998) 4725; M. Venturini, M.Reiser Generalized KV beam in a Dispersive Channel, Particle Accelerator Conference 1997, Proceedings.
3. J.J. Barnard et al., 1992 Linear Accelerator Conference Proc., AECL-10728 Vol. 1, p.229; Barnard et al., Emittance Growth in Heavy Ion Rings Due to the Effects of Space Charge and Dispersion, these Proceedings.
4. F.J. Sacherer, IEEE Transactions on Nuclear Science NS-18 (1971) 1105.
5. M. Reiser, Theory and Design of Charged Beams, Wiley & Son, New York (1994).
6. A. Garren, Proc. Heavy Ion Fusion Workshop, 1979 Reports LBL-10301/SLAC-PUB 2575, UC-28, p.397.
7. M. Reiser, S. Bernal, A. Dragt, M. Venturini, J.G. Wang, H. Onishi and T.F. Godlove, Fus. Eng. and Design, 32-33 (1996) 293-298; J.G. Wang et al., Study of Space-Charge Physics in Beams for Advanced Acceleration Applications, these Proceedings.
8. R.A. Kishek, M. Venturini, M.Reiser, I. Haber, PIC Code Simulations of Space-Charge Dominated Beams in the University of Maryland Electron Ring, these Proceedings.
9. D.P. Grote et al., Fus. Eng. and Design 32-33, (1996) 193-200; A. Friedman, Overview of the WARP Code and Studies of Transverse Resonance Effects, these Proceedings.

THE SPACE-CHARGE IMPEDANCE OF RF-SHIELDING WIRES

Tai-Sen F. Wang

Los Alamos National Laboratory, Los Alamos, NM 87545, USA

Abstract

We studied the electrostatic fields due to the longitudinal and transverse perturbations of a charged particle beam with a uniform distribution propagating inside an rf-shielding cage constructed from evenly-spaced conducting wires. Simple formulae are derived for estimating the space-charge impedances. Numerical examples are given for illustration.

1 Introduction

An rf-shielding cage, or an rf-cage, used in an accelerator or storage ring is a cage-like structure made of conducting wires stretched in parallel to the direction of the circulating charged particle beam.[1] The conducting wires on the cage are arranged to surround the beam to create an electromagnetically shielded environment for the beam. This kind or the similar kinds of devices together with ceramic beam pipes have been implemented[1] and planned[2-4], or is being planned[5] in some high-intensity rapid cycling proton synchrotrons.

There are two reasons for choosing the rf-cage instead of solid beam pipe. The first reason is to avoid excess eddy current that may be induced on the beam pipe by the fast-changing magnetic field. The second reason is that, for reducing the coupling impedance, it is easier to vary the cross-section of an rf-cage than varying the cross section of a solid pipe. In the long wavelength region, an appropriately designed rf-cage can provide electromagnetic shielding near that of a solid beam pipe.

Although an rf-cage has been built and implemented in an existing proton synchrotron,[1] a serious study of the electromagnetic field of a charged particle beam propagating in an rf-cage has never been documented until the relative recent.[6] In Ref. 6, a rigorous formalism was established to investigate the electrostatic field of a charged particle beam with a uniform distribution inside an rf-shielding cage constructed from evenly-spaced conducting wires. However, mistakes were later found in that work. The purpose of this present work is to revisit the old problem considered in Ref. 6. We will present the correct solution of the electrostatic fields of a perturbed beam traversing inside of an rf-shielding cage. Simple formulae will be derived for both the longitudinal and transverse coupling impedances in the long wavelength regime. Numerical examples will be given to illustrate the effect of rf-shielding.

CP448, *Workshop on Space Charge Physics in High Intensity Hadron Rings*
edited by A. U. Luccio and W. T. Weng
© 1998 The American Institute of Physics 1-56396-824-X/98/$15.00

2 The Electrostatic Potential of a Perturbed Beam

The system considered here is shown in Fig. 1. A beam having a circular cross-section of radius r_b and a uniform charge distribution is propagating inside of an rf-cage composed of N conducting wires extended in the direction parallel to the beam. For simplicity, we shall limit our discussion to the geometry in which wires are evenly distributed over a circle and the rf-cage is positioned concentric with the beam. The radius of the rf-cage, measured from the center of the cage to the centers of wires, is r_c. We assume that wires are electrically grounded and all wires have the same circular cross-section of radius ρ_w. The discussions in the following will be restricted to the regime of $\rho_w \ll r_c$ and $N \gg 1$.

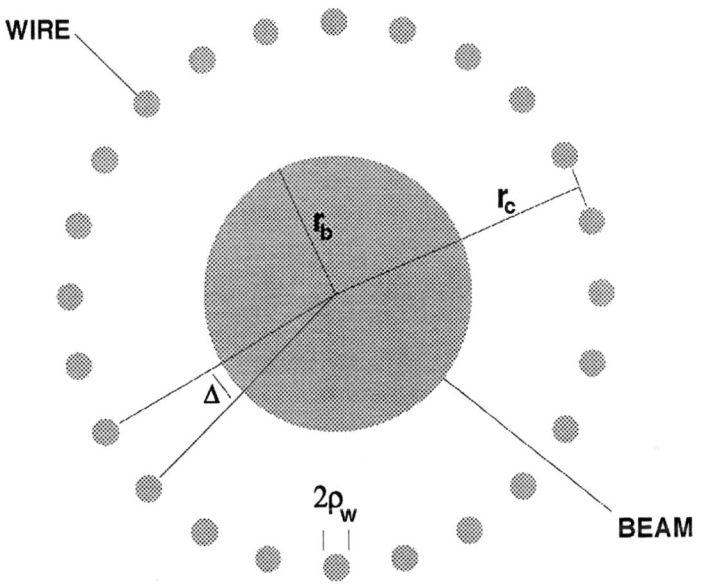

Fig. 1. Cross-sectional view of a beam inside an rf-cage and beam pipe. r_c and r_b are the radii of the rf-cage and the beam, respectively. Δ is the angle subtended by two adjacent wires, and ρ_w is the radius of a wire.

We choose a cylindrical coordinate system (r, θ, z) such that the z-axis coincides with the central axis of the beam. We shall call this coordinate system the "beam coordinate system" or the "global coordinate system". In order to make it convenient to describe the electric field near an individual wire, we shall also use another cylindrical coordinate system (ρ, ψ, z) in which the z-axis coincides with the central axis of a wire as shown in Fig. 2. This "local coordinate system" will also be referred to as the "wire coordinate system" in the following.

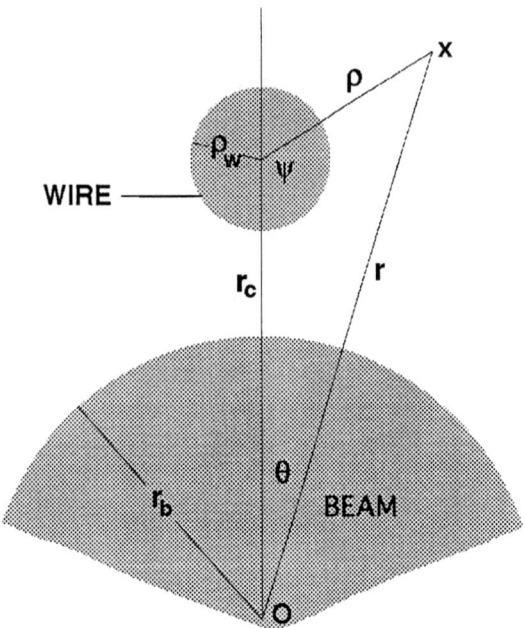

Fig. 2. The local and the global coordinates adopted in this study. The origins of the local and the global coordinates are located at the center of beam and the center of a wire, respectively. The positive directions of ψ and θ go counterclockwise and clockwise, respectively.

2.1 The Potential due to a Longitudinal Perturbation

Here, we shall concentrate on the electrostatic potential due to the longitudinal charge-density perturbation that varies in the z-direction according to e^{ikz}, where k is the wave-number of the perturbation. The Poisson equation we want to solve is

$$\frac{1}{r}\frac{\partial}{\partial r}\left(r\frac{\partial \Phi}{\partial r}\right) + \frac{1}{r^2}\frac{\partial^2 \Phi}{\partial \theta^2} + \frac{\partial^2 \Phi}{\partial z^2} = \begin{cases} 0, & \text{for } r > r_b, \\ -(\sigma/\epsilon_o)e^{ikz}, & \text{for } r \leq r_b, \end{cases} \quad (1)$$

where σ is the volume charge density associated with the perturbation, ϵ_o is the permittivity of free space, and r_b is the radius of the beam. We are interested in the solution of Eq. (1) in the region where $kr_c \ll 1$.

In the absence of any boundary, we assume $\Phi = 0$ at infinity. The solution of Eq. (1) in this case is

$$\Phi = \phi_b = \begin{cases} b_\| K_0(kr)e^{ikz} , & \text{for } r > r_b, \\ \dfrac{\sigma}{\epsilon_o k^2}\left[1 - kr_b K_1(kr_b)I_0(kr)\right]e^{ikz} , & \text{for } r \leq r_b, \end{cases} \qquad (2)$$

where

$$b_\| = \left(\frac{\sigma r_b}{k\epsilon_o}\right)I_1(kr_b) ; \qquad (3)$$

$I_n(x)$ and $K_n(x)$ are the nth order modified Bessel functions of the first kind and the second kind, respectively.

Moving into the discussion of solving Eq. (1) in the presence of wires, we consider the solution in the region of $r \geq r_c$ first. In the presence of wires, the beam field will induce electric charges on the surfaces of wires. These induced charges also have an electric potential associated with them. We then assume that each wire induces a field which can be expanded into a Fourier series of $e^{in\psi}$ in the local coordinate as

$$\phi_w = \sum_{n=-\infty}^{\infty} C_n K_n(k\rho)e^{in\psi}e^{ikz} , \qquad (4)$$

where C_n is an unknown quantity to be solved. Applying the addition theorem of Bessel functions,[7] ϕ_w can also be expressed in the global coordinate variables as

$$\phi_w = \sum_{l=-\infty}^{\infty}\sum_{n=-\infty}^{\infty} (-1)^n C_n I_{n+l}(kr_c) K_l(kr)e^{il\theta}e^{ikz} . \qquad (5)$$

Note that Eq. (4) is virtually the multipole expansion of the induced field in the wire coordinate system.

Since all wires are electrically identical and are evenly distributed, one can study the field around wires by considering the electric potential around any individual wire. Thus, we call the wire under consideration the 0th wire and number all others by their relative locations with respect to the 0th wire. If S_w is the sum of the induced potentials from all other wires nearby the 0th one, we have

$$S_w = \sum_{l=-\infty}^{\infty}\sum_{n=-\infty}^{\infty} C_l \left[\sum_{\mu=1}^{N-1} e^{i(n+l)(\pi-\mu\Delta)/2} K_{l-n}(kd_\mu)\right] I_n(k\rho)e^{in\psi}e^{ikz} , \qquad (6)$$

where $\Delta = 2\pi/N$ is the angular separation between two adjacent wires; d_μ is the distance between the centers of the 0th and the μth wires.

The total electric potential around the 0th wire is

$$\Phi = \phi_b + \phi_w + S_w . \qquad (7)$$

On the surface of each wire, the potential due to the induced charge should cancel the potential due to the beam and the potential contributed by all other wires, i.e., $\Phi = 0$ at $\rho = \rho_w$. Thus, substituting Eqs. (2), (4) and (6) into Eq. (7), we derive that on the surface of the 0th wire,

$$\sum_{l=-\infty}^{\infty} C_l \left[\sum_{\mu=1}^{N-1} e^{i(n+l)(\pi - \mu \Delta)/2} K_{l-n}(kd_\mu) \right] + \frac{C_n}{h_n(k\rho_w)} + b_\| K_n(kr_c) = 0 \quad, \quad (8)$$

where
$$h_n(x) = I_n(x)/K_n(x) \quad . \quad (9)$$

The closed-form solution of Eq. (8) appears to be inaccessible. However, in the regime of $k\rho_w \ll kr_c \ll 1$, it is possible to find a solution expanded in the power series of $h_n(k\rho_w)$. We define a quantity η_n such that

$$C_n = -b_\| \eta_n / [N I_n(kr_c)] \quad . \quad (10)$$

Then, if the multipole coupling is neglected, the lowest order solution of η_n is

$$\eta_n \approx N h_n(k\rho_w) I_n(kr_c) K_n(kr_c) D_n \quad , \quad (11)$$

where
$$D_n = \left[1 + (-1)^n h_n(k\rho_w) \sum_{\mu=1}^{N-1} e^{-in\mu\Delta} K_0(kd_\mu) \right]^{-1} \quad . \quad (12)$$

The lowest order solution of η_n that includes the multipole coupling is

$$\eta_n \approx N h_n(k\rho_w) I_n(kr_c) D_n \left\{ K_n(kr_c) - \sum_{l=-\infty}^{\infty} (1 - \delta_{ln}) h_l(k\rho_w) \right.$$
$$\left. \times K_l(kr_c) \sum_{\mu=1}^{N-1} e^{i(n+l)(\pi - \mu\Delta)/2} K_{n-l}(kd_\mu) \right\} \quad . \quad (13)$$

The contribution of the potential from all wires now can be expressed as

$$\phi_w + S_w = -b_\| \sum_{p=-\infty}^{\infty} \sum_{n=-\infty}^{\infty} \frac{(-1)^n \eta_n}{I_n(kr_c)} I_{n+pN}(kr_c) K_{pN}(kr) e^{ipN\theta} e^{ikz} \quad . \quad (14)$$

In arriving at Eq. (14), use has been made of the relation (holds good only when $\Delta = 2\pi/N$)

$$\sum_{j=0}^{N-1} e^{-ijl\Delta} = N\delta_{l,pN} = \begin{cases} N, & \text{for } l = pN, \, p = 0, \pm 1, \pm 2, \cdots, \\ 0, & \text{for } l \neq pN. \end{cases} \quad (15)$$

Substituting the potentials in Eqs. (2) and (14) into Eq. (7) yields the following total potential in the region of $r \geq r_c$

$$\Phi_\| = b_\| K_0(kr)e^{ikz} - b_\| \sum_{p=-\infty}^{\infty} B_p K_{pN}(kr)e^{ipN\theta}e^{ikz} \ , \tag{16}$$

where

$$B_p = (1-\delta_{N0}) \sum_{n=-\infty}^{\infty} \frac{(-1)^n \eta_n}{I_n(kr_c)} I_{n+pN}(kr_c) \ , \tag{17}$$

Next, we consider the solution in the region of $r \leq r_c$. In this region,

$$\phi_w + S_w = -b_\| \sum_{p=-\infty}^{\infty} \sum_{n=-\infty}^{\infty} \frac{\eta_n}{I_n(kr_c)} K_{n+pN}(kr_c) I_{pN}(kr) e^{ipN\theta} e^{ikz} \ . \tag{18}$$

From this, we can derive

$$\Phi_\| = b_\| K_0(kr)e^{ikz} - b_\| \sum_{p=-\infty}^{\infty} \hat{B}_p I_{pN}(kr)e^{ipN\theta}e^{ikz} \ , \tag{19}$$

for $r_c \geq r \geq r_b$, and

$$\Phi_\| = \frac{\sigma}{\epsilon_o k^2}\left[1 - kr_b K_1(kr_b)I_0(kr)\right]e^{ikz} - b_\| \sum_{p=-\infty}^{\infty} \hat{B}_p I_{pN}(kr)e^{ipN\theta}e^{ikz} \ , \tag{20}$$

for $r \leq r_b$, where

$$\hat{B}_p = (1-\delta_{N0}) \sum_{n=-\infty}^{\infty} \frac{\eta_n}{I_n(kr_c)} K_{n+pN}(kr_c) \ . \tag{21}$$

2.2 The Potential due to a Transverse Perturbation

We now consider the electrostatic potential due to a sinusoidal transverse perturbation in a beam propagating inside an rf-cage. The model of the perturbation to be studied here is a shell with surface charge density varying according to $e^{ikz}\cos\theta$. Here we assume the beam coordinate is oriented in such a way that the maximal charge density is at the angle of $\theta = 0$. The Poisson equation is

$$\frac{1}{r}\frac{\partial}{\partial r}\left(r\frac{\partial \Phi}{\partial r}\right) + \frac{1}{r^2}\frac{\partial^2 \Phi}{\partial \theta^2} + \frac{\partial^2 \Phi}{\partial z^2} = -\frac{\sigma \bar{d}}{2\epsilon_o}\delta(r-r_b)\left(e^{i\theta} + e^{-i\theta}\right)e^{ikz} \ , \tag{22}$$

where $\bar{d}$ is the averaged displacement of the beam.

Because the system under consideration is symmetric with respect to θ, we shall solve the equation

$$\frac{1}{r}\frac{\partial}{\partial r}\left(r\frac{\partial \Phi}{\partial r}\right) + \frac{1}{r^2}\frac{\partial^2 \Phi}{\partial \theta^2} + \frac{\partial^2 \Phi}{\partial z^2} = -\frac{\sigma \bar{d}}{2\epsilon_o}\delta(r - r_b)e^{i\theta}e^{ikz} \;, \qquad (23)$$

first. The solution of Eq. (22) then can be obtained from the solution of Eq. (23) by the substitution of $e^{in\theta}/2 \to \cos\theta$ appropriately. In the absence of any boundary, Eq. (23) has the solution

$$\Phi = \phi_b = \begin{cases} b_\perp K_1(kr)e^{i\theta}e^{ikz} \;, & \text{for } r > r_b, \\ b_\perp \left[K_1(kr_b)/I_1(kr_b)\right]I_1(kr)e^{i\theta}e^{ikz} \;, & \text{for } r \leq r_b, \end{cases} \qquad (24)$$

where

$$b_\perp = \left(\frac{\sigma \bar{d} r_b}{2\epsilon_o}\right) I_1(kr_b) \;. \qquad (25)$$

For the solution including the rf-shielding wires, we again consider the region of $r \geq r_c$ first. Different from the case of longitudinal perturbation, the axial symmetry does not exist here and the angular positions of wires need to be taken into account when describing the potential around each wire. Take the wire located at the angle of $m\Delta$, the mth wire, as an example. When applying the addition theorem of Bessel functions, one has to make a change of variable from θ to $\theta - m\Delta$ for translating the description of the potential from the global coordinate variables to the local coordinate variables around the mth wire. Therefore, the multipole expansion coefficients of the field induced on each wire depend on the wire's location.

The analysis will follow the same path as in Section 2.1. Thus, we assume the mth wire induces a potential which can be described in terms of the wire coordinate system variables as

$$\phi_w = \sum_{n=-\infty}^{\infty} P_{m,n} K_n(k\rho) e^{in\psi} e^{ikz} \;, \qquad (26)$$

where $P_{m,n}$ is the unknown we want to solve. The contribution from all other $N-1$ wires, S'_w is

$$S'_w = \sum_{l=-\infty}^{\infty} \sum_{n=-\infty}^{\infty} \sum_{j=0}^{N-1} (1-\delta_{j,m}) P_{j,l} e^{i(n+l)[\pi-(j-m)\Delta]/2}$$
$$\times K_{l-n}(kd_{j,m}) I_n(k\rho) e^{in\psi} e^{ikz} \;, \qquad (27)$$

where $d_{j,m}$ is the distance between the centers of the mth and the jth wires. Using some algebra, one can prove that $P_{m,n} = e^{im\Delta} P_n$, where $P_n = P_{0,n}$ is the multipole expansion coefficient for the 0th wire. Applying this relation, the requirement of the total potential be zero on the surfaces of all wires can be reduced to a single equation similar to Eq. (8):

$$\sum_{l=-\infty}^{\infty} P_l \left[\sum_{\mu=1}^{N-1} e^{i(n+l)(\pi-\mu\Delta)/2} e^{i\mu\Delta} K_{n-l}(kd_\mu) \right] + \frac{P_n}{h_n(k\rho_w)} + b_\perp K_{n+1}(kr_c) = 0 \ . \tag{28}$$

Eq. (28) can be solved in the same procedure as solving Eq. (8). Thus, defining η'_n according to

$$P_n = -b_\perp \eta'_n \Big/ \left[N I_{n+1}(kr_c) \right] \ , \tag{29}$$

then the lowest order solution of η'_n without the multipole coupling is

$$\eta'_n \approx N h_n(k\rho_w) I_{n+1}(kr_c) K_{n+1}(kr_c) D'_n \ . \tag{30}$$

where

$$D'_n = \left[1 + (-1)^n h_n(k\rho_w) \sum_{\mu=1}^{N-1} e^{-i(n-1)\mu\Delta} K_0(kd_\mu) \right]^{-1} \ . \tag{31}$$

Including the multipole coupling, the lowest order solution of η'_n is

$$\eta'_n \approx N h_n(k\rho_w) I_{n+1}(kr_c) D'_n \Bigg\{ K_{n+1}(kr_c) - \sum_{l=-\infty}^{\infty} (1-\delta_{ln}) h_l(k\rho_w)$$

$$\times K_{l+1}(kr_c) \sum_{\mu=1}^{N-1} e^{i(n+l)(\pi-\mu\Delta)/2} e^{i\mu\Delta} K_{n-l}(kd_\mu) \Bigg\} \ . \tag{32}$$

Using these results, we can obtain the following solution of Eq. (22):

$$\Phi_\perp = 2b_\perp K_1(kr)\cos\theta e^{ikz} - 2b_\perp \sum_{p=-\infty}^{\infty} B'_p K_{pN+1}(kr)\cos\theta e^{ipN\theta} e^{ikz} \ , \tag{33}$$

for $r \geq r_c$,

$$\Phi_\perp = 2b_\perp K_1(kr)\cos\theta e^{ikz} - 2b_\perp \sum_{p=-\infty}^{\infty} \hat{B}'_p I_{pN+1}(kr)\cos\theta e^{ipN\theta} e^{ikz} \ , \tag{34}$$

for $r_c \geq r \geq r_b$, and

$$\Phi_\perp = 2b_\perp \left[\frac{K_1(kr_b)}{I_1(kr_b)} \right] I_1(kr)\cos\theta e^{ikz} - 2b_\perp \sum_{p=-\infty}^{\infty} \hat{B}'_p I_{pN+1}(kr)\cos\theta e^{ipN\theta} e^{ikz} \ , \tag{35}$$

for $r \leq r_b$, where

$$B'_p = (1-\delta_{N0}) \sum_{n=-\infty}^{\infty} \frac{(-1)^n \eta'_n}{I_{n+1}(kr_c)} I_{n+pN+1}(kr_c) \ . \tag{36}$$

293

and
$$\hat{B}'_p = (1 - \delta_{N0}) \sum_{n=-\infty}^{\infty} \frac{\eta'_n}{I_{n+1}(kr_c)} K_{n+pN+1}(kr_c) \ . \tag{37}$$

3   The Space-Charge Impedance

Before proceeding to the derivation of space-charge impedance, it is worthwhile to discuss the limits of our solutions when $N \to \infty$ and $\rho_w \to 0$. At these limits, the summation over wires can be replaced by an integration to show that the values of both η_n and η'_n approach one. Then from Eqs. (19), (20), (34), and (35), one obtains the perturbed potentials for a beam inside a conducting pipe without wires. Thus, for an rf-cage made of infinite number of thin wires, the shielding effects is the same as that of a conducting beam pipe.

We now consider the space-charge impedance in the long wavelength region, i.e. the region where $k\rho_w \ll kr_c \ll 1$. We shall concentrate our discussions to the field at the center of the beam, i.e. at $r = 0$. From Eqs. (20) and (35), we can derive the perturbed electric field at the center of the beam:

$$E_z = -\frac{\partial \Phi_\parallel}{\partial z} \approx \frac{-i\sigma}{\epsilon_o k} \left\{ 1 - kr_b K_1(kr_b) - \frac{\eta_0 kr_b I_1(kr_b) K_0(kr_c)}{I_0(kr_c)} \right\} e^{ikz} \ , \tag{38}$$

and

$$E_y = -\frac{\partial \Phi_\perp}{\partial r}\bigg|_{\theta=0} \approx \frac{-\sigma k \bar{d} r_b}{2\epsilon_o} \left\{ K_1(kr_b) - \frac{\eta'_0 I_1(kr_b) K_1(kr_c)}{I_1(kr_c)} \right\} e^{ikz} \ . \tag{39}$$

where we have retained only the most significant terms (the $n = 0$ or the monopole terms) in the summations, and we have left off the couplings among the multipole fields since they are negligibly small in the parameter range of our interest. Then, applying the small argument expansions of Bessel functions, we derive from Eqs. (11) and (30) that for $N \gg 1$

$$\eta_0 \approx \frac{N \ln(kr_c)}{N \ln(kr_c) - \ln[r_c/(N\rho_w)]} \approx \frac{N \ln(kr_c)}{N \ln(kr_c) + \ln(\pi f_w)} \ , \tag{40}$$

and

$$\eta'_0 \approx \frac{N}{N + 2\ln[r_c/(N\rho_w)]} \approx \frac{N}{N - 2\ln(\pi f_w)} \ . \tag{41}$$

where the "wire filling factor" f_w is defined as the ratio between the angle subtended by a wire in the beam coordinate system θ_w and Δ, i.e.

$$f_w = \frac{\theta_w}{\Delta} \approx \frac{N\rho_w}{\pi r_c} \ . \tag{42}$$

Making use of the approximations in Eqs. (40) and (41), we obtain from Eqs. (38) and (39) that

$$E_z \approx \frac{-i\sigma k r_b^2}{4\epsilon_o}\left[1 + 2\ln\left(\frac{r_c}{r_b}\right) + W_\|\right]e^{ikz} , \qquad (43)$$

and

$$E_y \approx \frac{-\sigma \bar{d} r_b^2}{2\epsilon_o}\left(\frac{1}{r_b^2} - \frac{1}{r_c^2} + \frac{W_\perp}{r_c^2}\right)e^{izk} , \qquad (44)$$

where

$$W_\| = -2(1 - \eta_0) \approx \frac{-2\ln(kr_c)\ln(\pi f_w)}{\ln(\pi f_w) + N\ln(kr_c)} , \qquad (45)$$

and

$$W_\perp = 1 - \eta_0' \approx \frac{-2\ln(\pi f_w)}{N - 2\ln(\pi f_w)} . \qquad (46)$$

Note that in Eqs. (43) and (44), when $W_\| = 0$ and $W_\perp = 0$, one obtains the electric field inside a solid beam pipe. In a crude way, the quantities $W_\|$ and $W_\perp$ can be linked to the electric flux leaks out from the rf-cage; the quantities η_n and η_n' can be linked to the electric flux contained by the rf-cage. Thus, the approximations in Eqs. (43) and (44) allow one to estimate how the usual impedances of a solid beam pipe are modified when using an rf-cage for shielding. For example, in a hypothetical case of the rf-cage occupies the whole ring, the longitudinal and the transverse space-charge impedances are

$$Z_\| = \frac{inZ_o}{2\beta\gamma^2}\left[1 + 2\ln\left(\frac{r_c}{r_b}\right) + W_\|\right] , \qquad (47)$$

and

$$Z_\perp = \frac{iRZ_o}{\beta^2\gamma^2}\left(\frac{1}{r_b^2} - \frac{1}{r_c^2} + \frac{W_\perp}{r_c^2}\right) , \qquad (48)$$

respectively, where

$$W_\| \approx \frac{-2\ln(nr_c/R)\ln(\pi f_w)}{\ln(\pi f_w) + N\ln(nr_c/R)} , \qquad (49)$$

$W_\perp$ is the same as in Eq. (46), n is the azimuthal harmonic of the perturbation around the ring, $Z_o = 377\Omega$, β is equal to the averaged particle speed divided by the speed of light, $\gamma = (1 - \beta^2)^{-1/2}$, and R is the machine's effective radius.

Numerical examples of $W_\|$ and $W_\perp$ are given in Figs. 3 and 4 for $kr_c = 0.1$ (or $nr_c/R = 0.1$). These results indicate that in the long wavelength regime, the diameter of wires is a less important parameter when large number of wires are used for shielding. Also note that $W_\| = W_\perp = 0$ when $r_c = N\rho_w$, or $f_w = 1/\pi$ as indicated in Eqs. (45) and (46).

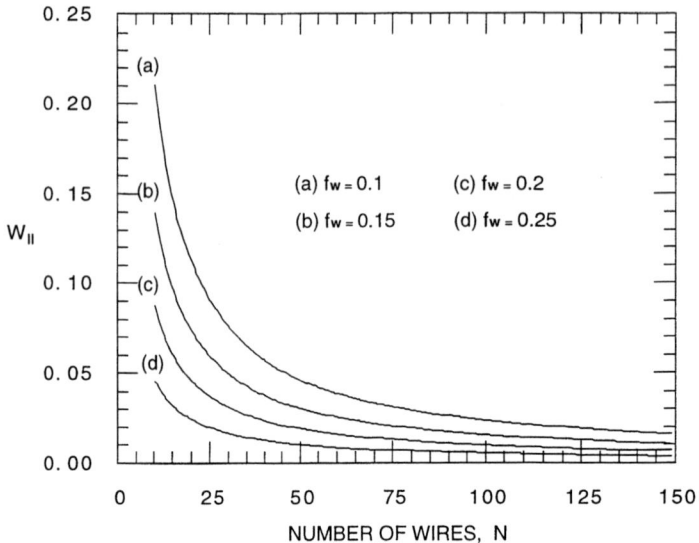

Fig. 3. $W_\parallel$ as a function of the total number of wires N for (a) $f_w = 0.1$, (b) $f_w = 0.15$, (c) $f_w = 0.2$, and (d) $f_w = 0.25$. Where $W_\parallel$ is calculated using Eq. (45) for $kr_c = 0.1$.

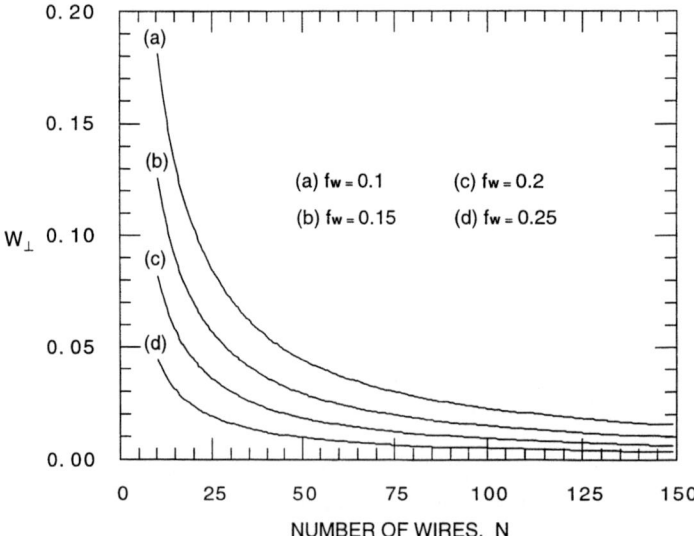

Fig. 4. $W_\perp$ as a function of the total number of wires N for (a) $f_w = 0.1$, (b) $f_w = 0.15$, (c) $f_w = 0.2$, and (d) $f_w = 0.25$. Where $W_\perp$ is calculated using Eq. (46) for $kr_c = 0.1$.

4 Conclusions

For a charged particle beam propagating inside of an rf-shielding cage made of evenly-spaced conducting wires, the electrostatic fields due to sinusoidal longitudinal and transverse perturbations have been solved analytically for the case that both the cage and wires have circular cross sections. Only the dipole mode has been treated for the transverse perturbation. We have assumed that the beam has a uniform charge distribution and the unperturbed system is azimuthally symmetric. Using the calculated fields, we have derived simple formulae for the coupling impedances in the long wavelength region. Numerical examples were given to show the shielding effects of the rf-cage.

Acknowledgements

The initial study of this topic was sponsored by Accelerator Technology (AT) Division of Los Alamos National Laboratory and AUSTRON Project. The author would like to thank AT Division and AUSTRON Project for the support he received during the course of this study. Many thanks to Dr. Phil Bryant for his encouragement and comments as well as his suggestions. The author also like to thank Dr. R. Gluckstern for correcting some mistakes made in the earlier version of this paper; Drs. B. Zotter and D. Mohl for their comments.

References

[1] Dr. G. Rees private communication. An rf-cage has been implemented at ISIS.

[2] *The Physics and Plan for a 45 GeV Facility That Extends the High-Intensity Capability in Nuclear and Particle Physics*, Los Alamos National Laboratory Report, LA-10720-MS, May 1986.

[3] *Kaon Factory Study: Accelerator Design Report*, TRIUMF-Kaon Project, TRIUMF, 1990.

[4] *AUSTRON Feasibility Study*, Eds. P. Bryant, M. Regler, M. Schuster, Im Auftrage des Bundesministeriums für Wissenschaft und Forschung, Wein, Österreich, Nov. 1994.

[5] *Proposal for Japan Hadron Facility*, JHF Project Office, High Energy Accelerator Research Organization, KEK Report 97-3, JHF 97-1, May 1997.

[6] T. F. Wang, CERN Internal Report, "Electric Field of a Perturbed Beam with RF-Screening Wires", CERN/PS-94-08, April 1994.

[7] The addition theorem of Bessel functions can be found in almost any textbook or mathematical table discusses about Bessel functions. See, for example, Chapter 9 of *Handbook of Mathematical Functions*, by M. Abramowitz and I. A. Stegun, US National Bureau of Standards, 1964.

Tune Shift Due to Non-linear Space Charge Effect

Joanne Beebe-Wang

AGS Department, Brookhaven National Laboratory, Upton, NY 11973

Abstract. Betatron tune shift due to space charge effect is investigated by solving the equation of motion of particles including total space charge (linear and non-linear part). We introduce a binomial representation for 4-dimensional phase space distribution, which can characterize the important factors to tune shift while largely reducing the mathematical difficulties. With this representation, a numerically solvable equation of motion is obtained. From its solution, betatron tune is deduced by FFT (Fast Fourier Transformation). We derive the tune shift as function of beam current, amplitude of betatron oscillation and the particle distribution. We also present the differences between tune shifts calculated by including total space charge effect and including only its linear term.

I INTRODUCTION

The commonly used analytical formula for transverse tune shift due to space charge in a circular accelerator of circumference $2\pi R$ is [1]:

$$\Delta \nu_\xi = \nu_\xi - \nu_{\xi 0} = -\frac{K_\xi R^2}{2\nu_{\xi 0}} \quad (1)$$

where ν_ξ and $\nu_{\xi 0}$ are the betatron tunes with and without space charge, the subscript ξ representing either horizontal x or vertical y direction. It is obtained by solving equation of the motion [1]:

$$\xi'' + (\frac{\nu_{\xi 0}}{R})^2 \xi = \frac{F_\xi}{m\gamma\beta^2 c^2} \quad (2)$$

with an assumption that the space charge force F_ξ is small enough to be treated as a perturbation that is linear in ξ, therefore the right side of Eq.(2) is replaced by $K_\xi \xi$. However, this assumption becomes invalid for beams with high charge densities or unusual distributions, such as the one in SNS accumulator ring. Since the perturbation is not so small, taking only the linear term may no longer be a good approximation. Also, a linear perturbation can not describe the nature of a more complicated distribution, such as smoke ring. Furthermore, it is important for us to determine the tune shift produced by the particles with large betatron oscillation where ξ is not small, so non-linear terms are not small compared to the linear term.

In this study, the total space charge effect (linear and non-linear part) is included in the equation of motion, and the beams may have non-Gaussian (for example, smoke ring) distribution. In Section 2, we introduce a binomial representation for 6-dimensional phase space distribution and show how this representation can effectively characterize the majority types of distributions while largely reducing the mathematical difficulties of solving the differential equations. In Sections 3, we present how total charge in the beam, amplitude of betatron oscillation and the particle distribution affect the transverse tune shift. The conclusions of this study will be given the Section 4.

II PHYSICAL MODEL AND EQUATIONS OF MOTION

Transverse tunes are the characteristic of particle motion which is given by the solution of the equation of motion (2). For $F_\xi = 0$, the solution describes the particle motion without space charge. While for small F_ξ, the right side of Eq. (2) can be treated as perturbation, and when only linear teams of F_ξ is included it yields to tune shift given by Eq. (1). Generally Eq. (2) is difficult to solve since the space charge force $\vec{F}$ is a complicated integral:

$$\vec{F}(\vec{r}) = \frac{\mu_0 e^2 c^2 N}{4\pi\gamma^2} \int \rho(\vec{r}_s) \frac{\vec{r} - \vec{r}_s}{|\vec{r} - \vec{r}_s|^3} d\vec{r} \tag{3}$$

where $\mu_0 = 4\pi \times 10^{-7}$H/m is the permeability of free space, e is electron charge, c is speed of light, N is total number of particles, γ is Lorentz factor, and $\rho(\vec{r}_s)$ is particle density. The goal of this study is to create a simple and clean physical model which characterizes the factors important to the investigations of tune shift due to space charge effect and ignore the rest. With this model, the equation of motion should be analytically or numerically solvable. From the solution we derive the tune shift as function of beam current, amplitude of betatron oscillation and the particle distribution.

The physical model we give here is based on the conceptual design for the accumulator ring of the Spallation Neutron Source (SNS) [4]. With proper modification, this model may also by suitable for other synchrotron/storage rings. There are following major features in the model:

a. The external focusing force is uniform and linear to the transverse displacement, and longitudinally homogeneous in the ring. Since horizontal and vertical focusing forces are equal, the particles oscillate in x and y directions with the same betatron frequency. Therefore the horizontal tune ν_x and vertical tune ν_y without space charge are also equal. In this study we choose

TABLE 1. SNS Accumulator ring parameters

Beam Average Power	1.0 MW
Kinetic Energy	1.0 GeV
Average Current	1.0 mA
Repetition Rate	60 Hz
Ion Source Peak Current	35 mA
Linac Peak Current	27.7 mA
Beam Duty Cycle	6.18 %
Linac Pulse Length	1.03 msec
Beam Loss (Controlled/Uncontrolled)	< 4 / 0.02 %
Number of Turns Injected	1158
Revolution Period	0.8413 μsec
Revolution Frequency	1.1886 MHz
Circumference	220.688 m
Space-Charge Tune-Shift	< 0.1
Bunching Factor (Dual RF Systems)	0.405
Number of Protons/Ring	1.04×10^{14}
Beam Emittance (Transverse, norm.)	217 πmm-mr
Tunes ν_x/ν_y	5.82 / 5.80
β_{max}(x/y)	19.2 / 19.2 m
Dispersion X_p(max/min)	4.1/0.0 m
Transition Energy γ_t	4.93
RF Voltage (1st Harmonic)	40 kV
RF Voltage (2nd Harmonic)	20 kV
RF bucket	17 eV-sec
Beam Emittance (Longitudinal)	10 eV-sec
Beam Gap (injection)	295 nsec
Injected Pulse Length	546 nsec
Extracted Pulse Length	591 nsec
Vacuum	10^{-9} Torr

$\nu_x = \nu_y = \nu_0 = 5.81$ which is the average value of designed horizontal and vertical tunes of SNS accumulator ring. The SNS Accumulator ring parameters are listed in Table 1.

b. The beam have only one long bunch which covers about 2/3 of the circumference (see Table 1). Considering the bunch length $L_{Bunch} \gg$ beam size a, and the bunch has a very flat longitudinal distribution, for the particles not located at the end of the bunch, it is reasonable to assume that the particle density $\rho(x, y, z)$ does not change longitudinally in the bunch:

$$\rho(x, y, z, t) = \rho_2(x, y)\rho_z(z, t) \qquad (4)$$

and

$$\rho_z(z, t) = \lambda(t) = \text{constant to } z \qquad (5)$$

c. In the current design of Spallation Neutron Source (SNS), the particles are injected one pulse per turn into the accumulator ring during 1225 turns. The particle density increases linearly with time $t - t_0$ during the injection:

$$\lambda(t) = \lambda_{final} \frac{t - t_0}{t_f - t_0} \propto t - t_0 \tag{6}$$

where λ_{final} is the linear particle density at the end of injection, t is the time during the injection, t_0 and t_f are the time when the injection begins and finishes respectively.

d. Transverse particle distribution is cylindrically symmetric and the shape of the distribution does not change during injection. So, in cylindrical coordinates we can express the transverse particle density as:

$$\rho_2(x, y) = f(r) \tag{7}$$

where $f(r)$ is not a function of angle θ.

III BINOMIAL REPRESENTATION FOR 4-D PHASE SPACE DISTRIBUTION

In order to obtain the transverse tune shift due to space charge by solving the differential equation (2), it is essential to formalize F_ξ to the form that can be solved analytically or numerically. Here, we introduce a normalized 4-dimensional phase space distribution:

$$\rho_4(x, y, x', y') = \begin{cases} C_M (1 - \tilde{r}_4^2)^{M-1} & \text{for } 0 \leq \tilde{r}_4 \leq 1 \\ 0 & \text{for } \tilde{r}_4 > 1 \end{cases} \tag{8}$$

where M is a real number which may take any real value in the range of $[0, +\infty)$, C_M is normalization factor and

$$\tilde{r}_4^2 = \tilde{x}^2 + \tilde{y}^2 + \tilde{x}'^2 + \tilde{y}'^2 \tag{9}$$

here $\tilde{x}, \tilde{y}, \tilde{x}', \tilde{y}'$ are reduced phase space coordinates defined as:

$$\tilde{x} = \frac{x}{a}, \quad \tilde{y} = \frac{y}{b}, \quad \tilde{x}' = \frac{a p_x}{\varepsilon_x}, \quad \tilde{y}' = \frac{b p_y}{\varepsilon_y} \tag{10}$$

where a and b are horizontal and vertical beam envelopes, while ε_x and ε_y are horizontal and vertical beam emittances. In the simplest model, we have $a = b$ and $\varepsilon_x = \varepsilon_y$.

C_M is determined by normalization condition:

$$\int d\tilde{y}' \int d\tilde{x}' \int d\tilde{y} \int d\tilde{x} \rho_4(x, y, x', y') = 1 \qquad (11)$$

The integral is over a 4-dimensional sphere with boundary $\tilde{x}^2 + \tilde{y}^2 + \tilde{x}'^2 + \tilde{y}'^2 = 1$. The multiple integral in Eq.(11) yields:

$$C_M = \frac{1}{I_{2M-1} I_{2M} I_{2M+1} I_{2M+2}} = \frac{M(M+1)}{\pi^2} \qquad (12)$$

where I_n is the definite integral of $\cos^n \alpha$ over $[-\frac{\pi}{2}, \frac{\pi}{2}]$ and it takes form of [3]:

$$I_n = \int_{-\frac{\pi}{2}}^{\frac{\pi}{2}} \cos^n \alpha \, d\alpha = \begin{cases} \frac{\sqrt{\pi}\Gamma(\frac{n+1}{2})}{\Gamma(\frac{n}{2}+1)} & n > -1 \\ \frac{\pi(n-1)!!}{n!!} & n \text{ even} \\ \frac{2(n-1)!!}{n!!} & n \text{ odd} \end{cases} \qquad (13)$$

In order to observe the characteristics of the particle distributions expressed by Eq. (8), five different cases of transverse particle densities are illustrated in Figure 1. 4-dimensional phase space particle densities as functions of $\tilde{r}_4$ expressed by Eq.(8) with M=0, 0.5, 1, 1.5, 3 are presented by Fig. 1(A-E) respectively.

The normalized real space particle density can be obtained by integrating $\rho_4(x, y, x', y')$ as defined in Eq.(8) over x' and y' in the 4-dimensional sphere:

$$\rho_2(x, y) = \int_{-\sqrt{1-\tilde{x}^2-\tilde{y}^2-\tilde{x}'^2}}^{\sqrt{1-\tilde{x}^2-\tilde{y}^2-\tilde{x}'^2}} d\tilde{y}' \int_{-\sqrt{1-\tilde{x}^2-\tilde{y}^2}}^{\sqrt{1-\tilde{x}^2-\tilde{y}^2}} d\tilde{x}' \rho_4(x, y, x', y') \qquad (14)$$

It is convenient to express the real space particle density in cylindrical coordinates $(\theta, \tilde{r})$ because of it's cylindrical symmetry:

$$\rho_2(\tilde{r}) = \begin{cases} \frac{M+1}{\pi}(1-\tilde{r}^2)^M & \text{for } 0 \leq \tilde{r} \leq 1 \\ 0 & \text{for } \tilde{r} > 1 \end{cases} \qquad (15)$$

where $\tilde{r}^2 = \tilde{x}^2 + \tilde{y}^2$ or $r = a\tilde{r}$. Five different cases of transverse particle densities expressed by Eq. (15) are illustrated in Figure 1. Cross sections of normalized real space particle densities as functions of $\tilde{r}$ expressed by Eq.(15) in cylindrical coordinates with M=0, 0.5, 1, 1.5, 3 are presented by Fig. 1(F-J) respectively.

This distribution has following properties:

a. The particle distribution of a general case is characterized by a single parameter M. The binomial representation and M values for distributions of K-V, smoke ring, uniform, Gaussian and some other distributions are listed in Table 2.

b. The distribution has a finite range for a finite value of M, i.e. there are no particles outside a given limiting ellipse.

c. The real space particle densities for M 0 have continuous derivatives.

TABLE 2. Properties of binomial representations for particle density distributions

M	Distribution	4-D Phase Space Density $\rho_4(x,y,x',y')$ $\tilde{r}_4^2 = \tilde{x}^2 + \tilde{y}^2 + \tilde{x}'^2 + \tilde{y}'^2 \leq 1$	Real Space Density $\rho_2(x,y)$ $\tilde{r}^2 = \tilde{x}^2 + \tilde{y}^2 \leq 1$
0	K-V	$\frac{1}{\pi^2}\delta(1-\tilde{r}_4^2)$	$\frac{1}{\pi}$
0.5	smoke ring	$\frac{3}{4\pi^2}(1-\tilde{r}_4^2)^{-0.5}$	$\frac{3}{2\pi}(1-\tilde{r}^2)^{0.5}$
1	uniform	$\frac{2}{\pi^2}$	$\frac{2}{\pi}(1-\tilde{r}^2)$
1.5	elliptical	$\frac{15}{4\pi^2}(1-\tilde{r}_4^2)^{0.5}$	$\frac{5}{2\pi}(1-\tilde{r}^2)^{1.5}$
2	parabolic	$\frac{6}{\pi^2}(1-\tilde{r}_4^2)$	$\frac{3}{\pi}(1-\tilde{r}^2)^2$
3		$\frac{12}{\pi^2}(1-\tilde{r}_4^2)^2$	$\frac{4}{\pi}(1-\tilde{r}^2)^3$
4		$\frac{20}{\pi^2}(1-\tilde{r}_4^2)^3$	$\frac{5}{\pi}(1-\tilde{r}^2)^4$
5		$\frac{30}{\pi^2}(1-\tilde{r}_4^2)^4$	$\frac{6}{\pi}(1-\tilde{r}^2)^5$
⋮	General	$\frac{M(M+1)}{\pi^2}(1-\tilde{r}_4^2)^{M-1}$	$\frac{M+1}{\pi}(1-\tilde{r}^2)^M$
∞	Gaussian	$\frac{1}{4\pi^2}\exp(-\frac{\tilde{r}_4^2}{2})$	$\frac{1}{2\pi}\exp(-\frac{\tilde{r}^2}{2})$

d. Projection of a binomial distribution in N-dimensional space into the $(N-1)$-dimensional space increases the parameter M by 0.5. This means that the distribution remains binomial under projection.

To further simplify the mathematics, we consider the particle's canonical angular momentum p_θ is zero. The restriction $p_\theta = 0$ implies that the particles are launched from a magnetically shielded source (i.e., $B = 0$ at source) with $d\theta/dt = 0$. Finally, by substituting $\rho(x,y,z)$ in Eq. (2) with Eq. (15), we have the equation of motion which includes total space charge effect:

$$\tilde{r}'' + (\frac{\nu_0}{R})^2 \tilde{r} = \frac{2r_0}{\gamma^3 \beta^2 a^2} \frac{s\lambda_{final}}{2\pi R N_t} \frac{1}{\tilde{r}} [1 - (1-\tilde{r}^2)^{(M+1)}] \qquad 0 \leq \tilde{r} \leq 1. \qquad (16)$$

The left side of equation is the space charge effect. The factor $\frac{s\lambda_{final}}{2\pi R N_t}$ on the left side is the $\lambda(t)$ as expressed by Eq.(6). If $\tilde{r} \ll 1$, Eq.(16) can be written as a linear differential equation:

$$\tilde{r}'' + (\frac{\nu_0}{R})^2 \tilde{r} = \frac{2r_0}{\gamma^3 \beta^2 a^2} \frac{s\lambda_{final}}{2\pi R N_t} (M+1)\tilde{r} \qquad \text{for } \tilde{r} \ll 1. \qquad (17)$$

This equation of motion only include the linear part of space charge effect. It is valid for particles oscillate with small amplitude. In the following section we will present the tune shifts deduced from both Eq.(16) and Eq.(17) as a comparison of including total space charge effect and including only its linear term.

IV TRANSVERSE TUNE SHIFT

The motions of particles including total/linear space charge fields may be obtained by solving Eq.(16)/Eq.(17) with any given initial conditions. Generally, the solutions are sine-like oscillations accompanied with frequency reduction and amplitude growth due to the non-zero term of space charge forces. From its solution, betatron tune can be deduced by FFT (Fast Fourier Transformation). The tune shift is obtained by subtracting the tune without space charge from the calculated betatron tune. In order to compare results, the initial conditions are $\tilde{r}(s=0) = 0.6$ and $\tilde{r}'(s=0) = 0.0$ in all the cases presented in this study.

A Tune Shift vs the Increase of the Beam Current

In this section we investigate, for a given particle distribution, how tune shift increases while increasing charge density during the injection. Fig. 2 shows transverse particle motions and the tunes at five different stages of injection. Sub-plots (A-E) present the motion of particle obtained by solving Eq.(16) at the times 100, 300, 600, 900 and 1225 turns of injection are completed. The FFT of the particle motion at the corresponding times are shown in (F-J), and from which the tunes are found to be 5.81, 5,78, 5.75, 5.72, 5.68, respectively. Fig. 3 shows tune shifts due to linear and non-linear space charge effects as functions of number of turns during injection. Eight different cases with M=0, 0.5, 1, 1.5, 2, 3, 4, 5 are presented for comparison. 3(A) gives the tune shifts when only linear space charge effect is included. 3(B) gives the tune shifts total (linear and non-linear) space charge effect is included. In both figures, the cases of M=0, 0.5, 1, 1.5, 2, 3, 4, 5 are shown by the 8 curves from top to bottom respectively.

B Tune Shift vs the Amplitudes of Betatron Oscillations

In this section we investigate, for a given particle distribution, how large tune shifts are for particles with different betatron oscillation amplitude. Figure 4 shows transverse particle motions and the tunes of five different cases with amplitude of betatron oscillation $\tilde{r}_2$=5, 10, 15, 20, 25mm. Sub-plots 4(A-E) present the motion of particle obtained by solving Eq.(16) at the last turn of the injection. The FFT of the particle motion for the corresponding amplitude of betatron oscillation are shown in 4(F-J), and from which the tunes are found to be 5.65, 5,66, 5.68, 5.71, 5.73, respectively. Figure 5 shows tune shifts due to linear and non-linear space charge effects as functions of amplitude of betatron oscillation. Eight different cases with M=0, 0.5, 1, 1.5, 2, 3, 4, 5 are presented for comparison. 5(A) gives the tune shifts when only linear space charge effect is included. While 5(B) gives the tune shifts when total (linear and non-linear) space charge effect is included. In both figures, the cases of M=0, 0.5, 1, 1.5, 2, 3, 4, 5 are shown by the 8 curves from top to bottom respectively.

C Tune Shift vs Different Particle Distributions

In this section we investigate, for given betatron oscillation amplitudes at given stages of the injection, how large the tune shifts would be for different particle distributions. Figure 6 demonstrates transverse particle motions and the tunes of five different cases with M=0, 0.5, 1, 1.5, 3. Sub-plots 6(A-E) present the motion of particle obtained by solving Eq.(16) at the last turn of the injection. The FFT of the particle motion in the corresponding space charge fields are shown in 6(F-J), and from which the tunes are found to be 5.76, 5,73, 5.72, 5.70, 5.66, respectively. In figure 7 we present the tune shifts due to linear and non-linear space charge effects as functions of transverse particle distribution characterized by M parameters. Seven different cases with amplitude of betatron oscillation r_0=0.25, 5, 10, 15, 20, 22.5, 25mm are presented for comparison. 7(A) shows the tune shifts when only linear space charge effect is included. All the six curves are merged to one since tune is not r_0 dependent in a linear approximation. 7(B) gives the tune shifts when total (linear and non-linear) space charge effect is included. The curves from bottom to top give the cases of r_0=0.25, 5, 10, 15, 20, 22.5, 25mm respectively. In figure 8 we present tune shifts due to linear and non-linear space charge effects as functions of transverse particle distribution characterized by M parameters. functions of number of turns during injection. Seven different cases at 100, 300, 500, 700, 900, 1100, 1225 turns during injection are presented for comparison. 8(A) shows the tune shifts when only linear space charge effect is included. 7(B) gives the tune shifts when total (linear and non-linear) space charge effect is included. In both figures, the 7 curves from top to bottom represent 7 the cases of different cases at 100, 300, 500, 700, 900, 1100, 1225 turns during injection respectively.

V DISCUSSIONS AND CONCLUSIONS

The results of this study give us a clear over-all insight of tune shift due to linear and non-linear space charge effect. The quantitative relationship between the tune shift and beam current, amplitude of betatron oscillations and the particle distributions deduced from this idealized error-free model may be used as guide lines to check our computer simulation results. The methods developed here may also be implemented in a new computer code to be developed in the future.

REFERENCES

1. A. W. Chao, "Physics of Collective Beam Instabilities in High Energy Accelerators", John Wiley &ons, Inc. 1993.
2. W. Joho, "Representation of Beam Ellipses for Transport Calculations", SIN-REPORT TM-11-14, 8.5.1980.

3. I.S. Gradshteyn and I.M. Ryzhik, "Table of Integrals, Series, and Products" Academic Press, Inc.,1980.
4. Weng, W.T. et al., "Accumulator Ring Design for the NSNS Project", Proc. PAC97, Vancouver, Canada, May 12-16, 1997.

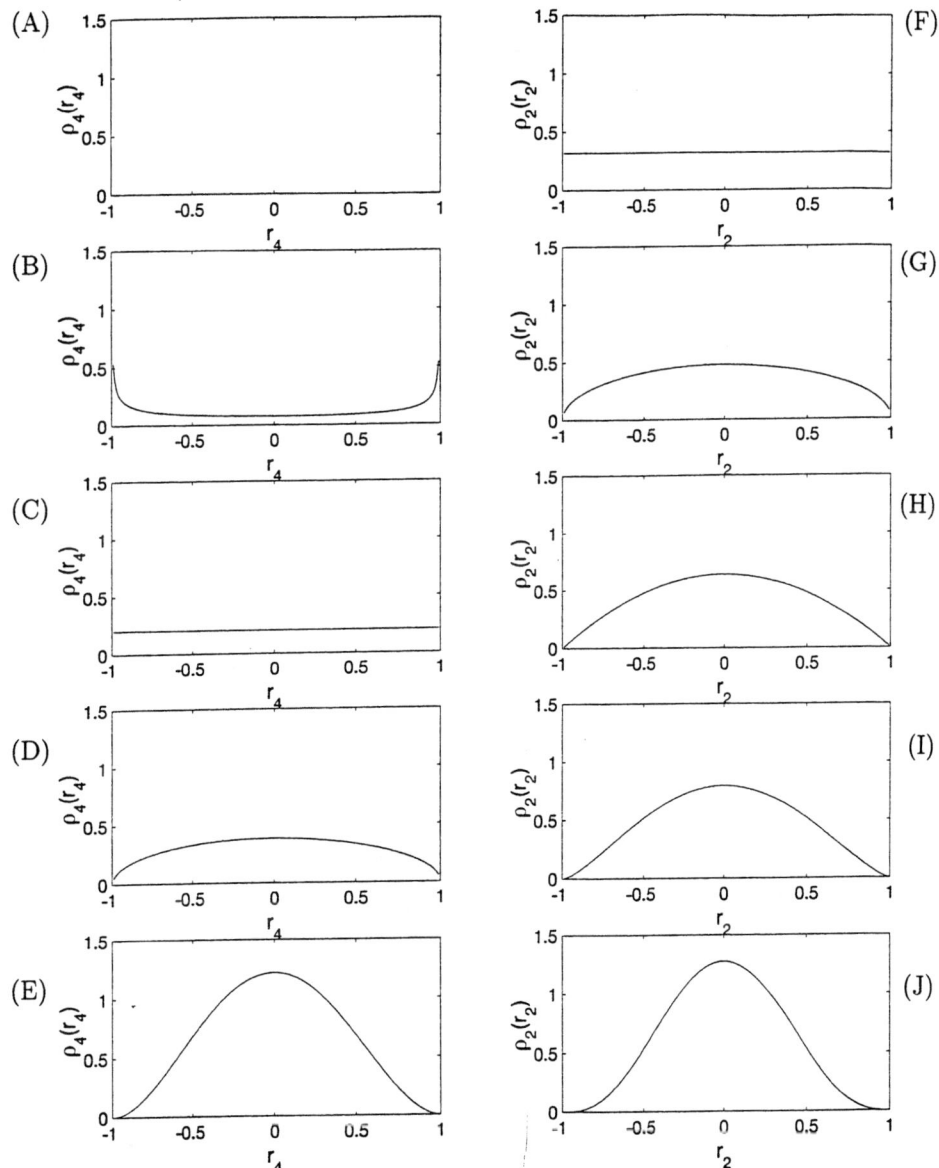

FIGURE 1. Five different cases of 4-D phase space particle densities and 2-D transverse real space particle densities. Sub-plots (**A-E**) present normalized 4-D phase space particle densities as functions of $\tilde{r}_4$ expressed by Eq.(8) with M=0, 0.5, 1, 1.5, 3, respectively. Correspondingly, with the same M values, normalized 2-D transverse real space particle densities as functions of $\tilde{r}_2$ expressed by Eq.(14) are plotted in (**F-J**) respectively.

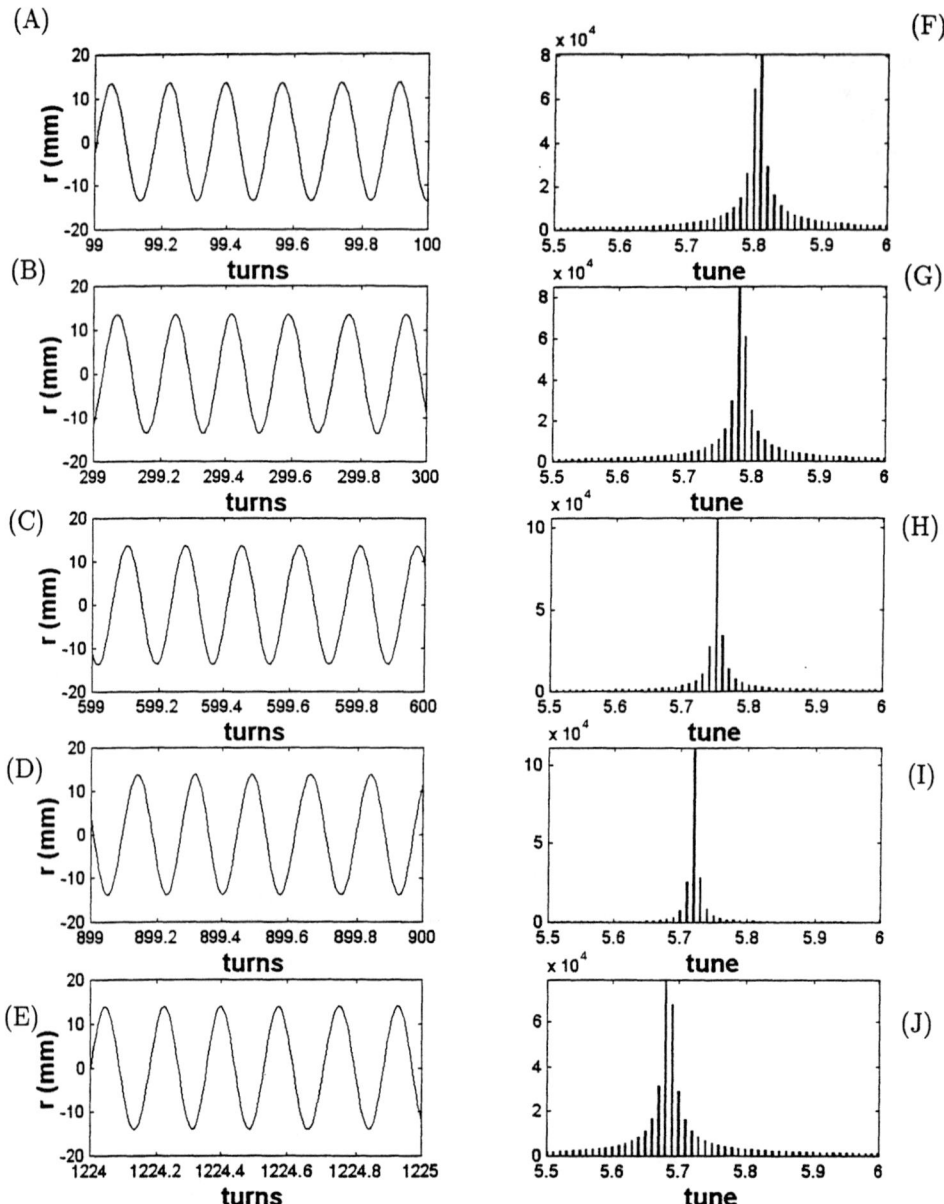

FIGURE 2. Transverse particle motions and the tunes at five different stages of injection. Sub-plots (**A-E**) present the motion of particle obtained by solving Eq.(16) at the times 100, 300, 600, 900 and 1225 turns of injection are completed. The FFT of the particle motion at the corresponding times are shown in (**F-J**), and from which the tunes are found to be 5.81, 5,78, 5.75, 5.72, 5.68, respectively.

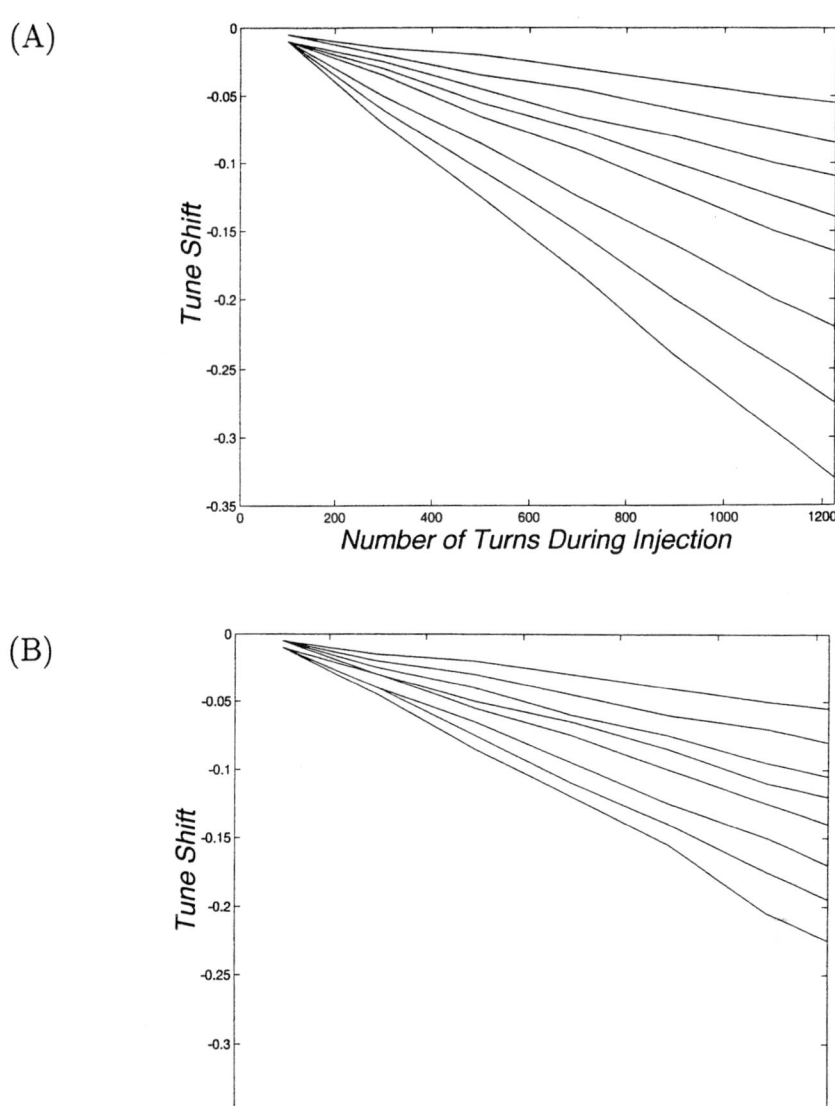

FIGURE 3. Tune shifts due to linear and non-linear space charge effects as functions of number of turns during injection. Eight different cases with M=0, 0.5, 1, 1.5, 2, 3, 4, 5 are presented for comparison. **(A)** only linear space charge effect is included. **(B)** total (linear and non-linear) space charge effect is included. In both figures, the cases of M=0, 0.5, 1, 1.5, 2, 3, 4, 5 are shown by the 8 curves from top to bottom respectively.

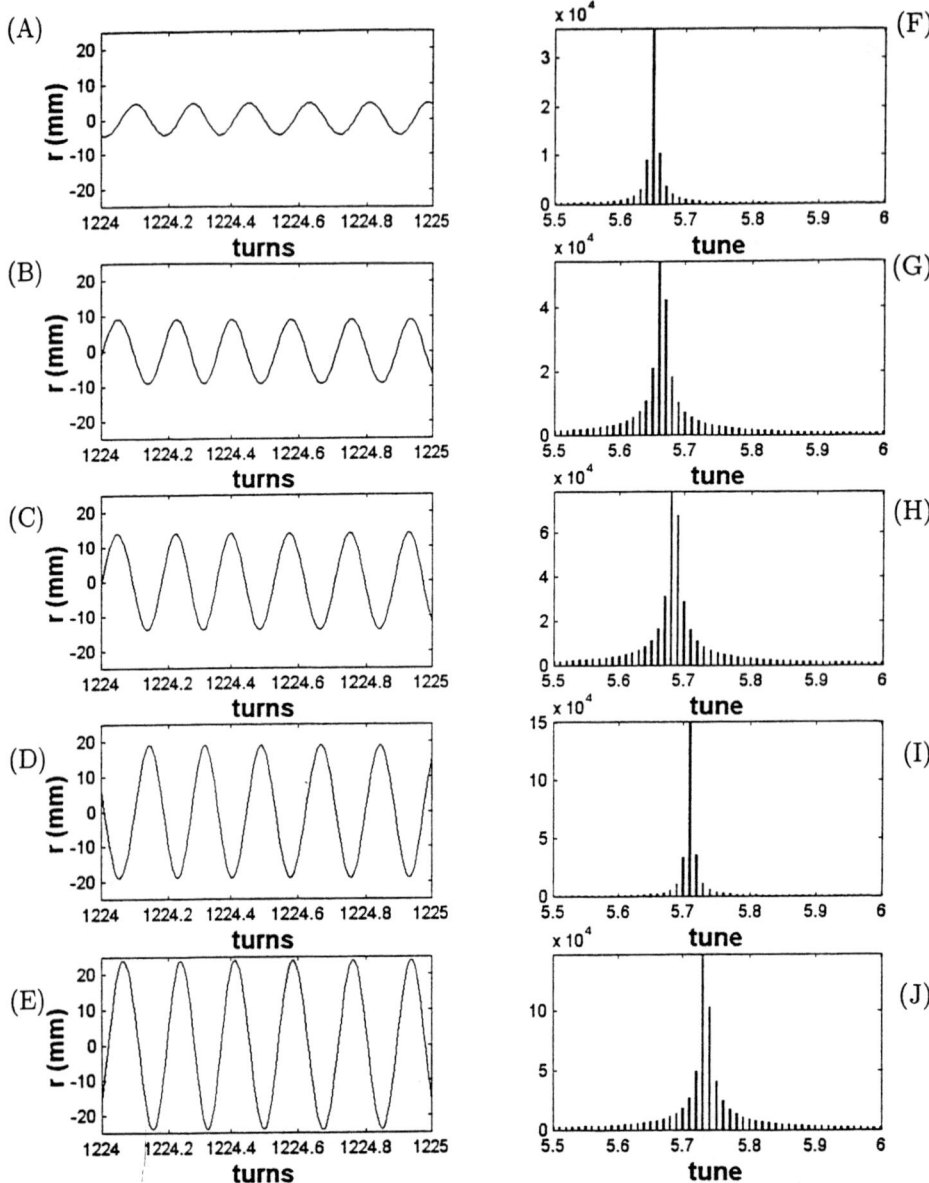

FIGURE 4. Transverse particle motions and the tunes of five different cases with amplitude of betatron oscillation $\tilde{r}_2$=5, 10, 15, 20, 25mm. Sub-plots (**A-E**) present the motion of particle obtained by solving Eq.(16) at the last turn of the injection. The FFT of the particle motion for the corresponding amplitude of betatron oscillation are shown in (**F-J**), and from which the tunes are found to be 5.65, 5,66, 5.68, 5.71, 5.73, respectively.

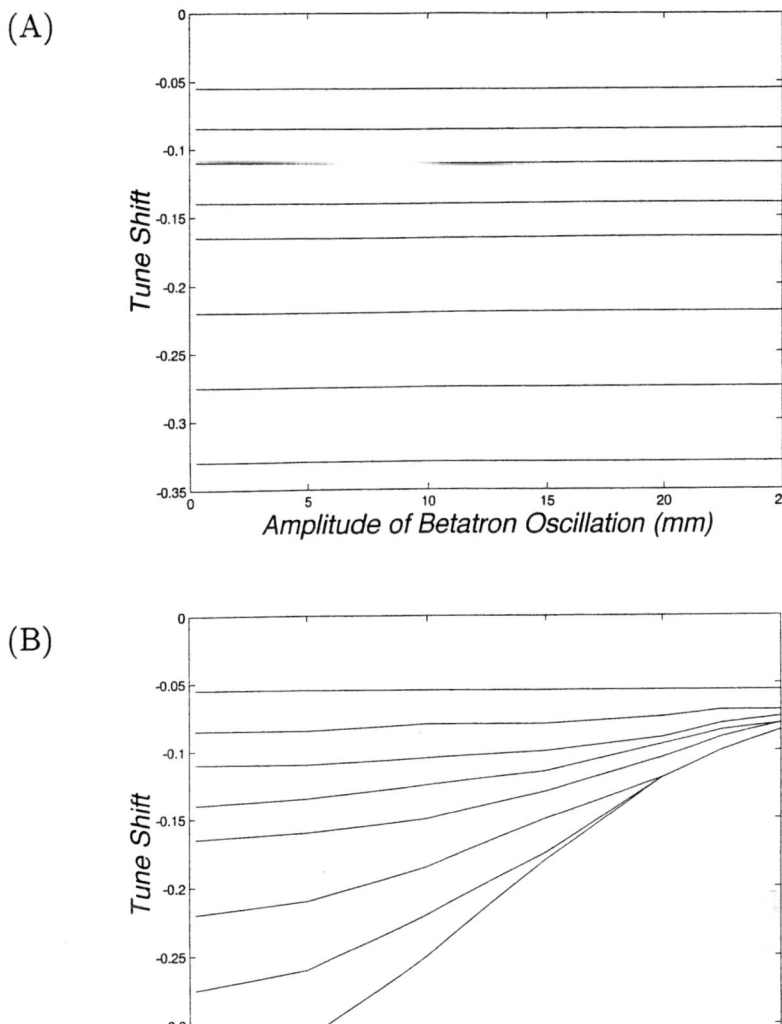

FIGURE 5. Tune shifts due to linear and non-linear space charge effects as functions of amplitude of betatron oscillation. Eight different cases with $M=0$, 0.5, 1, 1.5, 2, 3, 4, 5 are presented for comparison. **(A)** only linear space charge effect is included. **(B)** total (linear and non-linear) space charge effect is included. In both figures, the cases of $M=0$, 0.5, 1, 1.5, 2, 3, 4, 5 are shown by the 8 curves from top to bottom respectively.

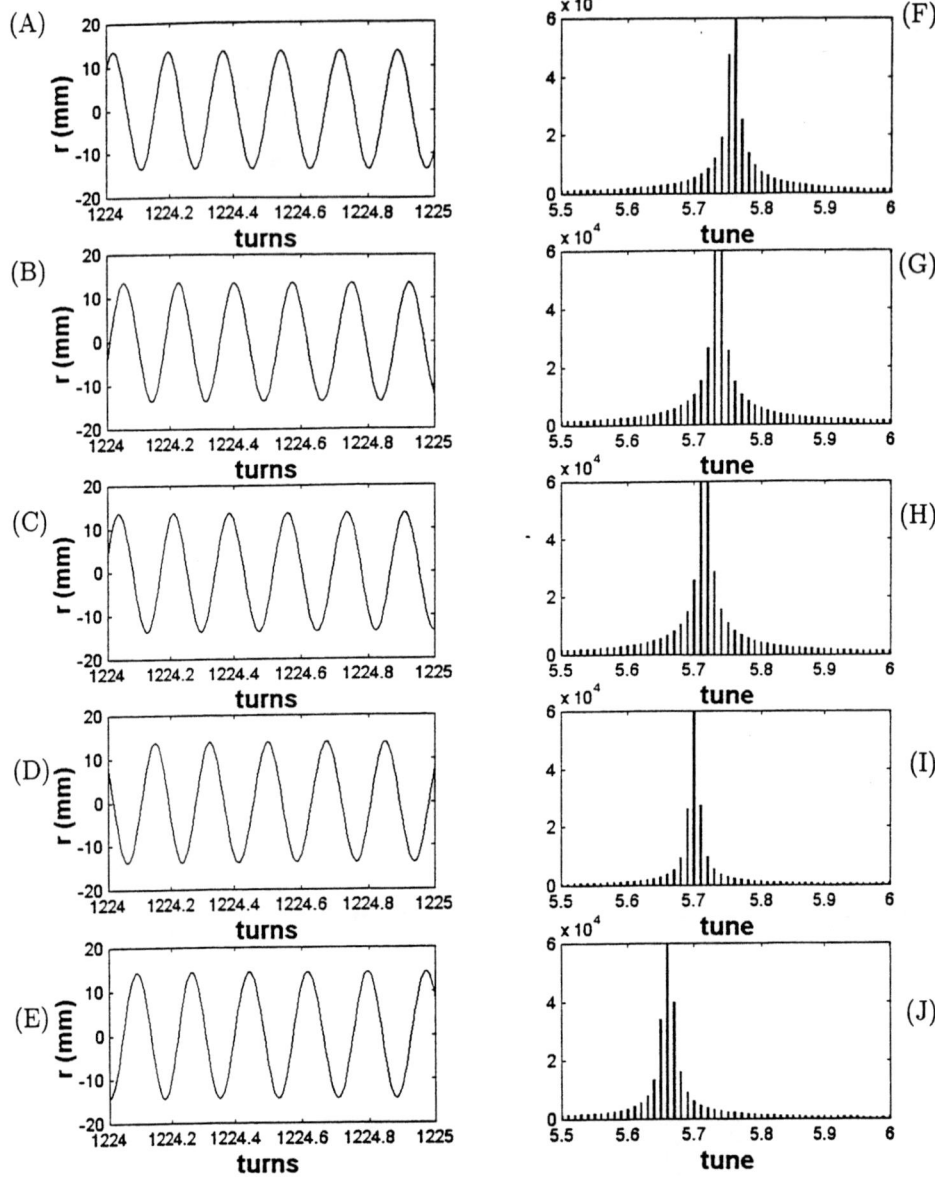

FIGURE 6. Transverse particle motions and the tunes of five different cases with M=0, 0.5, 1, 1.5, 3. Sub-plots (**A-E**) present the motion of particle obtained by solving Eq.(16) at the last turn of the injection. The FFT of the particle motion in the corresponding space charge fields are shown in (**F-J**), and from which the tunes are found to be 5.76, 5,73, 5.72, 5.70, 5.66, respectively.

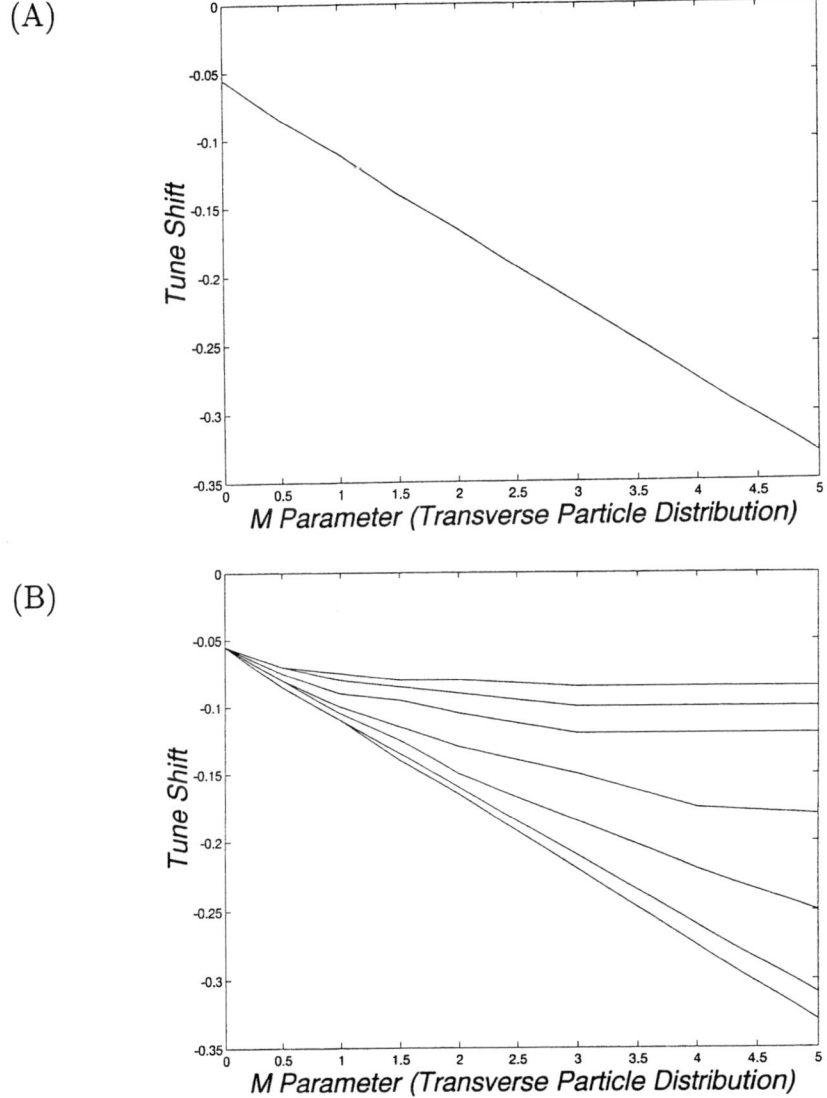

FIGURE 7. Tune shifts due to linear and non-linear space charge effects as functions of transverse particle distribution characterized by M parameters. Seven different cases with amplitude of betatron oscillation r_0=0.25, 5, 10, 15, 20, 22.5, 25mm are presented for comparison. **(A)** only linear space charge effect is included. All the six curves are merged to one since tune is not r_0 dependent in a linear approximation. **(B)** total (linear and non-linear) space charge effect is included. the curves from bottom to top give the cases of r_0=0.25, 5, 10, 15, 20, 22.5, 25mm respectively.

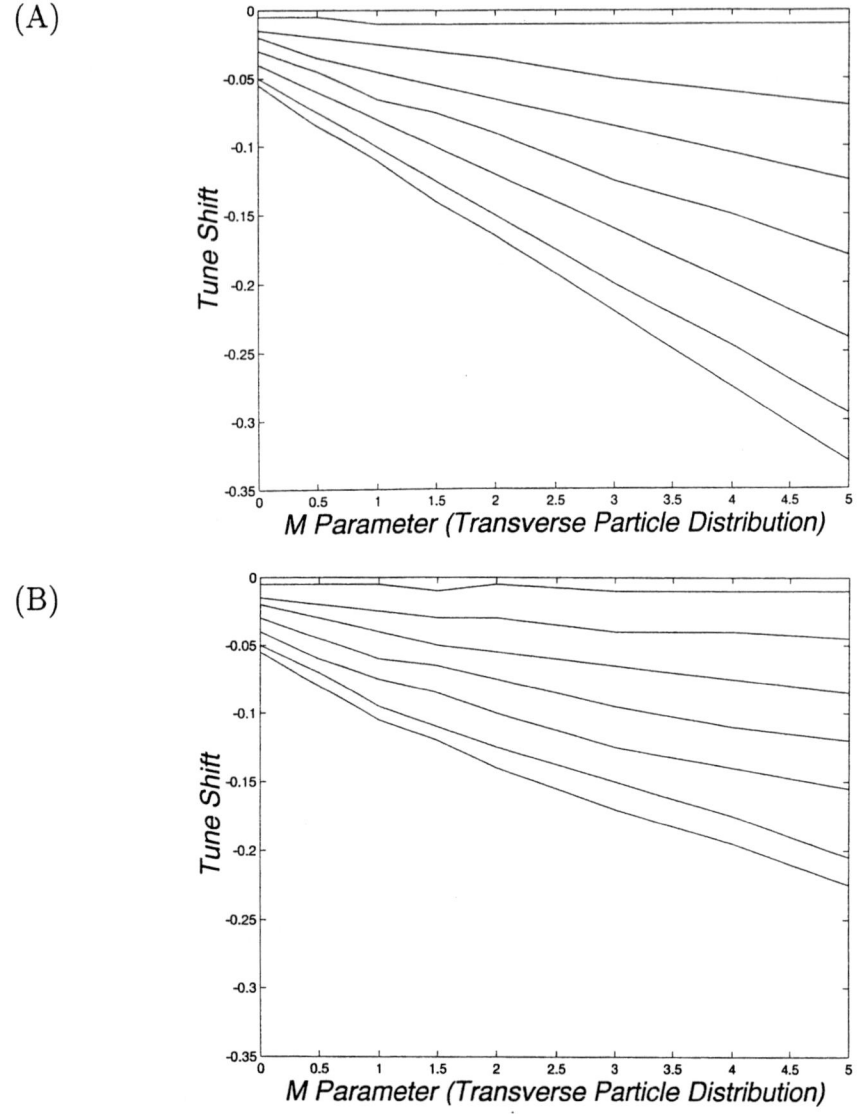

FIGURE 8. Tune shifts due to linear and non-linear space charge effects as functions of transverse particle distribution characterized by M parameters. functions of number of turns during injection. Seven different cases at 100, 300, 500, 700, 900, 1100, 1225 turns during injection are presented for comparison. **(A)** only linear space charge effect is included. **(B)** total (linear and non-linear) space charge effect is included. In both figures, the 7 curves from top to bottom represent 7 the cases of different cases at 100, 300, 500, 700, 900, 1100, 1225 turns during injection respectively.

CONTRIBUTED PAPERS: SIMULATION

Analytical and Numerical Treatment of Halo-Free Beam Transport

Yuri K. Batygin

The Institute of Physical and Chemical Research (RIKEN), Saitama, 351-01, Japan

Abstract. High brightness beam transport in a non-linear focusing channel, while avoiding beam emittance growth and halo formation, is considered. Invariance of beam distribution function is treated as a problem of proper matching of the beam with focusing channel. Matched conditions for intense beam are obtained from the stationary Vlasov's equation for beam distribution function and Poisson's equation for electrostatic beam potential. Two approaches to self-consistent problem for matched beam distribution are treated: (i) focusing field is determined by given beam distribution, and (ii) matched beam distribution is defined by focusing potential of the structure. Attained solutions provide theoretical basis for choosing parameters of the space charge dominated beam transport with suppressed halo.

MATCHED BEAM DISTRIBUTION

Distortion of beam emittance in particle accelerators results in serious limitations in achieved value of phase space density of the beam and particle losses. Prevention of emittance growth and halo formation in high intensity beams is a key problem for next generation of particle accelerators. Importance of the problem is connected with the development of particle accelerators for heavy ion fusion, spallation neutron sources, radioactive waste transmutation and other applications. In most of the cases the beam emittance growth have to be suppressed to provide small beam losses in accelerator or to produce small beam size at the target.

Conventional focusing structures employ quadrupole lenses and axial - symmetric lenses (both electrostatic and magnetostatic) with linear focusing field. Kapchinsky-Vladimirsky (KV) distribution is a solution of the self-consistent problem for the intense beam in transport channel with linear focusing field. In KV beam particles are uniformly distributed on the surface of 4-dimensional hypervolume which rotates in phase space during beam transport. Realistic beam distribution is far from the KV distribution. In most of the realistic beams particle distribution is close to truncated Gaussian distribution. Transportation of realistic beam in linear focusing channel results in strong emittance growth due to beam mismatch with the channel. In Fig. 1 results of particle-in-cell simulation of high brightness beam transport utilizing code BEAMPATH are presented. Intrinsic mismatch of non-uniform beam profile with linear focusing channel is a reason of appearance of halo around the beam core. Let us define beam core as an ellipse

$$\left(\frac{x}{2\sqrt{<x^2>}}\right)^2 + \left(\frac{y}{2\sqrt{<y^2>}}\right)^2 = 1 , \qquad (1)$$

where $2\sqrt{<x^2>}$ and $2\sqrt{<y^2>}$ are x - and y - beam envelopes, respectively. Particles outside ellipse, Eq. (1) will be considered as particles generating beam halo. In Fig. 2 fraction of particles in beam halo as a function of longitudinal position, z, for beam transport of Fig. 1 is presented. As seen, number of particles in beam halo oscillates around beam core.

CP448, *Workshop on Space Charge Physics in High Intensity Hadron Rings*
edited by A. U. Luccio and W. T. Weng
© 1998 The American Institute of Physics 1-56396-824-X/98/$15.00

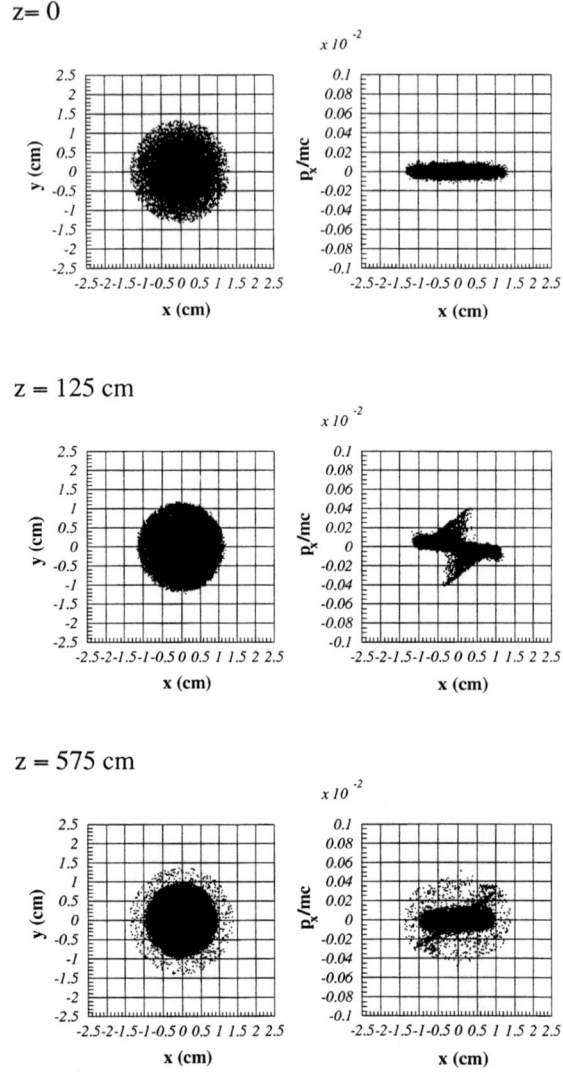

FIGURE 1. Emittance growth and halo formation of 150 keV, 0.1 A, 0.06 π cm mrad proton beam with truncated at $2.6\sqrt{<x^2>}$ Gaussian distribution in linear focusing channel.

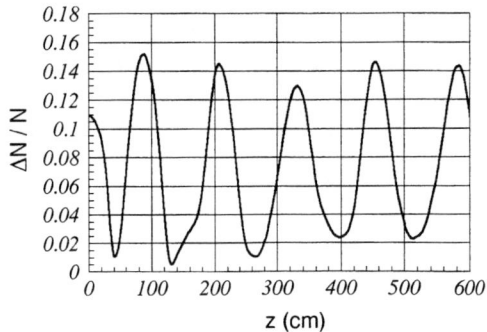

FIGURE 2. Fraction of particles in beam halo for transport of Gaussian beam in linear focusing channel, presented in Fig. 1.

To prevent emittance growth and halo formation, beam has to be matched with the channel. Beam distribution function, $f(x, p_x, y, p_y)$, expressed as a function of Hamiltonian, H, is a constant of motion in a z - uniform, time-independent focusing channel

$$f = f(H), \quad H = \frac{p_x^2 + p_y^2}{2m\gamma} + q U_{ext} + q \frac{U_b}{\gamma^2}, \quad (2)$$

where p_x and p_y are components of transverse particle momentum, q and m are charge and mass of the particles, respectively, γ is a particle energy, U_{ext} is a potential of focusing field, and U_b is a space charge potential of the beam. Matched beam distribution function, Eq. (2), obeys self-consistent set of Vlasov-Poisson equations:

$$\begin{cases} \frac{1}{m\gamma}(\frac{\partial f}{\partial x} p_x + \frac{\partial f}{\partial y} p_y) - q(\frac{\partial f}{\partial p_x}\frac{\partial U}{\partial x} + \frac{\partial f}{\partial p_y}\frac{\partial U}{\partial y}) = 0 \\ \frac{1}{r}\frac{\partial}{\partial r}(r\frac{\partial U_b}{\partial r}) = -\frac{q}{\varepsilon_0}\int_{-\infty}^{\infty}\int_{-\infty}^{\infty} f(H) \, dp_x \, dp_y \end{cases}, \quad (3)$$

where $U = U_{ext} + U_b \gamma^{-2}$ is a total potential of the structure. The most interesting are two formulations of the self-consistent problem.

1. Assume, the beam distribution function is known. The problem is to find the required focusing potential, which maintains this distribution in the channel:

$$f(x, p_x, y, p_y) \rightarrow U_{ext}(x,y) \,. \quad (4)$$

2. Assume, the potential of the focusing structure is given. The problem is to find the beam distribution function, matched with this focusing structure:

$$U_{ext}(x,y) \rightarrow f(x, p_x, y, p_y) \,. \quad (5)$$

First formulation of the problem is connected with search for focusing field to keep in equilibrium the beam with known distribution, while the second problem is to find beam equilibrium in given focusing structure. Below we consider both problems.

BEAM EQUILIBRIUM IN CONTINUOUS FOCUSING CHANNEL

General method to solve the first problem [1] is to substitute the given beam distribution function into Vlasov's equation and to solve it for unknown total potential of the structure, U. Required focusing field is then found as a difference between the total potential and known space charge potential of the beam

$$U_{ext} = U - U_b \gamma^{-2} . \tag{6}$$

The same relationship is valid for electrical field

$$E_{ext}(r) = E(r) - E_b(r) \gamma^{-2} . \tag{7}$$

Consider the following class of beam distribution functions

$$f = f(T), \quad T = \frac{p_x^2 + p_y^2}{p_0^2} + G(x, y) . \tag{8}$$

Calculation of derivatives of the distribution function give:

$$\frac{\partial f}{\partial x} = \frac{\partial f}{\partial T} \frac{\partial G}{\partial x}, \quad \frac{\partial f}{\partial y} = \frac{\partial f}{\partial T} \frac{\partial G}{\partial y},$$

$$\frac{\partial f}{\partial p_x} = 2 \frac{\partial f}{\partial T} \frac{p_x}{p_0^2}, \quad \frac{\partial f}{\partial p_y} = 2 \frac{\partial f}{\partial T} \frac{p_y}{p_0^2} . \tag{9}$$

Substitution of derivatives, Eq. (9), into Vlasov's equation, provides relationship for unknown total potential of the structure:

$$\frac{1}{m\gamma} (\frac{\partial G}{\partial x} p_x + \frac{\partial G}{\partial y} p_y) - 2q (\frac{p_x}{p_0^2} \frac{\partial U}{\partial x} + \frac{p_y}{p_0^2} \frac{\partial U}{\partial y}) = 0 . \tag{10}$$

Particular solution of the equation (10) can be obtained, if components of Eq. (10) are equal termwise:

$$\begin{cases} \frac{\partial U}{\partial x} = \frac{p_0^2}{2qm\gamma} \frac{\partial G}{\partial x} \\ \frac{\partial U}{\partial y} = \frac{p_0^2}{2qm\gamma} \frac{\partial G}{\partial y} \end{cases} . \tag{11}$$

In this case total potential is proportional to function G(x,y):

$$U(x,y) = \frac{p_0^2}{2qm\gamma} G(x,y) . \tag{12}$$

Therefore, given beam distribution function uniquely defines required total potential of the structure.

Beam Equilibrium with Elliptical Symmetry in Phase Space

Let the function G(x,y) in beam distribution, Eq. (8), to be a quadratic function of radius, r:

$$G = \frac{x^2 + y^2}{R^2}, \quad r^2 = x^2 + y^2 \ . \tag{13}$$

Equations (8) and (13) define the class of distribution functions with elliptical symmetry in phase space. Particle trajectories of the equilibrium beam are given by family of ellipses:

$$\frac{x^2}{R^2} + \frac{p_x^2}{p_o^2} = \text{const} \ . \tag{14}$$

Let us introduce normalized beam emittance

$$\varepsilon = \frac{R \, p_o}{mc} \ . \tag{15}$$

Total potential of the structure for this class of distributions from Eq. (12) is also quadratic function of radius

$$U = \frac{mc^2}{q} \frac{1}{\gamma} \left(\frac{\varepsilon}{R}\right)^2 \left(\frac{x^2 + y^2}{2R^2}\right) \ . \tag{16}$$

Therefore, total field has to be linear function of radius:

$$E = -\frac{mc^2}{qR} \frac{1}{\gamma} \left(\frac{\varepsilon}{R}\right)^2 \left(\frac{r}{R}\right) \ . \tag{17}$$

This result has a simple explanation. Particles rotate in total linear field along elliptical phase space trajectories, Eq. (14), and beam distribution with elliptical symmetry is, therefore, conserved.

Equilibrium of the KV beam

One of the distributions of this type is the KV distribution

$$f = f_o \, \delta \left(\frac{p_x^2 + p_y^2}{p_o^2} + \frac{x^2 + y^2}{R^2} - T_o \right) \ . \tag{18}$$

Distribution, Eq. (18), is projected at configuration space (x,y) as uniformly populated beam of radius R with space charge potential

$$U_b = -\frac{I}{4\pi\varepsilon_o \beta c} \frac{(x^2 + y^2)}{R^2} \ , \tag{19}$$

where I is a beam current and β is a longitudinal particle velocity. Required focusing potential from Eq. (6) gives the following equilibrium condition for KV beam:

$$U_{ext} = \frac{mc^2}{q} \frac{1}{\gamma} \left(\frac{r}{R}\right)^2 \left[\frac{1}{2} \left(\frac{\varepsilon}{R}\right)^2 + \frac{I}{\beta\gamma I_c}\right], \text{ or} \quad (20)$$

$$E_{ext} = -\frac{mc^2}{qR} \frac{1}{\gamma} \left(\frac{r}{R}\right) \left[\left(\frac{\varepsilon}{R}\right)^2 + 2\frac{I}{\beta\gamma I_c}\right], \quad (21)$$

where $I_c = 4\pi\varepsilon_0 mc^3/q = 3.13 \cdot 10^7 (A/Z)$ Amp is characteristic value of beam current. Let us note, that $I_A = \beta\gamma I_c$ is the Alfven current.

The same relationships, Eqs. (20) and (21), follow directly from KV envelope equation for the axial-symmetric beam in focusing channel with applied potential $U_{ext} = G_f (r^2/2)$

$$\frac{d^2R}{dz^2} + \mu_o^2 R - \frac{\varepsilon^2}{(\beta\gamma)^2 R^3} - \frac{2I}{R (\beta\gamma)^3 I_c} = 0, \quad (22)$$

where μ_o is a frequency of particle oscillation in a linear focusing channel without space charge forces

$$\mu_o^2 = \frac{q\, G_f}{m\gamma\beta^2 c^2} = \frac{q}{m\gamma\beta^2 c^2} \frac{1}{r} \frac{dU_{ext}}{dr}. \quad (23)$$

If the KV beam is in equilibrium with external focusing field, beam radius R remains constant and the envelope equation, Eq. (22), is reduced to

$$\frac{q}{m\gamma\beta^2 c^2} \frac{R}{r} \frac{dU_{ext}}{dr} - \frac{\varepsilon^2}{(\beta\gamma)^2 R^3} - \frac{2I}{R (\beta\gamma)^3 I_c} = 0. \quad (24)$$

Equation (24) is the same equilibrium condition for KV beam, as Eq. (20).

Equation (20) consists of two terms: one is proportional to square of beam emittance, ε^2, and another one is proportional to beam current, I. Ratio of two terms defines significance of beam current on beam dynamics with respect to beam emittance:

$$b = \frac{2}{\beta\gamma} \frac{I}{\varepsilon^2} \frac{R^2}{I_c}. \quad (25)$$

Dimensionless parameter b is a ratio of beam brightness, I/ε^2, and normalized value of I_c / R^2, therefore, it has a meaning of dimensionless beam brightness. Regime of b>>1 corresponds to high-brightness, space-charge-dominated beam transport. Another case of b<<1 corresponds to emittance dominated beam transport.

Let us rewrite Eq. (20) as follow

$$U_{ext} = \frac{mc^2}{q} \frac{1}{\gamma} \left(\frac{r}{R}\right)^2 \left(\frac{I}{\beta\gamma I_c}\right) \left(1 + \frac{1}{b}\right). \quad (26)$$

In case of high brightness beam transport the term 1/b is small as compare with unity and required focusing potential is mainly defined by the value of beam current and almost independent on beam emittance.

Equilibrium of the Gaussian Beam

Let us check, how equilibrium condition is changed for non-uniform beam distribution with elliptical symmetry [1]. Consider beam with Gaussian distribution function

$$f = f_o \exp\left(-2\frac{p_x^2 + p_y^2}{p_o^2} - 2\frac{x^2 + y^2}{R^2}\right). \qquad (27)$$

Space charge field of the beam with Gaussian distribution function is given by

$$E_b = \frac{I}{2\pi \varepsilon_o \beta c} \frac{1}{r}\left[1 - \exp\left(-2\frac{r^2}{R^2}\right)\right]. \qquad (28)$$

Equilibrium condition for Gaussian beam is a combination of total field, Eq. (17) and space charge field, Eq. (28):

$$E_{ext} = -\frac{mc^2}{q R}\frac{1}{\gamma}\frac{r}{R}\left[\frac{\varepsilon^2}{R^2} + 4\frac{I}{\beta\gamma I_c}\frac{(1 - \exp(-2r^2/R^2))}{2(r^2/R^2)}\right]. \qquad (29)$$

Condition to maintain Gaussian beam, Eq. (29), is different from that of KV beam, Eq. (21). For non-uniform beam external focusing field has to be nonlinear function of radius.

Beam Equilibrium without Elliptical Symmetry in Phase Space

Consider another class of particle distribution, with function $G(x, y)$ in Eq. (8) as

$$G = \frac{(x^2 + y^2)^2}{2R_o^4}. \qquad (30)$$

Example of this kind of beam distribution is

$$f = f_o \exp\left(-2\frac{p_x^2 + p_y^2}{p_o^2} - \frac{(x^2 + y^2)^2}{R_o^4}\right). \qquad (31)$$

Phase space trajectories of the equilibrium beam in this case are not ellipses anymore. According to Eq. (12), total potential of the structure for this class of particle distributions is

$$U(x,y) = \frac{p_o^2}{qm\gamma}\frac{(x^2 + y^2)^2}{4 R_o^4}. \qquad (32)$$

Total field of the structure is

$$E = -\frac{1}{\gamma}\frac{mc^2}{q R_o}\left(\frac{\varepsilon^*}{R_o}\right)^2\left(\frac{r^3}{R_o^3}\right), \qquad (33)$$

where $\varepsilon^* = R_o p_o/(mc)$ is an effective normalized beam emittance. In contrast with distribution functions with elliptical symmetry, total field, Eq. (33) is not a linear function of radius, but is an essentially nonlinear function $\sim r^3$.

Space charge density of the beam is attained after integration of beam distribution function over particle momentum:

$$\rho(r) = \int_{-\infty}^{\infty} \int_{-\infty}^{\infty} f_o \exp\left(-2\frac{p_x^2 + p_y^2}{p_o^2} - \frac{(x^2+y^2)^2}{R_o^4}\right) dp_x\, dp_y = \rho_o \exp\left(-\frac{r^4}{R_o^4}\right). \quad (34)$$

Space charge field of the beam is obtained from Poisson's equation

$$E_b = \frac{1}{\varepsilon_o r} \int_o^r \rho(r')\, r'\, dr' = \frac{I}{2\pi\varepsilon_o \beta c\, r} \operatorname{erf}\left(\frac{r^2}{R_o^2}\right). \quad (35)$$

Combination of total field, Eq. (33), and space charge field, Eq. (35), gives the expression for the required focusing field of the structure to maintain beam distribution:

$$E_{ext} = -\frac{1}{\gamma}\frac{mc^2}{q R_o}\left[\left(\frac{\varepsilon^*}{R_o}\right)^2 \left(\frac{r^3}{R_o^3}\right) + \frac{2I}{I_c \beta \gamma}\left(\frac{R_o}{r}\right) \operatorname{erf}\left(\frac{r^2}{R_o^2}\right)\right]. \quad (36)$$

In Fig. 3 results of numerical study of high-brightness beam transport with distribution function, Eq.(31), in focusing field, Eq. (36), are presented. As seen, beam remains in equilibrium in contrast with mismatched beam, presented in Fig. 1.

Important point is stability of beam equilibrium in nonlinear focusing field. Sufficient condition for stability is given by Newcomb-Gardner theorem [2], which states, that monotonically decreasing equilibrium distribution function of Hamiltonian $\partial f/\partial H < 0$ is stable with respect to perturbations. Distributions, Eqs. (27), (31) as well as most of realistic beam distributions satisfy stability condition.

SELF-CONSISTENT BEAM DISTRIBUTION IN CONTINUOUS FOCUSING CHANNEL

Inverse self-consistent problem is to find unknown beam distribution function via given focusing potential of the structure. In continuous channel with linear focusing field all self-consistent solutions tend to uniformly populated beam in the limit of high brightness beam [3]. In Ref. [4] this result was generalized for the case of an arbitrary applied focusing field. It was found that the self-consistent stationary particle distribution has such a shape that the space charge beam potential U_b is opposite to the arbitrary external focusing potential U_{ext}:

$$U_b = -\gamma^2 \frac{kb}{1+kb} U_{ext}, \quad (37)$$

where k is a beam profile parameter (k = 1 for uniform beam and k = 2 for Gaussian beam).

Space charge density of the matched high-brightness beam is determined by Poisson's equation $\rho_b = -\varepsilon_o \Delta U_b$. Matched beam profile is defined mostly by focusing potential function and is a weak function of particle distribution in phase space. Time-

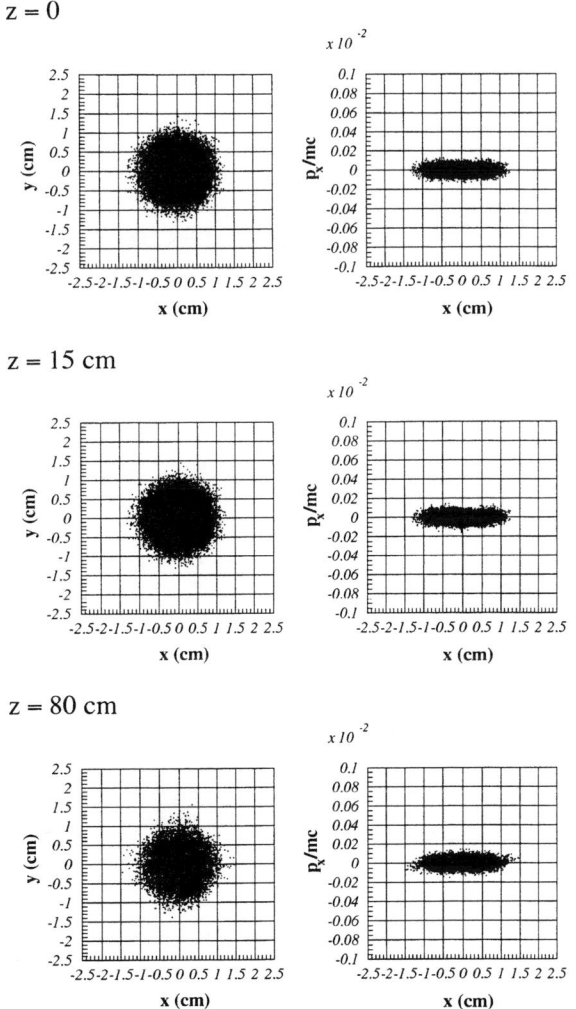

FIGURE 3. Conservation of 150 keV, 0.5A, 0.07 π cm mrad proton beam with distribution, Eq. (31), in nonlinear focusing field, Eq. (36).

dependent focusing potential $U(r,\varphi,t) = U_0(r,\varphi)\cdot\cos\omega_0 t$ can be substituted by an effective potential

$$U_{ext}(r,\varphi) = \frac{qE^2(r,\varphi)}{4\,m\,\gamma\,\omega_0^2}, \qquad (38)$$

if phase advance of particle oscillations per period is small enough (smooth approximation).

Beam Transport in a Quadrupole Channel with Duodecapole Field Component

Matched conditions for nonuniform beam requires the focusing field to be highly nonlinear function of radius. Incorporation of a duodecapole component into quadrupole focusing channel is a way to improve matching of a nonuniform beam with the channel [1, 4]. Consider a four-vanes quadrupole focusing structure with electrostatic potential

$$U(r,\varphi,t) = (\frac{G_2}{2} r^2 \cos 2\varphi + \frac{G_6}{6} r^6 \cos 6\varphi)\sin\omega_0 t, \qquad (39)$$

where G_2 is a quadrupole field gradient and G_6 is a duodecapole field component. Oscillating field, Eq.(39), creates the effective scalar potential

$$U_{ext}(r,\varphi) = \frac{mc^2}{q}\frac{\mu_0^2}{\lambda^2}[\frac{1}{2} r^2 + a_6 r^6 \cos 4\varphi + \frac{a_6^2}{2} r^{10}], \qquad (40)$$

where $\mu_0 \equiv q\, G_2\, \lambda^2/(\sqrt{8}\,\pi\, mc^2)$ is a smoothed transverse oscillation frequency, $\lambda = 2\pi c/\omega_0$ is a wavelength, and $a_6 = G_6/G_2$ is a ratio of two field components. Equipotential lines, $U_{ext}(r,\varphi) = $ const, are close to square.

Space charge density of the matched high-brightness beam in the considered channel is given by [4]:

$$\rho_b = \rho_0 (1 + 10\, a_6\, r^4 \cos 4\varphi + 25\, a_6^2\, r^8). \qquad (41)$$

Beam with distribution, Eq. (41), is maintained in the channel with effective potential, Eq. (40). Due to the term $\cos 4\varphi$, space charge density, Eq. (41), is an azimuthal periodic decreasing-increasing function of radius.

Realistic beam distribution has monotonically decreasing density function with radius. In Fig. 4 results of numerical study of transport of the Gaussian beam with the same parameters, as in Fig. 1, are presented. Beam boundaries were truncated to be close to the matched beam, Eq. (41). The value of quadrupole gradient, $G_2 = 50$ kV/cm^2, was kept constant along the channel. The duodecapole component was gradually reduced from the value of $G_6 = 1.5$ kV/cm^6, as required by matching conditions, to zero for the distance of 100 cm. It provided adiabatic transformation of the initial nonuniform beam distribution into distribution, matched with the quadrupole channel.

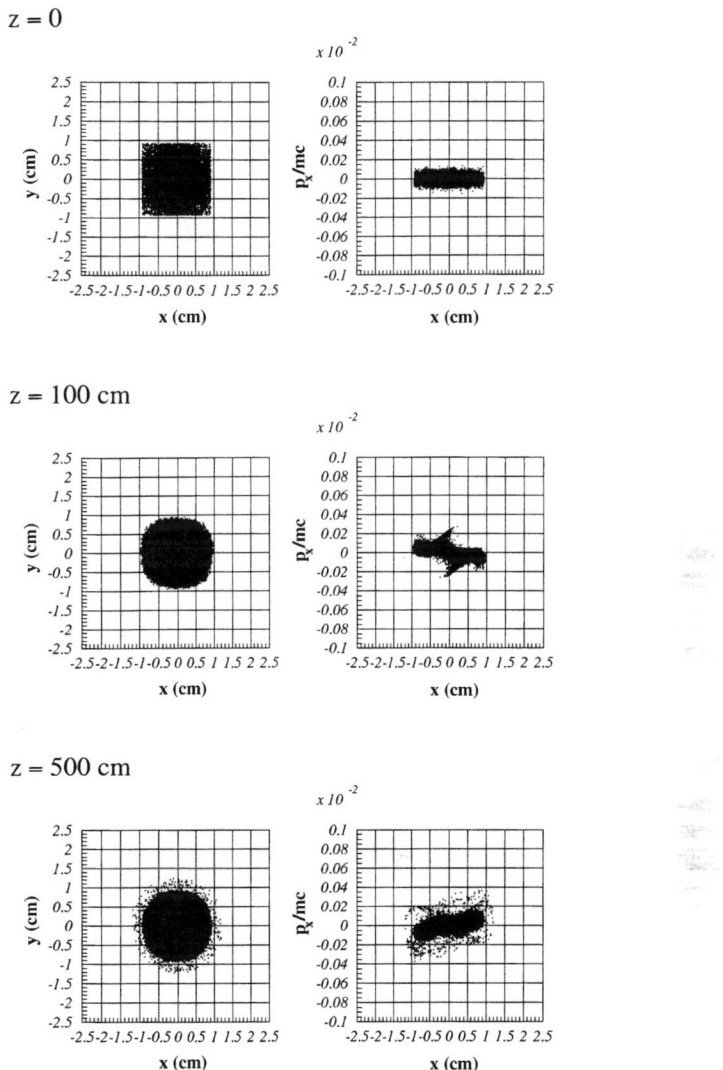

FIGURE 4. Transport of 150 keV, 0.1 A, 0.06 π cm mrad proton beam with truncated at $2.6\sqrt{<x^2>}$ Gaussian distribution in quadrupole focusing channel with field gradient $G_2 = 50$ kV/cm^2 and adiabatic decline of duodecapole field component from $G_6 = 1.5$ kV/cm^6 to zero for the distance 100cm.

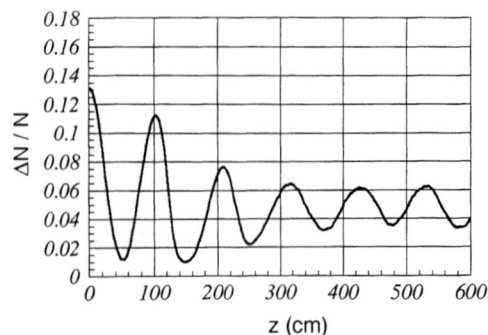

FIGURE 5. Fraction of particles in beam halo for transport of Gaussian beam in quadrupole-duodecapole channel, presented in Fig. 4.

As seen, such a beam is better matched with the focusing structure, than that, presented in Fig. 1. Beam mismatch still remains, because Gaussian beam distribution is different from the matched beam distribution, Eq. (41). It is expressed in emittance growth and appearance of halo, which are, however, smaller than that of Fig. 1. Fraction of particles generating halo in quadrupole-duodecapole channel is 6% (see Fig. 5) instead of 15% in Fig. 2. Therefore, utilizing of duodecapole field component with beam square truncation result in suppression of beam halo.

REFERENCES

1. Batygin, Y.K., *Phys. Rev. E* **53**, 5358 - 5365 (1996); **54**, 5673 - 5680 (1996).
2. Davidson, R.C., *Theory of Nonneutral Plasmas*, Benjamin, Reading, Massachusetts, 1974.
3. Kapchinsky, I.M., *Theory of Resonance Linear Accelerators*, Atomizdat, Moscow, 1966, Harwood, 1985.
4. Batygin, Y.K., *Phys. Rev. E* **57**, 6020 -6029 (1998).

Table 1. Nominal assumptions for the injection calculations.

2-MW beam, 2.07E14 protons per pulse at the end of injection
1158 turns, 10 macro-particles injected per turn, 11580 macro particles injected
3-sided, 300-μg/cm², carbon foil
1% of injected particles miss foil
65% chopping factor
Linac beam coordinates w.r.t. ring: x = 80 mm, x' = 0, y = 30 mm, y' = 0
Linac beam horizontal and vertical 2σ emittance ε = 0.276 π mm mrad, truncated Gaussian
Linac beam β_x = 15.7, α_x = -1.95, β_y = 5.2, α_y = 0.77
Nominal lattice tune: v_x = 5.82, v_y = 5.80
Alternate lattice tune: v_x = 5.85, v_y = 5.70
Dual harmonic RF system, first harmonic 40 kV, second harmonic 20 kV
480 space charge updates per turn, 20 space charge updates per FODO cell
32 x 32, horizontal x vertical, grid
Space charge smoothing parameter = 4.42 x 10⁻³ m

Table 2. Closed orbit bump coordinates at the foil.

Closed orbit for correlated x-y painting
$x_{co}(mm) = 41.7 + 32.0 e^{-7.07 t (m\,sec)}$
$x'_{co}(mrad) = -4.79 + 4.00 e^{-7.07 t (m\,sec)}$
$y_{co}(mm) = -0.20 + 26.1 e^{-1.34 t (m\,sec)}$
$y'_{co}(mrad) = 4.62 - 3.99 e^{-1.34 t (m\,sec)}$
Closed orbit for x only painting (smoke ring)
$x_{co}(mm) = 27.2 + 42.1 e^{-1.23 t (m\,sec)}$
$x'_{co}(mrad) = -6.60 + 45.263 e^{-1.23 t (m\,sec)}$

CORRELATED HORIZONTAL AND VERTICAL PAINTING

With the Schonauer space-charge model

An example of injection painting using the Schonauer space charge model is shown in Fig. 1. This figure is a "snapshot" of the transverse phase spaces at the end of injection (turn 1158). In this case, there is a continual bumping of both the horizontal and vertical closed orbits during the injection process. The horizontal direction is bumped with emittance > 180 π-mm-mrad with space charge effects included. We are presently working on understanding the numerical convergence of these results, and investigating other scenarios to mitigate the space-charge effects.

A more detailed desription of this work is given in Ref. 5.

SPACE CHARGE MODEL AND NOMINAL INJECTION ASUUMPTIONS

New transverse space charge models [4] have been implemented in an alternate version of ACCSIM. In this technical note, we exercise the PIC option of these new models to more generally calculate dynamic transverse space charge effects. This is a straightforward calculation, which dynamically evaluates the space charge forces by binning the macroparticle distribution. In these preliminary calculations we employ a 32 x 32 grid in the transverse plane to bin the macroparticles. Space charge kicks are calculated 480 times per turn, 20 times per FODO cell. This degree of longitudinal resolution is needed to track the horizontal and vertical real space modulation of the beam envelope and to resolve effects of structure resonances. At each of the 480 longitudinal locations the macroparticles are rebinned, and an FFT method is applied to calculate the space charge force giving kicks to each macroparticle. Ten macroparticles are injected each turn to give a total of 11,580 injected macroparticles. Other injection parameters are taken from previous optimizations.

The code includes longitudinal effects, but is not a full three-dimensional calculation. In particular, the beam is segmented longitudinally (into 40 segments here). The overall transverse charge distribution, used for all segments, is modulated by a longitudinal weighting factor. This factor is calculated from a separate one-dimensional model of longitudinal transport (including space charge effects) and applied to include the effect of varying charge density along the longitudinal coordinate. This factor, calculated once per turn, is included multiplicatively in each particle kick, according to the longitudinal coordinate of the macroparticle. The protons undergo about one synchrotron oscillation per injection cycle of 1158 turns.

Some nominal injection assumptions for these calculations are listed in Table 1. The calculations are performed for 2.0 MW of beam power, and the baseline-design dual harmonic RF system is employed. The closed orbit bumps used here were determined by optimization studies [6] which minimized both the foil traversals per particle and the maximum extent of the transverse tune spread *neck-tie* predicted by the simple Schonauer transverse space charge model. Exponential forms are assumed for the bumps, as shown in Table 2. In this technical note, we simply repeat injection scenarios arrived at via this optimization procedure, but use the new PIC space charge model in place of the Schonauer model. The intent is to investigate the impact of a more realistic transverse space charge model on nominal SNS injection scenarios. A detailed injection optimization with the PIC space charge model has not been performed.

Progress Towards Understanding Transverse Space Charge Effects on SNS Ring Injection

J. Galambos, J. Holmes, D. Olsen, J. Whealton
Oak Ridge National Laboratory, Oak Ridge TN, 37831, USA
M. Blaskiewicz, A. Luccio, J. Beebe-Wang
Brookhaven National Laboratory, Upton, NY 11973, USA

INTRODUCTION

The Spallation Neutron Source, SNS, is the next US neutron source for condensed matter physics. The SNS consists of a 35-mA peak-current H⁻ front end, a 1.0-GeV proton linac operating at a 6% duty factor, and a 220.67-m-circumference accumulator ring with injection, HEBT, and extraction, RTBT, beam transport lines. The four-fold symmetric ring lattice [1] consists of a total of 16 arc FODO cells and 8 straight-section FODO cells. A total of 1158 turns will be injected for each pulse. Various injection geometries and schemes have been studied. The SNS Ring injection process is facilitated by *painting* the linac beam phase space into the ring beam phase space by dynamically bumping the ring equilibrium orbit at the injection foil. The ring equilibrium orbit can be independently bumped in both the horizontal x and vertical y directions. Two painting options with bumping magnets are studied here: (1) directly correlated horizontal and vertical bumping away from the foil so that large horizontal and vertical betatron amplitudes occur together, and (2) horizontal bumping only, which produces a *smoke ring* in the vertical phase space.

Previous injection calculations for the SNS have employed the Schonauer model [2] in ACCSIM [3] for transverse space charge effects, which are limited to shifting the macroparticle tunes. No other space charge effects on the transverse spatial charge distributions are calculated in this model. As such, to date the final ring emittance is determined by the choice of the linac beam distribution, by the closed orbit bumps, and by scattering of the beam by the foil. In this model only foil interactions produce beam halo. In this technical note preliminary results are reported using a full Particle-In-Cell, PIC, space charge calculation [4] for SNS injection. The impact of transverse space charge on emittance growth and halo are examined by repeating the previous Schonauer model optimized injection scenarios, here using the PIC model. Although the initial operation of the SNS will be at 1.0 MW of beam power, these calculations are done for 2.0 MW of beam power, since the ring is being designed to allow for the 2.0 MW upgrade.

We find that the inclusion of the transverse space charge effects has a large impact on the final transverse beam distributions. Painting scenarios which result in final painted

REFERENCES

1. C. K. Birdsall and A. B. Langdon, *Plasma Physics via Computer Simulation*, ISBN 0-07-005371-5, Adam Hilger / IOP, Bristol, 1991.
2. R. W. Hockney and J. W. Eastwood, *Computer Simulation Using Particles*, ISBN 0-85274-392-0, Adam Hilger / IOP, Bristol, 1988.
3. The U.S. Heavy Ion Fusion World-Wide Web pages can be found at: http://fusion.lbl.gov/ ; see also the links and references therein.
4. *Energy from Inertial Fusion*, STI/PUB/944, ISBN 92-0-100794-9, W. J. Hogan, scientific editor, IAEA, Vienna, 1995.
5. *Proceedings* of the 10th International Symposium on Heavy Ion Inertial Fusion (Frascati): *Nuovo Cimento* 106 A Nos. 11-12 (1994).
6. *Proceedings* of the 11th International Symposium on Heavy Ion Inertial Fusion (Princeton): *Fusion Engineering and Design* 32-33 (1996).
7. *Proceedings* of the 12th International Symposium on Heavy Ion Inertial Fusion (Heidelberg): *Nuclear Instruments and Methods A*, in press (1998).
8. A. Friedman, "Induction-Accelerator Heavy-Ion Fusion: Status and Beam Physics Issues," *Proc. IUPAP / Int.'l Committee for Future Accelerators' Workshop on Space Charge Dominated Beams and Applications of High Brightness Beams*, Bloomington IN, Oct. 10-13, 1995 (S. Y. Lee, Ed., AIP Conference Proceedings No. 377, 1996)
9. A. Friedman, D. P. Grote, and I. Haber, "Three-dimensional particle simulation of heavy-ion fusion beams," *Phys. Fluids B: Plasma Physics* 4, 2203 (1992).
10. D. P. Grote, "Methods Used in WARP3d, a Three Dimensional PIC/Accelerator Code," *Proceedings of the Computational Accelerator Physics Conference*, 1996, AIP conference proceedings 391, pp 51-58.
11. D. P. Grote, A. Friedman, I. Haber, and S. S. Yu, "Three-Dimensional Simulations of High Current Beams in Induction Accelerators with WARP3d", *Fusion Engineering and Design* 32, 193 (1996).
12. S. M. Lund et. al., "Numerical Simulation of Intense-Beam Experiments at LLNL and LBNL," *Proc. 12th International Symposium on Heavy-Ion Inertial Fusion*, Heidelberg Germany, September 23-27,1997; accepted for publication in *Nuclear Instruments and Methods A*, 1998.
13. D. P. Grote I. Haber, A. Friedman, and S. M. Lund, "WARP3d, A Three-Dimensional PIC Code For High-Current Ion-Beam Propagation Developed For Heavy-Ion Fusion," *Lawrence Livermore National Laboratory Quarterly Report* for Fourth Quarter 1996, available on the Web at:
http://lasers.llnl.gov/lasers/pubs/jul-sep96/jul96.html
14. D. P. Grote, A. Friedman, I. Haber, W. Fawley, and J.-L. Vay, "New Developments in WARP: Progress Toward End-to-End Simulation," *Proc. 12th International Symposium on Heavy-Ion Inertial Fusion*, Heidelberg Germany, September 23-27,1997; accepted for publication in *Nuclear Instruments and Methods A*, 1998.
15. J. J. Barnard, F. J. Deadrick, A. Friedman, D. P. Grote, L. V. Griffith, H. C. Kirbie, V. K. Neil, M. A. Newton, A. C. Paul, W. M. Sharp, H. D. Shay, R. O. Bangerter, A. Faltens, C. G. Fong, D. L. Judd, E. P. Lee, L. L. Reginato, S. S. Yu, and T. F. Godlove, "Recirculating Induction Accelerators as Drivers for Heavy Ion Fusion," *Physics of Fluids B: Plasma Physics* 5, 2698 (1993).
16. A. Friedman et. al., "Progress Toward a Prototype Recirculating Ion Induction Accelerator," *Lawrence Livermore National Laboratory Quarterly Report* for April-June 1995, available on the Web at:
http://lasers.llnl.gov/lasers/pubs/apr-june95/Friedman.pdf
17. A. Friedman, J. J. Barnard, D. P. Grote, and I. Haber, Simulation Studies of Transverse Resonance Effects in Space-Charge Dominated Beams, *Proc. 12th International Symposium on Heavy-Ion Inertial Fusion*, Heidelberg Germany, September 23-27,1997; accepted for publication in *Nuclear Instruments and Methods A*, 1998.

elements are separated by a bend and so are described with reference to different frames. The user specifies a coordinate system which describes the beamline as it is laid out "in the laboratory," and need not specify a reference orbit. In the 3-D and r,z models, no paraxial approximation is invoked.

The code offers an interactive, interpreter-driven user interface for code steering and scripting-language control. This capability allows significant changes in the code's operation to be effected by the user without requiring recompilation. It provides a rich simulation environment which offers ready access to the code's internal database, algebraic transformations, and interactive visualization. WARP has been parallelized to run on supercomputers, symmetric multiprocessors, and workstation clusters using message-passing techniques. While already quite capable, the code is still under development; it has not yet "settled down" to a version suitable for unlimited release.

TRANSVERSE RESONANCE EFFECTS

In the recirculating induction accelerator approach to an HIF driver (15,16), the beam is rapidly accelerated "through resonances" — it is "different on every lap" and errors add much as in a long linac. Beam steering is used to correct the centroid motion. In some RF linac / storage ring approaches, resonance crossing is driven by rapid bunching of the beam, which shifts the resonances as a result of the increasing space charge density and image effects. We are carrying out studies (17) of transverse resonance effects in rings with space charge. The beams we consider are highly space-charge-dominated, with an undepressed phase advance per lattice period of $\sigma_0 \sim 75°$ reduced by the transverse self-fields to $\sigma \sim 15\text{-}20°$. These parameters are similar to those of the "small recirculator" (16) under development at LLNL. We consider periodic quadrupole strength errors which can resonantly drive symmetric and antisymmetric mismatch oscillations of the beam core. Interactions with those core oscillations can drive particles into the halo and degrade the emittance. Using a 2D transverse "slice" model in the WARP PIC code, we study how aggressive acceleration or bunching, leading to rapid passage through a single applied resonance, affects the emittance. Power spectra are computed and their many features identified. We confirm that rapid passage through the resonance significantly reduces the attendant emittance growth. The growth rate is seen to be strongly correlated with the instantaneous degree of mismatch.

As part of this study, the convergence of the PIC method employed was investigated. Runs using 20,000 particles showed some jaggedness of particle orbits and enhanced emittance growth (resulting from the nonphysically-large collisionality associated with the small particle number), but were generally deemed reliable enough for studies such as these. Runs using 80,000 and 320,000 particles were well-behaved, and produced essentially converged results.

ACKNOWLEDGMENTS

*This work was performed under the auspices of the U.S. Department of Energy by LLNL under contract W-7405-ENG-48.

The author gratefully acknowledges the contributions of his colleagues and most frequent co-authors, John J. Barnard, Debra A. Callahan-Miller, David P. Grote, Irving Haber, and Steven M. Lund, to the work listed in the References.

Overview of the WARP Code and Studies of Transverse Resonance Effects*

Alex Friedman

Lawrence Livermore National Laboratory, Livermore, CA 94551, USA

Abstract. Two papers presented at the Shelter Island workshop are very briefly summarized here, in view of recent publications elsewhere. The WARP code, developed for Heavy-Ion beam-driven inertial confinement Fusion (HIF) accelerator studies, combines features of a particle-in-cell plasma simulation and an accelerator tracking program. Its methods and architecture have been developed for efficiency both in detailed simulation of individual machine sections and in long-time beam tracking. The transverse "slice" model in the code has been applied to the study of transverse resonance effects associated with quadrupole strength errors. These simulations confirm that rapid passage through a resonance can reduce the associated mismatch and emittance growth. References to published details and to other sources of information are supplied.

WARP CODE OVERVIEW

WARP is a particle-in-cell (1,2) beam simulation code offering 3-D, axisymmetric (r,z), and transverse (x,y) geometries. The code was developed for Heavy-Ion beam-driven inertial confinement Fusion (HIF) accelerator studies (3-8). In this application, and particularly in the U.S. approach based on induction technology, space charge forces dominate over the beam thermal pressure.

WARP (9-14) incorporates detailed descriptions of various accelerator and beamline elements, and was designed to follow beams efficiently over long path lengths. A hierarchy of models affords increasing levels of detail in the description of the accelerator lattice. While some of these models are based on an expansion in powers of the off-axis separation, the code can use models which are good to "all orders" (such as field specification on a full 3-D grid) when sufficient knowledge of the field is available. Several Poisson equation solvers are available, optionally incorporating internal conducting elements from "first principles" with subgrid-scale resolution of the conductor boundary locations.

WARP uses time (rather than path length) as the independent coordinate for particle motion; this facilitates the treatment of longitudinally-extended beams. The code gets its name from its use of "warped" coordinates, that is, a sequence of Cartesian grid sections aligned with the local beamline (so that each lattice element is described in its own natural coordinate system) linked by sections of "polar coordinate" grid. Coordinate transformations account for the bends while preserving the symplectic nature of the underlying advance. The fields from neighboring elements can overlap, even when those faster than the vertical direction, as listed in Table 2, and both bumps include a slight initial offset, giving hollow phase space distributions. The x-y projection is rectangular since there are no horizontal-vertical correlations. Table 3 lists some global results from these calculations. The present case is listed in row 1 for the SNS design tunes. The bump scheme results in a final distribution with a maximum emittance near 120 π mm-mrad for both the horizontal and vertical directions. We note that there is very little spreading of the beam off of the original injected ellipses. For example, the hollow centers of the horizontal and vertical phase space distributions persist throughout the injection interval and the beam edges in the horizontal and vertical phase spaces remain sharp. The final beam has no particles with emittance > 120 π mm mrad

With the PIC space charge model

The calculation discussed above was repeated with the PIC space charge model. Snapshots of the phase-space distributions at the end of injection are shown in Fig 2. There is a noticeable diffusion of the beam: (1) the hole in the center of the horizontal and vertical phase spaces fills in, and (2) the outer edge of the distribution is much less sharp. The increased diffusion of the outer extremities of the beam is evident in the increased fraction of the beam at large emittances giving halo growth. These fractions are listed in percentages for emittances greater than 120, 150, and 180 π mm mrad in Table 3, case 2.

The reason there is more halo growth in the horizontal direction than in the vertical direction is because the horizontal bump is ramped more quickly than the vertical bump (Table 2). That is, the closed orbit in the horizontal direction quickly approaches its final coordinates, and the injection takes place at the outer horizontal region of the beam during most of the injection time, similar to a smoke ring. On the other hand, in the vertical direction, the bump is more gradual, and there is a continual painting over the expanding vertical beam edge as the injection proceeds.

In the real space x-y plane, a waist develops in the vertical direction. This is due to x-y coupling introduced by the space charge. Namely, for a perfect lattice, the space charge force introduces many non-linear terms into the betatron motion. For example, in the horizontal plane the space charge terms can be approximated on the RHS of Hill's equation as the series:

$$x'' + K_x x = c_1 x + c_2 x^2 + \ldots + d_1 y + d_2 y^2 + \ldots + e_{11} xy + \ldots.$$

The $d_1 y$ term is equivalent to a skew quadrupole, which gives rise to x-y coupling independent of any magnet errors. Because of the closeness of the horizontal and vertical tunes of the design SNS lattice, $v_x = 5.82$, $v_y = 5.80$, the coupling time scale is relatively short. The corners of the distribution begin to rotate, in both directions, forming the waist in the middle of the distribution. Eventually this rotation would transform the rectangular beam cross section into an ellipse. However, since the x and y closed orbits are continually bumping, the injection paints over the x-y ellipses, so only the effect at the corners is evident. Also, since the horizontal direction is ramped more quickly than the vertical direction, the ellipse shape is more evident in the horizontal direction.

Table 3. Calculated results with the Schonauer and the PIC space charge models with a painted total emittance of 120 π-mm-mrad.

Case	Tune (1)	PIC Model Used	Foil Hits/ Part,	RMS emittance (π-mm-mrad)		Particles with emittance > 120 π-mm-mrad (%)		Particles with emittance > 150 π-mm-mrad (%)		Particles with emittance > 180 π-mm-mrad (%)	
				x	y	x	y	x	y	x	y
Correlated x & y painting											
1	A	No	2.53	37.8	26.1	0.0	0.0	0.0	0.0	0.0	0.0
2	A	Yes	2.52	40.7	20.9	20.7	0.48	8.20	0.10	1.95	0.03
3	B	No	2.48	37.7	26.1	0.0	0.0	0.0	0.0	0.0	0.0
4	B	Yes	2.78	37.5	25.1	17.0	0.58	7.89	0.03	2.69	0.00
Only x painted (y smoke ring)											
5	A	No	5.00	27.7	46.2	0.0	0.0	0.0	0.0	0.0	0.0
6	A	Yes	6.32	35.1	44.3	10.5	20.6	2.02	8.56	0.59	3.71
7	B	No	4.98	27.7	46.5	0.0	0.0	0.0	0.0	0.0	0.0
8	B	Yes	5.47	27.1	49.2	0.89	29.1	0.06	12.6	0.01	4.60

(1) – Lattice tune "A" has a horizontal/ vertical tunes of 5.82/5.80, and lattice tune "B" has a horizontal/ vertical tunes of 5.85/5.70.

In addition, for this case the beam is asymmetric in phase, due to a lump that forms in the distribution (evident by the higher density of particles in the lower half of the horizontal phase space in Fig. 2). Monitoring the average horizontal position of all particles indicates that the lump rotates close to the betatron tune as shown in Fig. 3. Although the lump is more obvious in the horizontal direction there is also a small rotation of the average y at the betatron frequency, which is also shown in Fig. 3. The magnitude of this oscillation is modulated. The origin of this oscillating lump is not yet understood, but may be related to the asymmetry introduced by painting in a single "spot" in the Ring phase space.

With alternate tunes away from the $v_x = v_y$ coupling resonance

The design SNS lattice has closely spaced horizontal and vertical tunes, 5.82 and 5.80, leading to significant x-y coupling. The sensitivity to this coupling was investigated by repeating the calculation with the tunes of $v_x = 5.85$ and $v_y = 5.70$, accomplished by changing the straight section quadrupole strengths by a few percent. With the

Schonauer model, case 3, there is minimal impact with the change of tunes. With the PIC model, case 4, the effect of changing the tune is more pronounced. Separating the horizontal and vertical tunes from each other reduces the horizontal-vertical coupling. As discussed above, for this case, the space charge emittance growth is more pronounced in the horizontal direction, which is painted faster. With reduced coupling, there is less transfer of the larger horizontal emittance growth to the vertical direction. Also, the waist in real space is less pronounced, as seen in Fig. 4.

PAINTING IN THE HORIZONTAL PLANE ONLY (SMOKE RING)

With the Schonauer model

Painting in the horizontal direction only, with an initial offset in the vertical equilibrium orbit results in a smoke ring distribution in the vertical phase space. Typical results for this scenario are shown in Fig. 5 at the end of injection. (Note that the initial offset of the closed orbit in the horizontal direction also results in a hollow distribution in the horizontal phase space.) As indicated in Table 3 for case 5, there is no emittance growth above $120\ \pi$ mm-mrad, and the smoke ring shape remains sharp throughout the injection cycle. The average foil traversals per particle are about two times higher when painting in only one plane compared to painting in both planes.

With the PIC model

Phase space plots are also shown in Fig. 5 for this same case using the PIC model. There is a large distortion of the distribution due to space charge effects. The center of the smoke ring in the vertical direction has quickly filled in and there is significant emittance growth above $120\ \pi$ mm-mrad compared to the Schonauer model shown in Fig. 7. Table 3, case 6, lists large fractions of particles in the halo region for this case. Even though the horizontal direction is being continually painted over, particles appear in the horizontal halo relatively early in the injection cycle due to coupling with the vertical direction.

With alternate tunes away from the $v_x = v_y$ coupling resonance

The case using the alternate tunes of $v_x = 5.85$ and $v_y = 5.70$ with painting in only the horizontal direction is also examined. With the Schonauer model, there is little difference from the results obtained using the design lattice tunes, cases 5 and 7 in Table 3. With the PIC model and the alternate tunes, case 8, less emittance and halo growth is observed compared to the corresponding case using the design lattice tunes, case 6. With reduced x-y coupling in this case, there is less spread of the emittance growth from the vertical direction to the horizontal direction.

Discussion

- Space charge effects for the SNS may result in very appreciable emittance and halo growth.
- For the 2 MW case investigated here, where the beam is not stored longer than the injection time, painting the ring emittance from the inside of the distribution towards the outside of the distribution will probably minimize halo growth. In essence, the painted beam will cover up the emittance growth from previously painted beam.
- With the design tunes of $v_x = 5.82$ and $v_y = 5.80$ the x-y coupling may or may not cause less halo growth than tunes away from the $v_x = v_y$ coupling resonance. Coupling is an open question.
- Inclusion of the PIC space charge model has a major impact on the final painted distributions, which will affect injection hardware and collimation optimization.

This technical note is a first quick look at space charge effects in the SNS ring injection. Clearly more work needs to be done. In particular: (1) calculations need to be done with more macroparticles (2) the dependence of the calculated halo with the smoothing parameter needs to be studied (3) the injection scheme needs to be reoptimized with space charge, perhaps painting the initial beam into less than 120 π mm mrad and employing less rapid bumps so that more time is spent painting the center of the distribution (4) the tune space needs to be explored (5) the effects of horizontal coupling need to be better understood (6) collimation configuration effectiveness needs to be studied with space charge and (7) transverse distributions on the neutron producing target need to be investigated with space charge.

Acknowledgements

Research on the Spallation Neutron Source is sponsored by the Division of Materials Science, U.S. Department of Energy, under contract number DE-AC05-96OR22464 with Lockheed Martin Energy Research Corporation for Oak Ridge National Laboratory.

References

1 Lee Y.Y. 1997, "The 4-Fold Symmetric Lattice for the NSNS Accumulator Ring", BNL/NSNS TN 026.
2 Schonauer H. 1989, "Addition of Transverse Space Charge to ACCSIM Code", TRIUMF Design Note TRI-DN-89-K50.
3 Jones F. 1990, "Users Guide to ACCSIM", TRIUMF Design Note TRI-DN-90-17.
4 Holmes J. A., Galambos J. D. and Olsen D. K. 1998, "PIC and Particle-Core Model Space Charge Modules for Tracking Codes", SNS/ORNL/AP TN 006.
5 Galambos J. D., Holmes J. A. and Olsen D. K, J. Whealton, M. Blaskiewicz, "Preliminary SNS H- Injection Calculations with Space Charge using a PIC Model", SNS/ORNL/AP TECHNICAL NOTE, Number 005
6 Galambos J. D., Holmes J. A. and Olsen D. K. 1998, "Accumulator Ring H-Injection Optimization Studies", Particle Accelerator Conference, May 1997, Vancouver (In Press).

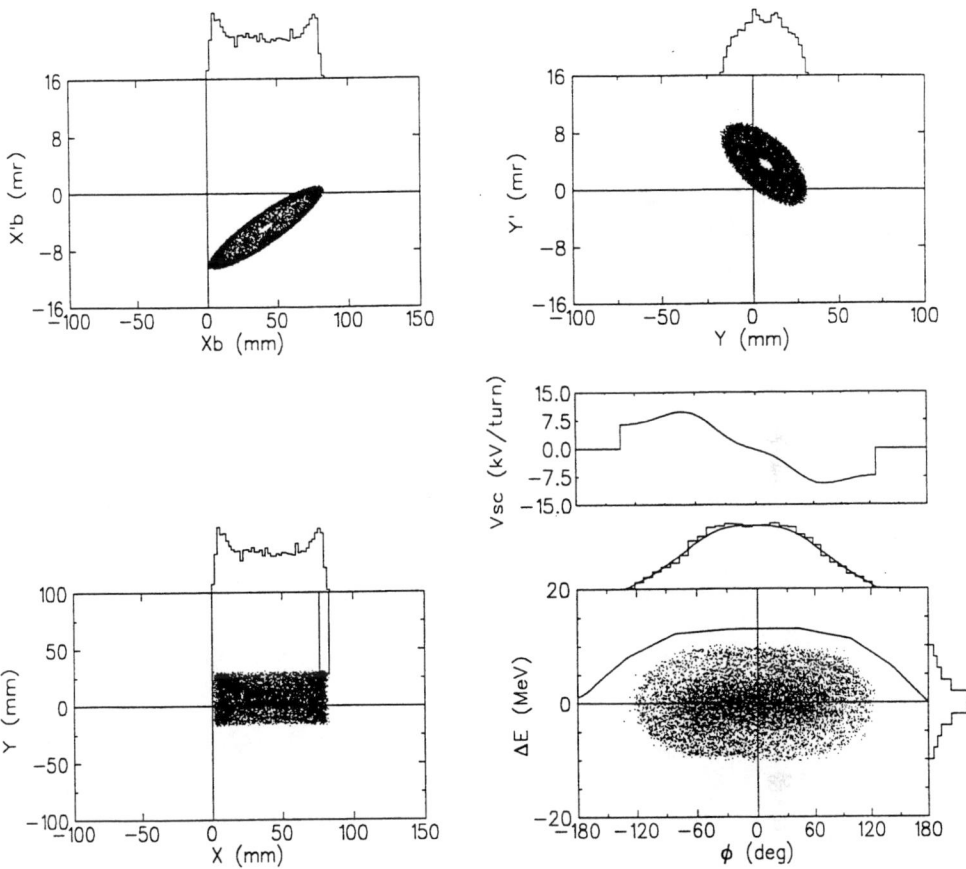

Fig. 1. Phase space plots at the end of injection for x and y correlated painting, with lattice x/y tunes of 5.82/5.80, using the Schonauer space charge model.

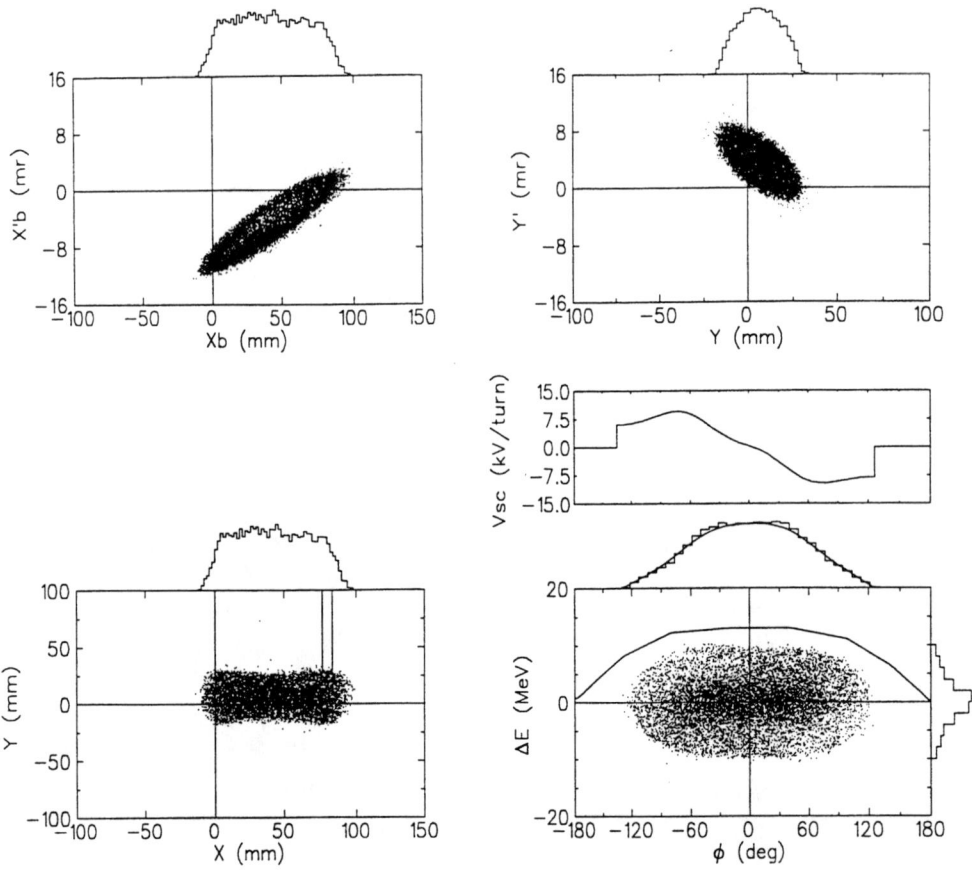

Fig. 2. Phase space plots at the end of injection for x and y correlated painting, with lattice x/y tunes of 5.82/5.80, using the PIC space charge model.

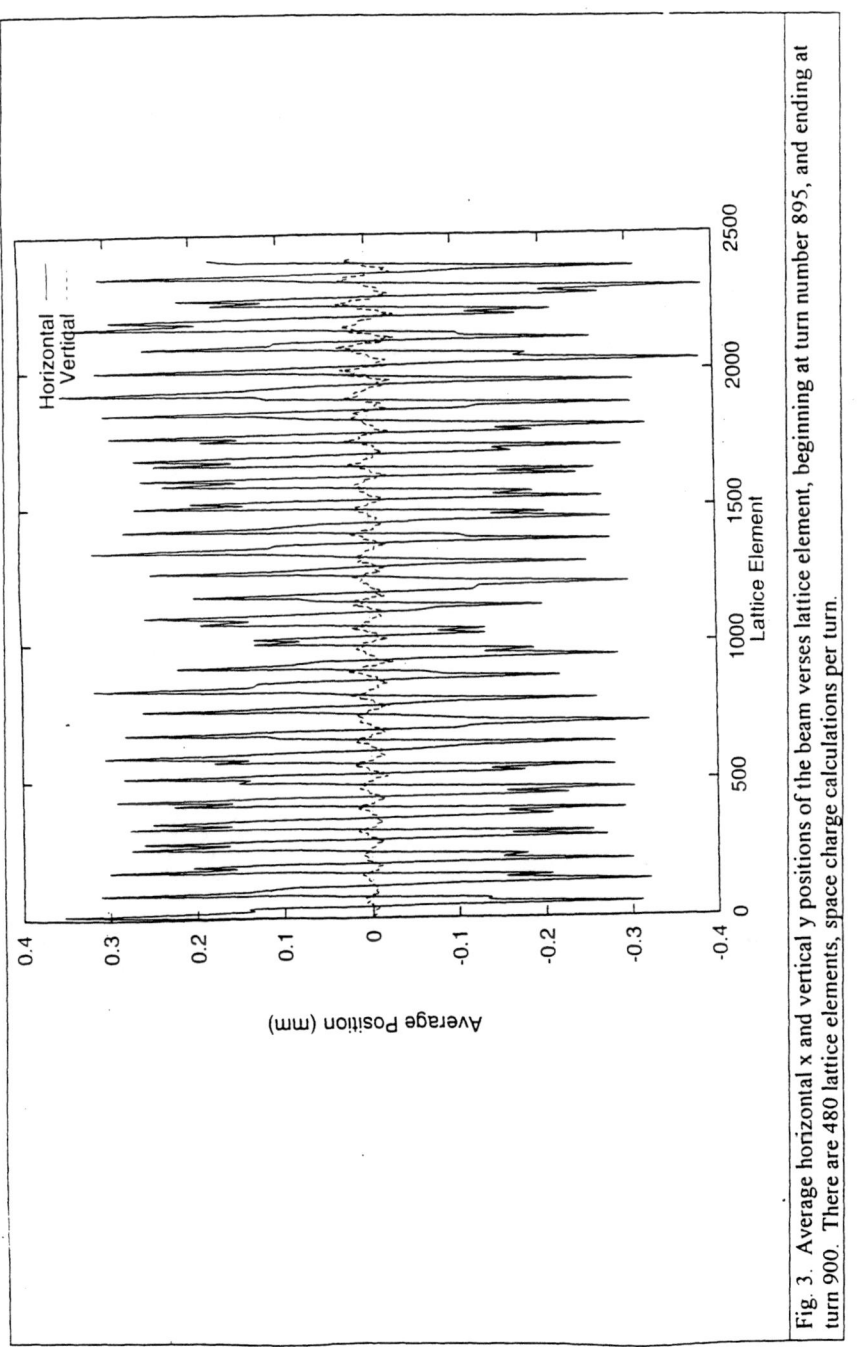

Fig. 3. Average horizontal x and vertical y positions of the beam verses lattice element, beginning at turn number 895, and ending at turn 900. There are 480 lattice elements, space charge calculations per turn.

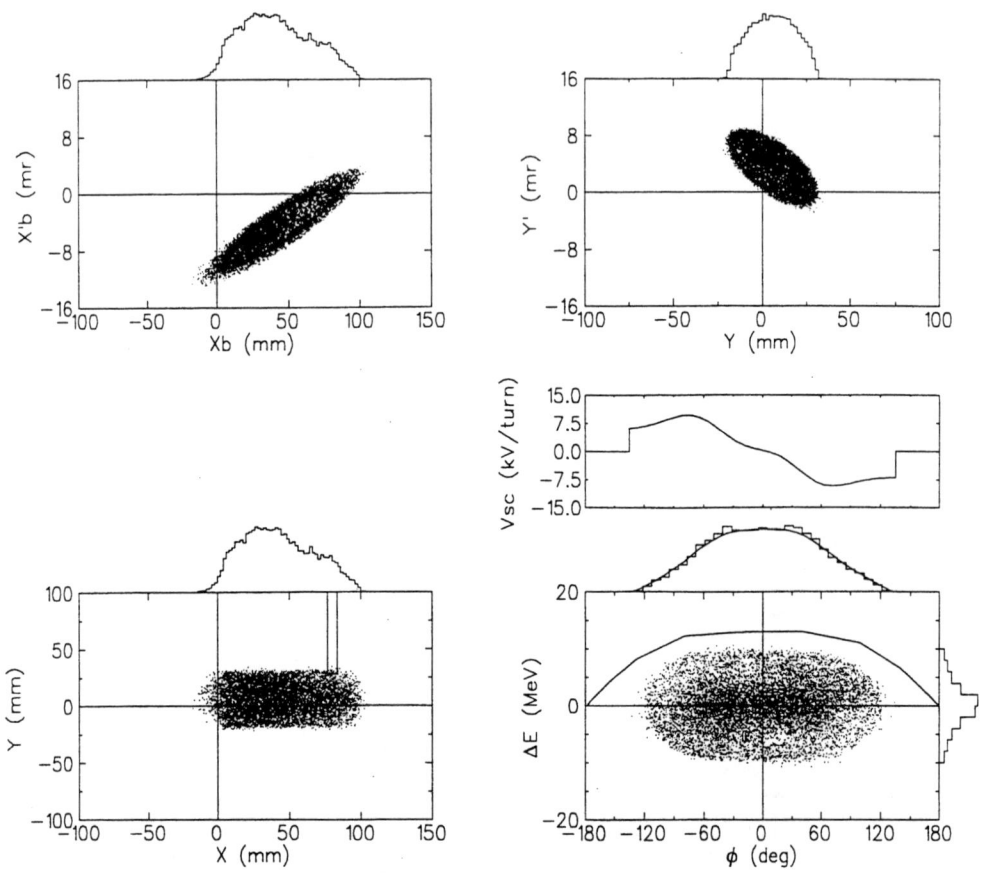

Fig. 4. Phase space plots at the end of injection for x-y correlated painting, with lattice x/y tunes of 5.85/5.70, using the PIC space charge model.

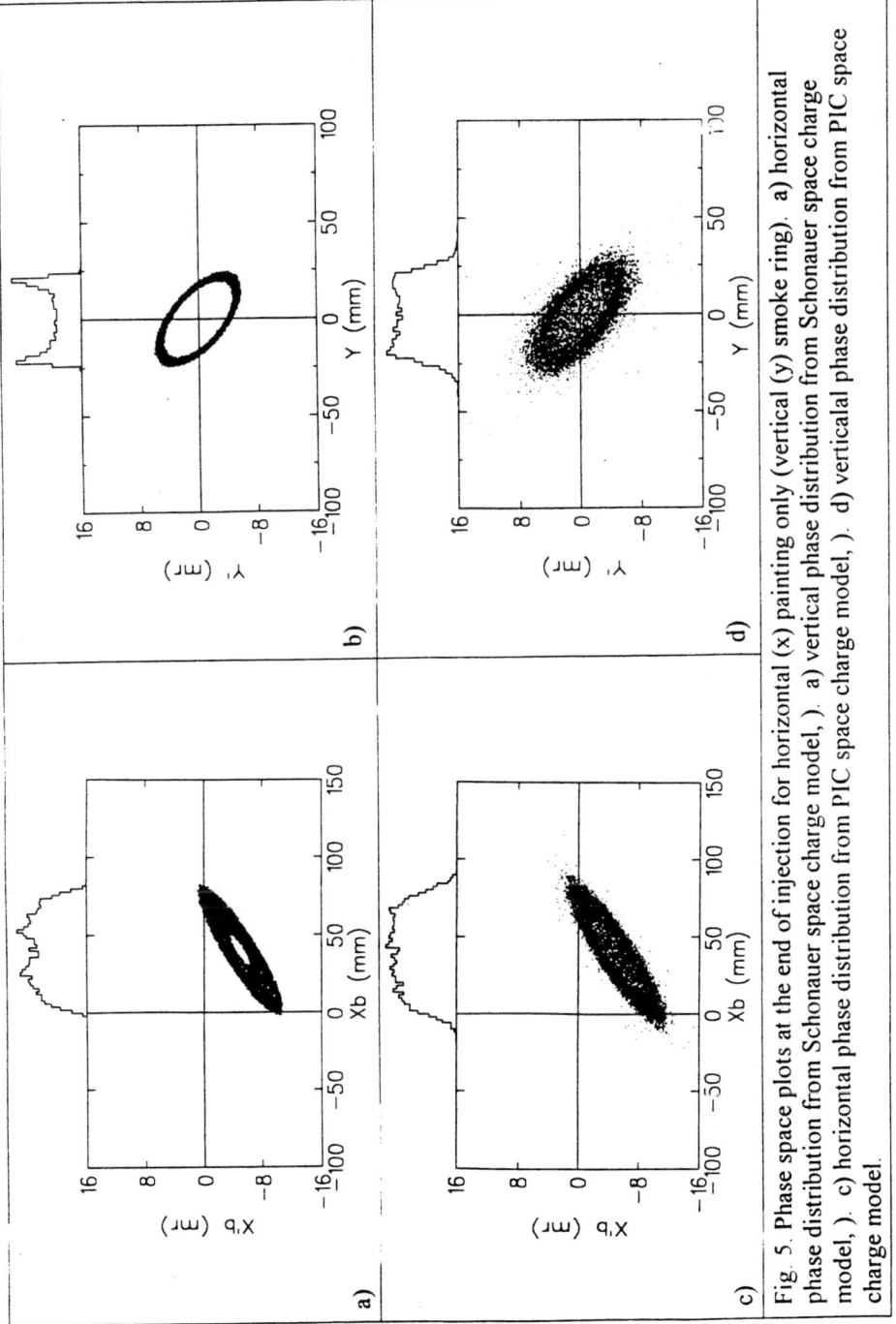

Fig. 5. Phase space plots at the end of injection for horizontal (x) painting only (vertical (y) smoke ring). a) horizontal phase distribution from Schonauer space charge model,). a) vertical phase distribution from Schonauer space charge model,). c) horizontal phase distribution from PIC space charge model,). d) vertical phase distribution from PIC space charge model.

Space Charge Calculations in Rings for Uniform Focusing, FODO, and Doublet Lattices

J.A. Holmes*, J.D. Galambos*, J.H. Whealton*, D.K. Olsen*,
M. Blaskiewicz**, A. Luccio**, J. Beebe-Wang**, and S.Y. Lee***

*Oak Ridge National Laboratory, Oak Ridge, TN 37831
**Brookhaven National Laboratory, Upton, NY 11973
***Indiana University, Bloomington, IN 47408

Abstract. This paper presents the results of calculations of the transverse effects of space charge of coasting beams in rings for three globally comparable lattice configurations: uniform focusing, FODO, and doublet. The parameters of the lattices and the H⁻ beam are chosen similar to those of the SNS accumulator ring. Three models for space charge are considered: 1) a particle core model based on rms beam parameters, 2) a self-consistent particle-in-cell (PIC) model, and 3) a phase-averaged PIC model. In all cases both matched and mismatched K-V distributions are considered by randomly initializing, and then tracking, collections of macroparticles representing the beam. In the particle core model the initial rms values of the macroparticle distributions are used as initial values for solving the envelope equations, including space charge forces and dispersion effects. For matched beams the calculations, performed using a modified version of the injection and tracking code, ACCSIM, reveal only a slight emittance growth and no halo generation with the particle core model. However, the self-consistent PIC model yields greater emittance growth and halo generation, particularly for the doublet lattice in the vertical plane. When the calculations are performed with the phase-averaged particle-in-cell (PIC) model, the results agree substantially with the particle core model, suggesting that the observed self-consistent PIC results are not a consequence of numerical truncation. As further confirmation, we have performed numerical convergence studies using the doublet lattice, and have observed the behavior of the self-consistent PIC model to persist. The tendency for emittance growth and halo generation in the doublet lattice is not surprising, as the fluctuations in beam area, which are an excellent indicator of space charge forces, are larger and more rapid than for the other lattices. For mismatched beams, only the particle core model has been applied at present, and good beam transport is obtained with up to 25% envelope oscillation amplitudes for all three lattices.

INTRODUCTION: LATTICES AND MODELS

Space-charge-induced emittance growth and halo generation are potential sources of beam loss in high intensity rings, such as the SNS. In such accelerators, uncontrolled losses to the walls as small as 10^{-4} would lead to activation (1), making maintenance difficult. For this reason it is important to study the effects of space charge on beam dynamics and halo generation in particular in high intensity rings. One important question is the impact of space charge effects on choice of lattice configuration. Although such effects may be studied in detailed lattice design, they are not usually considered in selecting the basic configuration.

In this paper we address this question by comparing and contrasting the effects of transverse space charge forces on beam emittance growth and halo generation for three globally comparable lattice configurations: uniform focusing, FODO, and doublet. Table 1 lists the parameters for these lattices, which are chosen to follow closely the design values of the SNS (1), except for the tunes, which were moved farther apart than the design values of $\nu_x = 5.82, \nu_y = 5.80$ in order to reduce possible resonant $x - y$ coupling.

Transverse space charge forces are sensitive to the beam size, due to its effects on both charge density and proximity of interacting particles. Figure 1 plots the quantity $A_f = \pi\sqrt{\beta_x \varepsilon_x \beta_y \varepsilon_y}$, the "transverse focusing" area of the lattice, versus the azimuthal distance for each of the lattices with the assumption that the emittances are $\varepsilon_x = \varepsilon_y = 100\pi$ mm-mrad. The transverse focusing area is the area of a K-V beam with the given rms emittances if only the linear focusing forces are present. The transverse focusing area displays the largest and most rapid variations for the doublet lattice, whereas it is constant in the uniform focusing case.

TABLE 1. Lattice Parameters

Lattice	Uniform focusing	FODO	Doublet
Superperiods	Circular	4	4
Circumference	220.668 m	220.668 m	220.668 m
x tune	5.85	5.85	5.85
y tune	5.70	5.70	5.70
β_x^{max}	6.004	18.806	15.766
β_x^{min}	6.004	2.397	2.000
β_y^{max}	6.162	19.571	16.940
β_y^{min}	6.162	1.336	1.527
η^{max}	1.026	4.101	3.804

To study the transverse space charge effects, a computational particle tracking approach is chosen. This is performed using a modified version of the injection and tracking code, ACCSIM (2). The details of the numerical models are described in Ref. (3), so only a brief description will be presented here. In particular, the particle dynamics can be summarized by Hill's equations for betatron oscillations:

$$x'' + K_x(s)x = F_x^{sc} + \frac{1}{\rho}\frac{\delta p}{p_0}$$
$$y'' + K_y(s)y = F_y^{sc}, \tag{1}$$

where derivatives are taken with respect to the azimuthal lattice distance s; $K_x(s)$ and $K_y(s)$ represent the periodic linear focusing forces from the lattice magnets; $\frac{1}{\rho}\frac{\delta p}{p_0}$ is the dispersion term; and $F_x^{sc}(s)$ and $F_y^{sc}(s)$ are space charge forces. In all cases the Hill's equations are solved for linear transport, including dispersion, using the matrix routines in ACCSIM. The only additional terms in the tracking calculations are the nonlinear space charge terms, which are evaluated as kicks to the transverse momenta before and after the matrix transports to make the tracking integration scheme second order symplectic. Wall effects and image charges are not yet included.

All calculations presented here perform particle tracking identically except for their treatment of the space charge terms, for which three alternative models are considered: 1) a particle core model based on rms beam parameters (4), 2) a self-consistent particle-in-cell (PIC) model, and 3) a betatron phase-averaged PIC model. In all three cases, the force terms are calculated as if the tracked particles represent line charges at the same azimuthal location, as in a coasting beam. Of the three models, the self-consistent PIC model is conceptually the most straightforward. It calculates the space charge forces directly from the actual distribution of tracked macroparticles, according to the following expressions:

$$F_x^{sc}(x,y) = \frac{K}{N}\sum_{i=1}^{N}\frac{x-x_i}{(x-x_i)^2+(y-y_i)^2+\varepsilon^2}$$
$$F_y^{sc}(x,y) = \frac{K}{N}\sum_{i=1}^{N}\frac{y-y_i}{(x-x_i)^2+(y-y_i)^2+\varepsilon^2}, \tag{2}$$

where the summation is over the macroparticles, $K = \frac{2Z^2 r_p \lambda}{\gamma^3 \beta^2 A}$, Z is the charge number, A is the mass number, r_p is the classical proton radius, γ and β are the relativistic mass and velocity, λ is the number of beam particles per unit length, N is the number of macroparticles, and ε is a numerical smoothing parameter designed to eliminate the singularities of short range macroparticle collisions. For reasons of computational speed, the implementation of the PIC force model is not so straightforward as shown here. The actual computation is made using fast Fourier transform (FFT) techniques, and is discussed in detail in Ref. (3). However, it is important to note that computational results depend on the number of macroparticles, N, the FFT resolution, and the smoothing parameter, ε.

The phase-averaged PIC model differs from the self-consistent PIC calculation in just one important respect: in the self-consistent PIC calculation each macroparticle contributes to the space charge force as a line source charge at its (x, y) location; in the phase-averaged PIC calculation each macroparticle contributes to the force by distributing its charge evenly over all phase angles of its betatron oscillation. This force model is based on the notion that the tune spreads in rings will cause the betatron phases to be populated uniformly as the beam circulates. Computationally, phase averaging yields a smooth space charge force distribution with relatively few macroparticles, but the validity of this model is limited to cases of the basic assumption that the beam is uniformly distributed over betatron phase angles. Dynamics that depend on breaking that symmetry are unobservable in the phase-averaged model.

The particle core model is the least self-consistent, but most theoretically tractable, of the three space charge representations. In this model the space charge force is obtained analytically from an assumed elliptical K-V distribution with x and y radii obtained from rms envelope equations for radii, a and b, respectively:

$$a'' + K_x(s)a = \frac{\varepsilon_x^2}{a^3} + \frac{2K}{\alpha + b}$$
$$b'' + K_y(s)b = \frac{\varepsilon_y^2}{b^3} + \frac{2K}{\alpha + b}.$$
(3)

Here ε_x and ε_y are the x and y emittances, and $\alpha = \{a^2 + 4D_x^2 <(\frac{\delta p}{p_0})^2 >\}^{\frac{1}{2}}$ contains a correction to the x radius for dispersion. D_x is the space-charge-corrected closed orbit dispersion function, which satisfies

$$D_x'' + K_x(s)D_x = \frac{1}{\rho} + \frac{2P}{\alpha + b}\frac{D_x}{\alpha},$$
(4)

and $<>$ denotes an average over the distribution. The space charge forces are calculated assuming a K-V distribution with radii α and b. Matched envelope solutions are obtained by solving Eq. (3) with periodic boundary conditions on a, a', b, and b' around the ring. Such solutions represent the beam core in its lowest energy state, and provide no energy to excite the beam particles. Other solutions give rise to core oscillations which not only are capable of exciting tracked particles, but also of undergoing their own collective dynamics with the lattice. A detailed derivation of the envelope equations is given in Ref. (4), and a thorough discussion of the particle core model is given in Refs. (3, 4).

In all calculations presented here K-V distributions of particles, representing the beam, are initialized randomly and then tracked. With the particle core model the envelope equations are initialized to the rms parameters of the initial particle distribution. Cases in which the beam is matched to the lattice and cases in which it is not matched are considered. For matched beam cases in the absence of collisions, the K-V distribution provides an equilibrium solution of the dynamic equations. Subsequent sections describe the calculations and computational assumptions, the results of the calculations, and our conclusions.

CALCULATIONS AND COMPUTATIONAL PARAMETERS

To understand the effects of space charge forces in the three lattices, calculations using the space charge models listed in Table 2 were performed for each lattice. The linear calculations, used as default reference cases, contain no space charge effects and the particles are simply tracked linearly through the lattice. The matched particle core model utilizes periodic closed solutions of Eq. (3) to generate the space charge forces. In the self-consistent PIC model the full space charge dynamics are present. The phase-averaged PIC calculation imposes a smoothing and symmetrizing (vertically about the x-axis and also horizontally about the y-axis for zero dispersion cases) effect on the charge distribution. The intent is to provide a good representation of the charge distribution of circulating beams using relatively few macroparticles. Because the K-V distribution is an equilibrium solution of the Vlasov equation for matched beams, to the extent that collisions are suppressed and numerical fluctuations in the macroparticle K-V distributions are small, we anticipate quiescent behavior in all the matched beam cases, with little emittance growth and no halo generation. In contrast, the mismatched beam calculations could result in both emittance growth and halo generation.

As with the lattice configurations, the physical beam parameters for these calculations are similar to the SNS baseline design. Specifically, as listed in Table 3, an H^- beam of energy 1 Gev and maximum energy spread ± 9.4 Mev is assumed. The number of beam particles, $3.08*10^{14}$ for a nonbunched circulating beam, corresponds to $2.0*10^{14}$ particles for a bunched beam with bunching factor equal to 0.65. We use a coasting beam and neglect azimuthal variations in charge density. The rms beam emittances are chosen to be 100 mm-mrad in both the x and y planes. The tracking calculations are carried out for a total of 1250 turns; but, unlike the SNS, we track the full beam intensity through the entire calculation rather than increase the intensity through the injection process. In Ref. (5) we presented results, specific to the SNS, that assume bunched beams, azimuthal charge density variation, and injection

Table 2. Space Charge Calculations Used for Each Lattice

Space charge model	Comments
Matched beams:	
Linear calculation - no space charge	Reference case
Particle core model	Space charge tune shifts, no energy source to drive particles
Self-consistent PIC	Full dynamic space charge interaction for "steady state" configuration
Phase-averaged PIC	Less fluctuation than self-consistent PIC, beam symmetry, similar to matched particle core model
Mismatched beams:	
Linear calculation- no space charge	Reference case
Particle core model	Fluctuating space charge forces due to envelope dynamics can enhance emittances and generate halo
Self-consistent PIC	Full dynamic space charge interaction for evolving configuration - not yet completed

for beam accumulation.

The numerical/computational parameters are also listed in Table 3. The number of azimuths used to represent the lattice for particle tracking was determined by requiring convergence of the periodic solutions of the envelope equations (3) which, as nonlinear equations, share many features with the betatron oscillation equations (1). The size of the FFT grid, the number of macroparticles N, and the smoothing parameter ε were selected as follows: in the FFT algorithm the linear size of the spatial grid is twice the greatest linear extent of the beam, so the beam occupies no more than one fourth of the Fourier spatial grid area. We want to have at least several macroparticles particles in each grid cell, choosing 30 particles/cell in the beam. The smoothing parameter is chosen to be at least as big as the maximum cell size, but small compared to the beam. Given these criteria, the values shown in Table 3 were selected to provide reasonable computational times for this initial study. Computational studies of the effects of varying the size of the FFT grid, N, and ε are underway, and we present partial results here. Finally, in the phase-averaged PIC model, the distributed charge of each phase-averaged particle is approximated by a weighted rectangular array of charges, with the weights summing to unity. The choice of a 5 x 5 representation for the array was made by carrying out convergence studies with the other parameters fixed as described.

Table 3. Physical and Computational Parameters

Parameter	Value	Comments
H^- beam energy	1 Gev	$\gamma = 2.066, \beta = 0.875$
Energy spread	± 9.4 Mev	RMS energy deviation is 4.7 Mev, elliptical distribution in longitudinal phase space
Number of beam particles - N_B	3.08×10^{14}	Corresponds to 2.0×10^{14} bunched beam with 0.65 bunch length
Emittances - $\varepsilon_x, \varepsilon_y$	100π mm-mrad	Initial matched K-V beam and matched core
Number of turns tracked	1250	Major output each 250 turns
Number of azimuths for tracking	488	
FFT grid	32 x 32	
Number of macroparticles - N	7680	7.5 particles/FFT cell, effectively 30 particles/FFT cell
Smoothing parameter - ε	4.42 mm	$\sqrt{2}$ *FFT cell size for 5 cm beam diameter
Phase-averaged resolution	5 x 5	Discretization of distributed charge in phase averaging

COMPUTATIONAL RESULTS FOR MATCHED BEAMS

Linear calculations for matched beams with no space charge were carried out for reference purposes. For all three lattices the results of these calculations reveal that each particle independently performs betatron oscillations in response to the focusing forces of its lattice. The action variables, J_x and J_y, are conserved for each macroparticle tracked, where J_x and J_y are given by

$$J_x = \frac{1}{2}\{\frac{x_\beta^2}{\varepsilon_x \beta_x} + \frac{\beta_x p_x^2}{\varepsilon_x}\}$$
$$J_y = \frac{1}{2}\{\frac{y_\beta^2}{\varepsilon_y \beta_y} + \frac{\beta_y p_y^2}{\varepsilon_y}\}.$$
(5)

In Eq. (5) $x_\beta = x - \eta_x \frac{\delta p}{p_0}$, $y_\beta = y$, $p_x = x_\beta' + \frac{\alpha_x}{\beta_x} x_\beta$, $p_y = y_\beta' + \frac{\alpha_y}{\beta_y} y_\beta$, η_x is the closed orbit dispersion function, and $\alpha_x, \beta_x, \alpha_y, \beta_y$ are the Courant-Snyder parameters. Also, the tunes v_x and v_y of each particle are calculated to be precisely 5.85 and 5.70, respectively. Furthermore, the rms emittance values stay fixed at 100 mm-mrad; the particle distributions remain matched K-V distributions; and the beam cross section remains unchanged.

The particle core model with matched boundary conditions provides a periodic analytic representation of the space charge forces for a matched K-V beam, and thus is expected to yield quiescent behavior for the tracked particles. For all three lattices, the calculations reveal exactly this behavior: almost no beam spreading and essentially no halo generation. The main effect of the space charge forces is to provide tune shifts, observed for all three lattices to be in the range 0.1 ± 0.02, which agrees well with the calculated Laslett tune shift of 0.105 for the uniform focusing lattice. Examination of the distribution of actions of the individual macroparticles reveals only a slight spreading due to the space charge force, but the macroparticle beam remains essentially a K-V distribution.

A different picture emerges from the self-consistent PIC calculations, particularly for the doublet lattice. For the uniform focusing and FODO lattices, there is somewhat more spreading of the beam cross section than with the particle core model, but there is little to no halo generated and the results are substantially the same as for the particle core model. However, for the doublet lattice significant halo generation is observed in the $y - y'$ phase plane. This is evident in Figure 2, which shows the final particle distribution in real $x - y$ space for the doublet lattice, using the self-consistent PIC and linear (no space charge) models, respectively. A modest beam core growth, comparable to that observed for the doublet and uniform focusing cases, is accompanied by a sizable number of halo particles sprouting out along the positive and negative y axes. Other quantities corroborate this halo growth. Figure 3 plots the particle actions for the three lattices after 1250 turns as well as the initial actions in the K-V distribution. The spread in the actions away from the initial K-V line is entirely due to the space charge forces. Furthermore, the actions diverge mostly in the doublet lattice, primarily for particles with small x and large y motion, and the divergence is mostly in y. Figure 4 shows the tunes of the particles for the three lattices, and the doublet tunes have the greatest spread, with the y tune spread exceeding the x tune spread. The Laslett maximum tune shift for the K-V distribution in the uniform focusing case is calculated to be 0.105, which leads to predicted tune values of $v_x = 5.745$ and $v_y = 5.595$, consistent with Fig. 4.

Consider now the time histories of the beams. Figure 5 plots the histories of the rms y emittances of the beams for the three lattices. For the uniform focusing and FODO lattices there is little increase, but for the doublet lattice there is an increase of about 3%, occurring mostly between 400 and 800 turns. The rms values of the x emittances remain basically unchanged for all three lattices, so again the growth is observed in the y motion in the doublet lattice. To gain a somewhat different picture of the halo, we arbitrarily define halo particles to be those macroparticles that acquire betatron oscillation amplitudes in either x or y greater than some constant factor times the maximum betatron oscillation amplitude of the matched K-V beam in the corresponding direction. We choose the constant halo factor to be 1.25, so that halo particles are those with either x or y oscillation amplitudes exceeding the corresponding maximum K-V amplitudes by at least 25%. Using this definition, Fig. 6 plots the number of halo particles versus turns for the three lattices. While the uniform focusing and FODO lattice beams display $\leq 0.1\%$ halo, the halo growth for the doublet lattice is $> 1\%$, and it occurs mostly between 400 and 800 turns.

Because matched K-V distributions are equilibrium solutions of the Vlasov equation and because the initial beams in these calculations are randomly generated matched K-V distributions, the beam spreading and halo generation in the PIC model calculations must be caused either by the deviation of the macroparticle beam from an exact K-V distribution or by macroparticle collisions resulting from the numerical treatment of the beam in this model. In order to distinguish between these possibilities, we consider the results of the phase-averaged PIC model. In the phase-averaged PIC model the numerical calculation of the space charge force is identical to that in the full PIC model: macroparticle charges are binned to a spatial grid and treated by FFTs to obtain the forces. All numerical prescriptions, including grid choice, FFT representation, and smoothing parameter, are identical in the two models. The only difference between the phase-averaged PIC model and the full PIC model is the method of distributing the charge of each macroparticle to the FFT gridpoints. A K-V distribution has the property of being phase-averaged, i.e. it yields itself upon phase averaging. Therefore, to the extent that the macroparticle distribution is a K-V distribution, the process of phase averaging will not effect it. Phase averaging will smooth some of the fluctuations between the numerical and analytic K-V distributions. Thus, the behavior of the phase-averaged PIC model in comparison with the full PIC and particle core models is revealing. If the beam spreading and halo generation using the phase-averaged PIC model are comparable to those obtained with the full PIC model, then the numerical aspects of the calculations may be the cause. However, if the phase-averaged PIC results are comparable to those of the particle core model, then the PIC model halo generation is likely to be caused by fluctuations in the macroparticle distribution.

The results of the phase-averaged PIC model, including phase space puncture plots, actions, tunes and emittances, are essentially identical to those of the matched particle core model. The agreement of the phase-averaged PIC model and the matched particle core model both with each other and with the anticipated behavior for a matched K-V beam suggest that fluctuations in the macroparticle distribution cause the full PIC model to generate beam spreading and halo for the doublet lattice. Calculations using more macroparticles, finer FFT resolution, and reduced smoothing parameter values are underway to investigate this notion, and partial results will be given here for the doublet lattice.

Figure 7 plots the fraction of halo particles over 1250 turns for three cases varying only in the number of macroparticles: 7680, 30720, and 122880 macroparticles. In all three cases the halo saturates between 1.0% and 1.5% of the beam by 800 turns. The only observable variation is that the case with 122880 macroparticles destabilizes later, but more suddenly than the beams with fewer particles. In Fig. 8, 30720 particles are used and the grid and smoothing parameters are varied. Again all cases result in saturated halo fractions of between 1.0% and 1.5% of the beam by 800 turns. It is also seen that, for a given number of macroparticles, increasing the FFT resolution or decreasing the smoothing parameter increase the fraction of halo particles. All PIC model calculations carried out thusfar for the doublet lattice corroborate the results given here. Indeed, the beam cross section shown in Fig. 2 was calculated using 122880 macroparticles.

Although we have not yet identified the exact cause of the instability in the doublet lattice, it is interesting to return to the beam focusing area, $A_f = \pi\sqrt{\beta_x \varepsilon_x \beta_y \varepsilon_y}$, presented in Fig. 1. This area undergoes the greatest and most rapid fluctuations for the doublet lattice. Because the space charge forces depend on the beam area through the charge density, Fig. 1 shows that these forces undergo the greatest, most rapid fluctuations in the doublet lattice. Thus, it is perhaps not surprising that we observe halo growth in this configuration.

COMPUTATIONAL RESULTS FOR MISMATCHED BEAMS

Calculations with mismatched beams have been carried out for all three cases using the particle core model. The initial envelope radii were increased by 25% over their matched values and the macroparticle distributions were made consistent with the envelopes. For all three lattices the behavior over 1250 turns was benign, with no significant emittance growth nor halo generation. PIC calculations of these cases are underway and complete results will be given in a future paper.

CONCLUSIONS

This paper is the start of a detailed study of space-charge-induced halo and emittance growth in the SNS ring. Matched K-V beams were tracked for 1250 turns through three different lattices. Both beams and lattices were given global parameters similar to the SNS. All calculations were carried out identically except for the treatment of space charge. Compared to calculations without space charge, both particle core model and phase-averaged PIC model calculations yield only slight beam spreading and no halo generation for all three lattice configurations. Hence, the results using both particle core and phase-averaged PIC models are consistent with the expected steady state behavior of matched K-V beams.

The results using the self-consistent PIC model yield somewhat more beam spreading and, primarily for the doublet lattice, significant halo creation. Because the same parameters were used in the numerical algorithms for the self-consistent and phase-averaged PIC calculations, it is likely that the observed behavior in the self-consistent PIC calculations is a manifestation of fluctuations in the randomly initialized K-V distribution of macroparticles, coupled to the lattice. The behavior of the doublet calculations is maintained throughout the parameter variation of convergence studies.

Calculations with mismatched beams have been carried out using the particle core model, and so far no unstable behavior has been observed.

In all these calculations, observed space charge tune shifts were found to be in good agreement with the calculated Laslett tune shift of 0.105, and because of the use of matched K-V distributions the tune spreads were found to be small (< 0.05).

* Research on the Spallation Neutron Source is sponsored by the Division of Materials Science, U.S. Department of Energy, under contract number DE-AC05-96OR22464 with Lockheed Martin Energy Research Corporation for Oak Ridge National Laboratory.

REFERENCES:

1. National Spallation Neutron Source Conceptual Design Report, Volumes 1 and 2, NSNS/CDR-2/V1,2, (May, 1997).
2. Jones, F., *Users' Guide to ACCSIM*, TRIUMF Design Note TRI-DN-90-17, (1990).
3. Holmes, J.A., Galambos, J.D., Blaskiewicz, M., Lee, S.Y., and Olsen, D.K., *Dynamic Space Charge Models for Computational Applications in Rings*, SNS/ORNL/AP TECHNICAL NOTE Number 006, (May, 1998).
4. Holmes, J.A., Galambos, J.D., Olsen, D.K., and Lee, S.Y., "RMS Envelope Equations for Rings," in *Proc. of the Workshop on Space Charge Physics in High Intensity Hadron Rings*, Shelter Island, NY, May 4-7, 1998, to be published by American Institute of Physics.
5. Galambos, J.D., Holmes, J.A., Olsen, D.K., Whealton, J., and Blaskiewicz, M., "Preliminary SNS H⁻ Injection Calculations with Space Charge using a PIC Model," in *Proc. of the Workshop on Space Charge Physics in High Intensity Hadron Rings*, Shelter Island, NY, May 4-7, 1998, to be published by American Institute of Physics.

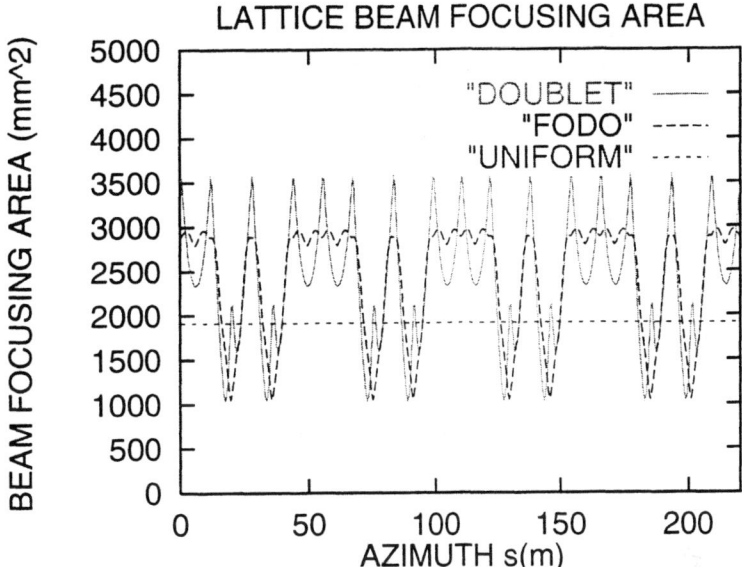

FIGURE 1. Beam focusing area $A_f = \pi\sqrt{\beta_x \varepsilon_x \beta_y \varepsilon_y}$ for doublet, FODO, and uniform focusing lattices.

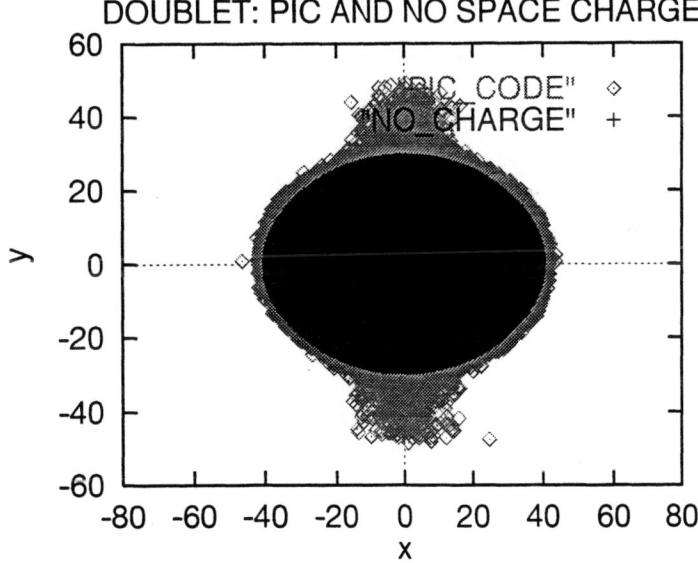

FIGURE 2. Final beam cross sections in $x-y$ plane for PIC and linear calculations in doublet lattice.

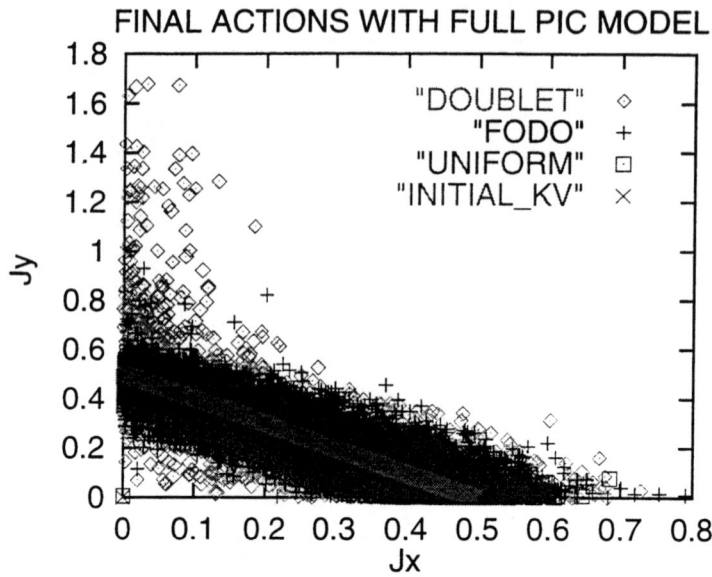

FIGURE 3. Final macroparticle actions for PIC calculations of doublet, FODO, and uniform focusing lattices, and initial K-V actions.

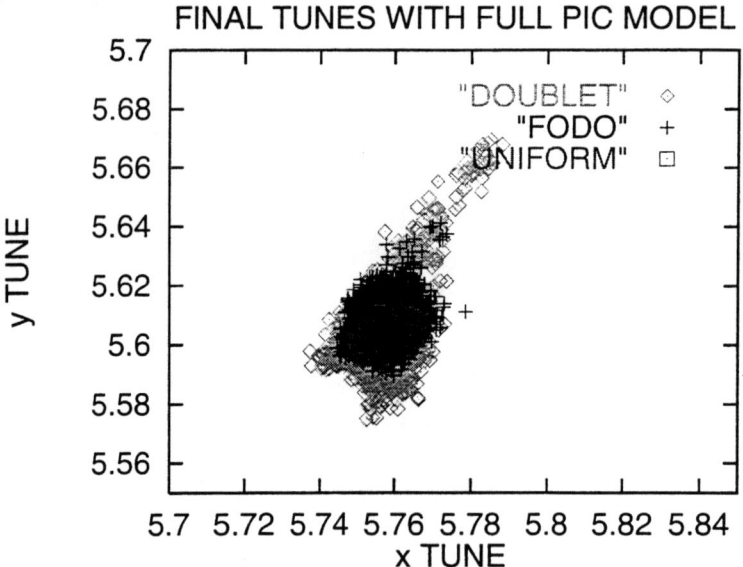

FIGURE 4. Final macroparticle tunes for PIC calculations of doublet, FODO, and uniform focusing lattices.

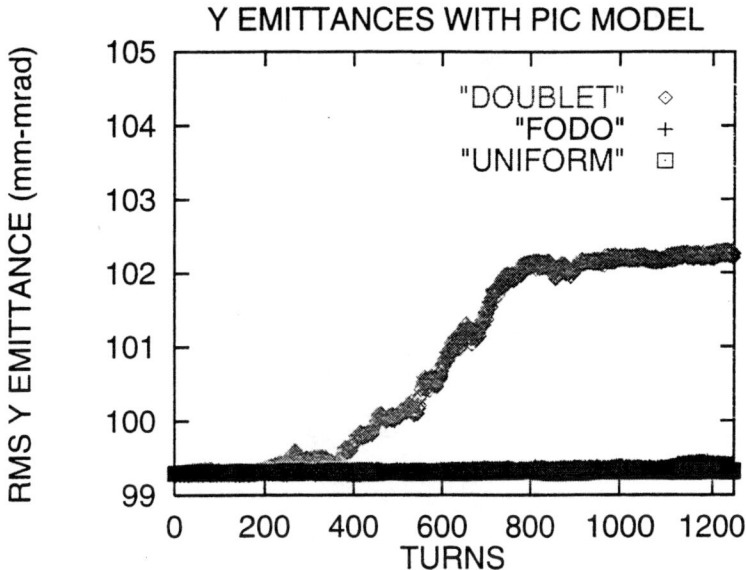

FIGURE 5. y emittance versus turns for PIC calculations of doublet, FODO, and uniform focusing lattices.

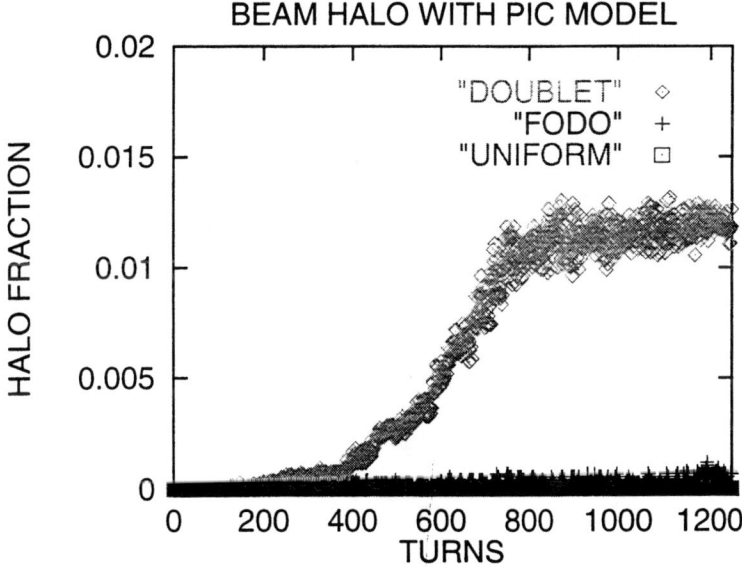

FIGURE 6. Beam halo fraction versus turns for PIC calculations of doublet, FODO, and uniform focusing lattices.

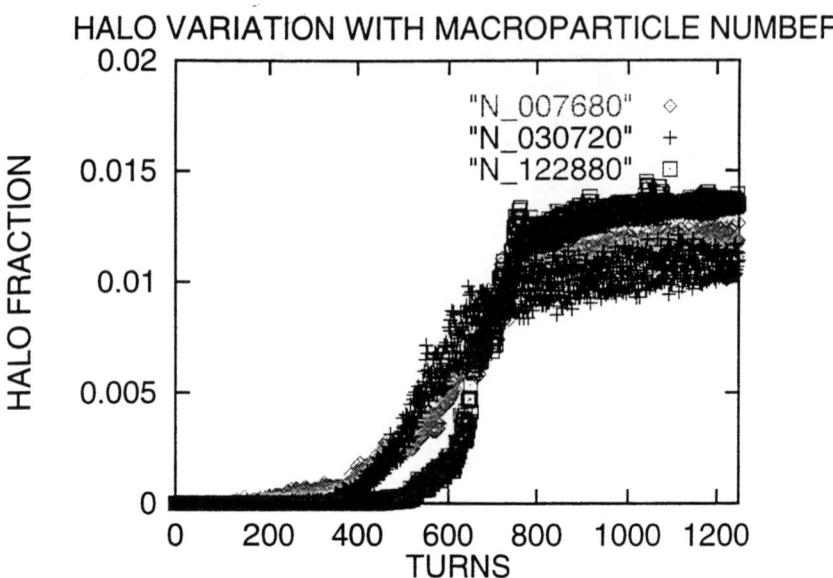

FIGURE 7. Beam halo fraction versus turns for PIC calculations with varying numbers of macroparticles in doublet lattice.

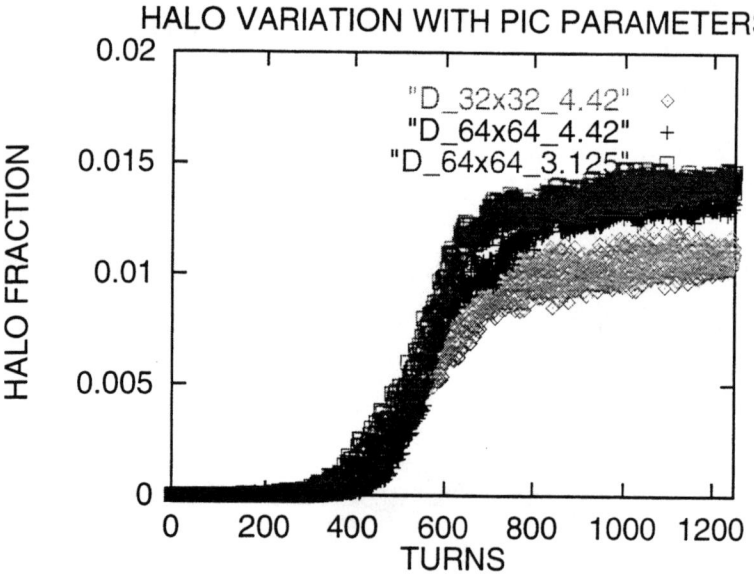

FIGURE 8. Beam halo fraction versus turns for PIC calculations with varying FFT grid and smoothing parameter in doublet lattice.

A Hybrid Fast-Multipole Technique for Space-Charge Tracking With Halos

F.W. Jones

TRIUMF
4004 Wesbrook Mall, Vancouver, Canada V6T 2A3

Abstract. The simulation of injection and accumulation of intense proton beams in synchrotrons and accumulator rings requires a flexible and robust treatment of space-charge effects. In particular, the simulation must be able to correctly incorporate the large-amplitude beam halo which is known to arise from a variety of mechanisms.

This paper describes a technique for the efficient and accurate evaluation of transverse space-charge forces in a simulated beam that may include halo particles at arbitrarily large amplitudes. The technique uses the Fast Multipole Method (FMM), together with some elements of Particle-In-Cell (PIC) methods, to evaluate the space-charge field at the location of each macroparticle. Timing comparisons show that when a large-amplitude halo is present this hybrid technique is significantly faster than plain FMM and also offers performance advantages over PIC/FFT field calculations with equivalent spatial resolution.

The hybrid FMM method has been implemented and tested in the multiparticle tracking and simulation code ACCSIM and some initial results are presented.

INTRODUCTION

For the design and operation of the next generation of intense proton rings for spallation sources and other applications, there is a need for comprehensive large-scale simulation programs to study the injection, capture, accumulation, collimation, and loss management processes in these rings. The 3D multiparticle tracking and simulation code Accsim [1] was developed to bring together various simulation models for these processes into a single ready-to-run package. It provides a relatively easy and efficient way to explore and optimize different options and parameters for injection schemes [3], collimation systems, and other aspects of the machine design.

Because of stringent loss requirements and other constraints of high-intensity operation, there is a need to improve the accuracy of Accsim and in particular to upgrade the treatment of space-charge effects. Especially important is the correct treatment of beam halo: the evolution of halo particles, their interactions with the beam core, and their large-amplitude motion. To cite just one of several sources of beam halo, in typical H^- injection schemes a small fraction of the beam will be scattered to large angles in the stripping foil and will occupy a halo that spans the entire aperture of the machine.

For Accsim, which has traditionally provided relatively fast algorithms and short turnaround times on typical desktop machines, the incorporation of a self-consistent space charge model presents a problem: to find a sufficiently fast simulation with acceptable accuracy. Such simulations are notorious consumers of cpu and memory resources and are often run on supercomputers. In the literature, one sometimes sees encouraging statements like "we have developed a fast method of solving ..." but they are usually followed later by "we implemented and tested the method on a Cray ...".

The following sections describe some methods that are available for calculating space-charge fields and some initial experience with one such, the Fast Multipole Method, which is able to easily and accurately accommodate large-amplitude halo particles. As will be shown, for beams the FMM method can be dramatically speeded up by hybridizing with a PIC-style grid covering the dense core of the beam, to the point where it outperforms a conventional PIC/FFT calculation of the same spatial resolution when a large halo (e.g. 6σ) is present.

REQUIREMENTS FOR SPACE-CHARGE MODEL

Accsim, which is typically run with ensembles of 10^4–10^5 particles, performs kick/matrix tracking (through an extended TRANSPORT formalism) in 6-dimensional phase space. Matrices represent the basic magnet lattice and kicks represent longitudinal space-charge effects, rf cavities, and thin multipoles. The program provides a variety of injected beam distributions and allows many quantities to be user-programmed as time-varying functions to facilitate multiturn injection, phase-space painting, orbit bumps, rf capture and acceleration. There are also more exotic features such as multiple rf systems, barrier rf, and mini-wire septum skimmers for collimation. Simulation of particle interactions with matter (injection foil, collimators, vacuum chamber walls, etc.) includes plural or multiple Coulomb scattering, energy loss, and an elementary treatment of nuclear interactions.

During tracking, the calculation of longitudinal space charge effects is done at regular intervals around the ring in the standard way by binning particles in rf phase (with optional digital smoothing), deriving the longitudinal space-charge potential, and applying the appropriate energy kick to each particle (the algorithm constitutes a symplectic first-order integrator).

For transverse space charge, until the currently-reported work there has been no general self-consistent treatment, although the code does incorporate a package [2] which evaluates, through an analysis of the betatron amplitudes and longitudinal charge density, the distribution of transverse space-charge tune shifts in the beam.

Such a feature-laden code as Accsim entails that any upgrade to the space-charge treatment: (1) must coexist with many other computational elements for stepwise tracking and simulation; (2) must allow beam distributions of arbitrary shape, size, and position, without any assumed symmetries; (3) must be reasonably efficient, in keeping with the rest of the program; and (4) should not have too many calculation parameters that must be chosen by the user.

Without ascending to the realm of supercomputers, a self-consistent fully 3D calculation is out of reach for the present, but Accsim lends itself well to a "$2\frac{1}{2}$D" model with semi-independent transverse and longitudinal integration steps and force calculations.

The term "$2\frac{1}{2}$D" refers to the mixed 2D and 3D aspects of the model: the model is 2D in the sense that the nominal transverse space-charge field is evaluated by viewing all macroparticles as 2-dimensional (line) charges. Accsim's transfer-matrix/kick formalism tracks in distance steps rather than time steps. Each particle carries a time-difference (rf phase) coordinate locating it in the bunch, and the stored transverse coordinates are in fact a "snapshot" in space rather than in time. However, the model is 3D in the sense that the space-charge force on a given macroparticle is scaled according to the longitudinal charge density at its position in the bunch, thus coupling the longitudinal motion into the transverse tune space.

The $2\frac{1}{2}$D model assumes that the bunch length is much larger than the transverse beam size, and that there is no significant correlation between transverse and longitudinal distributions (in typical Accsim applications, any such correlations in the injected linac beam will be masked in the ring due to the multiturn injection and phase-space painting).

N-BODY PARTICLE SIMULATION METHODS

As well as in Beam Physics, N-Body particle simulation methods [4,5] have been developed and utilized in disciplines such as Astronomy, Plasma Physics, Molecular Dynamics, and Computer Science. The solution methods can be characterized as either "Particle-Mesh" or "Particle-Particle" or some combination of the two.

Particle-Mesh methods

In space-charge simulations for beams, the use of mesh-based methods seems almost universal. On a 2D or 3D mesh, charges are assigned to the mesh points according to Nearest-Grid-Point, Cloud-In-Cell, or other schemes. Poisson's equation is solved (usually by FFT and/or cyclic reduction) for the potential and, by differentiation, the electric field components E_x and E_y at the mesh points. The fields at each particle location are interpolated by applying the inverse operation of the charge assignment.

Although there is pre-processing and post-processing overhead, the computation time is usually dominated by the Poisson solver and therefore depends on the number of grid points N_g, for example with FFT the time is $\mathcal{O}(N_g \log N_g)$. For modelling beams with large halos, this dependence has negative implications because maintaining the spatial resolution of the grid in the core region while including large-amplitude halo particles may entail a dramatic uplift in the number of grid points. For example, if a 100×100 grid is used to model the core of a beam, extending the grid to cover a halo extending to 3 times the core size requires a factor of ~10 increase in computation time.

Tree codes

For beams with large halos, the Particle-Particle "Tree code" methods at first look more promising. In these methods, there is no discretization via mesh points with assigned charge values and potential values. Rather, the potential and/or field at each particle location is derived by surveying the other charges and invoking the 2D or 3D force law. For N_p particles this is fundamentally an $\mathcal{O}(N_p^2)$ operation, but significant economies can be obtained by lumping together charges at various distance scales. The tree codes utilize quad-tree subdivision of the domain of charges, whereby the solution for each particle consists of traversing the tree. Tree-code computation times depend on the number of *particles* and are typically between $\mathcal{O}(N_p)$ and $\mathcal{O}(N_p \log N_p)$.

Because of the quad-tree subdivision and direct force evaluation, these methods can easily accommodate distributions with sparse populations of large-amplitude (and growing) halo particles. One drawback of such particle-particle methods is that of "close encounters", where one must intervene with an appropriate physical model when the distance between particles becomes small. For beams, there is also the issue of imposing boundary conditions at chamber walls in order to correctly account for image forces, by the placement of an appropriate set of fictitious charges. In the few-GeV regime of typical Accsim applications these forces are of reduced significance [7] and in the present study are not included.

Fast Multipole Method

The adaptive Fast Multipole Method (FMM) of Greengard [6] and Rokhlin is one of the best-performing and most accurate tree-code algorithms. It uses adaptive subdivision to break down the problem domain into a quad-tree in which the "leaf" regions each contain a small number (e.g. 40) of charges and uses the properties of multipole expansions and of "well-separated" groups of charges to optimize the computation of the potential and/or field at each particle location.

The FMM method offers a number of benefits: (1) the computation time scales as $\mathcal{O}(N_p)$; (2) there is a choice of computing field, potential, or both; (3) it is not sensitive to the shape, size, or smoothness of the charge distribution; (4) there are few free parameters to be chosen, and the main thing the user has to worry about is setting the desired accuracy of the computation.

A well-documented and versatile 2D implementation of FMM, the DAPIF2 package of Fortran subroutines, was obtained from author Greengard. The package is self-contained and was easily added to Accsim, where two options for using it were implemented: (1) a stand-alone test mode, where the potential and field for the current particle ensemble are calculated for benchmarking and visualization purposes, and (2) a full-tracking mode where the FMM field calculation is done at

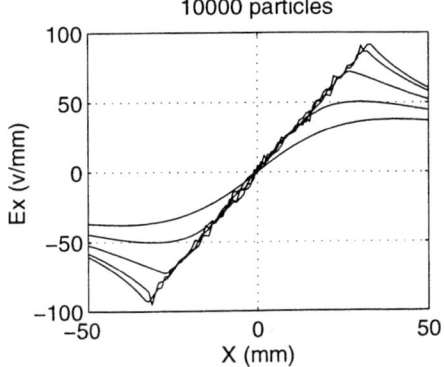 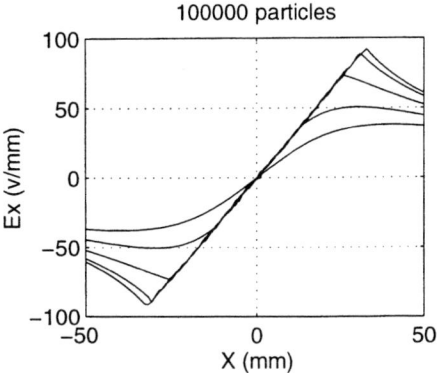

FIGURE 1. FMM computed curves of electric field component E_x along horizontal lines at $y=0, 10, 20, 30, 40$mm crossing a K-V beam of 32.6mm radius.

user-specified intervals in the ring, the force on each particle is derived, and angular kicks representing the force integral are applied.

To support space-charge testing a 4-dimensional K-V beam generation option was added to Accsim by modifying the distribution generator to populate x–y correlated phase-space shells. Figure 1 shows results of the FMM field calculation for such a K-V distribution, of circular cross section, representing a DC beam of $2 \cdot 10^{14}$ protons in a ring of 190.4m circumference.

For $\sim 10^4$ particles the FMM method exhibits field noise due to charge density fluctuations in the simulated beam. Raising the number of particles to $\sim 10^5$ reduces the noise to a respectable level. At this point one should examine computation times, as shown by the "FMM" curve in Figure 2. For up to 100000 particles, the FMM algorithm as implemented by Greengard does not stray too far from $\mathcal{O}(N_p)$ dependence, but the timings indicate that it is not practical for "production" use in Accsim, which may be doing hundreds of such calculations per turn and accumulating beam for thousands of turns.

A HYBRID FAST MULTIPOLE TECHNIQUE

To gain relief from the N_p-dominated computation, it is natural to think of overlaying a grid, not over the whole domain (because of the "halo factor" objections described above), but just on the dense core region of the beam. Then one could assign PIC-style charges to the grid points and invoke an $\mathcal{O}(N_g \log N_g)$ Poisson solver. The difficulty is that Poisson's equation cannot be solved without specifying boundary conditions at the edges of the grid, where varying charge densities exist and for which no boundary information is available beforehand.

At this point an experiment comes to mind: the FMM algorithm accepts charges of arbitrary location and magnitude, so give the PIC-style charges to FMM as "superparticles" located at the grid points and representing the nearby macroparticle

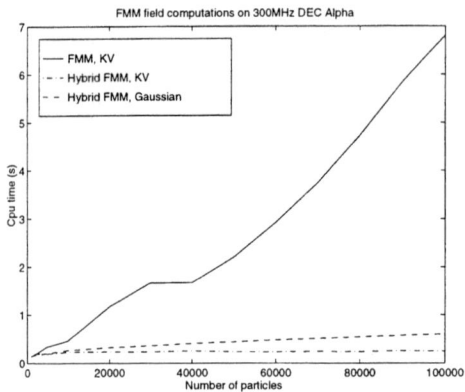

FIGURE 2. Timings for FMM field solution for K-V beam, and hybrid FMM solution for K-V beam and Gaussian beam.

charges. After assembling the charge/location data for the grid points, throw in all the charges *outside* the grid region (i.e. the halo particles) as discrete unit charges. Then, give the whole ensemble of charges to FMM which will return the field (1) at each grid point, and (2) at each halo particle. The efficiency benefit of this scheme is dramatic, as shown in the "Hybrid FMM" curves in Figure 2, where a 1mm grid ($65 \times 65 = 4225$ grid points) is used. The cpu time requirement is reduced to:

$$\text{cpu time} \propto \mathcal{O}(N_\text{g})_\text{core} + \mathcal{O}(N_\text{p})_\text{halo}$$

Although the FMM algorithm is designed to handle arbitrary charge locations, it does a reasonably quick job of finding the fields on a regular grid of charges.

Note: the hybrid technique described here is different from, and in a sense opposite to, the Particle-Particle Particle-Mesh (P^3M) [4,5] and Tree-Code-Particle-Mesh (TPM) [4] hybrids wherein a Particle-Mesh framework is supplemented by direct-sum or tree-code calculations to get more accurate short-range forces.

The Algorithm

In the simplest first-order form of the algorithm, particle charges are assigned to the nearest grid point and after the FMM solution the field seen by the particle is taken from the nearest grid point.

1. For each macroparticle...
   - If the particle lies within the grid, add its charge to the nearest grid point.
   - If the particle lies outside the grid, add it to the list of particles to be passed to the FMM method, and record its position (index) in the list.

2. For each grid point...

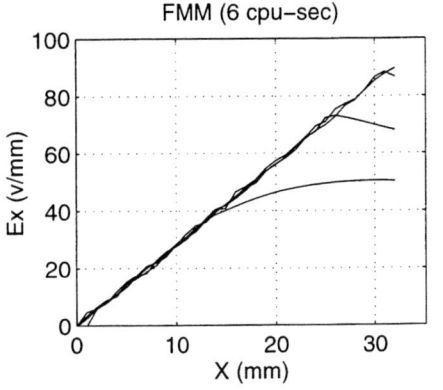 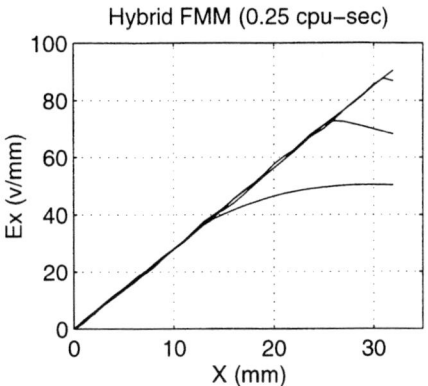

FIGURE 3. Field curves (E_x for $y=0, 10, 20, 30, 40$mm) of K-V beam solution for 90000 particles by FMM and by hybrid FMM on a 65×65 grid.

- If the grid point has zero charge, skip it.
- If the grid point has nonzero charge q, add it as a "particle" of charge q to the list of particles for the FMM routine. After doing this, replace the charge quantity in the grid array with the grid point's index in the FMM particle list.

3. Call the FMM routine, which returns arrays of field components E_x and E_y.

4. For each macroparticle...
   - If the particle lies within the grid, find its nearest grid point, get the grid point's list-index from the grid array, and use the list-index to recover the FMM field components.
   - If the particle lies outside the grid, use the particle's recorded FMM list-index to recover the field components directly.
   - Apply angular kicks $\Delta x'$ and $\Delta y'$ representing the force integral for the particle.

Figure 3 shows E_x field curves for the K-V beam described above, on a 1mm regular grid, by plain FMM and by the above algorithm. The hybrid FMM provides not only an accurate solution that agrees closely with plain FMM, but also provides a smoothing of density fluctuations and consequent noise in the field values.

Extending the algorithm to Cloud-In-Cell charge assignment and field interpolation requires slightly more book-keeping. For each particle within the grid, the charge is divided among the nearby grid points and the location (array indices) of the nearest grid point to the particle is recorded. If necessary, storage can be economized by using one-dimensional indexing into the grid array. As before, each grid point storage location does double duty, first to accumulate charge and then to record the grid point's index into the FMM particle list. As before, particles outside the grid record their individual indices in the FMM list.

After FMM solution, the outside particles get their field value directly through the list index. The inside particles, via their recorded grid indices, get field values from the neighbouring grid points (via the grid points' list indices), and apply the relevant interpolation formula.

The overhead for pre- and post-processing is similar to PIC, and tests with up to 10^5 particles showed that it was insignificant compared to the field solution time.

Timing Comparison with PIC/FFT Solution

Hybrid FMM was compared with a PIC/FFT calculation for an ensemble of 50000 macroparticles representing a beam with a dense core of parabolic profile, extending to $\sim 2\sigma$ in each transverse dimension, and a sparse halo containing approximately 5% of the beam and extending to $\sim 6\sigma$. Timings were done for the electric field solution only, not including the pre- and post-processing overheads which are quite small for both methods.

Grid Size		CPU time (s)	
Core	Total	PIC/FFT	Hybrid FMM
32×32	128×128	0.16	0.16
64×64	256×256	0.73	0.30
128×128	512×512	4.97	0.94

TABLE 1. Timings on 300MHz DEC Alpha for 50000-particle field solutions

As shown in Table 1, different grid resolutions for the core region were chosen and field-solves were done for the core+halo region using both methods. It should be noted that in the PIC/FFT solutions, the periodic boundary conditions at the edges of the grid require the addition of a "guard band" of $\sim 2\sigma$ in order to avoid distortion of the fields in the solution region. Inside this guard band, the two methods agree well in both the core and halo regions, as seen in Figure 4.

Given the variety and sophistication of fast Poisson-solvers available, it is not claimed that these tests are definitive or that hybrid FMM always gives the best performance for any core+halo problem. A more conservative conclusion is that hybrid FMM can provide a robust and accurate solution of these problems, at high spatial resolution in both core and halo, with a low level of programming complexity and with performance that exceeds or is competitive with Poisson solvers.

Advantages of Hybrid FMM

To summarize some advantages of the hybrid FMM technique:

1. A dramatic speed increase over plain FMM. For beams with a populous core and a sparsely populated halo, better performance that an equivalent PIC/FFT solution.

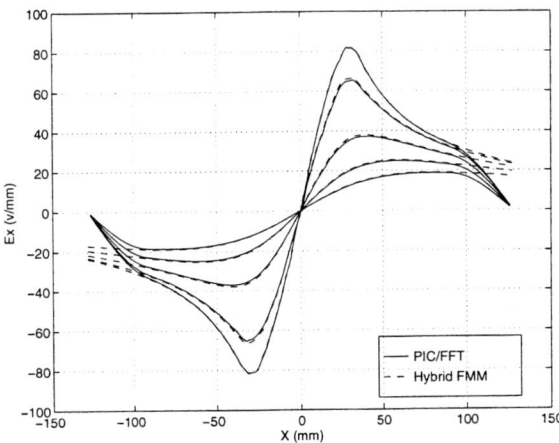

FIGURE 4. Field curves E_x for $y=0, 10, 20, 30, 40$mm for beam with 6σ halo, by PIC/FFT and by hybrid FMM, with core grid of 64×64 and core+halo PIC grid of 256×256

2. The spacing of the gridded charges smooths much of the field noise due to density fluctuations in the core region.
3. Unlike Poisson solvers, grid points with zero charge are left out of the computation.
4. The grid size and shape can be chosen freely, as long as it encloses enough particles to achieve the desired computation speed and the desired smoothness in the core region. The grid does not have to be regularly-spaced or to follow any particular geometry, as long as there is a consistent way to assign charges and interpolate fields.
5. The problem of "close approach" of particles is much reduced. In the sparse halo it can be dealt with by imposing a cut-off radius.

TRACKING TESTS IN ACCSIM

To evaluate the accuracy and utility of the hybrid FMM algorithm some initial tracking tests with a K-V beam were conducted in Accsim. The first set of tests comprised tune and emittance measurements in a realistic lattice (a 190m proton driver ring for a spallation source), as shown in Figure 5. Runs were conducted with up to 100000 macroparticles representing a beam of $2 \cdot 10^{14}$ protons.

For the stepwise force integration, a minimal first-order configuration was used: field evaluation and integration steps were performed at the center of each quadrupole and at the center of each drift, with angular kicks $\Delta x'$ (and similarly $\Delta y'$) derived for each particle by:

$$\Delta x' = \frac{E_x L_s}{\beta^2 \gamma^3 m_0 c^2}, \tag{1}$$

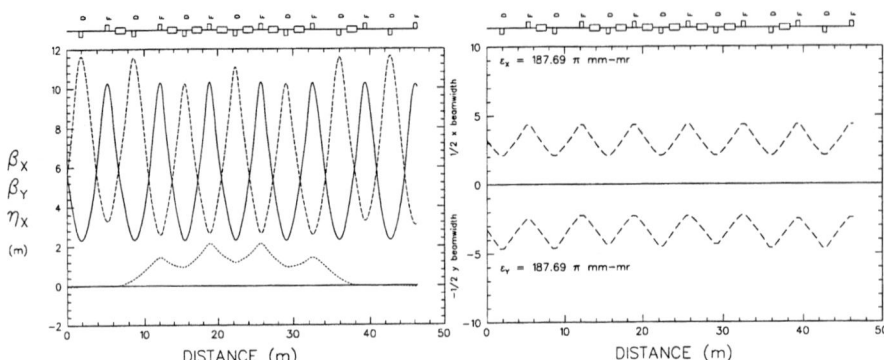

FIGURE 5. Optical functions and envelopes for one superperiod of Accsim test lattice

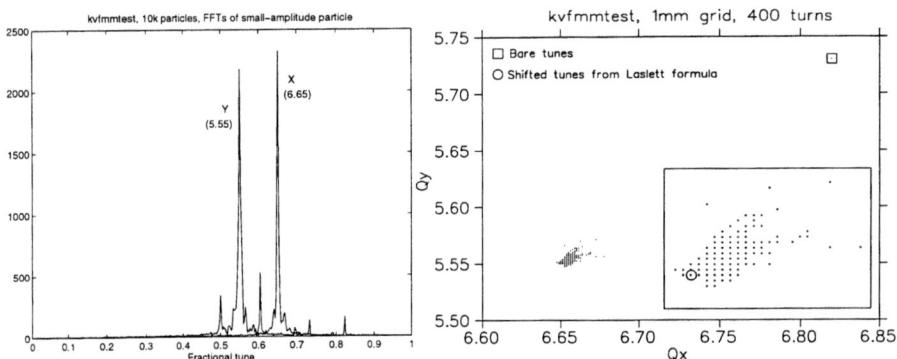

FIGURE 6. Tune measurements: (a) Superimposed x and y FFTs of small-amplitude test particle, (b) Scatterplot, with magnified inset, of tunes for 10000 particles measured by counting zero-crossings (many dots are superimposed due to the limited precision).

where L_s is the distance to the next integration point.

For comparison with theory, a DC beam with $\Delta p/p = 0$ was chosen. Figure 6 shows particle tune measurements for a single small-amplitude test particle (by FFT) and for all particles in the beam (by counting zero-crossings during tracking). For this case, which involved 10000 macroparticles tracked over 400 turns (400 × 28 = 11200 FODO cells) the K-V distribution is well modelled by Accsim tracking, which shows only ~0.3% tune spread over the entire macroparticle population.

The measured single-particle tunes of 6.65 and 5.55 are in good agreement with the theoretical shifted tunes 6.652 and 5.547 as predicted by the generalized Laslett tune-shift formulas [8]

$$\Delta \nu_{x,y} = \frac{r_p N_I (q^2/A) F_{x,y} G_{x,y} \bar{H}_{x,y}}{\pi \varepsilon_{x,y} \beta^2 \gamma^3 B_f} \quad (2)$$

where F, G, $\bar{H}$ are form factors describing the image forces, transverse distribution, and beam aspect ratio. For the case in point, $F=G=1$, $\bar{H}_x=0.478$, and $\bar{H}_y=0.522$.

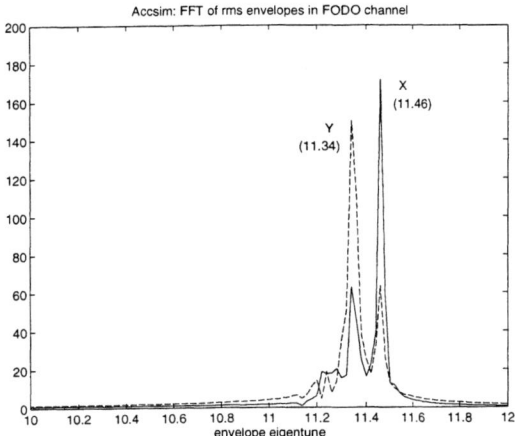

FIGURE 7. FFTs of RMS envelopes of mismatched K-V beam in FODO channel

The bare tunes $(6.820, 5.730)$ are also seen in the FFT, indicating a weak coherent dipole mode which is not subject to space-charge detuning [7] and probably arises from small numerical asymmetries in the charge distribution

Envelope tests used a pure FODO structure obtained by removing the bends from the above-mentioned lattice. The quadrupole strengths were adjusted to achieve a small tune split and a nearly round beam in the smooth approximation. For a round beam Baartman [7] expresses the envelope eigenfrequencies in terms of the tune-shift $\Delta\nu_x$ and the unperturbed tunes ν_{0x} and ν_{0y} as

$$\nu^2 = 2\nu_{0x}^2 + 2\nu_{0y}^2 - 5\nu_{0x}\Delta\nu_x \pm \sqrt{(2\nu_{0x}^2 - 2\nu_{0y}^2)^2 + (\nu_{0x}\Delta\nu_x)^2} \qquad (3)$$

Figure 7 shows FFT's of the RMS envelopes for a 10000-macroparticle K-V beam tracked for 1120 cells in the FODO channel. The spectrum shows the expected pair of strongly-coupled symmetric and antisymmetric transverse modes, with one mode being dominant in the horizontal plane and the other in the vertical. The measured envelope frequencies of 11.46 and 11.34 are in good agreement with the frequencies 11.450 and 11.320 predicted by the above formula.

SUMMARY

The requirements have been outlined for an efficient and robust transverse space-charge simulation in Accsim and similar codes, and the issue of the large-amplitude halo has been identified as a major concern that may not be economically handled with conventional Particle-In-Cell methods.

The adaptive Fast Multipole Method, one of a class of N-Body particle simulation methods, has been found to be a reliable and accurate method to evaluate space-charge fields in charge distributions with widely-dispersed halo particles, as it is not particularly sensitive to an evolving mix of distance scales.

While pure FMM is not suitable (at least in the context of Accsim) for large-scale beam simulations, it can be effectively combined with elements of the PIC method by overlaying a suitable grid on the more densely-populated core region of the beam, assigning composite charges to the grid points, and letting FMM solve the whole system of core (grid) + halo charges. In this scenario, halo particles of abritrarily large amplitude can be accurately modelled without difficulty or significant computation penalty, and for core sizes of 64×64 and 128×128 factors of 2 and 5 performance gains were observed over a conventional PIC/FFT calculation on the same charge domain.

The hybrid FMM technique has been implemented and tested in Accsim where it has accurately reproduced the expected single-particle and envelope tunes for a K-V beam. The efficiency of this method, and its ability to meet the requirements and time constraints of the code, warrants its further testing and eventual inclusion in a new Accsim release, possibly as one of a range of field-evaluation options for transverse space charge.

ACKNOWLEDGMENTS

Thanks to L. Greengard for his FMM subroutines and documentation, to J. Holmes and J. Galambos for providing PIC/FFT routines, and to M. Craddock, M. Dyachkov, D. Kaltchev, and S. Koscielniak for their advice and encouragement.

REFERENCES

1. F.W. Jones, "Developments in the Accsim Multiparticle Tracking and Simulation Code," *Proc. 1997 Particle Accelerator Conference*, Vancouver, 1997, p. 2597.
2. H.O. Schönauer, "Addition of transverse space charge to Accsim code," TRIUMF Design Note TRI-DN-89-K50 (1989).
3. A. Luccio, D. Maletic, F.W. Jones, "Proton Injection and RF Capture in the National Spallation Neutron Source," *Proc. 1997 Particle Accelerator Conference*, Vancouver, 1997, p. 1882.
4. Amara Graps, "N-Body / Particle Simulation Methods," http://www.amara.com/-papers/nbody.html.
5. R.W. Hockney and J.W. Eastwood, *Computer Simulation Using Particles*, Bristol: Adam Hilger, 1988.
6. Leslie F. Greengard, *The Rapid Evaluation of Potential Fields in Particle Systems*, Cambridge, Mass: MIT Press, 1988.
7. R. Baartman, "Betatron Resonances with Space Charge," *Emittance in Circular Accelerators*, International Workshop on Particle Dynamics in Accelerators, KEK Proceedings 95-7, September 1995.
8. K.H. Reich, K. Schindl and H. Schönauer, "An approach to the design of space-charge limited high intensity synchrotrons," *Proc. 12th International Conference on High-Energy Accelerators*, Fermilab, 1983, p. 438.

PIC Code Simulations of the Space-Charge-Dominated Beam in the University of Maryland Electron Ring

Rami A. Kishek [*], Irving Haber [†], Marco Venturini [*], and Martin Reiser [*]

[*] Institute for Plasma Research, University of Maryland, College Park, MD 20742
[†] Naval Research Laboratory, Washington, DC 20375

Abstract. Numerical simulations using the WARP particle-in-cell code are applied to study the evolution of the space-charge-dominated beam in the University of Maryland Electron Ring (1). The self-consistent simulations play a special role because the nonlinear nature of the dynamics makes accurate analytic predictions difficult. Simulations of a matched beam at the nominal design parameters show negligible degradation in beam quality after 10 turns. The role of lattice element nonlinearities on the beam evolution has been investigated. An rms mismatch is shown to lead to bounded oscillations and an acceptable level of emittance growth. A mismatch in the dispersion function, however, is shown to lead to a higher levels of emittance growth, and dispersion matching is currently under investigation.

INTRODUCTION

The Maryland Electron Ring project (1), which is currently in the design/early construction stages, employs many innovative features to investigate, at modest cost, the fundamental physics of a space-charge-dominated beam undergoing bending. In view of the nonlinear nature of many features of the beam evolution, numerical simulation is a versatile tool for both the design of the experiment and the subsequent interpretation of experimental measurement. Further, the Maryland E-Ring is being developed as a low-cost testbed for the experimental benchmarking of theories and computer codes. The WARP particle-in-cell (PIC) code (2) is specifically designed to efficiently follow a space-charge-dominated beam propagating in an alternating gradient focusing lattice that includes bending elements. The particle orbits are integrated using the fully non-linear self-consistent electrostatic self-field. The WARP code is an attractive choice for the Maryland ring simulations especially because it has been used in simulating the similar Heavy Ion Recirculator at Lawrence Livermore

National Lab (2), and is therefore well-suited to such geometries and already contains a variety of lattice element representations.

In ref. (3) we had presented results of preliminary simulations in a straight channel of nonlinear quadrupoles. This paper extends the on-going ring simulations to follow the beam along the bent lattice, examining the consequences of lens non-linearities, mismatches, and dispersion on the beam quality. Because of space limitations in the mechanical design, the printed-circuit quadrupoles and dipoles used in the ring (4) have a short aspect ratio and consist almost entirely of fringe fields. Although the focusing fields were designed to be linear in an integrated sense, they may be non-linear when integrated over the actual particle orbits, necessitating the use of PIC code simulations that self-consistently include the space charge. Additionally, limited resolution of the diagnostics in the matching section of the injector may result in a mismatch in the rms size of the beam injected into the ring, while a non-zero energy spread in the beam may lead to a dispersion mismatch. WARP simulations have been applied in order to assess the effects of such nonlinearities and mismatches.

We present evidence that the nominal design of the entire ring lattice employing realistic lens elements will be adequate to propagate the beam without substantial degradation, even in the presence of realistic mismatches. The dispersive effects of the ring lattice, on the other hand, do lead to an emittance growth. Ref. (5) presents a set of envelope equations modified to include dispersion, and compares this analytic theory with the WARP simulations mentioned here. The rms envelope-dispersion equations from ref. (5) can be used in matching dispersion so as to reduce the emittance growth from a dispersion mismatch.

Notice that all the simulations described here assume a perfectly aligned lattice and no injection errors. A systematic study of realistic errors in ring construction and operation is currently underway. Note also that we have benchmarked WARP against experimental measurements from a prototype injector at Maryland, and the excellent agreement is discussed in ref. (6). Before we present the results of the simulations, we briefly describe the ring lattice, the setup of the simulations, and the numerical testing. Then we discuss the effects of lens non-linearities and mismatches. Results of simulations of dispersion effects will also be presented, but a full discussion and comparison with analytical theory is deferred to Ref. (5).

RING LATTICE AND SIMULATION SETUP

The design and progress of the electron ring project have been described more fully elsewhere (1). For the purposes of this paper, it will be sufficient to briefly describe the lattice and nominal beam parameters. The Maryland Electron Ring is designed to operate as a scaled testbed for highly space-charge-dominated, non-relativistic, ion beams. The nominal beam current is 100 mA at 10 keV, resulting in a generalized perveance of 0.0015. A nominal (unnormalized 4*rms) emittance of 50 mm-mrad and nominal average beam radius of 10.2 mm ($\sigma_o = 72°$) results in a tune depression (ν / ν_o) of 0.16, placing the beam in the strongly space-charge-dominated regime.

The lattice consists of 36 FODO cells around the 11.52 m circumference ring. Each cell is therefore 32.0 cm long and contains two evenly-spaced printed-circuit quadrupoles (4) and, in between those, a printed-circuit dipole which bends the beam by 10°. As mentioned earlier, the quadrupoles and the dipoles are short relative to the pipe radius (effective length of both ~ 3.7 cm; pipe radius ~ 2.5 cm). This implies that the lattice can be thought of as a combination of bends and straight sections. The quadrupole gradient is 0.078 Tesla/m, while the bending dipole peak field is 0.001535 Tesla.

For these simulations of the transverse dynamics, we use the single-slice version of WARP, which solves for the self-fields on a 2-D mesh, but retains the velocity information in all 3 dimensions (2). The lattice is modeled from first principles by computing the lens magnetic field values on a 3D grid with the aid of a magnetics program, starting with the actual conductor geometry. At every step, WARP interpolates this field data to the particle locations. Thus the dipoles and quadrupoles used in the simulation include the nonlinearities caused by the fringe fields. We load a beam with an initial semi-Gaussian distribution into the field-free region between two quadrupoles, and then follow its evolution along s, the distance traversed by the beam, for a number of turns (typically 10). Except for the case investigating the effects of a mismatch, the beam is initially rms matched into the lattice using the standard rms envelope equations.

In the final simulations we use a mesh of 256×128 cells across the 5 cm-diameter pipe (symmetry in the vertical direction halves the number of cells needed), so that the resulting cell size is sufficient to resolve Debye-length-scale variations in the potential. The number of simulation particles used (100,000) is chosen to minimize effects of numerical collisions, producing a potential that is relatively smooth from cell to cell. The time step used corresponds to a beam advance by 0.4 cm. These numerical parameters, as well as others such as the gridding and truncation of the magnetic fields and the use of symmetry, have all been thoroughly tested. In the regime just described the simulation results are found to be insensitive to a change in the numerical parameters, except for a small numerical growth in emittance that can be reduced by further increasing the number of particles.

SIMULATION RESULTS AND DISCUSSION

The simulations of the perfectly-matched ring (with an energy spread of 1 eV) show no loss in the number of particles and very little emittance growth. Figure 1 shows the beam emittance in the transverse coordinates x and y as a function of

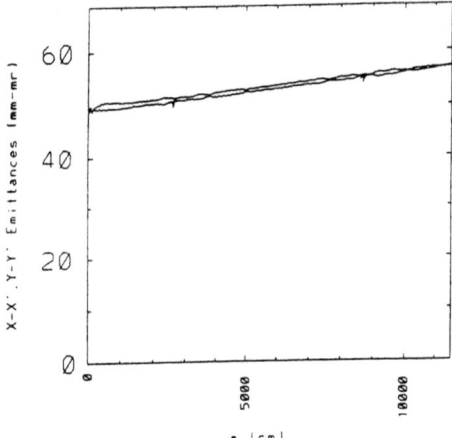

FIGURE 1. Evolution of a matched beam with the nominal design parameters over 10 turns (360 lattice periods). Value of 4*rms emittances strobed every lattice period in between quadrupoles.

distance, s, around the ring, for the nominal parameters described above.[1] The figure shows an increase in emittance from 50 mm-mrad to about 57 mm-mrad (or approximately 14 %) over 10 turns. From the numerical tests, which examine the scaling of the emittance growth with the number of particles, we can deduce that about a third of that growth is numerical in nature (i.e., can be reduced by increasing the number of simulation particles). This results in a long-term emittance growth that averages to less than 1 % per turn.

Since the lens nonlinearities are a strong function of the radial distance from the axis, we explore them by changing the size of the beam in the simulation. This is accomplished by appropriately scaling both the beam current and emittance in order to maintain both the matched condition and the tune depression. Also note that the energy spread of the beam was adjusted to maintain the same ratio of longitudinal to transverse thermal velocities [as we shall see, this ratio affects dispersive behavior, whereas in the present case we wish to keep the influence of dispersion unchanged]. Furthermore, the numerical parameters of the simulation (e.g., # particles, cell size) were appropriately adjusted in the instance where the beam was scaled significantly.

With the proper scaling, a bigger beam will sample more of the lenses' deviations from linearity, and we therefore expect to see a larger emittance growth. This indeed is the case, as Fig. 2 demonstrates. For a small beam, the long-term emittance growth is insensitive to changes in size, indicating that the growth is largely

---

[1] The split between ε_x and ε_y in Fig 1 arises because we launch an initially uniform beam into a channel where the focusing force is nonlinear. Therefore, the beam we specify is initially not in local equilibrium. This is analogous to a non-uniform beam in a linear focusing channel that also experiences a relaxation to equilibrium.

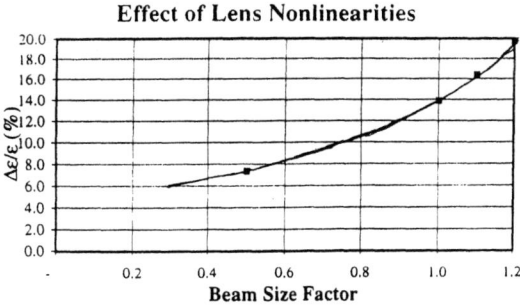

FIGURE 2. Effect of beam size on the long-term emittance growth (over 10 turns), as a percentage of the initial emittance.

numerical. However, as the beam size is further increased, the emittance growth is also observed to increase, indicating that part of that growth (about 2/3 of that seen with the nominal parameters) is caused by the lens nonlinearities. Even so, simulations show that a beam having a radius 10 or 20 % greater than the nominal design still propagates for the equivalent of 10 turns without any loss of particles or any significant degradation in beam quality.

In reality, experimental diagnostics used in matching the beam have an uncertainty that can be as much as 0.5 mm. Therefore, we have explored the effects of introducing a mismatch in the initial rms beam size by that amount. Fig. 3 compares the evolution of the emittance in the matched case to two mismatched cases: one with a 0.5 mm error in x, and the other with 0.5 mm errors in x and y. It is worth noting that in the mismatched cases, we observe a very slight loss of particles

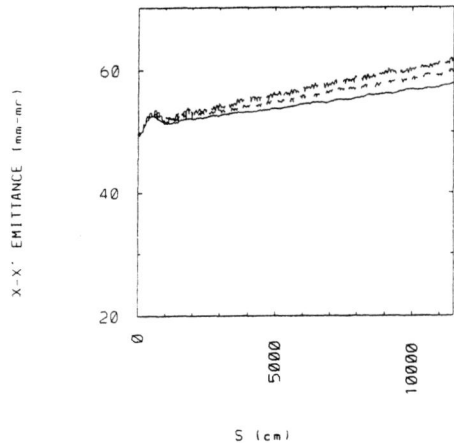

FIGURE 3. Effect of an rms mismatch on the 4*rms beam emittance. Shown is the emittance in x for 3 cases: (bottom) matched beam; (middle) 0.5 mm mismatch in x; (top) 0.5 mm mismatches in both x and y.

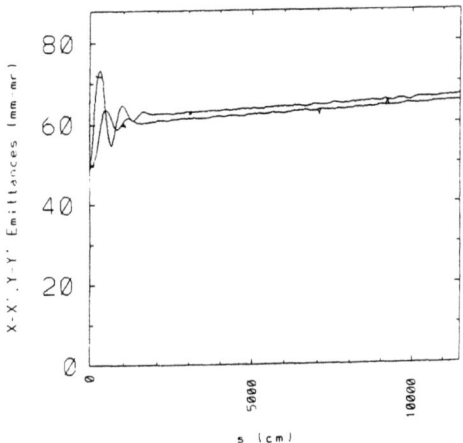

FIGURE 4. Effect of an energy spread of 50 eV (dispersion mismatch) on the 4*rms beam emittance in x and y.

(less than 0.1 %) over the 10 turns. In both cases we see mismatch oscillations, which do eventually decay in amplitude. In addition to those oscillations, we can see that the rms mismatch, in combination with the lens nonlinearities, leads to an additional long-term growth in emittance that is comparable to that caused by the nonlinearities alone. Even so, this growth may be acceptable if it cannot be resolved by the experimental diagnostics.

The next step is to increase the energy spread of the beam and explore the effects of dispersion. Since we make no attempts to match the initial distribution to the dispersion, the beam will also experience the effects of a dispersion mismatch. Shown in Fig. 4 is one such case where we assume an energy spread of 50 eV. This corresponds to the case where the longitudinal thermal velocity is equal to the transverse thermal velocity. [Since we are not yet doing full 3D simulations, we do not really model any transverse to longitudinal coupling in this case]. Notice that the

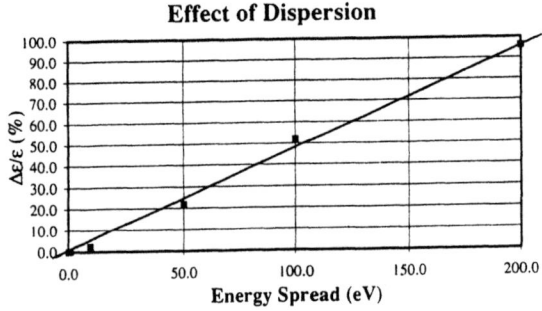

FIGURE 5. Effect of a non-zero energy spread on the net short-term growth in emittance, as a percentage of the initial emittance.

usual slow, gradual growth in emittance is preceded by a sharp rise in the x emittance and oscillations on a shorter time-scale (comparable to a betatron period). These oscillations are then damped by an exchange in energy between the two transverse directions. Similar results have been reported for simulations of the Livermore ring (7). The theoretical explanation for this behavior, which is due to the dispersion mismatch, is provided in ref (5).

Figure 5 displays the *short-term* emittance growth of the beam (net growth after the oscillations subside) as a function of the energy spread. The simulations show that this emittance growth is roughly proportional to the energy spread. In the case of the Maryland ring, the energy spread of the actual beam has not yet been measured, but is not expected to be any more than 50 eV. It is further hoped that matching of dispersion at injection into the ring will reduce the net growth in the emittance.

CONCLUSION

Extending the studies reported in ref. (3), simulations are being conducted of the full lattice of Maryland Electron Ring (both dipoles and quadrupoles) in the presence of dispersion. Nonlinearities of the lens elements are shown to lead to an acceptable growth of emittance over a period of 10 turns, and to cause very little degradation in beam quality, when operating close to the nominal design parameters. Although the level emittance growth remains acceptable in the presence of a realistic mismatch in the rms beam size, a dispersion mismatch can lead to much more rapid growth early on. In the future, we plan to consider the possibility of dispersion matching as well as explore other issues such as the sensitivity to errors in ring construction and operation.

ACKNOWLEDGEMENTS

We wish to thank John J. Barnard, Terry F. Godlove, J.G. Wang and Santiago Bernal for many fruitful discussions, and Alex Friedman, Dave Grote and Steve Lund for support with the WARP code. The code runs on computers provided by the National Energy Research Scientific Computing Center (NERSC). This research is sponsored by the U. S. Department of Energy (DOE).

REFERENCES

1. M. Reiser, *et. al.*, Fusion Engineering and Design **32-33**, 293 (1996); J. G. Wang, S. Bernal, P. Chin, *et. al.*, "Studies of Space-Charge Physics in Beams for Advanced Accelerator Applications," these proceedings.

2. D. P. Grote, *et. al.*, "Three-Dimensional Simulations of High Current Beams in Induction Accelerators with WARP3d," Fus. Eng. & Des. **32-33**, 193-200 (1996); A. Friedman "Overview of the WARP Code and Studies of Transverse Resonance Effects," these proceedings.
3. R. A. Kishek, S. Bernal, M. Reiser, *et. al.*, "Beam Dynamics Simulations of the University of Maryland Electron-Ring Project", NIM-A, accepted for publication (1998).
4. T. Godlove, S. Bernal, and M. Reiser, "Printed-Circuit Quadrupole Design," Proc. of 1995 Particle Accelerator Conference, 2117 (1995).
5. M. Venturini and M. Reiser, "RMS Envelope Equation in the Presence of Space Charge and Dispersion," Phys. Rev. Lett., in print (1998); Marco Venturini, Rami A. Kishek, and Martin Reiser, "Dispersion and Space Charge," these proceedings.
6. S. Bernal, P. Chin, R. A. Kishek, *et. al.*, "Transport of a Space-Charge Dominated Electron Beam in a Printed-Circuit Quadrupole Channel," submitted to Physical Review Special topics: Accelerators and Beams (1998)
7. J. J. Barnard, G. D. Craig, A. Friedman, *et. al.*, "Emittance Growth in Heavy Ion Rings due to the Effects of Space Charge and Dispersion," these proceedings.

Simulation of Longitudinal Micro-Bunch Dynamics

Patrick R. Knaus

European Laboratory for Particle Physics CERN, Division PS
1211 Geneva 23, Switzerland

Abstract. The longitudinal aspects of H$^-$ charge exchange injection into the CERN PS from a future Superconducting Proton Linac (SPL) are studied with due regard to space charge effects. The micro-structure of the beam introduced by the Linac RF is taken into account in the simulation. Algorithmic principles of the applied code are outlined. Recommendations are made on the characteristics of the beam to be provided by the new Linac in as far as the chopping factor is concerned.

INTRODUCTION

Most of the radiofrequency hardware of LEP-2 will be available after the year 2000 due to the decommissioning of the lepton collider. A proposal was made to use the large inventory of RF equipment for the construction of a 2 GeV Linac injecting directly into the PS [1,2], thus making obsolete the PS Booster. The brightness of the beam in the PS at low energy would double, helping the injector complex to satisfy the LHC requirements. The planned CERN physics programme with protons would benefit from such a machine and additional users could easily be accommodated thanks to the capability of the Linac to operate at much higher duty factor than strictly required for high energy physics.

The longitudinal aspects of the injection scenario from this future Linac into the PS have been simulated. A modified version of the multi-particle tracking code Track1d [3], extended to allow for the modelling of micro-bunches, was used.

SPL TO PS INJECTION SCHEME

The parameters of the SPL and PS machines relevant for this simulation are summarised in Table 1. The fundamental idea of reusing as much as possible of the LEP-2 equipment lead to the adoption of the LEP RF frequency of 352.2 MHz. This enables reutilisation of a significant number of klystrons with their power distribution system. The injection energy into the PS has to be as high as possible to improve transverse beam stability and minimise space charge problems. The

TABLE 1. Parameter of the SPL and PS machines for the chosen injection scenario.

SPL parameter	
Mean beam current	10 mA
Bunch current	56 mA
RF frequency	352.2 MHz
SPL macro-pulse	250 μs
Filling factor	1/3
Nb. of Linac bunches per PS bucket	7
SPL micro-pulse	59.6 ns
Chopping factor	54 %
Total bunch energy spread ($\sqrt{5}\sigma_{rms}$)	± 0.5 MeV
Energy jitter (pulse to pulse)	± 2 MeV
PS parameter	
Injection energy	2 GeV
Accumulated particles	$1.5\ 10^{13}$
Injected turns	110
RF voltage at injection	50 kV
RF frequency	9.5 MHz
Harmonic number	21
Machine radius	100 m
Bending radius	70 m
Magnetic field at injection	0.1327 T
Transition γ_t	6.086
Revolution period	2.212 μs
Synchrotron period	1.015 ms
Geometry factor $g_0 = 1 + 2\ln(b/a)$	6.45

beam energy assumed for this study is 2 GeV, but could eventually be raised to 2.2 GeV after a more detailed analysis of the LEP equipment. Energy ramping of the SPL during injection is presently not foreseen. For production of the LHC type beam an intensity of $1.5\ 10^{13}$ protons has to be reached in the PS, i.e. an equivalent of 110 injected turns. The macroscopic model of the SPL beam is represented in Figure 1. The Linac beam is chopped at the PS revolution frequency, to reduce longitudinal capture losses. Moreover, one pulse per machine revolution is omitted to allow for the ejection kicker rise time.

The microscopic situation at injection, that is the beam distribution within a single PS bucket, is represented schematically in Figure 2. The minimum micro-bunch separation is fixed to 2.8 ns by the RF frequency of the Linac. However, a few other combinations with integer multiples of this value are possible, resulting in different filling factors. Certain limits imposed by the chopper frequency have to be respected. A unity filling factor for example would require an illusory 352.2 MHz beam chopper. Two realistic cases with different filling factors have been considered, the 1/7- and 1/3-schemes, injecting into a PS bucket one bunch out of seven and one out of three respectively.

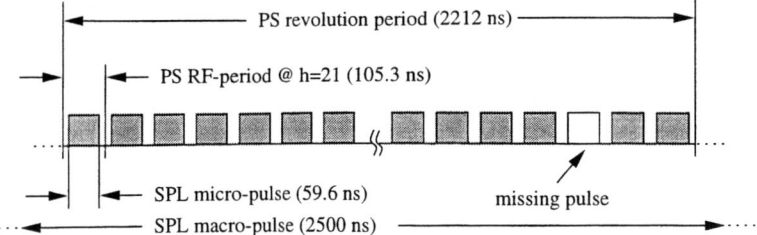

FIGURE 1. Macro-structure of the SPL beam as seen on the time scale of the PS revolution period. Every turn one SPL micro-pulse is injected into 20 of the 21 PS buckets. One bucket is left over to allow for the ejection kicker rise time.

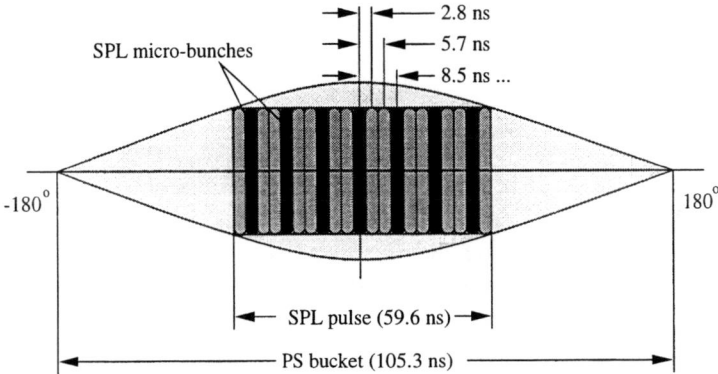

FIGURE 2. A zoom into a single SPL micro-pulse as injected into one of the 21 PS buckets reveals its micro-bunch structure. The injected pulse shows a filling factor of 1/3, where the filling factor is defined as the ratio between the RF frequencies of the RFQ and the Linac.

For the 1/3-scheme the spacing between two successive micro-bunches is three times the SPL RF period, that is 8.5 ns, demanding for a still reasonable 7 ns chopper rise time. Such a 117.4 MHz chopper is actually at the limit of the technically feasible. The peak bunch current reaches 56 mA. On the other hand the very relaxed 1/7-scheme with a chopping at 50.3 MHz would allow to reuse the present Linac2 hardware, but results in an intolerable peak bunch current of 140 mA.

SPACE CHARGE

An important issue for the injection of high intensity beam into the PS is the control of space-charge effects. This holds especially for the very short SPL micro-bunches of only a couple of nanoseconds, where the peak charge density in the bunches is particularly important. Space charge effects are driven by high particle densities and tend to defocus the bunch. In order to counteract the space charge defocusing, the applied RF voltage has to be high compared to the space charge voltage. The space charge voltage increases with increasing bunching and reaches very high local values in the case of micro-bunching.

The aim of this study was to determine whether injection of an SPL-type beam into the PS would be possible, or if it would lead to beam explosion. The simulation should allow to favour one out of the different injection schemes varying mainly in the number of micro-bunches injected per bucket and turn.

Computational Model Used in the Code

The underlying equations to numerically treat the synchrotron motion are

$$\frac{d}{dt}\frac{\Delta E}{\omega_0} = \frac{q}{2\pi}[V(\varphi) - V(\varphi_s) + V_{SC}(\varphi)] ,\quad (1)$$

$$\frac{d}{dt}\Delta\varphi = \frac{h\omega_0\eta}{\beta^2 E_s}\Delta E ,\quad (2)$$

where $\Delta E = E - E_s$, $\Delta\varphi = \varphi - \varphi_s$ and the slip factor $\eta = 1/\gamma_t^2 - 1/\gamma^2$. E and φ are the energy and the phase of the particle, $V(\varphi)$ is the applied RF-voltage and h the harmonic number. The suffix s refers to the synchronous particle and, since the equations govern only first-order variations from the synchronous values, the revolution frequency ω_0 is evaluated for the synchronous particle in the calculations. The space-charge voltage V_{sc} generated within the beam is given by the derivative of the line density $\lambda(\varphi)$ of the particle distribution

$$V_{SC}(\varphi) = -q\frac{d\lambda(\varphi)}{d\varphi}\left[\frac{Rg_0}{2\varepsilon_0\gamma^2} - L\beta^2 c^2\right] ,\quad (3)$$

where R is the radius of the machine. For a beam with assumed circular cross section of radius a in a circular pipe of radius b, the geometry factor g_0 is defined as $g_0 = 1 + 2\ln(b/a)$. L is the total inductance of the reactive vacuum chamber wall and may be neglected in the present problem.

The applied RF voltage is a standard sinusoidal of harmonic number $h=21$ given by

$$V(\varphi, t) = V_0(t)\sin\varphi .\quad (4)$$

A second order multi-step algorithm (leap-frog) is used to time-integrate the equations of motion, and space charge forces are calculated from the smoothed line density found by weighting the particles to a series of bins in phase.

To solve the dilemma of keeping the total number of particles that have to be tracked to a reasonable limit, while injecting a statistically meaningful particle distribution every turn into the machine, a special technique of charge accumulation has been applied: As well as phase and energy, the model particles carry different charges. Simulation of injection is achieved by distributing particles onto the nodes of an imaginary phase space grid and allowing charge to build up as more and more beam is accumulated. This Eulerian approach to the motion concentrating on values at specific points, is adopted from the theory of fluid dynamics and is used only in the course of injection. During a revolution in the machine, charges at these points move away from their nodes under the effect of Equations (1) and (2), but are brought back using specific weighting schemes. Newly injected beam is similarly weighted onto the nodes. In this way a charge density is build up, but the number of particles is bounded by the scaling on the grid. After the end of injection a more conventional Lagrangian approach is adopted. The nodal charges are regarded as macro-particles for the beam and their motion is followed up. The beam is represented by up to 150.000 of those macro-particles.

Self-consistency Test of the Code

In order to verify the correctness of the results obtained with a tracking code it is necessary to model test cases that can be solved analytically and then compared to the simulations. Such a test case is given by a self-consistent particle distribution. A code is said to be self-consistent if the overall particle distribution used to describe the beam is conserved as it is followed up turn by turn through the machine.

A self-consistent particle distribution for the longitudinal phase space was given by Hofmann and Pedersen [4]. This distribution describes bunches with elliptic energy distribution as observed in proton synchrotrons and makes several analytical calculations for bunched beams in longitudinal phase space possible. Its phase space density $g(W, \varphi)$ is given by

$$g(W, \varphi) \propto \sqrt{H_e - H} \tag{5}$$

where $W = \Delta E/\omega_0$. W and φ are canonical variables whose variation is governed by the Hamiltonian

$$H = \frac{h\omega_0^2 \eta}{2\beta^2 E_s} W^2 - \frac{q}{2\pi} U(\varphi). \tag{6}$$

$U(\varphi)$ is the potential

$$U(\varphi) = \int_{\varphi_s}^{\psi} [v(\varphi) - V(\varphi_s)] \, d\varphi \tag{7}$$

and H_e is the value of H for the particles at the beam extremities. For this distribution, the space charge potential mirrors the applied potential and the distribution is preserved during the motion.

In Figure 3 a matched Hofmann-Pedersen beam is shown right at injection and 5000 machine revolutions later. Since the distribution remains unchanged this test

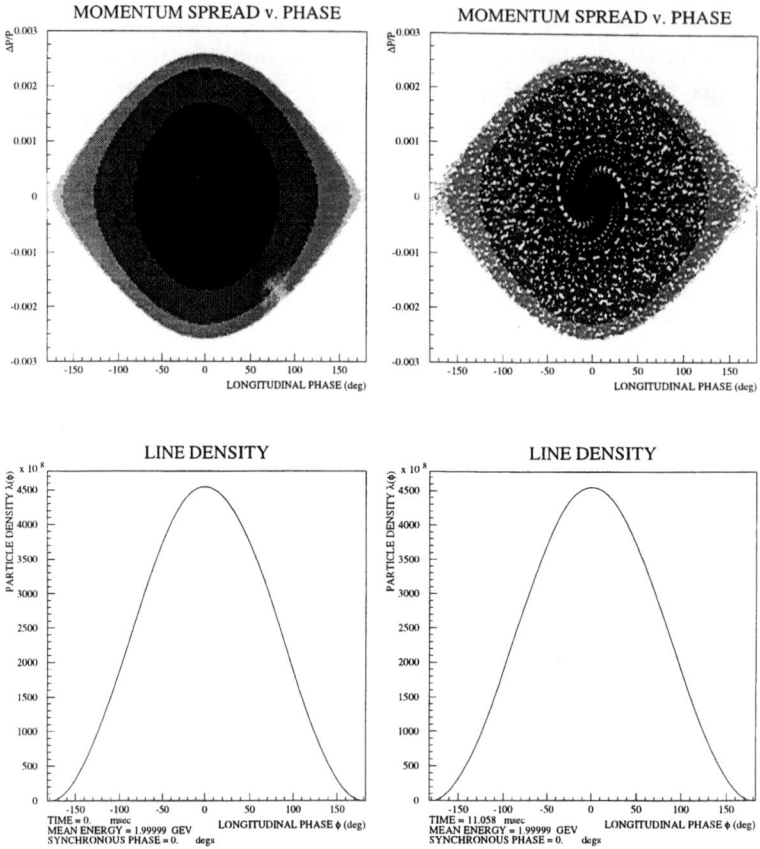

FIGURE 3. *Tracking of the self-consistent Hofmann-Pedersen distribution. The particle distribution in longitudinal phase space and its line density remain unchanged within numerical precision. This can be deduced from the phase space plots at injection and 11 synchrotron periods or 5000 machine revolutions later. The code may thus be considered as self-consistent and gives real physical results.*

confirms the self-consistency of the code. This is true up to the limiting intensity of the Hofmann-Pedersen model, when the induced space charge voltage finally cancels the applied RF voltage.

THE SIMULATION

The total number of protons in the ring should reach $N=1.5 \cdot 10^{13}$. Injection of a $I=10$ mA mean current SPL beam covers $n=110$ turns. In the simulation this beam

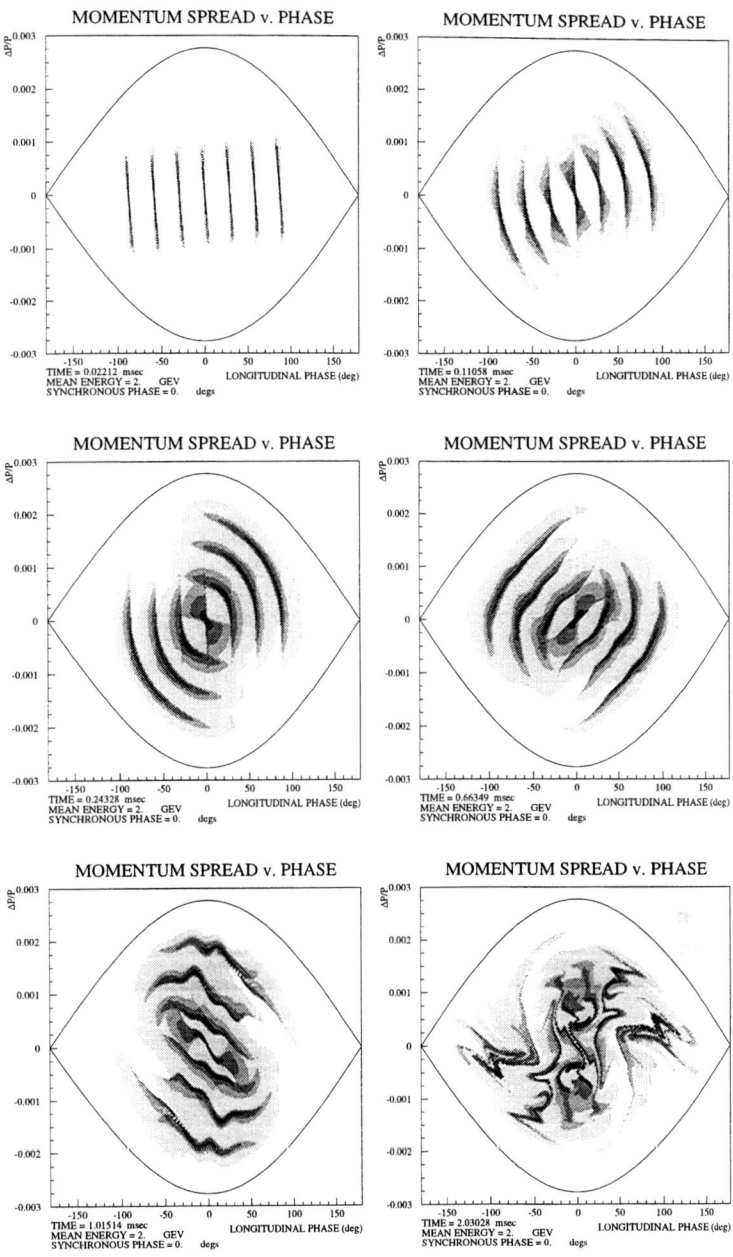

FIGURE 4. *Phase space distribution as it builds up during injection of 7 micro-bunches per bucket and turn using the 1/3-scheme. The beam after 10, 50, 110, 300, 459 and 918 machine revolutions is shown. Injection stops after the 110th turn.*

is represented by up to 150.000 macro-particles. The initial longitudinal particle distribution injected in every bucket consists of several micro-bunches of parabolic phase and momentum distribution. The initial total momentum spread of every micro-bunch is $\Delta p/p=1.9‰$. This includes the internal bunch momentum spread as well as the pulse to pulse jitter and corresponds to an absolute energy spread of $\Delta E=\pm 2.5$ MeV.

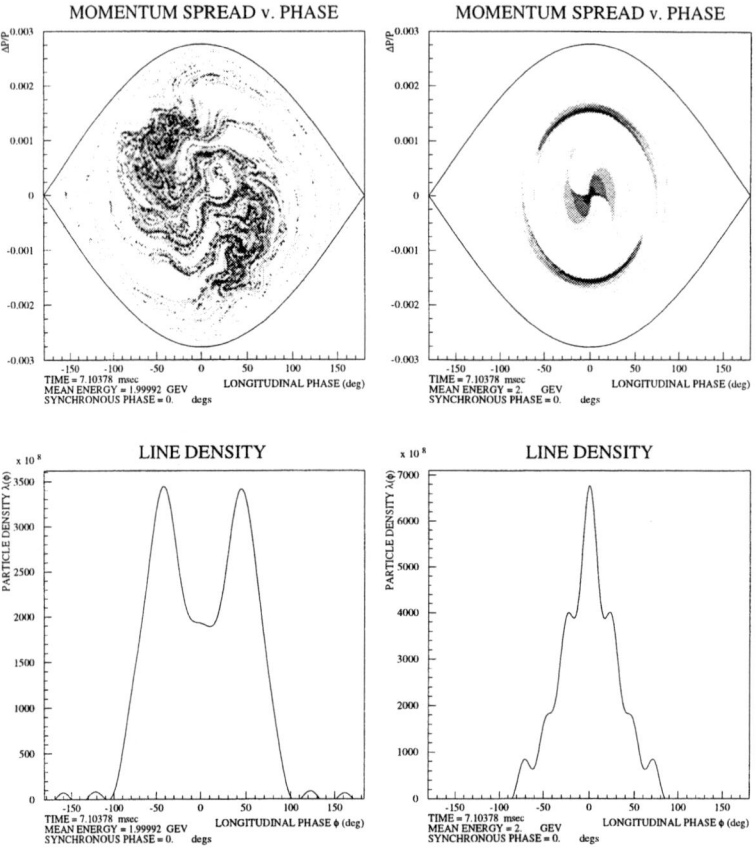

FIGURE 5. *Particle distribution in longitudinal phase space and the corresponding line densities for the 1/7-scheme. Clearly problems arise from the two low density islands in between the three micro-bunches that are still discernible even after several synchrotron periods. The two right-hand side plots represent the same situation, but tracking was performed neglecting space charge forces. In this case no mixing of the micro-bunches is observed and the resulting peak line density is significantly higher.*

The line density of the beam builds up during beam accumulation, reaches its peak value right at the end of injection when many bunches are still upright in longitudinal phase space. As time goes by the bunches rotate in the bucket and smear out filling the entire bucket area. The peak line density and the space charge voltage decrease. Therefore the first few turns right after injection are the most critical ones. The evolution of the phase space distribution during injection is represented in Figure 4. The zero-current RF bucket is drawn to guide the eye. It does not take into account the deformation due to the space charge voltage. The grey scale in the scatter plot corresponds to different particle densities. The scale goes from black (high density) over dark grey to light grey (low intensity).

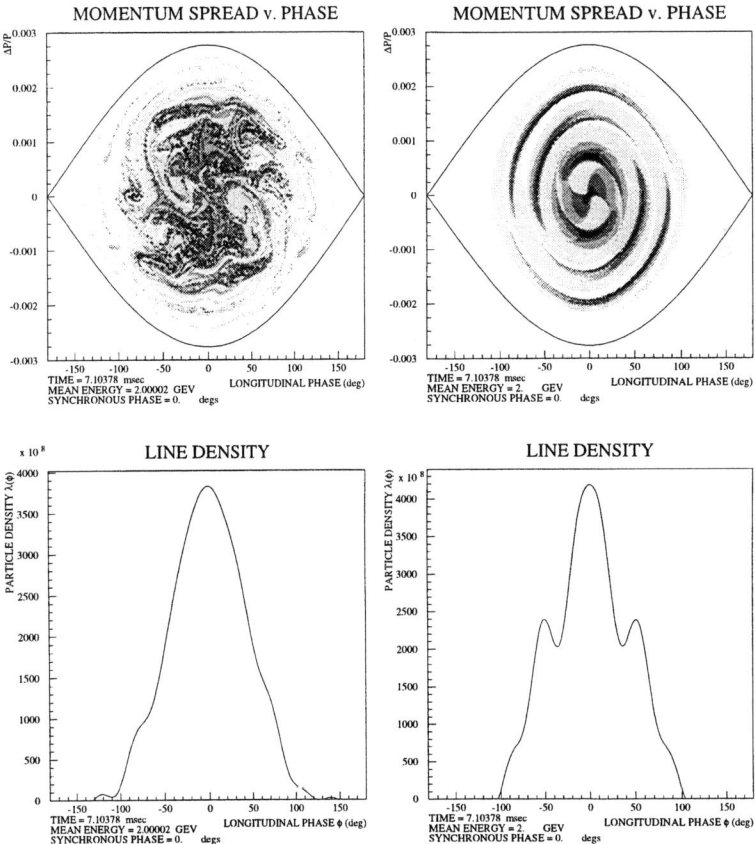

FIGURE 6. *Particle density in longitudinal phase space and the corresponding line densities for the 1/3-scheme. The resulting particle distribution shows no holes in phase space. The right-hand side plots correspond again to particle tracking neglecting space charge forces.*

Results of the Simulation

1/7-Scheme

The 1/7-scheme proposes a Linac filling factor of 1/7 thus injecting one bunch out of seven into the PS, resulting in 3 micro-bunches per PS bucket and turn. The micro-bunches are spaced by 67.95° at the PS RF of 9.5 MHz. The interest of this layout was that it would have allowed a reutilisation of the existing 50 MeV Linac2 as injector for the 352.2 MHz structure, just requiring a new RFQ at 50.3 MHz and a retuning of Linac2 at 201.3 MHz instead of the present 202.6 MHz. This arrangement would have made the chopper design trivial, but the low filling factor in the high frequency section leads to a 140 mA bunch current giving rise to space charge problems in the PS where capture became critical. This scheme was therefore discarded. Figure 5 shows the particle distribution in longitudinal phase space and the corresponding line density after about 7 synchrotron periods.

1/3-Scheme

The alternative 1/3-scheme proposes injection of 7 micro-bunches per PS bucket and turn. The distance between successive bunches is 29.12° at the PS RF frequency, demanding for a still reasonable chopper rise time of 7 ns. The resulting peak bunch current is only 56 mA, i.e. about 40 % of the current in the previous case. Furthermore it can be seen from Figure 6 that the resulting particle distribution in the PS bucket doesn't show the previously observed density holes. This gives a direct explanation of the much lower space charge peak voltage of 8 kV as compared to the previous 14 kV (cf. Figure 7).

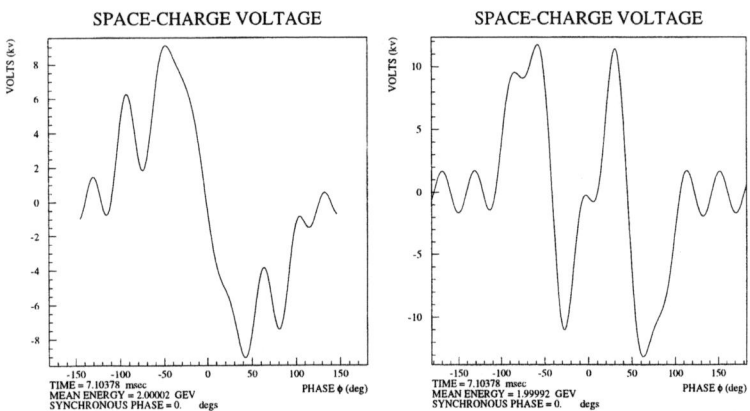

FIGURE 7. *The space charge voltage resulting from the 1/3- and 1/7-schemes after 7 synchrotron periods. The peak voltage is significantly higher for the 1/7-scheme.*

Acknowledgements

I am indebted to M. Lindroos for introducing me to the SPL Study Group and to R. Garoby and M. Vretenar for the illuminating discussions we shared. I should particularly like to acknowledge C. Prior for entrusting me Track1d, his willingness in answering may questions on the code and invaluable support when adapting the code to the needs of the SPL.

REFERENCES

1. Garoby R., Vretenar M., *Proposal for a 2 GeV Linac Injector for the CERN PS*, PS/RF/Note 96-27, October 1996
2. Autin B., Boussard D., Cappi R., Garoby R., Haseroth H., Hill C.E., Knaus P., Lombardi A. (Ed.), Martini M., Ostroumov P.N., Tessier J.M., Vretenar M. (Ed.), *Report of the Study Group on the Superconducting Proton Linac as a PS Injector*, CERN PS Report, to be published summer 1998
3. Prior C.R., *The Longitudinal Tracking Code Track1d*, Rutherford Appleton Laboratory, 1996
4. Hofmann A., Pedersen F., *Bunches with Local Elliptic Energy Distributions*, IEEE Trans. Nucl. Sci., NS-26, No. 3, June 1979

A Particle Simulation Code for a High Intensity Accelerator Ring[*]

A.U. Luccio, J. Beebe-Wang, M. Blaskiewicz
Brookhaven National Laboratory, Upton, NY 11973-5000, USA
J. Galambos, J. Holmes, D. Olsen
Oak Ridge National Laboratory, Oak Ridge, TN 37831, USA

Abstract. A computer code to simulate in 6 phase-space dimensions the dynamics of a high intensity particle beam in a circular accelerator is being created as a collaborative effort of two laboratories. The main and immediate purpose of the code is to study in detail the behavior of the 1 GeV proton accumulator ring for the Spallation Neutron Source to be built at Oak Ridge. The structure of the code and some of the basic guidelines are described.

PROLOGUE

A computer code, SAMBA (= Sensible Analysis Model for Beams in Accelerators), is being created by a collaboration between BNL and ORNL. We build pieces of the code, together or individually, and put them in a common cauldron. The code is steadily growing. This is a report on the status of the code with considerations on the physics and on the programming.

Six dimensional simulation computer codes, including space charge effects, are indispensable tools for the study of high intensity particle accelerators. A few codes that meet these requirements are being developed by different laboratories. The underlying physical principles are mostly known, the real problem is how to implement them in a way that the results could be trusted, and also in a way that the computer running time would be reasonably short, due to the large multitude of examples that are being run for any project.

1. THE SPALLATION NEUTRON SOURCE

There is a joint effort between National Laboratories in the USA to study and build a pulsed neutron spallation source, SNS [1]. The accelerator part of this facility, that will be ultimately built and operated at the Oak Ridge National Laboratory, consists of a 1 GeV proton linac followed by a constant energy accumulator ring, designed by a team at the Brookhaven National Laboratory The initial design beam power is 2 MW. This calls

---

[*] Sponsored by the Division of Material Sciences, U.S.Department of Energy, under contract number DE-AC05-96OR22464 with Lokheed Martin Energy Research Corp. for Oak Ridge National Laboratory.

CP448, *Workshop on Space Charge Physics in High Intensity Hadron Rings*
edited by A. U. Luccio and W. T. Weng
© 1998 The American Institute of Physics 1-56396-824-X/98/$15.00

for a very intense circulating current in the accumulator (2.10^{14} protons per turn). Uncontrolled beam losses should be limited to 10^{-4}.

The physics of the beam in the ring, both with theory and computer simulation, is being studied by two teams, at Oak Ridge and Brookhaven.

2. A NEW CODE

To study the beam dynamics in the accumulator ring the code ACCSIM [2], developed at TRIUMF was used for some time. Recently, we decided to write the new code, SAMBA. The reason is that we felt that the SNS project needed a specialized code, that could be developed in parallel by all the members of the joint team, and was more comprehensive than existing codes, yet devoid of things that were not so important for our project. In doing so, we have freely borrowed from other codes we know, trying to collect the best in each one of them. This paper is a progress report on the development of simulation codes by the joint team. The physical algorithms and methods are described, together with the principles of programming being followed. Some of the approaches are not yet well consolidated, since in many cases we are experimenting different ways.

2.1. Programming Language

SAMBA is written in C++. We did this for two main reasons
- Development of programming modules in parallel is more natural with C++ than with FORTRAN. This is why C++ has become the standard in the Industry, where a large group of programmers are often working on the same code, to bring it as early as possible to the market.
- The code is run under the supervision of a SuperCode that allows the coexistence of compiled and interpreted modules, making development and debugging very natural, as it will be explained later in detail.

2.2. Input From Machine Optics/Output To Graphics

- The machine optics code that produces the main accelerator descriptor is MAD [3]. This is a well developed and maintained code able to describe all of the features of the ring lattice, including higher order transformation matrices and the treatment of lattice errors.
- SAMBA is independent from any specific graphic packages. Output to graphics can be done through files or by direct pipe from the code to graphics, in order to allow for animation. An intriguing possibility is to use the graphic firmware tools of the workstation to perform some of the calculations.

3. BEAM DYNAMICS. PIC CODE

The real beam is represented by an ensemble of macroparticles in fully 6-dimensional phase space. The calculations include
- transport of the particles in the lattice using the first order and second order transformations calculated by MAD
- In the transverse space, space charge forces are calculated that apply an angle kick to the macros. Betatron tune distribution in the beam is calculated ("necktie"). Transverse impedances are used to calculate beam to wall coupling in a realistic vacuum chamber geometry.
- In the longitudinal space, we have cavities with RF voltage plus a realistic longitudinal impedance budget, that allows the representation of longitudinal space charge kicks and the description of the bunch and the bucket.

3.1. Orbit Transport

For the purpose of orbit transport, the lattice is subdivided in sections containing a number of machine modules or nodes. In order to save computer time, the number of section should be kept as small as possible. First order **R** 6x6 matrices and second order **T** transforms for each machine section are produced in output by MAD. Also, the values of the Twiss functions α, β are output at the end of a section. If the closed orbit is distorted because an orbit bump is applied, typically at injection, or because of lattice errors, the matrices are displaced and centered with respect to the distorted orbit

3.2. Transverse Motion

Call y the vertical macroparticle displacement with respect to the reference orbit, x the radial displacement and s the longitudinal coordinate along the orbit. Transverse motion is in (x,y) and is dealt to first order with the firs 4 rows and columns of **R** that transforms the spatial coordinates x and y and the transverse components of the momentum p_x and p_y. To calculate transverse space charge effects, at the end of each machine section the beam charge distribution ρ is calculated by binning and counting macroparticles over a regular mesh, and from ρ the space charge force is calculated by a three dimensional integration

$$\vec{F}(\mathbf{P}) = \frac{\mu_0 e c^2 Ne}{4\pi\gamma^2}\vec{g}, \quad \vec{g} = \int \frac{\rho(\mathbf{Q})}{r^3}\vec{r}d\mathbf{Q} \tag{1}$$

In this calculation the beam is seen as individual macros at **P**, acted upon by forces generated by elements of a continuum at **Q** at a distance r, where

$$r^2 = r_\perp^2 + \gamma^2 r_\parallel^2 \tag{2}$$

with a relativistic dilation factor γ^2 in the longitudinal direction. The factor $1/\gamma^2$ in the expression for the force is due to the partial compensation of the electric and magnetic fields. The longitudinal part of the integration extends only to a distance of the order of the radial size of the beam, and is almost a constant unless we are close to the head or tail of the bunch. For macroparticles close to both ends of the bunch, of coordinate s_0, if we factor out the longitudinal charge density as

$$\rho(\mathbf{Q}) = \rho_\parallel \rho_\perp \tag{3}$$

an approximation for the radial integration is

$$\frac{1}{\gamma} \int \frac{d(s_P - s_Q)}{r^3} = \frac{1}{r_\perp^2}\left[1 + \frac{s - s_0}{\gamma r(s_0)}\right] \tag{4}$$

This is equivalent to treat the beam more as a bundle of spaghetti than a cloud of points.

At the end of a machine section of length L a space charge kick is applied to every macro. The transverse momentum kick applied to a macro in **P** is calculated from the space charge force

$$\delta\mathbf{p}_\perp = \int_L \bar{F}(\mathbf{P})dt \approx \bar{F}\frac{L}{\beta c} \tag{5}$$

The transverse space charge kick can be calculated "brute force" by performing a 3-dimensional summation over the distribution of macros in the beam, or by FFT. We are experimenting with both, considering the accuracy of results vs. computer time. The betatron tune distribution within the beam is calculated by the partial derivatives of the force function. A tune value is attributed to each macro in the simulation. The tune shift distribution in the beam gives indications on the stability limit of the transverse motion, but is not directly used in any calculation.

The calculation of the transverse space charge force is the single most expensive operation as far as computer time is concerned. Many strategies are being experimented to cut this time. They all derive from the observation that, at least during one turn in the machine, the general shape of the beam charge profile would not change too much. This leads to the concept of "rubberbanding" of the distribution, i.e. a given shape is stretched or compressed by a radial scaling factors to cover the section of the beam at the next interface, once calculated, say, at the beginning of a turn. The beam section and rubberbanding strategies can be derived

- from a statistical evaluation of the emittance, averaged over the macro distribution
- by using the local Twiss beta functions as given by MAD. In this case the scaling factor is proportional to the square root of the twiss beta at the interface
- by using the envelope equation integrated in a Core model (see Section 4).
- by fitting the calculated force profile by FFT or with another basis (e.g. Chebycheff polynomials) and transferring the coefficients to the next interface with proper scaling.

Transverse space charge forces introduce coupling between the radial and vertical motion. The betatron oscillation differential equations become

$$\begin{cases} x'' + K_x^2 x = f_x = \dfrac{F_x(x,y)}{m_0 \gamma \beta^2 c^2} \\ y'' + K_y^2 y = f_y = \dfrac{F_y(x,y)}{m_0 \gamma \beta^2 c^2} \end{cases} \quad (6)$$

Eqs. (6) are coupled. The tune spectrum in one mode will contain side bands due to the other mode [5]. Polynomial expansion of the function at the r.h.s of the equation yields, to first order, the tune shift is given by

$$\Delta v = -\frac{r_0 N}{\beta^2 \gamma^3 v} \frac{g_r}{2r} \quad (7)$$

with r_0 the classical radius of the particle (say, proton). Eq. (7) gives the tune shift in x or y (r), that for common unimodal charge distributions (say, Gaussian) is the maximum tune shift in the center of the beam. To higher orders, Eqs. (6) lead to higher order nonlinear tune terms and coupling. If we expand or fit the function in the r.h.s of Eqs. (6) in harmonics, they can be interpreted and solved as coupled Hill equations and we may be able to find Floquet exponents for instability growth.

3.3. Emittance

SAMBA statistically calculates the transverse emittance of the beam in the macro population. The 4-dimensional emittance (to the 4th power), including coupling between the vertical and the radial mode, can be defined as the determinant of the covariance matrix

$$\varepsilon^4 = \det \begin{pmatrix} <x^2> & <xp_x> & <xy> & <xp_y> \\ <p_x x> & <p_x^2> & <p_x y> & <p_x p_y> \\ <yx> & <yp_x> & <y^2> & <yp_y> \\ <p_y x> & <p_y p_x> & <p_y y> & <p_y p_x> \end{pmatrix} \quad (8)$$

The radial emittance (squared) is defined as

$$\varepsilon_x^2 = \det\begin{pmatrix} <x^2> & <xp_x> \\ <p_x x> & <p_x^2> \end{pmatrix} = <x^2><p_x^2> - <xp_x>^2 \qquad (9)$$

and similarly the vertical emittance. It is then

$$\varepsilon^4 = \varepsilon_x^2 \varepsilon_y^2 + \text{coupling terms} \qquad (10)$$

If there is no coupling between the two radial modes, and the orbit transfer matrices are simplectic, the horizontal and radial emittances should be individually conserved during tracking. However, as noted in the previous section, space charge is producing coupling and the important quantity to watch is the complete 4-dimensional emittance.

3.4. Foil. Wall. Losses

Injection in the proton machines that interest us mostly is accomplished via the well studied scheme of multi-turn injection of H⁻ ions stripped in a foil. The phase-space acceptance of the machine is being populated by moving the equilibrium orbit at the foil with a time programmed magnetic orbit bump, that generally collapses to bring the final beam close to the axis of the accelerator vacuum chamber. Also, transverse painting can be achieved by moving the position of the injection spot.

The bump profile may assume exponential assumes profiles of the form

$$x(t) = x_0 + (x_0 - x_f)\left[\frac{1 - e^{\frac{-\tau_x t}{\Delta_t}}}{1 - e^{-\tau_x}}\right] \qquad (11)$$

where $x_{0(f)}$ is the initial (final) x bump position, Δ_τ is the time for the bump, and t_x is a normalized time constant. The x', y and y' bumps are specified in a similar manner

SAMBA, as most tracking codes for accelerator rings, contain provisions to simulate orbit bumps as well as for simulate ion conversion and scattering in the foil. We have freely copied from others. Lost particles by interactions with walls or collimators, or not being converted at the foil are accounted for and subtracted from the cycle. Also in this, our physics is not original, only the bookkeeping may be different

3.5. Longitudinal Motion. Impedance Budget

Macroparticle transport in the longitudinal direction is done in part by using the last two rows and columns of the **R** matrices, that take care of the linear part of the transformation and of the coupling with the transverse variables, and in part by applying the equations of synchrotron motion to the two variables, longitudinal momentum and phase. The second of these equations is intrinsically not linear

$$\begin{cases} \phi_{n+1} = \phi_n + \tau\omega_{RF}\left(\frac{\delta p}{p}\right)_{n+1} \\ \delta p_{n+1} = \delta p_n + \frac{q}{\beta^2 c}\sum_h V_h(\sin(h\phi_n) - \sin(h\phi_s)) \end{cases} \qquad (12)$$

In Eq. (12) Φ_n represent the synchronous phase and dp_n the momentum difference between a given particle and the synchronous particle at turn n. In the simulation, we are studying different combinations of RF voltages and frequencies. We are in particular interested in RF barrier voltages produced with two or more frequencies

$$V_{RF} = \sum_{i=1,n} V_i \sin(h_i\phi - \phi_i), \qquad (13)$$

where n is the number of RF harmonics used[1]. Like in the code ESME [6], SAMBA calculates longitudinal space charge effects by applying to the macros in the beam additional momentum kicks according to

$$\delta p_{SC,n} = \frac{q}{\beta^2 c}\sum_k I_k Z_k \qquad (14)$$

In Eq. (14), I_k are the Fourier components of the beam current and Z_k the amplitudes of the longitudinal coupling impedance at each beam current frequency. The I_k are calculated by a FFT of the beam at given positions in the lattice, or just once per turn. The impedances are given in input as an Impedance Budget independently calculated from the configuration of the accelerator chamber and of the devices present in it, such as kickers etc.

Commonly, the most important piece of the impedance budget is the lowest frequency impedance due to the coupling of the beam with its own space charge and to the image currents on the beam chamber walls [7]. This is a pure imaginary impedance, giving rise in a cylindrical chamber with perfectly conducting walls to a longitudinal electric field

---

[1] In the SNS we use two frequencies to produce a "pseudo barrier bucket" [11]

$$E_s = -\frac{1}{2\pi\beta\gamma^2} Z_0 \left[\frac{1}{2} + \ln\frac{b}{a} \right] \frac{\partial I}{\partial s} \tag{15}$$

with $Z_0 = \mu_0 c = 377$ ohm the impedance of free space. b and a are the chamber and beam radii, respectively, and $\partial I/\partial s$ the longitudinal gradient of the current in the beam. The $1/\beta\gamma^2$ factor represents the usual compensation between electric and magnetic forces. Eq. (15) shows that the longitudinal space charge field is particularly large at the head and tail of the bunch where the current gradient is commonly the largest, and has opposite signs. The beam current gradient can be numerical very noisy if calculated directly from the longitudinal macroparticle shape distributed in longitudinal bins, and some smoothing algorithm is needed. Since the use of a longitudinal budget impedance requires an FFT of the beam, a Fourier filtering where the high frequencies are eliminated naturally provides such smoothing.

4. BEAM DYNAMICS. CODE VALIDATION. CORE MODEL.

To be trustworthy, the calculated beam behavior must agree with experimental data and with established theories, when available, Thus the validation of the code is a very essential exercise. At least, comparison with the theory can be done on the following principles

- We believe that the beam will show a core behavior plus a halo, then a core calculation based on the solution of differential equations should show the main feature of beam propagation.
- The results of a CORE model [8] should continuously guide and control the PIC model results. PIC will best describe details and the formation of halo.
- The integration of CORE and PIC is one of the future development in store for SAMBA.

5. PROGRAMMING CONSIDERATIONS

5.1. General Code architecture

5.1.1. The SuperCode Driver Shell

The Track code is written in C++, and operates within the SUPERCODE [9] driver shell. Before discussing any specifics of the Track code itself, a few words are useful to explain the relationship between user provided "physics" modules and the driver shell. Conventional scientific programs typically have a prescribed flow of logic originating in a main program. Often there are some program flow choices, governed by appropriate choice of input variable settings. The program execution remains in compiled code until program completion. If the user desires some new flow logic, an edit-compile (debug) programming cycle is required.

On the other-hand, the SUPERCODE is a programmable driver shell, which can execute interpreted script files as well as compiled "physics" modules. Generally runs are done by reading in a "script" file (analogous to an input file). These script input files are more than typical "namelist" files which simply assign values to variables. Since SUPERCODE is a programmable interface, calling sequences to compiled code can be customized within a script "input" file (without recompiling). In fact there is no fully compiled set of logic to do a run. Rather the "general purpose" workhorse physics modules are compiled, and the calling sequences, looping, initial variable settings etc. are done through the shell. In fact, routines can be entered on-the-fly in script files.

Both interpreted and compiled code have their place. Compiled code is much faster in general. Compiling everything however leads to code clutter (i.e. one-off numerical experiments, or scans that are never used again, along with variables to control them). The general philosophy for the separation of interpreted / compiled code is: code that is execution intensive, and/or general purpose is compiled. Examples of compiled code would be particle transport through a matrix multiplication operation. Code that is problem specific and generally not called often during a run is interpreted. Interpreted code examples are variable initialization, setting up parametric scans, etc.

5.1.2. SUPERCODE Driver Shell Modules

SUPERCODE comes with a number of general purpose modules. These include:
- Optimization (calculus based and genetic-algorithm[2]),
- Probabilistic risk analysis (or uncertainty analysis),
- Mathematical tools (splines, B-splines, random numbers, Bessel functions, ...),
- Interactive plotting[3],
- Parallel processing[4].

Note also that a number of these features require special libraries to be installed on the system. All of these libraries are free, and generally run on most UNIX platforms. The Shell is also buildable on Mac and PC platforms.

5.1.3. Module / Shell Relationship

The procedure for actually constructing "physics modules" to be run with the SUPERCODE shell is described in Ref. [9][5]. When adding a new module in practice, it's

---

[2] Requires installation of the GALib library on your system. See http://lancet.mit.edu/galib-2.4/
[3] Requires installation of the PLPLOT and TCL/TK libraries. See
http://www.mech.ubc.ca/Students/Computer/Software/Plplotdoc/node4.html.
Note - the plotting capability provides rudimentary, X-Y and 3-D capabilities, and is intended to provide a quick diagnostic capability, not an exhaustive plotting capability).
[4] Requires installation of the PVM software library on your system (see
http://www.netlib.org/pvm3/book/pvm-book.html)
[5] FORTRAN modules are no longer supported.

usually sufficient to mimic the method of an existing physics model. The general relationship between the user supplied physics modules and the driver shell is shown schematically in Fig. 1.

The Module Descriptor File is a useful place to get information about a module. It lists all the variables and routines that the Driver shell knows about in that particular Module. This is the place where these quantities should be documented. Variable quantities in these files can be manipulated from the Shell. The routines in the Module descriptor files can also be called from the Shell. In fact, putting together a set of variable initializations, and routine calls in a script file is how a "run" is typically done.

Note that the driver shell can not directly access class members. To access class members from the Shell (manipulate, view, etc.), a Module routine must be written (and compiled) to perform the appropriate manipulation. Examples where this may be done would be a routine to create a macro-particle, or a routine to dump macro-particle information to a stream. These compiled routines could then be called from the Shell, and the actual class member manipulations would be performed when the compiled module is called.

5.2. Class Hierarchy for Ring and MacroParticles

In thinking about the tracking of macro-particles around a ring, two main ideas come to mind: the macro-particles themselves and the Ring elements which operate on them in various ways. These are each represented as classes, and much of the code is based on these. As such they are described in some detail here. We reiterate that the actual user implementation of these classes is done via "modules". These modules contain the user interaction mechanisms for instantiating objects, performing member function calls, etc. The modules and actual use of the classes are described later in Section 6..

5.2.1. MacroParticle Class

The MacroParticle class is a simple container class to hold information specific to each macro-particle. Most of the information held for each particle is self evident. Of particular interest is that each macro-particle object contains a reference to a "synchronous particle" object. (There is a separate SyncPart class to hold the synchronous particle information). The synchronous particle must be instantiated before any macro-particles can be created. The SyncPart[6] class contains information for the synchronous particle.

---

[6] Presently, only one synchronous particle is allowed, but it is straightforward to extend this to multiple synchronous particles, if ever needed.

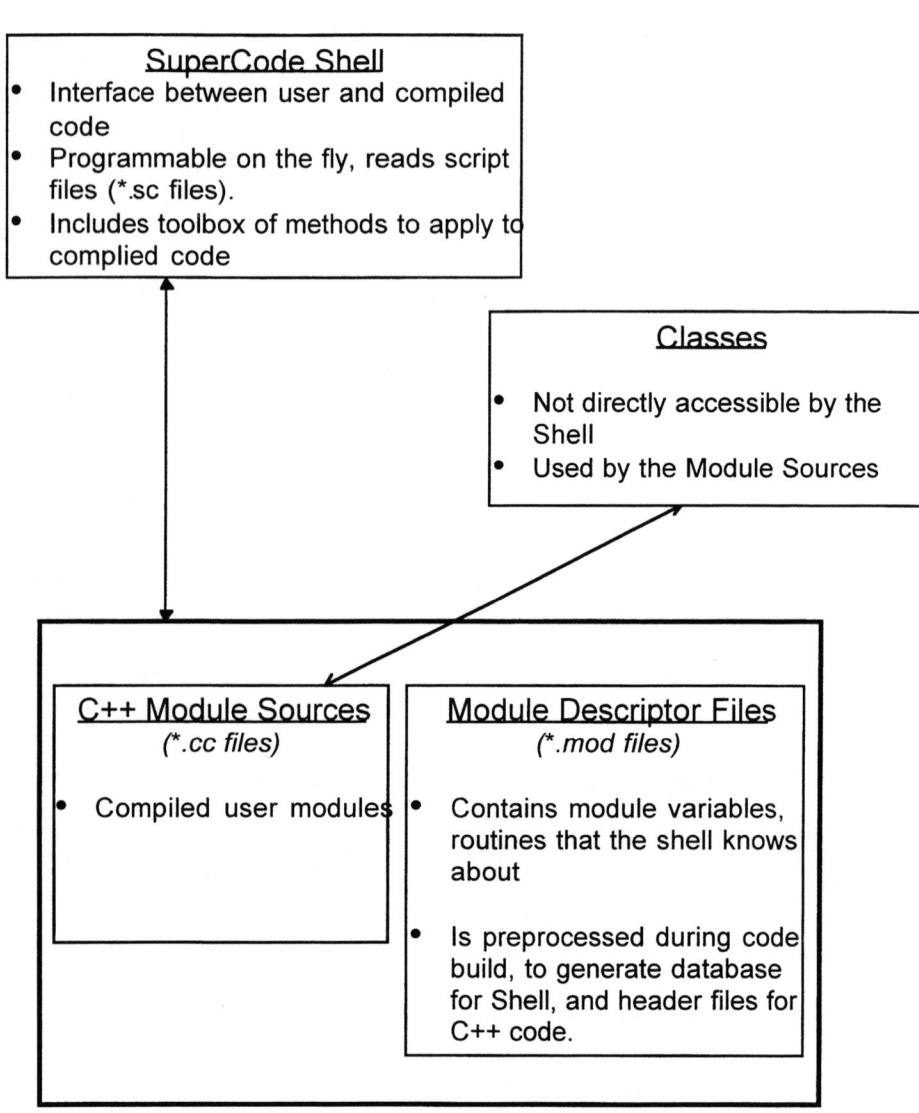

FIGURE 1. General relationship between the physics modules and the SuperCode driver Shell.

5.2.2. Node Class

As the macroparticles circulate around the ring, different operations will be performed on/with them. For instance they may undergo a matrix transfer operation, or a space-charge update, or a dump of information to a stream. Each of these operations will be performed at a "node" of the Ring. The Node class represents the common set of features such operators have. This is an abstract class, with the general purpose information, common to all nodes in the ring. As this is an abstract class, no Node objects will be created, but rather this class will be inherited by sub-classes that will have actual objects (for example a Transfer-Matrix class). Most of the Node members are self-evident, but a few require additional discussion. The oindex member is used for ordering the nodes into a calling sequence which the calculations will be done on/with macro-particles. The calling sequence is done in ascending order of oindex. Whenever a node is instantiated, an oindex value is required. Before any actual Ring calculations are done, the nodes are sorted into the order in-which the calculations should be done. The length member is the length of the node (which can be 0). Each node presumable should do some operation with the macro-particles. Two Node member function hooks are provided for this: (1) a nodeCalculator, and (2) an updatePartAtNode routine. The first routine is a place where preliminary calculations can be done, which depend on more than one macro-particle. For example, the potential resulting from the ensemble of macro-particles could be calculated here for a space-charge kick node. The second routine (updatePartAtNode) operates on an individual macro-particle (it is automatically looped trough all active macro-particles). For example, a "Transfer matrix" node could advance a single macro-particle through a transfer matrix here. Note that either of these routines can be omitted if not needed.

5.2.3. Derived Node Sub-Classes

As mentioned above, the Node class is an abstract class. It will be inherited by the actual Node sub-classes. These derived classes will perform the actual calculations needed to transport the macro-particles around the Ring, and performing other manipulations on them. The mechanism for implementing these classes is generally to create a module for each class. While the class contains the routines that do the calculations, the "module" members contain the mechanisms for creating an object from the Shell, and other actions requiring direct user access. Examples of Node sub-classes are: (1) a TransferMatrix class for transporting macro-particles around the Ring, (2) a LSpaceCharge class to give longitudinal space charge kicks, and (3) a RFCavity class to give RF voltage kicks.

When a Node sub-class is added, a constructor should be supplied that at a minimum accepts input for the Node name and the order number in-which the Node calculators are called. Also, at least one calculator should be supplied (either a nodeCalculator or updatePartAtNode). With these class members defined and declared, we know both "when" each node should be called as we work our way around the Ring and "what" it should do.

6. MODULES

6.1. The Macro-Particle and Ring Modules

6.1.1. Particles Module

This module contains the interfaces between the user/shell and the macroParticle class. Member functions in this module offer a window into the macroParticle class. The addMacroParticle routines allow direct addition of a macro-particle with specified values. This is the mechanism for actually adding a macro-particle. It can be called directly from the shell or from other modules[7]. An example application of this routine would be to read in a set of prescribed particle information - e.g., one could write a SUPERCODE script file to read data from an input file which calls the addMacroParticle routine. The macro-particle parameters are stored in the macroParticle class. Since the Shell does not have direct access to class members, routines are provided here to dump macro-particle information to forms accessible by the Shell. nMacroParticles specifies the number of macro-particles instantiated so far. The routines xVal(i), yVal(i), xpVal(i), ypVal(i), deltaEVal(i), and phiVal(i) return the x, y, x', y', ΔE and ϕ values for macro-particle i.

6.1.2. Ring Module

The Ring module is a general module which controls the overall execution flow for the particle tracking. Although it has no classes itself, it contains initialization routines to determine the order in-which the Node class actions are performed, and the looping routines that actually perform the calls to the Node class calculators. Some of the important routines from this module are described below.

The initRing routine performs some general initialization calls. In particular it sorts all Nodes which have been added to the ring by ascending order of the oindex value[8]. This routine does not need to be called by the user, and is automatically called before and calculations, if it is needed.

The doTurn routine is the routine that actually performs turns around the ring. It takes an Integer argument which specifies the number of turns to do. This is the routine that loops over all active nodes. For each node it first calls the appropriate nodeCalculator routine and then loops over all the macroparticles while calling the updatePartAtNode routine.

The calcBucket routine is a special routine to track "a bucket particle", which can be used to find the separatrix bucket boundary in longitudinal space. It takes two arguments, (1) an initial energy (GeV) and, (2) an initial angle (deg). The energy and angle coordinates

---

[7] For example, it is used in the Inject module, which has "Initializers" which sample from specified distributions.
[8] It's a good idea to include some "padding" when assigning an oindex value to a Node. For example you might use intervals of 10 or so, to allow "space" between elements to add more nodes later.

for the bucket are stored in the vectors bucketDE (GeV) and bucketPhi (deg) respectively. These vectors are automatically dimensioned as needed.

6.2. Modules Governing Derived Node Classes

6.2.1. Injection Module

This module includes capabilities for adding a foil, and automatically injecting particles at the foil at each turn. This module contains a Foil class which includes information about the foil. Generally this Node will be at the beginning of the Ring (but it doesn't have to be). A Foil node can be included with a call to the routine:

>Void addFoil(const String &name, const Integer &order,
>   const Real &xMin, const Real &xMax, const Real &yMin,
>   const Real &yMax, const Real &thick, const Real &zNum)

Here name is a name of the node, order is the oindex value, xMin(xMax) is the minimum (maximum) horizontal foil extent in mm, yMin(yMax) is the minimum (maximum) vertical foil extent in mm, thick is the foil thickness (mg/cm^2) and zNum is the atomic charge number. Each time the Foil node is encountered as the ring is looped around, the routine InjectParts() is called. This routine will pick nMacrosPerTurn macro-particles from a prescribed distribution, and add them to the macroParticle ensemble (until a maximum of nMaxMacroParticles macro-particles have been injected). The distribution which is sampled to pick the macro-particles is specified by the routines:

>Void addXInitializer(const String &n, Subroutine subX)
>Void addYInitializer(const String &n, Subroutine subY)
>Void addLongInitializer(const String &n, Subroutine subL)

These routines tell the InjectParts() routine what routines to use to sample the horizontal, vertical and longitudinal distributions respectively. For example, subX will be called to provide the x, x' values for a new macro-particle. Some compiled routines are available for this purpose. The

>Void JohoXDist(), and Void JohoXDist()

routines supply values sampled from "Joho" horizontal and vertical distributions respectively, where the Joho distribution is a general form taken from Ref. [2]. The Void UniformLongDist() routine supplies samples from a uniform longitudinal distribution in energy and phase. The values provided by these sampling routines are displacements (the $\delta x, dx\tilde{O}, dy, dy', dE$ and $\delta\phi$) which assigned to the variables dXInj, dXPInj, dYInj, dYPInj, deltaE, and phi for. These displacements are added to the central coordinates of the injected beam (x_0, x'_0, y_0, y'_0) for the actual macro-particle initial values. See the Injection.mod file for a complete description of the controlling variables for these routines. Note that the user can provide her own, say, horizontal distribution sampling routine in an input script file, and simply point to it with the addXInitializer call, without recompiling.

6.2.2. TransMat Module

Transfer matrices are a fundamental mechanism used to transport macro-particles from one point in a ring lattice to another. A TransMatrix class is provided to contain transfer matrix information. Presently, only first order transfer matrices are included. To add a transfer matrix to the list of ring Nodes, the following routine from the TransMat module can be used:

> Void addTransferMatrix(const String &name, const Integer &order,
> const RealMatrix &R, const Real &bx, const Real &by,
> const Real &ax, const Real &ay, const Real &ex, const Real
> &epx, const Real &l)

Here, order is the oindex value (i.e. where in the Ring it is situated), R is the 6x6 1st order transfer matrix, $\beta x(y)$ is the horizontal (vertical) beta value at the beginning of the element [m], $\alpha x(y)$ is the horizontal (vertical) alpha value at the end of the element, εx is the horizontal dispersion [m] value at the end of the element, εpx is the horizontal dispersion derivative [m] value at the end of the element, and l is the length of the element [m]. A transfer matrix can be added directly from the Shell with a call to this routine, but this is a rather cumbersome process. Two routines are provided which read files to get the transfer matrix information about an entire ring:

> Void readDIMADFile(const String &s), and
> Void readMADFile(const String &s).

The argument of these routines is the name of the file to read. Each of these routines reads the specified file, parses it appropriately to find the transfer matrices, and makes the calls to the addTransferMatrix routine. The readDIMADFile routine expects to read a file produced by DIMAD [10], in the same format as that used by ACCSIM [2] to enter the transfer matrices. Each transfer matrix desired to be used must be explicitly outputed in the DIMAD run.

The readMADFile routine expects to read a TWISS file produced by MAD [3]. This is a convenient way to read in the transfer matrices between all elements of a ring, without going through a cumbersome set up procedure.

Each transfer matrix member is added with an oindex value incremented by 5 from the previous transfer matrix (initial value is 5). Thus there is room between these transfer matrices to go back and add more Nodes (like a Foil, RF Cavities, etc.).

6.2.3. Bump Module

The closed orbit of the Ring can be artificially altered anywhere in the Ring by introduction of an ideal bump. This is typically done directly before (and after) a Foil Node to facilitate painting of the Ring distributions. The Node sub-class used to contain the bumps is the IdealBump class.

Bumps should be added in pairs: and "up" bump, and a "down" bump. These nodes will typically surround a Foil element. An IdealBump node can be added with the Bump module routine:

> Void addIdealBump(const String &name, const Integer &order,
> const Integer &upDown, const Subroutine sub)

Here order is the oindex value specifying where in the Ring the bump is. Sub is the name of a routine that will specify the bump values as a function of time, and upDown is a switch indicating whether the bump is moving the closed orbit up (==1) or down (!=1). Note that the same routine should be specified by "sub" when adding both the "up" and "down" bumps, to ensure no net movement of the closed orbit. The routine called to provide the bumps is expected to set the variables:

> xIdealBump   - "The x value of the ideal bump at a point in time (mm)",
> xPIdealBump   - "The x prime of the ideal bump at a point in time (mrad)",
> yIdealBump   - "The y value of the ideal bump at a point in time (mm)",
> yPIdealBump   - "The y prime of the ideal bump at a point in time (mrad)"

as a function of the variable "time" [msec]. These variables are then used to modify the closed orbit parameters, and accordingly the values of x, $x\tilde{O}$, y, and y' of each macro particle. The user can add a routine to specify any sort of bump profile, and simply refer to this routine in the addIdealBum() call. An exponentially decaying bump profile is provided in the routine eFoldBump(). The variable names for these quantities are described in the Bump.mod file.

6.2.4. RFCavity Module

This module is provided to add RF cavities. The Node sub class for this purpose is the RFCav. An RF Cavity with an arbitrary number of harmonics can be added with the routine:

> Void addRFCavity(const String &n, const Integer &o, Integer &nh,
> RealVector &v, RealVector &hn, RealVector &p)

The first argument is a name for the Node, the second argument is the order number where this RF cavity appears in the Ring, the third argument is the number of harmonic components the RF Voltage waveform has, and the fourth, fifth and sixth arguments are references to Real Vectors containing the RF voltage [kV], harmonic number, and phase offset [rad] of each harmonic component of the RF voltage, respectively. The Real Vectors referenced in this call can be created, and initialized, on the fly prior to this call.

6.3. Miscellaneous Modules

6.3.1. MiscCalcs Module.

This module performs miscellaneous calculations, which may have dependencies on both macro-particles and ring nodes. Presently, the only calculation done here is to find the emittance of the macro-particles, Eq. (8). The routine Void calcEmittance() can be called at anytime, and will calculate a distribution of macro-particle emittances. The maximum horizontal and vertical emittance are stored in emittanceXMax and emittanceXMax respectively [mm-mrad].

6.3.2. Output Module

Quantities from the various modules above can be output from the Shell as desired by the user. Output capabilities to the screen, plots and files are described in Reference [9]. These output capabilities can generally be done from the Shell, without requiring the need to add source code. However, several canned text output routines are provided to output Class members, which are not directly accessible from the Shell (as well as some other convenient outputs)

    Void showMacroPart(Ostream &os)
                           - "Routine to show MacroParticle info to stream os";
    Void showNodes(Ostream &os)
                           - "Routine to show Node info to stream os";
    Void showTransMatrix(Ostream &os)
                           - "Routine to show TransferMatrix info to stream os";
    Void showRing(Ostream &os)
                           - "Routine to show all the ring information to stream os";
    Void showTurnInfo(Ostream &os)
                           - "Routine to show misc. general turn information";

Note: the argument to these routines can either be:

                 cout   - long buffered output to the screen,
                 cerr   - short buffered output to the screen, or

Any user defined stream. A user defined stream fio to a file called fileout can be created on the fly by

OFstream fio("fileout", ios::out);

This stream will delete any previously defined file called fileout. The command

OFstream fio("fileout", ios::app);

is similar, but will append the prescribed output to whatever exists (if anything) in a file called "fileout". See Reference [9] for more information on inputting/outputting information from the Shell.

ACKNOWLEDGMENTS

We are indebted to the Authors of other codes, notably to Frederick Jones of TRIUMF and to his ACCSIM [6]. We have used that code for a long time for the simulation of the SNS accumulator in its various versions and stages, and we have derived much inspiration from it.

REFERENCES

1. SNS Collaboration. *"The National Spallation Neutron Source"* Conceptual Design Report Documentation. NSNS/CDR-5 1997
2. Jones, F. W. *"Users Guide to ACCSIM ",* TRI-DN-90-17, June 1990.
3. Iselin, F.Ch., *"The MAD Program, Version 8.7",* CERN/SL/92(AP), July 17, 1992
4. Gardner, C.J, Luccio, A.U. and Lee, Y.Y. *"Accumulator Ring Lattice for the National Spallation Neutron Source"* Proceedings of the PAC'97, Vancouver, B.C. Canada, May 1997 (in press)
5. Luccio, A.U. *"Numerical Calculation of the Tune Spread Induced by Transverse Space Charge in a Synchrotron",* Brookhaven National Laboratory BNL/NSNS Technical Note No. 023, January 29, 1997
6. Mac Lachlan, J.A. *"Longitudinal Phase Space Tracking With Space Charge and Wall Coupling Impedance".* Fermi National Accelerator Laboratory, Report FN-446, February 1987
7. Chao, A.W. *"Physics of Collective Beam Instabilities in High Energy Accelerators,* Wiley New York
8. Holmes, J., Galambos, J.D., Olsen, D.K. and Lee, S.Y. An RMS Particle Core Model for Rings", in These Proceedings
9. Hanoi, S.W. *"Using and Programming the SUPERCODE",* July 21, 1995
10. Servranckxz, R.V., Brown, K.L, Shachinger, L. and Douglas, D. *"Users Guide to the Program DIMAD".* SLAC Rept 285 UC-28(A), May 1985
11. Blaskiewicz, M. and Brennan, M. Proc. EPAC'96, p. 2373

Nonlinear δF Simulation Studies of High-Intensity Matched-Beam Propagation in Periodic-Focusing Transport Systems[1]

Peter H. Stoltz, W. Wei-li Lee and Ronald C. Davidson

Plasma Physics Laboratory
Princeton University
Princeton, New Jersey, 08543 USA

Abstract. The nonlinear δF formalism based on the Vlasov-Poisson equations is used to describe the propagation of a matched, high-intensity ion beam through a periodic-focusing solenoidal field in the thin-beam approximation ($r_b \ll S$). The distribution function F_b is divided into a zero-order part (F_b^0) plus a perturbation (δF_b) which evolve nonlinearly in the zero-order and perturbed field configurations. To illustrate the application of the technique to axisymmetric, matched-beam propagation, nonlinear δF simulation results are presented for the case of a periodic solenoidal focusing field in which the (oscillatory) coupling coefficient $\kappa_z(s)$ turns on adiabatically about a constant average value $\bar{\kappa}_z$. For a field oscillation which turns on gradually over about twenty lattice periods, the amplitude of the mismatch oscillation is reduced by about one order-of-magnitude compared to the case where the field is turned on suddenly.

1. Introduction

The influence of space-charge effects on the nonlinear dynamics and stability properties of high-intensity charged particle beams [1–3] is particularly significant at the high beam currents and charge densities envisioned for the next-generation accelerators and transport systems for tritium production, spallation neutron sources, and heavy ion fusion [4,5]. It is therefore increasingly important to develop an improved theoretical understanding of the equilibrium, stability and transport properties of intense nonneutral beams propagating in periodic focusing systems [6–9]. In this regard, advanced numerical simulations and analytical studies [6–13] are

---

[1] Presented at the Workshop on Space-Charge Physics in High-Intensity Hadron Rings (Shelter Island, New York, May, 1998).

playing an increasingly critical role in design optimization and in validating theoretical models for comparison with experiment. The present paper applies the nonlinear δF-formalism [12,13] to study *matched-beam* propagation through a periodic solenoidal focusing field for the case of a thermal equilibrium beam with intense space-charge fields. To obtain matched-beam solutions, the (oscillatory) coupling coefficient $\kappa_z(s)$ is turned on *adiabatically* about a constant average value $\bar{\kappa}_z$.

Nonlinear δF simulation techniques have been applied successfully to model the nonlinear dynamics and stability properties of magnetically confined fusion plasmas [14] and intense nonneutral beam propagation through a periodic quadrupole lattice [13]. For a field that turns on gradually over approximately twenty lattice periods, the mismatch is reduced by about one order-of-magnitude. Such δF schemes are found to be attractive in comparison with standard particle-in-cell simulations because they exhibit minimal noise and accuracy problems.

2. Theoretical Model and Nonlinear δF-Formalism

The present analysis assumes a thin, intense nonneutral ion beam with characteristic radius r_b and axial momentum $\gamma_b m \beta_b c$ propagating in the z-direction through a periodic solenoidal focusing field, $\mathbf{B}^{sol}(\mathbf{x}) = B_z(s)\hat{\mathbf{e}}_z - (r/2)B_z'(s)\hat{\mathbf{e}}_r$. Here, $B_z(s+S) = B_z(s)$ is the axial field component, s is the axial coordinate, $S = const.$ is the periodicity length, 'prime' denotes derivative with respect to s, and $r = (x^2 + y^2)^{1/2}$ is the radial distance from the beam axis. We assume a thin beam with $r_b \ll S$ and $\nu = Z_i^2 e^2 N_b/mc^2 \ll \gamma_b$, where ν is Budker's parameter, $\gamma_b mc^2$ is the characteristic energy of a beam particle, $\gamma_b = (1 - \beta_b^2)^{-1/2}$ is the relativistic mass factor, $v_b = \beta_b c$ is the axial velocity, c is the speed of light in *vacuo*, and $Z_i e$ and m are the ion charge and rest mass, respectively. The quantity $N_b = \int dx dy n_b$ is the number of beam particles per unit axial length, where $n_b(x,y,s)$ is the particle density. The thin-beam approximation ($r_b \ll S$) and the assumption of small Budker's parameter ($\nu \ll \gamma_b$) are consistent approximations provided the transverse momentum components of a beam particle, p_x and p_y, and the characteristic axial momentum spread, δp_z, are small in comparison with the directed axial momentum $\gamma_b m \beta_b c$.

In addition, the present analysis is carried out in the electrostatic approximation, where the self-electric field produced by the beam space-charge is $\mathbf{E}^s = -\nabla \phi^s$, and the electrostatic potential $\phi^s(x,y,s)$ is determined self-consistently from Poisson's equation. Furthermore, to determine the self-magnetic field $\mathbf{B}^s = \nabla \times A_z^s \hat{\mathbf{e}}_z$ produced by the beam current, we assume that the axial velocity profile $V_{zb}(x,y,s) \simeq \beta_b c$ is approximately uniform over the beam cross section, and the self-magnetic field is approximated by $\mathbf{B}^s = \beta_b \nabla \phi^s \times \hat{\mathbf{e}}_z$. Consistent with the assumptions described above, the nonlinear δF-formalism makes use of the nonlinear Vlasov-Poisson equations to describe the dynamics of the beam particles and their interaction with the field configuration $\mathbf{E}^s = -\nabla \phi^s$ and $\mathbf{B} = \mathbf{B}^{sol} + \beta_b \nabla \phi^s \times \hat{\mathbf{e}}_z$. For present purposes, it is convenient to introduce the focusing coefficient $\kappa_z(s)$ and the normalized electrostatic potential $\psi(x,y,s)$ defined by

$$\kappa_z(s) = \left(\frac{Z_i e B_z(s)}{2\gamma_b m \beta_b c^2}\right)^2,$$

$$\psi(x,y,s) = \frac{Z_i e}{\gamma_b^3 m \beta_b^2 c^2} \phi^s(x,y,s). \tag{1}$$

For a periodic focusing lattice with $\kappa_z(s+S) = \kappa_z(s)$, we express

$$\kappa_z(s) = \bar{\kappa}_z + \delta\kappa_z(s) \tag{2}$$

where $\bar{\kappa}_z \equiv S^{-1} \int_{s_0}^{s_0+S} ds\, \kappa_z(s)$, and $\delta\kappa_z(s+S) = \delta\kappa_z(s)$ oscillates about zero average value. In addition, we transform to a frame of reference rotating about the beam axis at the local (normalized) Larmor frequency $\Omega_L(s) = -\sqrt{\kappa_z(s)} = -Z_i e B_z(s)/2\gamma_b m \beta_b c^2$. Introducing the accumulated phase of rotation, $\theta_L(s) = -\int_{s_0}^{s} ds \sqrt{\kappa_z(s)}$, the transverse orbits, $X(s)$ and $Y(s)$, in the rotating frame are related to the transverse orbits, $x(s)$ and $y(s)$, in the laboratory frame by $X = x\cos\theta_L(s) + y\sin\theta_L(s)$ and $Y = -x\sin\theta_L(s) + y\cos\theta_L(s)$. Assuming that the beam particles have negligible axial momentum spread about the average value $\gamma_b m \beta_b c$, it can be shown that the distribution function $F_b(X,Y,X',Y',s)$ evolves according to the nonlinear Vlasov equation [8,9]

$$\frac{d}{ds} F_b = \frac{\partial F_b}{\partial s} + X'\frac{\partial F_b}{\partial X} + Y'\frac{\partial F_b}{\partial Y} - \left([\bar{\kappa}_z + \delta\kappa_z(s)]X + \frac{\partial \psi}{\partial X}\right)\frac{\partial F_b}{\partial X'}$$
$$- \left([\bar{\kappa}_z + \delta\kappa_z(s)]Y + \frac{\partial \psi}{\partial Y}\right)\frac{\partial F_b}{\partial Y'} = 0. \tag{3}$$

In Eq. (3), (X,Y,X',Y') are phase-space variables appropriate to the Larmor frame, and the normalized potential $\psi(X,Y,s)$ is determined self-consistently from Poisson's equation

$$\left(\frac{\partial^2}{\partial X^2} + \frac{\partial^2}{\partial Y^2}\right)\psi = -\frac{2\pi K}{N_b} \int dX' dY' F_b. \tag{4}$$

Here, $n_b(X,Y,s) = \int dX'dY' F_b$ is the particle density, $N_b = \int dXdY\, n_b$ is the number of particles per unit axial length, and $K = 2N_b Z_i^2 e^2/\gamma_b^3 m \beta_b^2 c^2$ is the self-field perveance. In Eq. (3), note that X' and Y' correspond to normalized velocity variables in the X-Y plane (i.e., X' denotes dX/ds and Y' denotes dY/ds), and the coefficients of $\partial F_b/\partial X'$ and $\partial F_b/\partial Y'$ correspond to the particle accelerations in the X- and Y-directions, respectively.

In the present application of the δF-formalism, we divide the distribution function $F_b(X,Y,X',Y',s)$ into a zero-order part (F_b^0) plus a perturbation (δF_b) according to [12,13]

$$F_b = F_b^0 + \delta F_b. \tag{5}$$

Here, $F_b^0(R, X', Y', s)$ is taken to be a *known* axisymmetric solution ($\partial/\partial\theta = 0$) to the nonlinear Vlasov-Poisson equations (3) and (4) with constant focusing coefficient $\bar{\kappa}_z$. To describe the nonlinear evolution of $\delta F_b(X, Y, X', Y', s) = F_b - F_b^0$, we introduce the weight function $w(X, Y, X', Y', s)$ defined by

$$w = \frac{\delta F_b}{F_b} = 1 - \frac{F_b^0}{F_b}. \tag{6}$$

From Eqs. (3) and (6), with $dF_b/ds = 0$, it then follows that

$$\frac{d}{ds}w = (1-w)\left[\delta\kappa_z(s)\mathbf{X} + \frac{\partial}{\partial \mathbf{X}}\delta\psi\right] \cdot \frac{1}{F_b^0}\frac{\partial F_b^0}{\partial \mathbf{X}'}, \tag{7}$$

where d/ds is the *total derivative* following the particle trajectory in the equilibrium plus perturbed field configuration, and $\delta\psi(X, Y, s)$ is the solution of

$$\left(\frac{\partial^2}{\partial X^2} + \frac{\partial^2}{\partial Y^2}\right)\delta\psi = -\frac{2\pi K}{N_b}\int dX'dY' \delta F_b. \tag{8}$$

The nonlinear δF-formalism described above has wide applicability to s-dependent focusing field configurations $\kappa_z(s) = \bar{\kappa}_z + \delta\kappa_z(s)$, even for large-amplitude variations in the oscillatory component $\delta\kappa_z(s)$. The nonlinear Vlasov-Poisson equations for the axisymmetric ($\partial/\partial\theta = 0$) background distribution F_b^0 and normalized self-field potential ψ^0 support a broad range of *equilibrium* solutions ($\partial F_b^0/\partial s = 0 = \partial \psi^0/\partial s$) in which the zero-order distribution function is of the form [9]

$$F_b^0 = F_b^0(H_\perp). \tag{9}$$

Here, $H_\perp$ is the single-particle constant of the motion

$$H_\perp = \frac{1}{2}(X'^2 + Y'^2) + \frac{1}{2}\bar{\kappa}_z(X^2 + Y^2) + \psi^0(R), \tag{10}$$

corresponding to the (normalized) kinetic plus potential energy of a beam ion, and $R = (X^2 + Y^2)^{1/2}$ is the radial distance from the beam axis. For the class of beam equilibria $F_b^0(H_\perp)$, the unnormalized beam emittance ϵ_0 and mean-square beam radius R_{b0}^2 are defined in the usual manner by

$$\epsilon_0^2 = 4\langle X'^2 + Y'^2\rangle_0 \langle X^2 + Y^2\rangle_0,$$

$$R_{b0}^2 = \langle X^2 + Y^2\rangle_0, \tag{11}$$

where the average over F_b^0 of a phase function ψ is defined by $\langle \psi \rangle_0 = N_b^{-1}\int dX dY dX' dY' \psi F_b^0$, and $N_b = \int dX dY dX' dY' F_b^0$ is the number of ions per unit axial length.

One such example of a background equilibrium distribution F_b^0 in a uniform focusing field $\bar{\kappa}_z$ is the thermal equilibrium distribution [9] specified by

$$F_b^0(H_\perp) = \hat{n}_b \left(\frac{\gamma_b m \beta_b^2 c^2}{2\pi \hat{T}_{\perp b}}\right) \exp\left\{-\frac{\gamma_b m \beta_b^2 c^2}{\hat{T}_{\perp b}} H_\perp\right\} . \tag{12}$$

Here, $\hat{n}_b$ and $\hat{T}_{\perp b}$ are positive constants with dimensions of density and temperature (energy units), respectively, and $H_\perp$ is the (dimensionless) Hamiltonian defined in Eq. (10). The transverse beam temperature $\hat{T}_{\perp b}$ and the transverse emittance ϵ_0 are related by $\epsilon_0^2 = (8\hat{T}_{\perp b}/\gamma_b m \beta_b^2 c^2) R_{b0}^2$ for the choice of distribution in Eq. (12). Without loss of generality, we take the on-axis self-field potential to be $\psi^0(R=0) = 0$, and identify $\hat{n}_b = n_b^0(R=0)$ with the on-axis beam density. The equilibrium density profile, $n_b^0(R) = \int dX' dY' F_b^0(H_\perp)$, calculated from Eq. (12) is given by

$$n_b^0(R) = \hat{n}_b \exp\left\{-\frac{\gamma_b m \beta_b^2 c^2}{2\hat{T}_{\perp b}} [\bar{\kappa}_z R^2 + 2\psi^0(R)]\right\} , \tag{13}$$

where $\psi^0(R)$ is determined self-consistently from Poisson's equation

$$\frac{1}{R}\frac{\partial}{\partial R} R \frac{\partial \psi^0}{\partial R} = -\frac{2\pi K}{N_b} n_b^0(R) . \tag{14}$$

The equilibrium density profile calculated from Eqs. (13) and (14) is bell-shaped [9] with density maximum on axis ($R = 0$). As the beam intensity is increased and $K/2R_{b0}^2$ approaches $\bar{\kappa}_z$, the density profile becomes increasingly *flat-top*, with $n_b^0(R) \simeq \hat{n}_b = const.$ in the beam interior.

3. Dynamics of Root-Mean-Square Beam Radius $R_b(s)$

For axisymmetric beam propagation ($\partial/\partial \theta = 0$), an *exact* consequence of the nonlinear Vlasov-Poisson equations (3) and (4) is that the rms beam radius $R_b(s) = \langle X^2 + Y^2 \rangle$ evolves according to [8]

$$\frac{d^2}{ds^2} R_b(s) + \left[\kappa_z(s) - \frac{K}{2R_b^2(s)}\right] R_b(s) = \frac{\epsilon^2(s)}{4R_b^3(s)} . \tag{15}$$

Here, $\kappa_z(s) = \bar{\kappa}_z + \delta\kappa_z(s)$ is the s-dependent focusing coefficient, statistical averages over the distribution function $F_b(X, Y, X', Y', s)$ are defined by $\langle \psi \rangle = N_b^{-1} \int dX dY dX' dY' \psi F_b$, and the unnormalized beam emittance $\epsilon(s)$ is defined by

$$\epsilon^2(s) = 4[\langle X'^2 + Y'^2 \rangle \langle X^2 + Y^2 \rangle - \langle XX' + YY' \rangle^2] . \tag{16}$$

For the special case of a uniform focusing field with $\delta\kappa_z(s) = 0$ and $\kappa_z(s) = \bar{\kappa}_z = const.$, and constant values of rms beam radius R_{b0} and emittance ϵ_0, Eq. (15) reduces to the radial force balance condition

$$\left[\bar{\kappa}_z - \frac{K}{2R_{b0}^2}\right] R_{b0} = \frac{\epsilon_0^2}{4R_{b0}^3} . \tag{17}$$

Equation (17), which is valid for general choice of $F_b^0(H_\perp)$ in a uniform focusing field $\bar{\kappa}_z$, represents a powerful constraint condition on equilibrium beam properties. As expected, Eq. (17) is similar in form to the familiar envelope equation [1,8] for the outer radius r_b of a uniform-density Kapchinskij-Vladimirskij (KV) beam in the smooth-beam approximation ($dr_b/ds = 0$) provided we make the identification $R_{b0} = r_b/\sqrt{2}$. For specified values of $\bar{\kappa}_z$, K and ϵ_0^2, note that Eq. (17) can be solved for the mean-square beam radius to give

$$R_{b0}^2 = \frac{K}{4\bar{\kappa}_z} + \left[\left(\frac{K}{4\bar{\kappa}_z}\right)^2 + \frac{\epsilon_0^2}{4\bar{\kappa}_z}\right]^{1/2}. \tag{18}$$

As expected, we find from Eq. (18) that R_{b0}^2 increases with increasing beam intensity (K), increasing beam emittance (ϵ_0), and decreasing field strength ($\bar{\kappa}_z$).

For application in the numerical simulations in Sec. 4, we consider a fully-developed periodic focusing field $\kappa_z(s+S) = \kappa_z(s)$ in which the axial field has a sinusoidal component with $B_z(s) = B_{z0}[1 + (\Delta_m/2)\sin(2\pi s/S)]$, where B_{z0}, Δ_m and S are constants. Thus, $\kappa_z(s) = [Z_i e B_z(s)/2\gamma_b m\beta_b c^2]^2 = \bar{\kappa}_z + \delta\kappa_z(s)$, where

$$\bar{\kappa}_z = \kappa_{z0}(1 + \Delta_m^2/8) ,$$

$$\delta\kappa_z(s) = \bar{\kappa}_z \left[\frac{\Delta_m}{1+\Delta_m^2/8}\sin\left(\frac{2\pi s}{S}\right) - \frac{\Delta_m^2/8}{1+\Delta_m^2/8}\cos\left(2\cdot\frac{2\pi s}{S}\right)\right] . \tag{19}$$

Here, $\kappa_{z0} \equiv (Z_i e B_{z0}/2\gamma_b m\beta_b c^2)^2$.

The nonlinear equation (15) for the rms beam radius $R_b(s)$ can be integrated numerically [8] for a wide range of system parameters and choices of periodic lattice function $\kappa_z(s+S) = \kappa_z(s)$. For the case of *small-amplitude* oscillations about the average beam radius R_{b0} defined in Eqs. (17) and (18), we express $R_b(s) = R_{b0} + \delta R_b(s)$, and linearize Eq. (15). Treating $\epsilon^2 \simeq \epsilon_0^2 = const.$, and approximating $\delta\kappa_z(s) = \bar{\kappa}_z \Delta_m \sin(2\pi s/S)$ for small values of $\Delta_m^2/8 \ll 1$, this gives

$$\delta R_b''(s) + k_e^2 \delta R_b(s) = -\bar{\kappa}_z \Delta_m \sin(k_s S) , \tag{20}$$

where $k_s \equiv 2\pi/S$, and the envelope-oscillation wavenumber k_e is defined by

$$k_e^2 \equiv \bar{\kappa}_z \left[1 + \frac{K}{2\bar{\kappa}_z R_{b0}^2} + \frac{3\epsilon_0^2}{4\bar{\kappa}_z R_{b0}^4}\right]$$

$$= 4\bar{\kappa}_z \frac{[(KS/\epsilon_0)^2 + 4\bar{\kappa}_z S^2]^{1/2}}{KS/\epsilon_0 + [(KS/\epsilon_0)^2 + 4\bar{\kappa}_z S^2]^{1/2}} . \tag{21}$$

From Eq. (21), we note that the envelope-oscillation wavenumber k_e varies from $k_e^2 = 4\bar{\kappa}_z$ for a low-intensity beam ($K/S\sqrt{\bar{\kappa}_z} \ll 1$) to $k_e^2 = 2\bar{\kappa}_z$ for a high-intensity beam ($K/S\sqrt{\bar{\kappa}_z} \gg 1$). Furthermore, the general solution to Eq. (20) is

$$\delta R_b(s) = \frac{\bar{\kappa}_z \Delta_m}{k_s^2 - k_e^2} \sin(k_s S) + \delta R_b(0) \cos(k_e s)$$
$$+ \frac{1}{k_e}\left[\delta R_b'(0) - \frac{k_s \bar{\kappa}_z \Delta_m}{k_s^2 - k_e^2}\right] \sin(k_e s) , \qquad (22)$$

where $\delta R_b(0)$ and $\delta R_b'(0)$ are the initial values at $s = 0$. It is evident from Eq. (22) that $\delta R_b(s)$ generally has oscillatory components at wavelength $\lambda_e = 2\pi/k_e$, and at wavelength $\lambda_s = 2\pi/k_s = S$ corresponding to the period of the applied focusing field. Only for the special initial conditions with

$$\delta R_b(0) = 0,$$
$$\delta R_b'(0) = k_s \bar{\kappa}_z \Delta_m/(k_s^2 - k_e^2), \qquad (23)$$

is the beam truly *matched*, with $\delta R_b(s)$ oscillating only at the period $S = 2\pi/k_s$ of the focusing field [the first term on the right-hand side of Eq. (22)].

In the nonlinear δF simulations presented in Sec. 4, we will find that both frequency components in Eq. (22) are generally present in $\delta R_b(s)$ for the case where $\delta\kappa_z(s)$ in Eq. (19) is *turned on suddenly* at $s = 0$. As a second approach, we adopt an *adiabatic turn-on model* in which $B_z(s) = B_{z0}[1 + (1/2)\Delta(s)\sin(k_s S)]$, where (for example) the coefficient $\Delta(s)$ is defined by

$$\Delta(s) = \Delta_m \left[1 - \exp\left(-\alpha \frac{s}{S}\right)\right] , \qquad (24)$$

where α is a positive constant. For $s = 0$, Eq. (24) reduces to $\Delta = 0$, and for $s \gg \alpha^{-1} S$, $\Delta(s)$ asymptotes at $\Delta_m = const$. That is, using Eq. (24) with sufficiently small value of α, $\delta\kappa_z(s)$ turns on adiabatically over many lattice periods, and achieves the constant-amplitude oscillatory form in Eq. (19) for $s \gg \alpha^{-1} S$. In this case, it is found in the nonlinear δF simulations presented in Sec. 4 that *matched-beam* solutions are readily obtained with $\delta R_b(s)$ oscillating with the same period S as the periodic focusing field.

4. Numerical Model and Simulation Results

In this section, we present the equations followed in the δF code, summarize the parameters chosen for a sample run, discuss initial conditions and conservation properties, and present numerical results for the rms beam radius oscillations for two different values of the adiabatic turn-on parameter, α. The main result of this section is that we are able to match the beam successfully. Turning the field on adiabatically over approximately twenty lattice periods reduces the beam mismatch by about one order-of-magnitude relative to the case where $\delta\kappa_z(s)$ is turned on suddenly.

The δF simulations follow the particle trajectories and weights as a function of s. One can derive the necessary equations using the Klimontovich representation for the distribution function [1]

$$F_b = \frac{N_b}{N_p} \sum_{i=1}^{N_p} \delta(\mathbf{X} - \mathbf{X_i}) \delta(\mathbf{X'} - \mathbf{X'_i}), \quad (25)$$

where N_p is the number of particles used in the simulations. For this representation of F_b, assuming perturbations about the thermal equilibrium distribution $F_b^0(H_\perp)$ in Eq. (12), the equations for $\mathbf{X_i}(s)$ and $w_i(s)$ reduce to [see Eqs. (3) and (7)]

$$\mathbf{X}''_i = -[\bar{\kappa}_z + \delta\kappa_z(s)]\mathbf{X}_i - \left(\frac{\partial}{\partial \mathbf{X_i}}\psi_0(\mathbf{X_i}) + \frac{\partial}{\partial \mathbf{X_i}}\delta\psi(\mathbf{X_i}, s)\right), \quad (26)$$

$$w'_i = (1 - w_i)\left[\delta\kappa_z(s)\mathbf{X_i} + \frac{\partial}{\partial \mathbf{X_i}}\delta\psi(\mathbf{X_i}, s)\right] \cdot \frac{\gamma_b m \beta_b^2 c^2}{\hat{T}_{\perp b}} \mathbf{X'}_i. \quad (27)$$

For the simulations presented here, we assume an axisymmetric beam, and use Eqs. (8), (25), and $\delta F_b = wF_b$ to write

$$\frac{\partial}{\partial R}\delta\psi(R, s) = -\frac{K}{RN_p} \sum_{R_i < R}^{N_p} w_i(\mathbf{X_i}, s), \quad (28)$$

where $R^2 = X^2 + Y^2$.

We present simulation results for $\sqrt{\bar{\kappa}_z}S = 1$, corresponding to a vacuum phase advance of 57.3 degrees, and for $K/\sqrt{\bar{\kappa}_z}\epsilon_0 = 3$, corresponding to a depressed phase advance, $\sigma = \epsilon_0 S/2R_{b0}^2$, of 17.3 degrees. We also take $\Delta_m = 0.2$ as suggested in [15], with 10,000 particles, 512 radial grid points, a time step of $\Delta s/S = 0.002$, and integrate to $s = 100S$. Results are presented for both $\alpha = \infty$ (sudden turn-on) and $\alpha = 0.1$ (adiabatic turn-on).

From the analysis in Sec. 3, the rms beam radius $R_b(s)$ is expected to oscillate at two distinct frequencies. Hence, we monitor the change in $R_b^2(s)$, calculated from

$$\delta R_b^2(s) = \delta\left\langle X^2 + Y^2\right\rangle = \left\langle X^2 + Y^2\right\rangle - \left\langle X^2 + Y^2\right\rangle_0$$
$$= \frac{1}{N_b} \int (X^2 + Y^2)(F_b - F_b^0) dXdYdX'dY'$$
$$= \frac{1}{N_b} \int (X^2 + Y^2) wF_b dXdYdX'dY'$$
$$= \frac{1}{N_p} \sum_{i=1}^{N_p} w_i(X_i^2 + Y_i^2). \quad (29)$$

The initial conditions at $s = 0$ for the particle positions and momenta are chosen consistently [12] with the thermal equilibrium distribution in Eq. (12). Although we do not do so here, Parker and Lee [14] have suggested a method for initializing with an arbitrary distribution, which could allow increased resolution in particularly interesting regions of the simulation, such as near the beam edge. The particle weights are chosen to be zero at $s = 0$. In the δF scheme, the particle weights determine how the beam properties deviate from equilibrium. Choosing zero for the initial value of the weights implies $\delta R_b(0) = 0$ and (because the thermal equilibrium distribution function is even in X' and Y') $\delta R_b'(0) = 0$. As shown in Sec. 3, these choices for $\delta R_b(0)$ and $\delta R_b'(0)$ lead to a mismatched beam for sudden turn-on of $\delta \kappa_z(s)$. Matching the beam by choosing $\delta R_b(0)$ and $\delta R_b'(0)$ as specified in Eq. (23) would require initializing the weights correctly, and it is not readily apparent how to accomplish this, as many choices for the initial particle weights would lead to the prescribed $\delta R_b(0)$ and $\delta R_b'(0)$. As an extreme example, one can imagine choosing all initial particle weights to be zero except for one, which is chosen by means of Eq. (29) to satisfy Eq. (23). To avoid this ambiguity, we take all initial particle weights to be zero, and instead match the beam by the adiabatic turn-on of $\delta \kappa_z(s)$.

Conservation of the total number of particles in the beam requires the sum of the particle weights be equal to zero for all s, i.e.,

$$\langle w_i(s) \rangle = \frac{1}{N_p} \sum_{i=1}^{N_p} w_i = 0. \tag{30}$$

This constraint is useful for testing how well the code is modeling the true beam dynamics. For the sample parameters given above, we find $\langle w_i \rangle \approx 0.001$ over a distance $s = 100S$. This deviation decreases approximately linearly with decreasing time step, so the deviation is due to integration error. In δF simulations of tokamak plasmas, similar behavior [16] has been observed. For the results presented in this paper, we adjust for this deviation by subtracting $\langle w_i \rangle$ from each particle weight at the beginning of each time step, thus ensuring particle number conservation.

Shown in Fig. 1 are the numerical results for sudden turn-on ($\alpha = \infty$). Figure 1(a) shows a plot of $\delta \kappa_z(s)$ as specified in Eq. (19). Only the final 10 periods (between $s = 90S$ and $s = 100S$) are presented so as to make the detailed structure clear. The change in rms beam radius is plotted in Fig. 1(b), again showing only the final 10 periods. Figure 1(c) shows the Fourier transform of $\delta R_b(s)$ over the entire 100 lattice periods. The peaks at the oscillation frequency of the focusing lattice, $k_s S = 2\pi$, and at the envelope frequency given by Eq. (21), $k_e S \simeq 1.54$, are clear in this figure. For the parameters chosen here, Eq. (22) predicts these peaks should have equal amplitude; the small discrepancy is due to integration error, and decreases with decreasing step size Δs. Also, there is a small DC offset in $\delta R_b(s)$ that appears in the $k = 0$ component of the Fourier transform in Fig. 1(c). This offset is also due to integration error, and decreases with decreasing Δs.

Figure 2 shows the numerical results for adiabatic turn-on ($\alpha = 0.1$). Figure 2(a) shows the full 100 periods of $\delta \kappa_z(s)$ to illustrate the adiabatic turn-on. Figure 2(b) again presents $\delta R_b(s)$ over the final 10 periods, and Fig. 2(c) shows the Fourier transform of $\delta R_b(s)$ over the entire 100 periods. From Fig. 2(c), it is clear the amplitude of the mismatched peak has decreased by approximately one order-of-magnitude, demonstrating that the adiabatic turn-on of $\delta \kappa_z(s)$ over about 20 lattice periods leads to a well-matched beam.

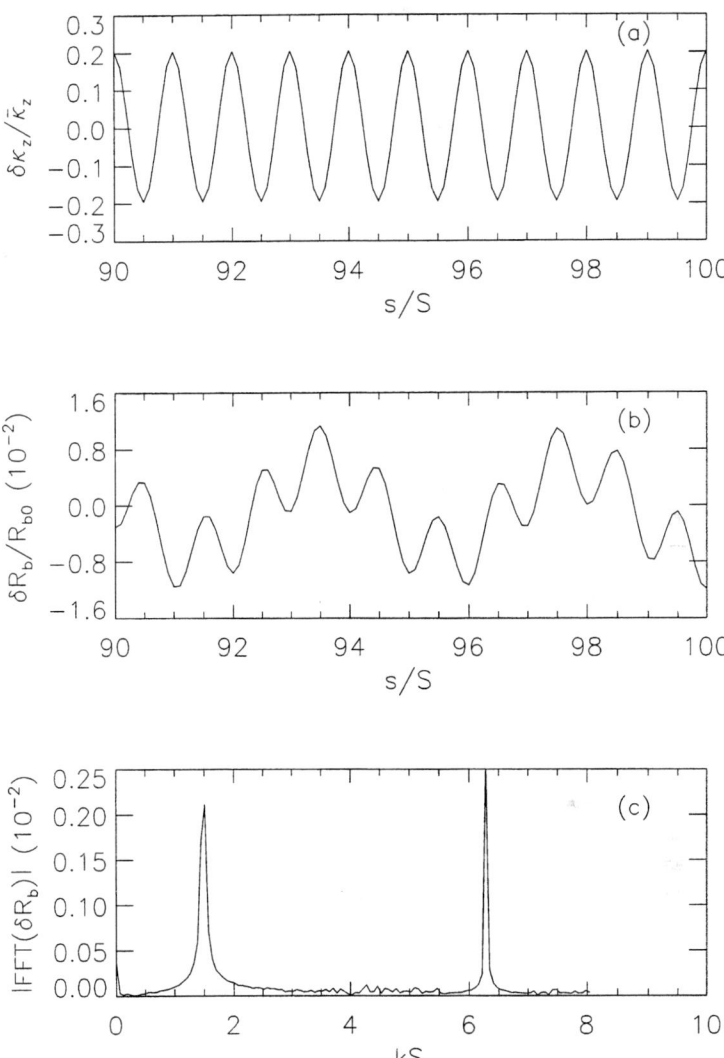

FIGURE 1. Mismatched beam with sudden turn-on ($\alpha = \infty$). Plotted versus s/S are: (a) $\delta\kappa_z(s)$ and (b) $\delta R_b(s)$ for the 10 lattice periods between $s = 90S$ and $s = 100S$. The Fourier transform of $\delta R_b(s)$ is shown in (c). The magnitude of the oscillations at $k_s S = 2\pi$ and $k_e S \simeq 1.54$ are nearly equal for this mismatched case.

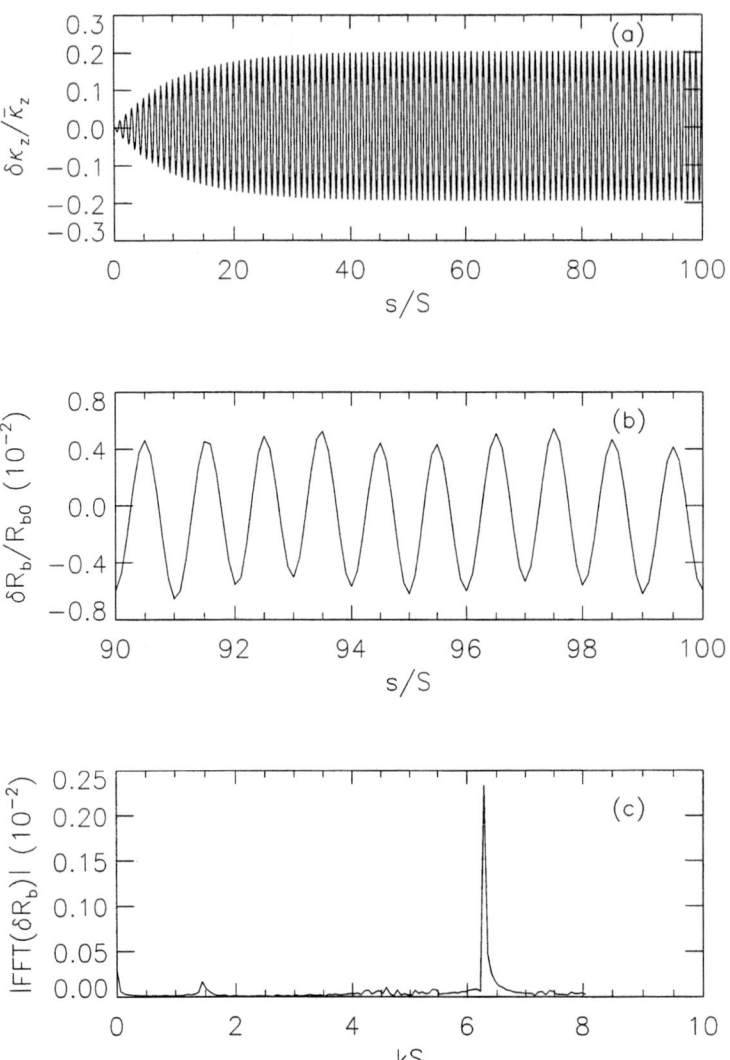

FIGURE 2. Matched beam with adiabatic turn-on ($\alpha = 0.1$). Plotted versus s/S are: (a) $\delta\kappa_z(s)$ between $s = 0$ and $s = 100S$, and (b) $\delta R_b(s)$ between $s = 90S$ and $s = 100S$. The Fourier transform of $\delta R_b(s)$ is shown in (c). The envelope oscillation peak at $k_e S \simeq 1.54$ has decreased by about one order-of-magnitude relative to the mismatched case in Fig. 1.

5. Conclusions

The nonlinear δF formalism has been described for intense nonneutral beam propagation through a periodic focusing solenoidal field $\kappa_z(s+S) = \kappa_z(s)$. Nonlinear δF simulation results were presented for a thermal equilibrium distribution F_b^0, illustrating application of the technique to axisymmetric, matched-beam propagation to the case where $\kappa_z(s)$ turns on adiabatically about a constant average value $\bar{\kappa}_z$. Adiabatic turn-on over about twenty lattice periods is found to produce a highly-matched beam in which the rms beam radius $R_b(s)$ oscillates with the same period, S, as the focusing field.

Acknowledgments

This research was supported by the U.S. Department of Energy and in part by the LANSCE Division of the Los Alamos National Laboratory.

REFERENCES

1. Davidson, R. C., *Physics of Nonneutral Plasmas* (Addison-Wesley Publishing Co., Reading MA, 1990).
2. Reiser, M., *Theory and Design of Charged Particle Beams*, (John Wiley & Sons, Inc., New York, 1994).
3. Wangler, T. P., *Principles of RF Linear Accelerators*, (John Wiley & Sons, Inc., New York, 1998).
4. Jameson, R. A., in *Advanced Accelerator Concepts*, edited by J. S. Wurtele, American Institute of Physics Conference Proceedings **279**, (American Institute of Physics, New York, 1993), p. 969.
5. Lee, E. P., and Hovingh, J., *Fusion Technology* **15**, 369 (1989).
6. Hofmann, I., Laslett, L. J., Smith, L., and Haber, I., *Part. Accel.* **13**, 145 (1983).
7. Struckmeier, J., and Hofmann, I., *Part. Accel.* **39**, 219, 1992.
8. Davidson, R. C., Lee, W. W., and Stoltz, P., *Phys. Plasmas* **5**, 279 (1998).
9. Davidson, R. C., and Chen, C., *Particle Accelerators*, in press (1998).
10. Friedman, A., and Grote, D. P., *Phys. Fluids* **B4**, 2203 (1992).
11. Haber, I., Callahan, D. A., Friedman, A., Grote, D. P., and Langdon, A. B., *J. Fusion Engineering and Design* **32**, 159 (1996).
12. Stoltz, P. H., Lee, W. W., and Davidson, R. C., *Nuclear Instruments and Methods in Physics Research*, in press (1998).
13. Lee, W. W., Qian, Q., and Davidson, R. C., *Physics Letters* **A230**, 347 (1997).
14. Parker, S. E., and Lee, W. W., *Phys. Fluids* **B5**, 77 (1993).
15. Lee, E. P., and Briggs, R. J., *The Solenoidal Field Transport Option: IFE Drivers, Near Term Research Facilities, and Beam Dynamics*, LBNL Report 40774 (1997).
16. Lin, Z., private communication (1998).

Nonlinear Self Consistent High Resolution Beam Halo Algorithm in Homomorphic and Weakly Chaotic Systems

J. H. Whealton, R. J. Raridon, D. K. Olsen, J. D. Galambos, J. A. Holmes

Oak Ridge National Laboratory, Engineering Technology, Physics, and Computational Physics and Engineering Divisions

ABSTRACT

A technique is described which enables high resolution of halo in beam dynamic studies by direct simulation. The method consists in first solving the beam dynamics problem using coarse initial data. The regions of the initial data, which result in beam halo, or extremums in phase space, are identified. The dynamics are resolved by continuing the calculation using initial data points slightly offset from those that result in halo formation, thus filling in the halo structure. The solution is repeated with appropriate scaling of such things as charge per orbit etc. This process may be continued indefinitely. The method can also shed some light on the halo generation in weakly chaotic systems. The scheme is essentially different from the Δf method in that no assumption is made about f_o. As an example, a bifurcation in a non-trivial space charge dominated homomorphic problem is resolved self-consistently using minor computational resources, rather than having to perform the calculation for 250 trillion effective particles.

INTRODUCTION

Recent accelerator systems require a very low beam-wall interception (as low as 1 part in 10^6). Modeling the beam dynamics including significant statistical sample in the halo would require considering on the order of a billion trajectories. In certain cases, which are not chaotic or weakly chaotic, a technique may be used which requires substantially fewer resources than the brute force approach.

The idea is easily understandable by reference to a specific example as shown in Fig. 1. Here is modeled an ion extraction from a plasma. The following Vlasov-Poisson equations

$$\nabla^2 \varphi(\mathbf{r},t) = \int f(\mathbf{r},\mathbf{v},t) d\mathbf{v} - \exp\left[-\varphi(\mathbf{r},t)\right] \tag{1}$$

$$\frac{\partial f(\mathbf{r},\mathbf{v},t)}{\partial t} + \mathbf{v} \bullet \nabla f(\mathbf{r},\mathbf{v},t) + \left\{ \mathbf{v} \times \mathbf{B} - \nabla \varphi \right\} \bullet \nabla_v f(\mathbf{r},\mathbf{v},t) = 0$$

provide an adequate description. Plasma electron density is described by a Boltzman distribution. Electrostatic equipotentials are shown in this 2-D example by dashed lines. Solid lines show ion trajectories (only on the top half of the figure). The extraction sheath, found self-consistently, is shown by the high field region in the middle of the figure. (Plasma electrons are abundant on the left side of this sheath and absent on the right side.) One electrode is shown in the middle of the figure on the top and bottom. The beam goes off the right side of the figure. The occupation in phase space of the exiting beam is shown in Fig. 2. Of particular interest are the beam aberrations shown in Fig. 2 due to the nonlinear transverse forces in the proximity of the electrode (shown in Fig. 1).

The scheme is basically to take those trajectories which are in the outer reaches of phase space, trace them back to their starting position (on the left side of Fig. 1) and then create, say ten trajectories, in the phase space neighborhood of the ten most aberrated trajectories. This is a very simple thing to do in the example illustrated in Fig. 1. First, the 10 most aberrational trajectories are replaced by 100 trajectories with a space charge weight of one-tenth the original. The self-consistent solution is shown in Fig. 3 (corresponding to Fig. 1) and Fig. 4 (corresponding to Fig. 3). This is denoted as the second major iteration. A comparison of Fig. 3 with Fig. 1 shows exactly what trajectories were selected. A comparison of Fig. 4 with Fig. 2 shows the corresponding region in phase space that was refined. Continuation of the process is done by again selecting the ten most aberrated trajectories in Figs. 3 and 4, replacing each of them with ten trajectories with a weighting of one-tenth of the replaced trajectory. This is the third major iteration. The result after 13 major iterations is shown in Figs. 5 and 6. As can be seen in Table 1, after 13 major iterations, only 10,306 cumulative trajectories were calculated, but if the resolution were uniform over all of the phase space 252 trillion trajectories would have been calculated – a saving of a factor of 25 billion in computation time. If all of these other 252 trillion trajectories were actually computed, the result would simply fill in the line in the central part of the phase space diagram – Fig. 6. While this factor may not be easily realizable in more complex problems with nontrivial occupation of 6D-phase space, large factors of resource savings in self-consistent halo calculations are still possible.

For non-chaotic, completely deterministic systems, there is no limit to the savings possible. For weakly chaotic systems, there will be a limit to the resolution obtainable, but information will be found on these chaotic regions of the phase space. For highly chaotic systems, the method will not be reliable.

Major Iteration	Actual	Apparent
1	126 x 2 = 252	252
2	252 - 10 + 100 = 342 Σ 342 + 252 = 594	2520
3	342 - 10 + 100 = 432 Σ = 1026	25,200
4	432 - 10 + 100 = 522 Σ = 1548	252,000
5	522 - 10 + 100 = 612 Σ = 2170	2,520,000 (M)
6	612 - 10 + 100 = 702 Σ = 2872	25,200,000
7	702 - 10 + 100 = 792 Σ = 3664	257,000,000
8	792 - 10 + 100 = 882 Σ = 4546	2,520,000,000 (B)
9	882 - 10 + 100 = 972 Σ = 5518	25,200,000,000
10	972 - 10 + 100 = 1062 Σ = 6580	252,000,000,000
11	1062 - 10 + 100 = 1152 Σ = 7732	2,520,000,000,000
12	1152 + 90 = 1242 Σ = 8974	25.2 T (T)
13	1242 + 90 = 1332 Σ = 10306	252.T

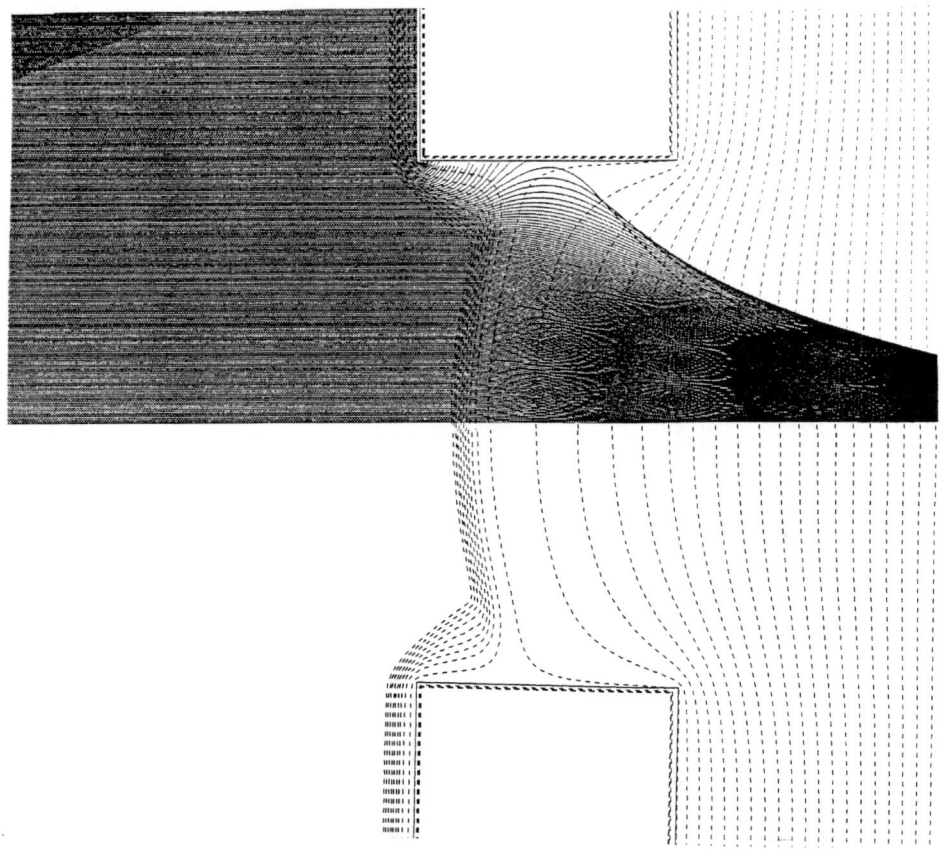

Figure 1. Ion orbits (solid lines starting at the left from inside the plasma) are accelerated by externally applied electric fields (the equipotentials indicated by dashed lines). The self-consistent plasma sheath is shown by the high field region (close spacing of equipotentials) shown. An electrode (constant equipotential) is shown at the middle top and bottom. The ion beam, which exists on the right, has a halo produced by aberrations caused by nonlinear fields near the electrode.

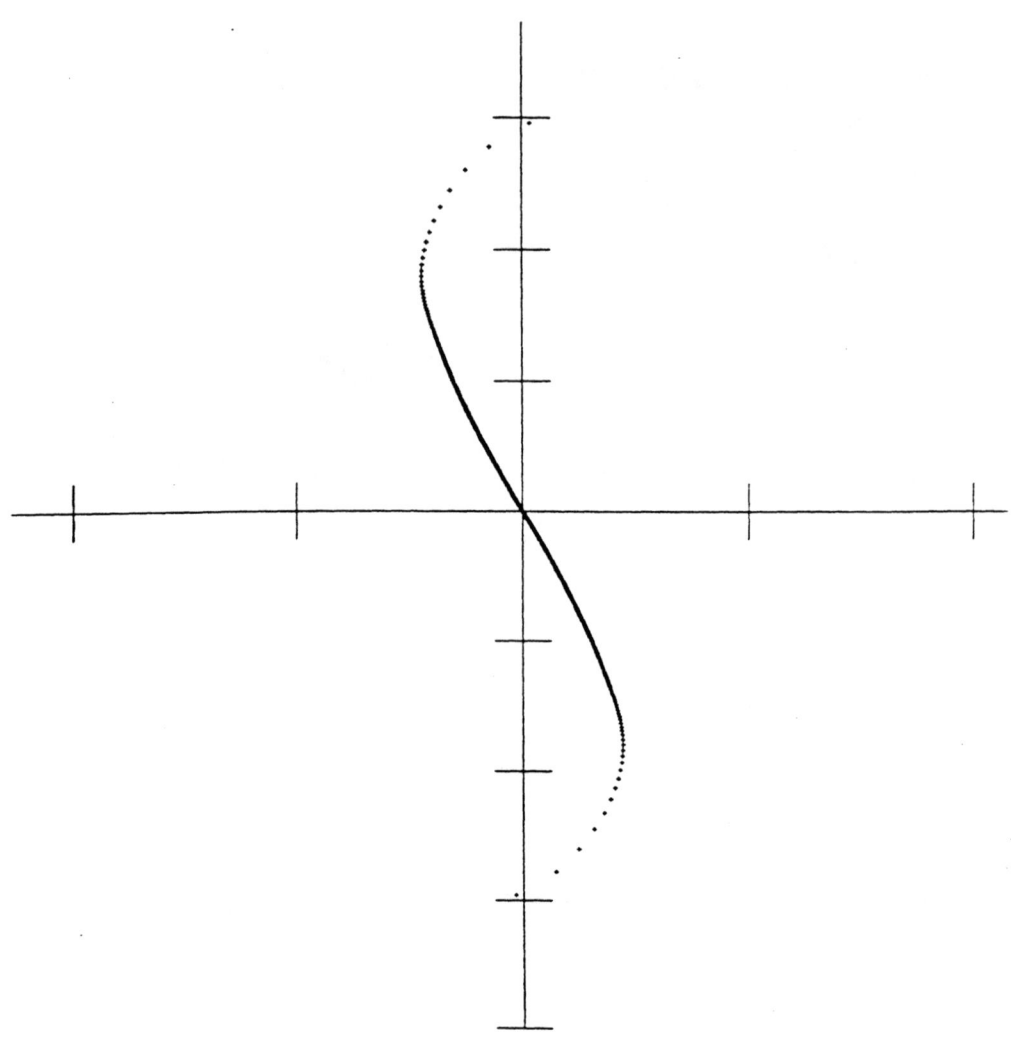

Figure 2. Phase space plot of the beam existing on the right hand side, Fig. 1, the transverse distance is the horizontal axis and the transverse velocity is the vertical axis. The points with large transverse speeds represent the aberrations (halo) of the beam, in this case.

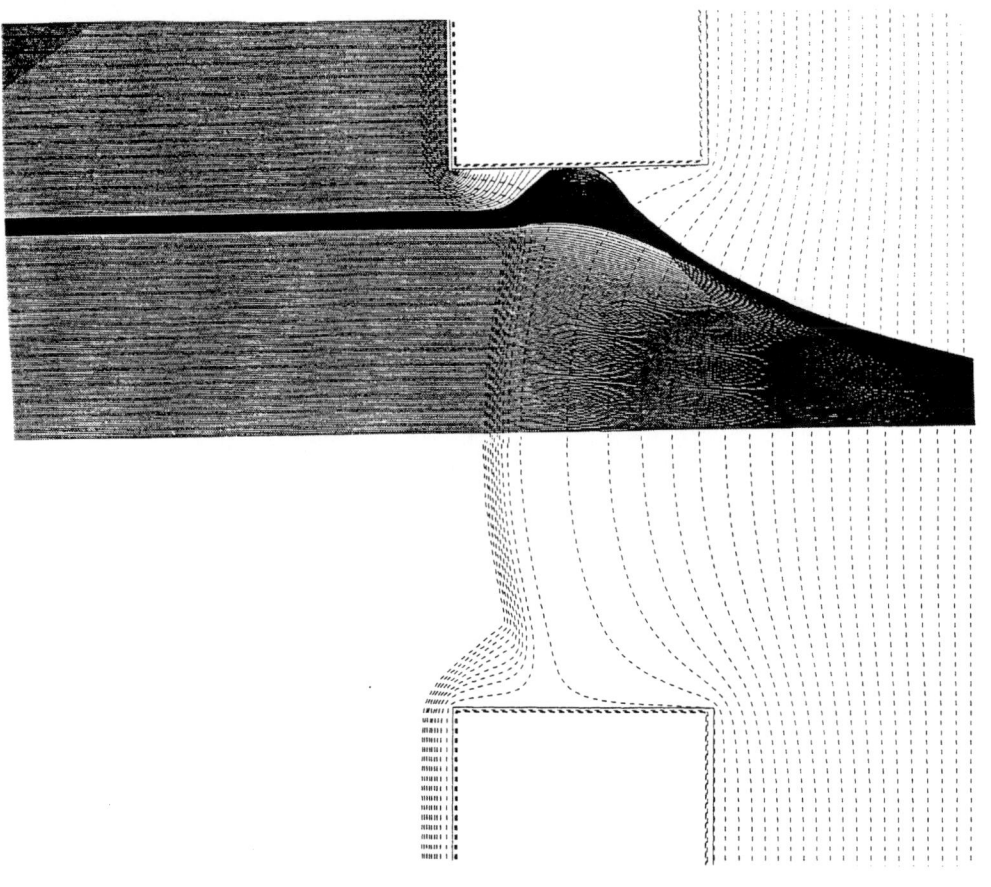

Figure 3. Same as Fig. 1 except the most aberrated 10 orbits are replaced by several more (100) for enhanced resolution.

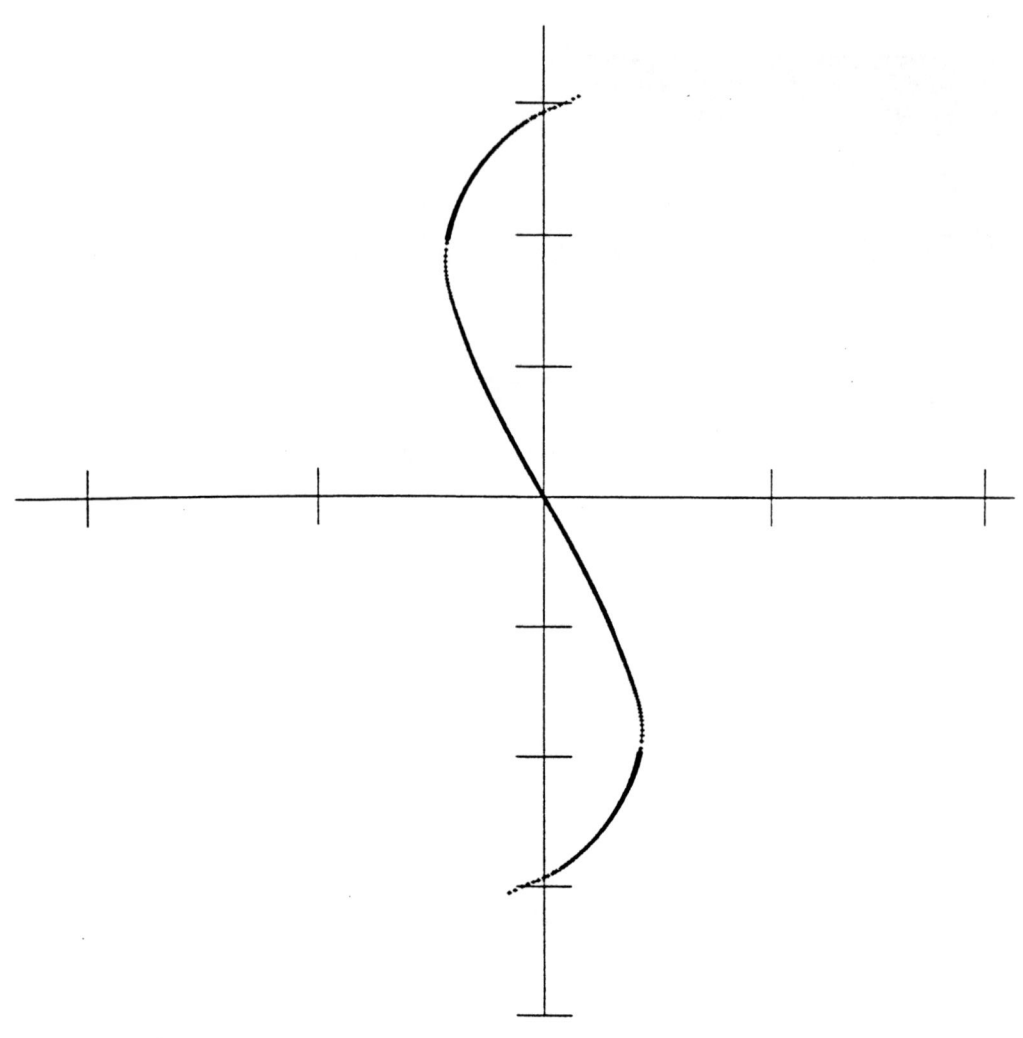

Figure 4. Same as Fig. 2 but now refers to the exiting beam of Fig. 3.

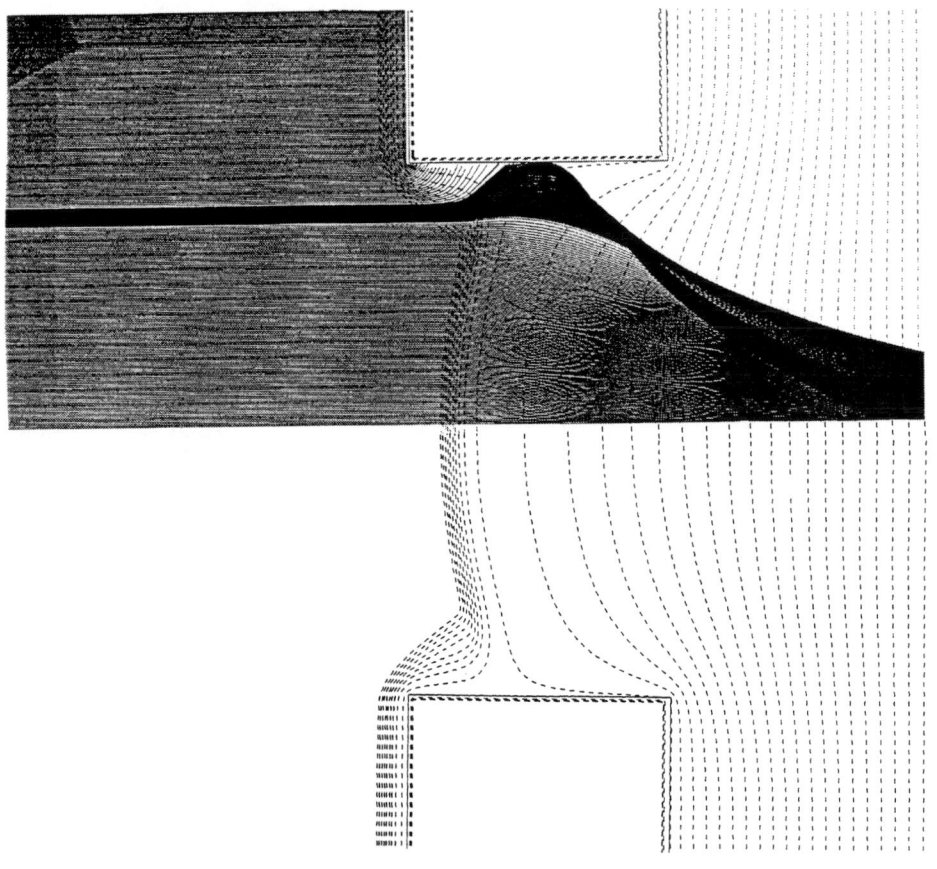

Figure 5. Same as Fig. 3 but the process has been continued 11 more times giving much greater resolution near the region of aberrations.

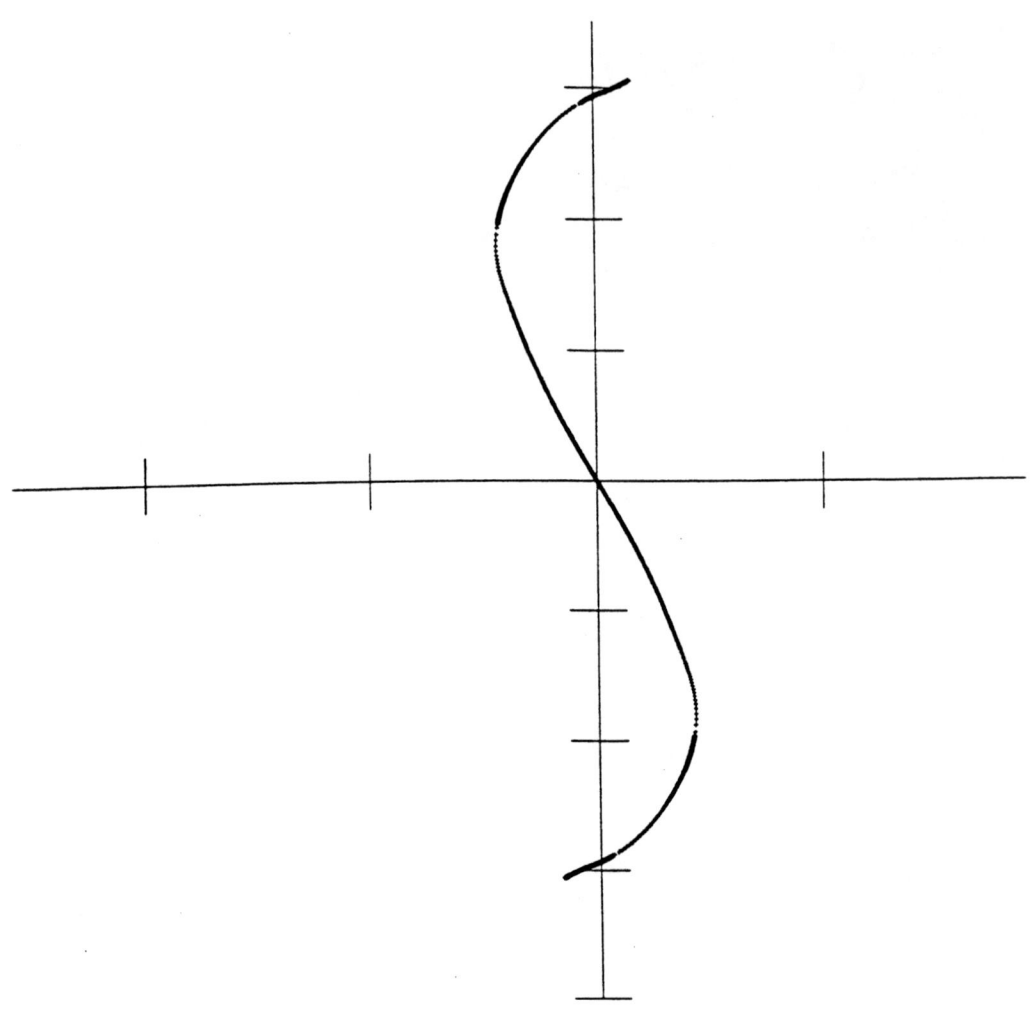

Figure 6. Same as Fig. 4, but now referring to Fig. 5. Now the aberrations are fully resolved.

SUMMARIES OF WORKING GROUPS

Summary of the Observation Group

Roberto Cappi

CERN, PS Division, 1211 Geneva 23, Switzerland

The main issues discussed in the 'Observation Group' group were:
- methods of observation of emittance growth
- strategies to design and place collimators to reduce halo in rings
- methods to observe particle losses and present experience
- space charge studies, etc.

EMITTANCE MEASUREMENTS

Some typical emittance measurement techniques were discussed comparing advantages and disadvantages. See Table 1.

Instrument	Advantages	Disadvantages
Wire scanners	Precise absolute value	Can burn for very high intensity beams. Slow (~3ms)
Ionisation Profile Monitors (collecting e-)	Fast (single turn measurements) Non destructive Can measure transverse mismatch	Relative measurements (sensitive to space charge fields) Poor radiation resistance
Extraction to a measurement line equipped with harps	Precise Reliable	Complicate (requires a beam extraction) and expensive. Does not measure the circulating beam.
Exotics (e.g. Lasers, scanning e- beams, etc.,)	To be studied	Puzzling? Reliable? Radiation resistance?

Table 1. A list of considered issues concerning transverse emittance measurements.

All these instruments are measuring the beam profile. The emittance is 'calculated' by using some optical parameters (e.g. the beta function, dispersion, etc.) which have to be known, that is measured, with good precision.
The suggestion of the group for a compression ring was to use an Ionization Profile Monitor in the ring and a measurement line with harps.

COLLIMATORS

Loss management is probably the main difficulty in the design and the operation of compressor rings as losses of 10^{-4} are already prohibitive.

After reminding the basic principle of a collimator implementation[1], see Fig.1,

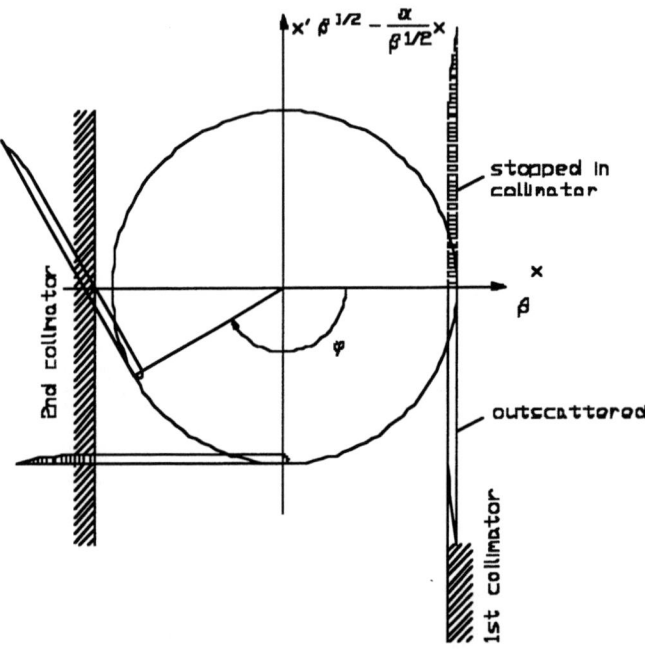

Fig.1 Basic of a collimating system using two collimators.

more sophisticated methods, were discussed, using for example an electrostatic wire septum (see Fig.2) replacing the first collimator [2-6].

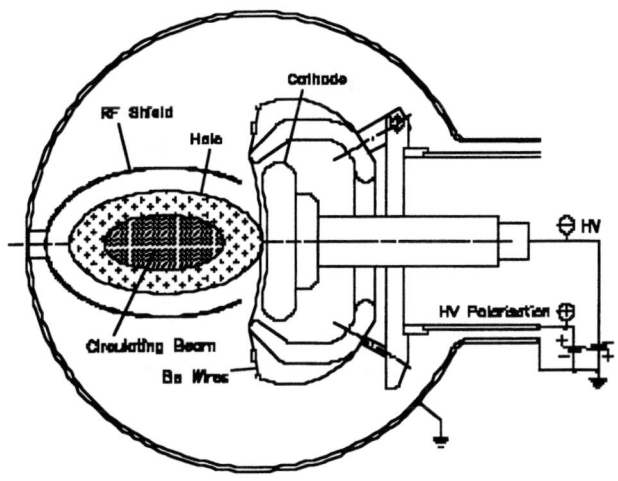

Fig.2 An electrostatic wire septum to replace the first collimator.

The working group conclusion was that this subject is of fundamental importance not only for halo collimation but also for concentration of any losses occurring repetitively or sporadically (e.g. : beam instabilities, hardware failures, timing faults, machine adjustments, setting-up, etc.). To obtain the best solution more effort should be devoted on this topic.

HALO MEASUREMENTS

Movable solid scrapers should anyway be able to 'digest' the total beam current (at least for one pulse) in case of a sudden beam instability or failure. This seems to be difficult to achieve.

On the other hand, one of the main collimators could be used. The local beam position should be adjusted with an orbit bump. The bump amplitude should be controlled (versus time) to take into account possible closed orbit distortions of injection / extraction bumps, kickers, etc.

The amount of the beam lost (normally less then 1%) on the collimator should be measured with scintillators properly placed and calibrated. It will be difficult anyway to obtain absolute values as the resolution of current transformers is probably not good enough.

BEAM LOSSES

A comparison of various machines by T. Wangler [7] indicated similar performance stressing the importance of mismatch as main source of particle losses. Even a small mismatch can produce a large emittance blow-up, while the dependence on space charge effects is relatively week. From simulations, the number of particles in the halo should be of the order of 1%.

The obvious conclusion was that the dynamic aperture of the machine must be as large as possible.

SPACE CHARGE STUDIES IN VARIOUS LABORATORIES

BNL

A careful study of S.Y. Zhang [8] on data from several machines, see Table 2, confirmed the Sacherer criterion that the beam envelope starts to resonate, producing emittance blow-up, when:

$$\chi \geq 1/4$$

where :

$$\chi = \frac{|\Delta v_y| - |\Delta v_{iy}|}{|\Delta v_{iy}|}$$

where : Δv_y is the distance of the coherent tune to the half integer (or integer) resonance (for y plane).
Δv_{iy} is the incoherent space charge tune shift.

For example: if $v_y = 6.22$ and $\Delta v_{iy} = -0.28$ then $\chi = 0.24$ and the beam starts to blow-up.

	FermiB	CPS	CPS	PSB	PSR	AGSB
v_y	6.80	6.22	6.28	5.45	2.19	4.80
$-\Delta v_{iy}$	0.38	0.28	0.37	0.66	<0.29	0.36
χ	0.21	0.21	0.24	0.32	<0.34	0.17

Table 2. The value of the parameter χ is ~ _ in several machines.

LANL

D.Neuffer and R.J Macek [9] presented some results from an experiment at Los Alamos PSR where approaching the integer resonance $v_y = 2.0$ the transverse emittance blows-up and Δv_{iy} increases less rapidly. See Fig.3.

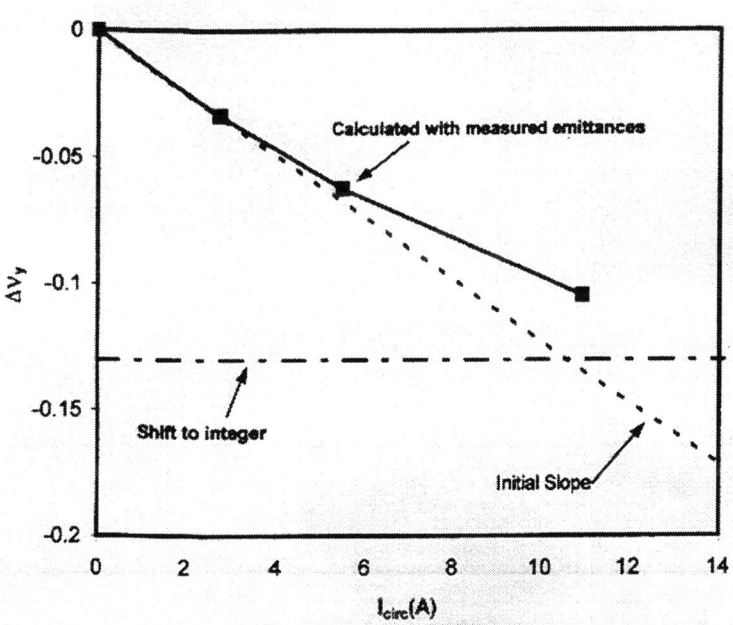

Fig.3. Tune shift versus beam intensity in PSR. Close to integer resonance the emittance becomes larger and the tune shift increases less rapidly.

F. Neri [10] reported on successful results of an experiment done at PSR to compensate the longitudinal space charge impedance (capacitive) by a ferrite inductance installed on the vacuum chamber. By biasing the ferrite it was possible to switch it on or off.

The total longitudinal impedance Z_{Ltot}/n seen by the beam is given by (excluding all other possible resonators):

$$\frac{Z_{Ltot}}{n} = j\left[\omega_0 L - \frac{Z_0 g_0}{2\beta\gamma^2}\right]$$

where: $Z_0 = 377 \, \Omega$

$g_0 = 1 + 2\ln\dfrac{b}{a}$ a and b are the beam and vacuum chamber size

β, γ are the usual relativistic factors

$\omega_0 = 2\pi$ x the revolution frequency (~ 2.9 MHz)

$n = \omega/\omega_0$

$L \approx 12 \, nH$ is the value of the ferrite inductance when switched on.

For a machine working below transition, the first term in the parenthesis has a focussing effects, or bunch shortening, while the second term (space charge capacitive impedance ~ 210 Ω in PSR) has a bunch lengthening effect. See Fig.4.

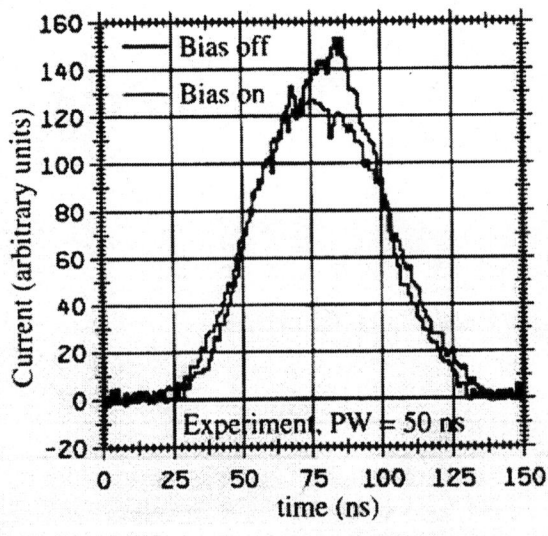

Fig.4 By switching on the inductance the bunch shortens.

GSI

I.Hofmann [11] recalled some considerations in the measurement of space charge detuning Δv_i using a quadrupole transfer function measurement. The method consists in exciting a coasting beam with a quadrupolar kicker and measure the beam response with a quadrupolar pick-up [12] .
In frequency domain one expects to see betatron lines at v_0 (dipole) and at $2v_0$ (quadrupole) for a low intensity beam. See Fig. 4a

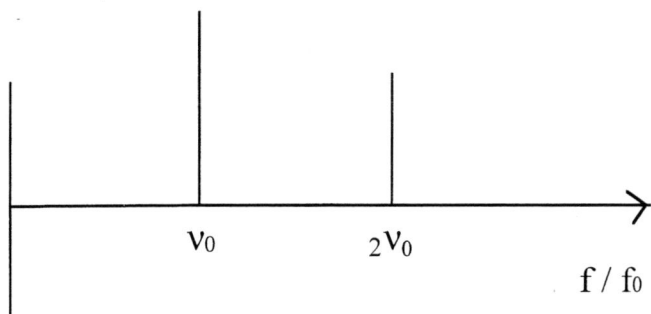

Fig. 4a Dipole and quadrupole betatron lines for a low intensity beam

Increasing the beam intensity the quadrupole frequency decreases (See Fig 4b) by space charge detuning as:

$$2\nu_{0,x} - \nu_{m,x} = \left(\frac{3}{2} - \frac{1}{2}\frac{a_x}{a_x + a_y}\right)\Delta\nu_{i,x}$$

where ν_m is the measured shifted quadrupole line, and a_x and a_y are the horizontal and vertical beam dimensions.

If $a_x \approx a_y$, then

$$\Delta\nu_{i,x} = \frac{4}{5}\left(2\nu_0 - \nu_{m,x}\right)$$

Calculating $\Delta\nu_{i,x}$ with the Laslett formula one can derive the transverse emittance.

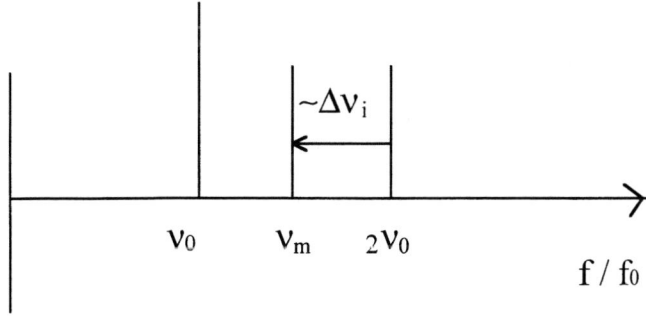

Fig.4b Increasing the beam intensity the quadrupole line shifts due to space charge to a smaller value.

University of Maryland

J.G. Wang [13,14] reminded us that the tradition at this place to study the behavior of space charge dominated beams continues with the construction of a mini accumulator ring of e-, despite the very limited manpower. The machine main parameters are : E_k=10 keV, R=1.8m, B~10 Gauss (!). The RF cavity is an induction cavity, similar to a resistive wall current monitor. A list of details and foreseen experiments will be too long for this summary and should be found in the Proceedings.

AGS

D. Trbojevic [15] proposes a missing magnet lattice to obtain a $\eta = \alpha - 1/\gamma^2 \approx 0$ (with no transition crossing). For a given RF voltage V_{RF} and longitudinal emittance ε_l, this will allow the produce a shorter bunch length τ_b for µ production, as

$$\tau_b \propto \varepsilon_l \sqrt{\frac{|\eta|}{V_{RF}}} .$$

ACKNOWLEDGMENTS

I wish to thank all the participants for their positive attitude and the enthusiasm shown in the discussions which would have deserved a longer time.

REFERENCES

1. R. Cappi, M. Chanel, F. Pedersen, U. Raich, K. Schindl, H. Schönauer, E. Wildner, *"GEB report on PSB consolidation"*, CERN, PS/ DI/ Note 95-25 (Tech.).
2. *"Outline Design of the European Spallation Neutron Source"*, I.S.K. Gardner, H. Lengeler,G.H. Rees, eds. , ESS 95-30-M, September 1995.
3. T. Trenkler, *"Collimation in the PS-Booster"*, PS/HI Note 95-15, August 1995.
4. T. Trenkler, *"Manual for the Medium Range Multi-Turn Tracking Code for Collimation"*, PS/HI Note 95-16, August 1995.
5. F. W. Jones, G. H. Mackenzie, H. Schönauer, *"ACCSIM - A Program to Simulate the Accumulation of Intense Proton Beams"*, Proc. 14th Int. Conf. High Energy Acc., Tsukuba 1989, p. [1409] /199.
6. H. Schönauer, *"Loss Concentration and Evacuation by Mini-Wire-Septa"*, Proc. of the 1995 PAC, Dallas, May 1995.
7. T. Wangler, these Proceedings.
8. S.Y. Zhang, these Proceedings.
9. D. Neufer and R.J Macek , these Proceedings.
10. F. Neri, these Proceedings.
11. I.Hofmann , these Proceedings.
12. M. Chanel, *"Study of beam envelope oscillations by measuring the beam transfer function with quadrupolar pick-up and kicker "*. Proc of EPAC '96 Sitges, Barcelona, Spain ; 10 - 14 Jun 1996, (1015-1017).
13. J.G. Wang, these Proceedings.
14. M. Reiser, these Proceedings.
15. D. Trbojevic, these Proceedings.

Summary: Theory Working Group

RICHARD BAARTMAN

TRIUMF, 4004 Wesbrook Mall, Vancouver B.C., V6T 2A3, Canada

I CHARGE TO WORKING GROUP

1. Relationship between single particle and coherent pictures of resonance in synchrotron. (II)

2. Methods and accuracy of emittance growth and particle losses by resonance model. (III)

3. Mechanisms of halo generation in synchrotron. (IVb)

4. Methods and accuracy of emittance growth by particle-core model. (IV)

5. How to minimize halo generation in synchrotron. (IVb)

II RELATIONSHIP BETWEEN SINGLE PARTICLE AND COHERENT PICTURES OF RESONANCE IN SYNCHROTRON

For a beam whose tune shift is dominated by direct space charge, the incoherent tunes for particles in the beam core are not relevant to betatron resonance. On the other hand, the halo incoherent tunes are relevant. Hence, both the coherent mode picture and the core-halo picture are valid, though in some sense they are opposite ends a spectrum. This picture is dependent upon the order of the resonance. At the lowest order (integer resonance), the incoherent tunes are not relevant for any amplitude. For half-integer resonance, we can apply the Sacherer prescription of paying attention only to the tune shift of the KV beam with the same rms size (the 'KV-equivalent beam'). For higher order, non-KV distributions are difficult to treat. However, experiments and simulations seem to indicate that the Sacherer prescription still applies. This leads to the general condition that there should be no uncorrected low-order resonances between the bare tune and the KV-equivalent tune.

III EMITTANCE GROWTH AND PARTICLE LOSSES BY RESONANCE MODEL

The rate of growth of amplitude on a betatron resonance can be calculated using standard (CERN 77-10) formulas, whether they be coherent core modes or incoherent halo modes. In fact, because of the amplitude dependence of the tune for a non-KV beam, there will be decoherence and this will limit the total growth, making this a conservative approach.

One can check the calculations using simulations, with the lattice errors contained in a high-order map, such as generated by COSY-∞. The calculations should ideally contain the effects of images; both the closed-orbit-dependent terms and the terms due to a modulated beam pipe.

IV HALO GENERATION, CONTROL AND MODELING

a Acceptance: There is a major difference between high intensity linac design and high intensity ring design: In the latter, we try to avoid halo generation, but in the former, we try to accept the halo. The ring approach is of course dictated by economics.

b Halo generation: The size of the halo relative to the beam size seems to be about the same in both cases ($\sim 6\sigma$), even though the tune depression in the ring is tiny compared with high intensity linacs. The mechanism of halo generation is core mismatch, i.e. time-dependent behaviour of the core causes halo. The H$^-$ injection scheme is very good in this regard; since accumulation is over many turns, the core can in principle be well-matched.

Therefore, it is important to investigate all possible sources of core mismatch: Power supply ripple, transverse quadrupole impedance, neutralizing electrons or dust particles, etc.

If core oscillations turn out to be a problem, there is still the possibility of active quadrupole damping, so there is no need to adopt the linac approach of providing very large acceptance.

It is important to ensure that the halo generated from other sources in the SNS is either negligible or efficiently removed by the collimation system.

c Lattice: Preliminary simulations for the SNS show a halo generated whose size is dependent upon the details of the lattice. Theoretically, this is not expected. More simulations are needed.

There is as yet no consensus on how properly to include both space charge and dispersion in envelope equations for rings. This is an important theoretical area, since it is related to the question of how to inject a matched beam at a point where the periodic dispersion is not zero.

Working Group on Simulation: Summary

A.U. Luccio
Brookhaven National Laboratory, Upton, NY 11973-5000, USA

1. CODES

Computer codes for multi dimensional beam simulation were presented and discussed in the Working Group:
- ACCSIM, developed by Fred Jones at TRIUMF, Vancouver, Canada,
- TRACK3D, developed by Chris Prior at the Rutherford Appleton Laboratory, in the UK, and being used in particular for studies on the European Spallation Source, ESS,
- SIMPSONS, developed by Shinji Machida at the SuperCollider Laboratory in Texas and then at the KEK Institute in Japan, being used for the Japanese Spallation Source Project,
- WARP3D, developed by Alex Friedman and Coll. at the Livermore National Laboratory in California, in conjunction with the Heavy Ion Fusion Program,
- BEAMPATH, developed by Yuri Batygin at RIKEN, in Tokyo,
- SAMBA, developed by a collaboration between Oak Ridge National Laboratory and Brookhaven National Laboratory, to be used for the Spallation Neutron Source studies.

2. METHODS

These codes are all characterized by full 6-dimensional simulation of beam propagating through a machine lattice, in the presence of strong space charge effects and beam-to-wall interaction. The lattice is represented by matrices and higher order transformations. The space charge forces or potential are obtained by solving -with different strategies- the Poisson Equation.

Simulation is done with Particle in Cell, PIC, algorithms. Water bag examples have been briefly discussed. Exact equation of motion for specific distributions (in particular the KV or Kapchinskij-Vladimirski distribution) can also be solved to study the evolution of the Core of the beam. Then, one can assume that the beam is simulated by its Core plus an Halo studied by PIC. This is also implemented in the δf method, in which the deviations from an equilibrium distribution are studied, solving the Vlasov Equation in a perturbative way.

Independent variable is most commonly the longitudinal coordinate of motion. Some codes (WARP) use time. In some cases the variable is just a lattice node counter.

3. LATTICE DESCRIPTORS

The machine lattice is commonly constructed by an external optics code. Examples are (1) MAD, maintained in different laboratories, mainly at CERN, (2) DIMAD, a transformation builder based on MAD, (3) SYNCH, (4) TRANSPORT. All the previous codes use thick building blocks to represent the machine elements. Another code used, (5) TEAPOT, uses a thin element representation. All these codes produce at the output matrices and higher order transformations for the transverse dynamical phase space coordinates x, x', y, y' and can provide sophisticated means of correcting distorted reference orbits due to lattice errors.

Some of the simulation codes being used have their own internal lattice descriptor to serve specific purposes.

4. PLATFORM

Multi-dimensional simulation is computer time hungry. One of the most important topics of discussion is always how to reduce computer run time, and what computer system (platform) to use. Most common are Unix Workstations maintained by a small number of users. Traditional Mainframes, maintained by computer centers are falling out of fashion. SuperComputers with parallel processing are being used in some cases. Personal Computers, PC's and Mac's, are a category fast growing in popularity.

To compare different platform, one has consider the following: a fast machine, mainframe, workstation or even supercomputer used by many customers not always favorably compares in speed with personal workstations or personal computers with one single user. Microprocessors are becoming faster and faster, so the differences between "computers" and PC's are smaller. PC's cost little money and Powerbooks are physically portable.

Comparison between machine is interesting. With SAMBA, preliminary studies seem to indicate that, in practical terms, PC's and Mac's are as fast as anything else. An interesting development is to put 20 or so PC's together and build inexpensively a parallel computer.

5. GRAPHICS

Graphics is important in simulation. We have of course to represent final data, but we increasingly want to use sophisticated representations, process data and monitor the evolution of beams with animation.

Some of the codes presented (e.g. ACCSIM) rely on some unique graphic packages, developed in house. The advantage are that graphics call are embedded in the program and can be easily customized. The disadvantages are that the code loses some of its portability, since in changing platform one should install the code and the graphics at the same time, and also that one is bound by the capabilities of the graphic packages. Using a commercial graphic package is the other option, with obvious advantages in portability and in the possibility to let the acquired software process and represent data in a more advanced way, ranging from 3-dimensional plots, contour maps, animation, etc.

Finally, the possibility of using graphic primitives and firmware graphic tools to do fast and efficiently some calculations was discussed.

6. LANGUAGE

Traditionally, for a long time, physicists have used FORTRAN as a high level language to code physical problems. Compiled programs are most commonly being run.

We had strongly opinionated discussions how this pattern may be changing. C++ has emerged as the language of choice in Industry, because its class structure and its modularity are considered by many particularly suitable to team work. Much of the development in software seem to be in C++. On the other hand, FORTRAN is simpler to learn and to use, is more compact in writing, has very extensive mathematical libraries, however it is arguably a one-person-at-a-time way of programming. SAMBA is written in C++, the other codes discussed are in FORTRAN. The debate goes on.

Compiling is another point. At an early time, programs in high level languages were interpreted, then compilers came in to save a large amount of time, actually making it possible to run large scientific codes with a lot of calculations. Now computers are so fast that the difference in run time between a compiled and an interpreted piece of code is smaller and smaller. This makes it a good alternative to mix compiled and interpreted code modules in the same program with great advantages in the development stage. We discussed how to do this, in particular how to run a program under the supervision of a SuperCode.

7. VALIDATION

Simulation codes must be validated, comparing their results with observations (still scarce for the problems of main interest of space charge effects in circular hadron machines) and available theories (say, the envelope equation). An issue that was discussed was how to create benchmark cases, and a recommendation that came out of our work was to establish a collaboration on the subject among all interested parties in various laboratories.

8. HIGHLIGHTS

Specific presentations to the Working Group (see the corresponding papers in the Proceedings) were

8.1. WARP3D

Alex Friedman from Livermore showed the use of their PIC WARP3 code on a Cray supercomputer applied to linear structures or to a low energy ring (recirculated linac). WARP3 is being run under a supervising code (PHYTON). The code moves a window or mesh (typically 64x64x256) following a warped environment. It typically uses 100,000 macro-particles. Independent variable is time, the coordinate system is absolute (no reference orbit). Poisson Equation for space-charge is solved with FFT or SOR to invert the operator ∇^2.

8.2. δf Method

Peter Stoltz from the Princeton Plasma Physics Lab showed application of his delta-f method, consisting of a PIC simulation study of δf or departure from equilibrium. The Vlasov equation was being solved for a function $f = f + \delta f$. The distribution was continuously renormalized on the corners of a mesh, by introducing statistical weights for the macro-particles. An example was given for the oscillations of δf for a mismatched beam in a channel.

8.3. Non-uniform Mesh

John Whealton of Oak Ridge showed on an example of plasma focus how to save computer time with a non uniform population (more macro-particles where there is more action).

8.4. Micro-map

Giuliano Franchetti, a collaborator of Ingo Hoffman from GSI presented a simulation based on the propagation of a beam through a space-charge modified transport map. It was a PIC calculation with an FFT of the density to calculate space charge effects. An example showed was on coupled resonances and the role of space charge to suppress them.

8.5. Mini Electron Ring

Rami Kishek and Marco Venturini, collaborators of Martin Reiser from the University of Maryland showed some experimental work on a space-charge dominated mini electron ring, plus the WARP3D simulation and the theory to model the observations, in particular the growth of the emittance. The Maryland mini ring is a small, very inexpensive machine built to experimentally simulate by scaling the conditions of much larger hadron machines. The model represented very well the observation with the theory using a space-charge modified definition of the emittance. We continued in an evening session to compare the discussion of this formulation with a similar one worked out by SY Lee of the University of Indiana.

8.6. Halo Suppression by Optics Matching

Yuri Batygin from Riken showed some hybrid PIC simulation with the code BEAMPATH of the propagation of a space-charge dominated beam through a line. The code performs a symplectic integration of the Vlasov Equation. A specialized Poisson solver is used for space-charge calculation. The beam was initial mismatched with respect to the line optics and halo clearly formed. However it was possible to use the results of the theory to re-establish the matching and modify the line optics with the use of multipoles. The simulation showed how this cure effectively suppressed the halo.

8.7. Injection in the SNS

John Galambos from Oak Ridge presented results of the simulation of the injection and RF capture of the protons in the Spallation Neutron Source accumulator ring, done with a version of ACCSIM modified with perfected space-charge kicks. In this version, the Poisson solver is FFT based. These results are part of the collaborative work with Brookhaven and represent progress in understanding how to inject in the ring in an optimized way, minimizing uncontrolled beam losses.

8.8. SAMBA

Alfredo Luccio from Brookhaven presented a progress report on the code SAMBA, developed in collaboration with Oak Ridge for the SNS. This is a six-dimensional simulation code with space charge, written in C++ working under the supervision of a SuperCode, containing the best we know, originally developed or freely borrowed from other codes.

8.9. FFM in ACCSIM

Fred Jones from TRIUMF presented improvements to ACCSIM in order to increase its speed and improve its resolution. The computer experiment consisted in introducing variable size cells tailored to the local particle density (Hybrid Fast Multipole Method). It was shown how to conveniently use PIC macro-particles in two categories: grid particles and halo particles.

8.10. Plenary Talks

Outside the Working Group there were two plenary talks on multi-dimensional simulation with space charge given by Shinji Machida from KeK (code SIMPSONS) and Chris Prior from the Appleton Rutherford Lab (code TRACK3D). They are both large and complex PIC codes, and are mentioned here for completeness.

Important points by Shinji were the concept of mode expansion over a variable mesh, the distinction between incoherent tunes of individual macro-particles and coherent tunes of modes, and the development of examples for validation.

Also Chris stressed the importance of code validation, presented an optimizer in examples for the European Spallation Source and raised the issue of code documentation and user's guides.

AUTHOR INDEX

A

Alessi, J. G., 271
Ankenbrandt, C., 128

B

Baartman, R., 56, 439
Bär, R., 15
Barnard, J. J., 221
Batygin, Y. K., 317
Beebe-Wang, J., 298, 332, 344, 390
Bernal, S., 189
Blaskiewicz, M., 213, 332, 344, 390
Boine-Frankenheim, O., 15

C

Cappi, R., 431
Chin, P., 189
Craig, G. D., 221

D

Davidson, R. C., 408

F

Fedotov, A. V., 245
Franchetti, G., 233
Friedman, A., 221, 329

G

Galambos, J. D., 254, 332, 344, 390, 420
Gluckstern, R. L., 245
Godlove, T., 189
Grote, D. P., 221

H

Haber, I., 189, 371
Hofmann, I., 15, 233
Holmes, J. A., 254, 332, 344, 390, 420

I

Ikegami, M., 73

J

Jones, F. W., 359

K

Kehne, D., 189
Kishek, R. A., 189, 278, 371
Knaus, P. R., 379
Kurennoy, S. S., 245

L

Lee, S. Y., 38, 344
Lee, W. W., 408
Lee, Y. Y., 271
Li, R. A., 189
Losic, B., 221
Luccio, A. U., 332, 344, 390, 441
Lund, S. M., 221

M

Macek, R. J., 116
Machida, S., 73
Merrill, F., 3

N

Neri, F., 171
Ng, K. Y., 178

O

Olsen, D. K., 254, 332, 344, 390, 420

P

Popovic, M., 128
Prior, C. R., 85, 141

R

Raparia, D., 271
Raridon, R. J., 420
Reiser, M., 189, 278, 371
Roser, T., 135
Rumolo, G., 15
Rybarcyk, L., 3
Ryne, R., 3
Ryne, R. D., 245

S

Stoltz, P. H., 408

T

Turchetti, G., 233

V

Venturini, M., 189, 278, 371

W

Wang, J. G., 189
Wang, T.-S. F., 286
Wangler, T. P., 3
Warsop, C. M., 104
Weng, W. T., 152, 271
Whealton, J. H., 332, 344, 420

Y

York, R. C., 189

Z

Zhang, S. Y., 198
Zhang, W. W., 189
Zotter, B. W., 26
Zou, Y., 189